The Mechanics of Ribbons and Möbius Bands

Roger Fosdick • Eliot Fried
Editors

The Mechanics of Ribbons and Möbius Bands

Previously published in *Journal of Elasticity* Volume 119, Issues 1–2, 2015

Springer

Editors
Roger Fosdick
Aeropsace Engineering and Mechanics
University of Minnesota
Minneapolis, Minnesota, USA

Eliot Fried
Mathematical Soft Matter Unit
Okinawa Institute of Science and Technology
Okinawa, Japan

ISBN 978-94-017-7299-0 ISBN 978-94-017-7300-3 (eBook)
DOI 10.1007/978-94-017-7300-3

Springer Dordrecht Heidelberg New York London

Printed on acid-free paper

Springer is part of Springer Science+Business Media (www.springer.com)

Contents

DOI 10.1007/978-94-017-7300-3_1
Reprinted from *Journal of Elasticity* Journal, DOI 10.1007/s10659-015-9516-7

Foreword

Special Invited Collection on the Mechanics of Ribbons and Möbius Bands

Roger Fosdick[1] · Eliot Fried[2]

This volume of the Journal of Elasticity contains a collection of papers dedicated to the historical development of and current research interests in the mechanics of the Möbius band. It contains four translations of landmark papers, originally written in German, that played a major role in the development of this and related topics in mechanics: three published by Sadowsky from the 1930's and one published by Wunderlich in 1962. In addition, there are twelve current research papers and reviews that provide insight into the intricate mechanics of stretchable and unstretchable elastic bands, their preferred equilibrium shapes as well as the geometry of surfaces and the representation of isometric mappings. A Möbius band need not be a ruled surface, but it may be a developable surface, which is a kind of ruled surface, with the additional property that it may be continuously flattened into a planar form while preserving its intrinsic lengths and angles, i.e., the result of an isometric mapping of a flat domain into a surface. The papers in this collection address mathematical and computational issues covering this wide range of possibilities.

The Möbius band was formally identified as an object of mathematical interest in the mid-nineteenth century. The first publications to include discussions of its topological properties were those of Listing in 1862 and Möbius in 1865. It has, though, been reported that both Listing and Möbius recognized the importance of the Möbius band a little earlier in 1858 and that in his 1847 study of topology Listing even made passing remarks concerning twisted ribbon-like surfaces.

✉ R. Fosdick
fosdick@aem.umn.edu

E. Fried
eliot.fried@oist.jp

1 Department of Aerospace Engineering and Mechanics, University of Minnesota, Minneapolis, MN 55455-0153, USA

2 Mathematical Soft Matter Unit, Okinawa Institute of Science and Technology, Onna, Okinawa 904-0495, Japan

The influence of the Möbius band now extends well beyond mathematics to encompass multiple branches of science and engineering, architecture, philosophy, psychology, and the musical, visual, literary, and performing arts. New directions for exploiting its intriguing topological properties in science and engineering have emerged in response to recent breakthroughs in the ability to fabricate objects with molecular-scale precision. Novel ideas of inductionless resistors and superconductors with high transition temperature, molecular engines, and helical magnetism have been proposed. It is the intriguing one-sided, one-edged, nonorientable nature of Möbius bands that is driving much of the modern work toward discoveries and applications of nanotechnological importance.

This volume is intended to enhance growth in, and provide insight for, the advancement of fundamental research and discovery in mechanics related to ribbons and the Möbius band. The contributions are wide in scope and they illustrate the important role that mathematical modeling and computation play in this novel area of research.

DOI 10.1007/978-94-017-7300-3_2
Reprinted from *Journal of Elasticity* Journal, DOI 10.1007/s10659-014-9490-5

Translation of Michael Sadowsky's Paper "An Elementary Proof for the Existence of a Developable MÖBIUS Band and the Attribution of the Geometric Problem to a Variational Problem"

Denis F. Hinz · Eliot Fried

Received: 8 August 2014 / Published online: 12 September 2014

Abstract This article is a translation of Michael Sadowsky's original paper "Ein elementarer Beweis für die Existenz eines abwickelbaren MÖBIUSschen Bandes und die Zurückführung des geometrischen Problems auf ein Variationsproblem." which appeared in *Sitzungsberichte der Preussischen Akademie der Wissenschaften, physikalisch-mathematische Klasse*, 17. Juli 1930.–Mitteilung vom 26. Juni, 412–415. Published on September 12, 1930.

Keywords Möbius band · Energy functional · Bending energy

Mathematics Subject Classification 74K20 · 74K10 · 53A04 · 74G55 · 01A75

Translation of the Original Paper

MÖBIUS [1] illustrated the band bearing his name by describing how one may be constructed by bending a rectangular sheet of paper. Subsequently, it has been asked whether this construction can be achieved solely as a consequence of the compliance of the sheet in bending or whether stretching is also required. In other words—the question has been raised as to whether the MÖBIUS band is developable in the strict sense.

MÖBIUS himself did not address this question, since it was irrelevant for his purposes. The developability of his band has been challenged by many, including, it is rumored, H.A. SCHWARZ. In the present work, the existence of a developable band is established on elementary geometric grounds.

Citations of this translation should refer also to Sadowsky's original paper, as cited in the Abstract.

D.F. Hinz · E. Fried (✉)
Mathematical Soft Matter Unit, Okinawa Institute of Science and Technology, Okinawa 904-0495, Japan
e-mail: eliot.fried@oist.jp

D.F. Hinz
e-mail: dfhinz@gmail.com

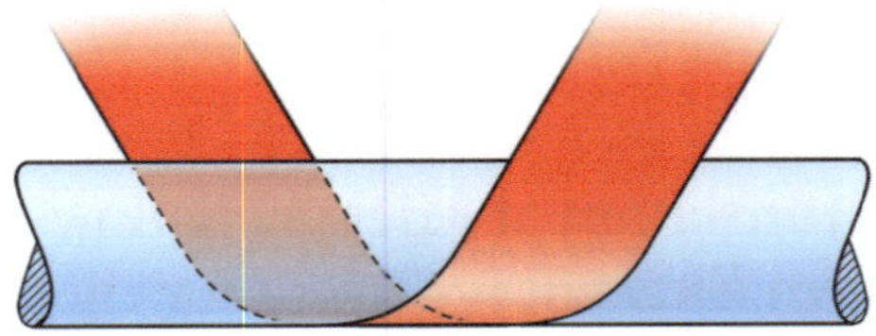

Fig. 1 Adaptation of Fig. 1 from the original version of the paper

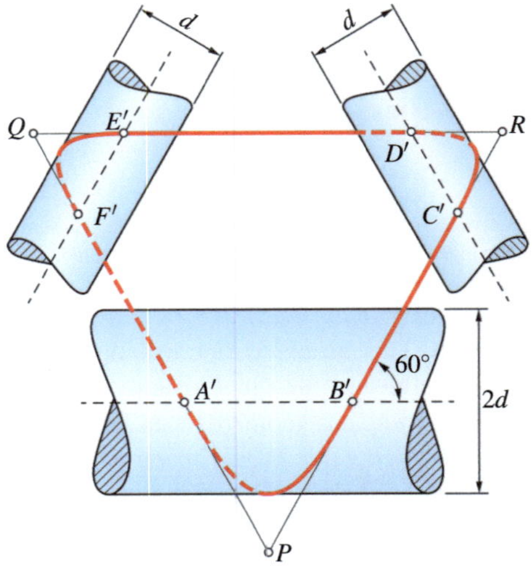

Fig. 2 Adaptation of Fig. 2 from the original version of the paper

Imagine an elongated rectangle of flexible but completely inextensible paper. Further, imagine a rigid cylindrical rod with circular cross-section. Construct two parallel planes tangent to the rod. A bendable but inextensible paper strip may then be positioned to lie within one of these planes, wrap halfway around the rod, and lie within the remaining plane. See Fig. 1 for a depiction of the described arrangement.

Consider now three cylindrical rods with circular cross-sections, one of which has diameter equal to the sum of the diameters of the remaining two. The rectangular strip of paper may then be threaded between the rods to form a MÖBIUS band. The resulting surface consists of three planar sections and three semicylindrical sections and, thus, is developable. An illustration of the construction for rods of diameters d, d, and $2d$ is provided in Fig. 2. The elementary connections between the particular angles and lengths needed to form such a band are not of interest here, since only the existence of the band matters.

The foregoing descriptive construction of a developable MÖBIUS band admits an equivalent analytical construction. If attention is restricted to the previously considered simple case in which the diameters of the rods are d, d, and $2d$, the projections onto the drawing plane of the midlines of the rectangular portions of the band lie on the edges of an equilateral triangle with side length L, and the axes of all three rods are parallel to the drawing plane. Let A, B, C, D, E, and F be points on the midline of the MÖBIUS band and let A', B', C', D', E', and F' be their projections onto the drawing plane. Let AB, BC, etc. denote the distances measured along the mid-line from point A to B, B to C, etc. and let $A'B'$, $B'C'$, etc. denote the lengths of the rectilinear connections between A' and B', B' to C', etc. in the drawing plane. Whereas the sections AF, BC, and DE of the midline remain rectilinear, the sections AB, CD, and EF are twisted along helical curves. These helical curves intersect

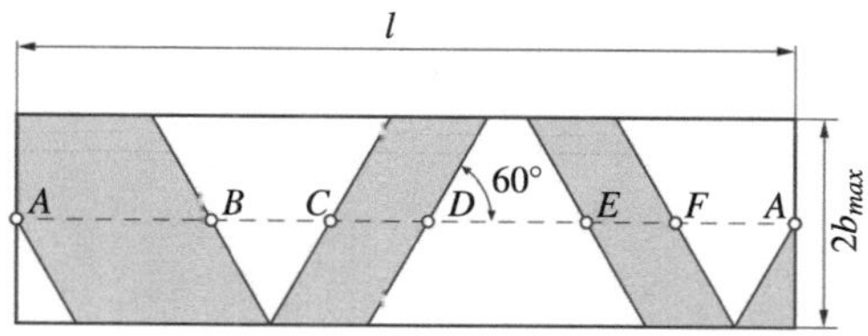

Fig. 3 Adaptation of Fig. 3 from the original version of the paper

the generators of the cylinder with an angle of 60°, from which it follows that (Fig. 2)

$$A'B' = AB\cos(60°), \qquad AB\cos(30°) = \frac{1}{2}\pi 2d = \pi d, \quad \text{etc.,}$$

$$\begin{aligned} AB = \frac{2\pi d}{\sqrt{3}}, \qquad & A'B' = \frac{\pi d}{\sqrt{3}}, \\ FE = CD = \frac{\pi d}{\sqrt{3}}, \qquad & F'E' = C'D' = \frac{\pi d}{2\sqrt{3}}, \\ PQ = QR = PR = L, & \\ AF = BC = L - \frac{\sqrt{3}\pi d}{2}, & \\ ED = L - \frac{\pi d}{\sqrt{3}}. & \end{aligned} \tag{1}$$

Moreover, the total length l of the midline obeys $l = 3L$.

Hereafter, l and d may be chosen arbitrarily (the latter within certain limits, cf. (2)) and, based on the computations above, the points A to F may be chosen to lie on the mid-line. These points also must be in the order A-B-C-D-E-F, which is the case for $3\sqrt{3}\pi d \leq 2l$ (cf. (1)). Next, the generators of the cylindrically twisted parts can be marked on the band with an angle of 60° (Fig. 3). The maximal band width $2b$ corresponding to one pair (l, d) of values in the construction is determined through the generators, because the portions of the band, which are wound around *different* cylinders may not overlap—at most they may come into contact at the boundary of the band. Since $BC = AF$ are the shortest of the unbent portions of the band, it follows that (Fig. 3)

$$b \leq \frac{BC}{2}\tan\left(60^o\right), \quad \text{i.e.} \quad b \leq \frac{l}{2\sqrt{3}} - \frac{3\pi d}{4}.$$

These inequalities identify the constraint

$$d \leq \frac{2l}{3\sqrt{3}\pi}. \tag{2}$$

For an infinitely narrow ($b = 0$) MÖBIUS band, we may set

$$BC = AF = 0.$$

For such a band, we also have

$$AB = \frac{4}{9}l, \qquad FE = CD = \frac{2}{9}l, \qquad DE = \frac{1}{9}l,$$

and (cf. the following considerations)

$$\int_0^l H^2 ds = \frac{15\pi^2}{l} \tag{3}$$

as an upper bound for the integral of the general variational problem (4) for a band that is infinitesimally narrow with respect to its length.

The existence of a developable MÖBIUS band is therefore established by construction. Consider, in more detail, the shape of the band so obtained: The band is assembled from pieces of individual flat and cylindrical surfaces; the actual band constructed with a piece of paper has a different smoothly curved shape. We may arrive at this form if we take into account that the band constructed out of planes and cylinder surface pieces may remain in equilibrium only under the influence of boundary and surface forces, respectively, along with moments. This follows, for example, on recognizing that the curvature of the band as constructed is discontinuous, and, moreover, that moments, as is well known, are associated with curvature. However, if freed from the influence of these forces the constructed band will deform and find an equilibrium configuration corresponding to a minimum of the internal energy associated with elastic deformation. The elastic bending energy density is proportional to the surface integral of the squared mean curvature, granted that the surface is developable. For the situation under consideration, the following observation holds: As initial shape we had a strictly developable band. Corresponding to the provision that the band is inextensible, the deformation is a pure deflection—without stretching. We thus conclude that any configuration resulting from such deformation must be strictly developable. In general, the shape of a the MÖBIUS band corresponds to a minimum of the bending energy. The determination of this ideal form is thus attributed to the variational problem

$$\iint H^2 dF = \text{Min}, \tag{4}$$

the integration being over the complete surface, where H is the mean curvature of the surface and the following conditions apply: the strip forming the band surface has rectangular shape and the band is both developable and possesses the proper one-sided spatial connectivity.

For a infinitely narrow band, that is a band with width b infinitesimally small compared to its length l, the variational problem (4) reduces to

$$\int_0^l \frac{(K^2 + W^2)^2}{K^2} ds = \text{Min}, \tag{5}$$

where K and W are the curvature and the torsion, respectively, of the mid-line of the band, and s is its arc length. For a further *exact* treatment of the problem, it seems that (5) is practically inappropriate, since the calculation becomes exceedingly complicated.

Addendum during the review: For further determination of the configuration of the band cf. the work of the author in the reports of the 3rd International Congress for Technical Mechanics (3. internationaler Kongress f. techn. Mechanik), Stockholm 1930.

References

1. Möbius, A.F.: Ges. Werke Bd. 2 or Leipziger Berichte Über die Bestimmung des Inhaltes eines Polyeders, 484–485 (1865)

DOI 10.1007/978-94-017-7300-3_3
Reprinted from *Journal of Elasticity* Journal, DOI 10.1007/s10659-014-9492-3

Translation and Interpretation of Michael Sadowsky's Paper "Theory of Elastically Bendable Inextensible Bands with Applications to the MÖBIUS Band"

Denis F. Hinz · Eliot Fried

Received: 9 August 2014 / Published online: 12 September 2014

Abstract This article is a translation of Michael Sadowsky's original paper "Theorie der elastisch biegsamen undehnbaren Bänder mit Anwendungen auf das MÖBIUS'sche Band" in 3. internationaler Kongress für technische Mechanik, Stockholm, 1930. The translation is augmented by an Appendix containing an interpretation of the last section of Sadowsky's original paper, including figures generated from recent numerical simulations.

Keywords Möbius band · Energy functional · Bending energy

Mathematics Subject Classification 74K20 · 74K10 · 53A04 · 74G55 · 01A75

Translation of the Original Paper

1 Definition of the Term *Band* Through Kinematic Properties

A body with distinguished midline that, at every point, behaves isotropically with respect to bending is termed a *rope* or a *wire* (or, alternatively, a one-dimensional body). The expression *isotropy with respect to bending* can be illustrated with the following example: If an initially straight and non-twisted section of rope or wire with elastic potential (that is, a conservative system) is bent circularly, then its bending moment depends only on the curvature of the circle and not on which rope or wire elements form the *outer* and *inner* fibers. Consequently, in such a rope or a wire no moment acts to counter eversion of the bent shape.

The circumstances for a *band* stand in contrast to those for a wire or rope. As an example of a band, imagine a MÖBIUS band made of a thin steel sheet. The surface of the band

Citations of this translation should refer also to Sadowsky's original paper, as cited in the Abstract.

D.F. Hinz · E. Fried (✉)
Mathematical Soft Matter Unit, Okinawa Institute of Science and Technology, Okinawa 904-0495, Japan
e-mail: eliot.fried@oist.jp

D.F. Hinz
e-mail: dfhinz@gmail.com

is a strictly (that is, exactly) developable surface. The initially planar steel sheet may thus only be bent but not stretched.[1] A band, no matter how broad or infinitesimally narrow, will respond differently to bending than will a isotropic rope or a wire. In particular, in a band there exists a inadmissible bending axis for relative cross-sectional torsion whereas in a rope or wire all directions have identical properties. In this sense the isotropic response of a rope or wire to bending was mentioned above and in the same sense, a band is anisotropic under bending. A band which has a plane rectangular shape in a stress-free configuration (for example, a MÖBIUS band) may be bent into circular shape but only in a way that the plane of the circle is everywhere cut perpendicularly by the band. The band thus lies on a cylindrical surface. An inside-out rotation of 90° by which the band surface would have to take the shape of a part of a circular annulus is impossible because a rectangular surface is not developable on a part of an annulus. Consequently, a band is subject to an eversional moment not present in an isotropic rope or wire.

The distinguishing features of isotropic ropes or wires and bands, approached intuitively here, become clear upon characterizing their midlines and the accompanying vector triads. The three vectors of such a triad will be denoted by **t**, **n**, and **b**.

1.1 Rope or Wire

We fundamentally require that the midline consists of the same particles at all times. It can thus be said that *the midline is material*. Since the tangent vector **t** connects two infinitesimally neighboring points on the midline, it is firmly connected with the material midline and, thus, we will term it *material* as well. The normal and binormal vectors **n** and **b** are unimportant here, because we require the rope or wire to be isotropic with respect to bending. The two latter vectors are thus not associated with the constitution of the rope or wire. Therefore, we may express the fundamental kinematical property of a rope or wire as follows:

The tangent vector **t** *of the accompanying vector triad of a rope or wire is material.*

1.2 Band

A band is part of the rectifying surface of its midline. (The rectifying plane of a point P on the midline is the plane through the tangent vector **t** and the binormal vector **b**, the rectifying surface is the envelope of all rectifying planes; cf. the work by the author cited in the footnote on the present page.) It is simultaneously required that the midline of the band appears as a straight line upon development of the band to a planar figure (namely a non-stressed configuration of the band); in other words, it is required that the midline coincides with a geodesic line on the band. The binormal vector **b** therefore always lies in the tangent plane of the band, which, however, is *material* in the above mentioned sense—that is, the tangent plane is firmly connected with the matter of the band. Naturally, the tangent vector **t** remains material as well, since the band is only a special case of an anisotropic rope or wire. In that case, the binormal vector **b** must be material as well: it lies in a material plane perpendicular to a material direction. Consequently the vectors **t** and **b** are both material. Moreover, the normal vector

$$\mathbf{n} = \mathbf{b} \times \mathbf{t}$$

is material as well, since it is uniquely determined trough the two material vectors **t** and **b**. Thus, the complete vector triad is material. This characteristic kinematical property of a band is encompassed by the following sentence:

For a band, the accompanying vector triad **t**, **n**, *and* **b** *of the midline is material.*

---

[1] On the questions of existence of an exactly developable MÖBIUS band, see Sadowsky [1].

Fig. 1 Adaptation of Fig. 1 from the original version of the paper

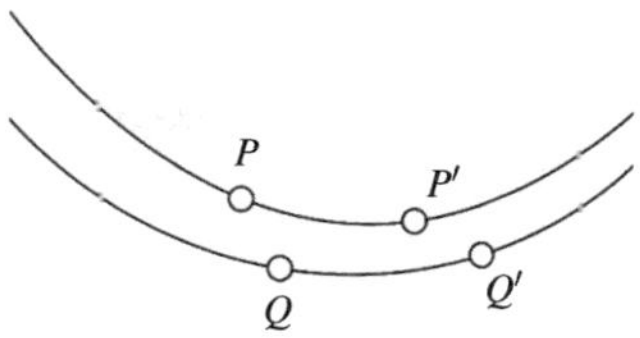

Remark 1 The statement *the vector triad is material* should not be misconstrued as *the vector triad is made of the same material particles* (which would anyway be senseless for the normal), but should rather be understood as *the vector triad is firmly connected to the matter of the band.*

This kinematical property, together with the requirement that the midline is inextensible (that is, the requirement that the variation of the arclength obey $\delta(ds) = 0$) will be viewed as the *definition of a band.*

2 Conclusions from the Kinematical Definition of a Band: Determination of the Virtual Torsion of the Accompanying Triad t, n, and b of the Midline Compatible with the Kinematical Properties of the Band

Let P be an arbitrary point on the midline, $\mathbf{r}$ the position vector of that point, s the arclength along the midline, K its curvature (at the point P), and W its torsion (at the same point). Let the equation describing the midline be

$$\mathbf{r} = \mathbf{r}(s). \tag{1}$$

The Frenet–Serret formulas from the geometry of space curves are then

$$\left.\begin{aligned} \frac{d\mathbf{t}}{ds} &= K\mathbf{n}, \\ \frac{d\mathbf{n}}{ds} &= -K\mathbf{t} + W\mathbf{b}, \\ \frac{d\mathbf{b}}{ds} &= -W\mathbf{n}. \end{aligned}\right\} \tag{2}$$

Let $\delta\boldsymbol{\vartheta}$ denote the virtual twist of the accompanying vector triad of the midline at the point P. The quantity $\delta\boldsymbol{\vartheta}$ should be interpreted as an infinitesimal vector, dependent on the arclength s. We now seek to determine the general virtual twist $\delta\boldsymbol{\vartheta}$ compatible with the kinematic condition discussed in Sect. 1.

To this end, we consider two neighboring points P and P' on the midline. The distance between these two points is denoted by ds. Since the band is capable only of undergoing bending without stretching, this distance is invariant under the virtual deformation. If the band is subject to an admissible (that is, compatible with the area and band conditions) virtual deformation, then the points P and P' will undergo a displacement taking, say, P to Q and P' to Q'. However, the distance QQ' between Q and Q' will remain equal to that ds, between P and P' (Fig. 1).

Upon application of the virtual deformation, the entire shape of the band changes; specifically, at each point along the midline, the vectors of the triad, as well as the curvature K and the torsion W, change. Let

$$K + \delta K, \quad \text{and} \quad W + \delta W \tag{3}$$

denote the curvature and torsion after application of the virtual deformation (both evaluated at the point Q, which originates from point P through the virtual deformation). Letting

$$\mathbf{t}, \quad \mathbf{n}, \quad \text{and} \quad \mathbf{b} \tag{4}$$

denote the accompanying vector triad at the point P, we seek to determine the accompanying vector triads at the points P', Q, and Q'.

To get from P to P', hold the midline fixed and proceed along the curve. For this traversal the Frenet–Serret formulas (2) hold and, consequently, we obtain expressions,

$$\left.\begin{aligned} &\mathbf{t} + K\mathbf{n}ds, \\ &\mathbf{n} + (-K\mathbf{t} + W\mathbf{b})ds, \\ &\mathbf{b} - W\mathbf{n}ds, \end{aligned}\right\} \tag{5}$$

for the three vectors of the triad at the point P'.

The point Q is reached after imposing the virtual deformation. For each band element, this deformation may be decomposed into two parts, one involving a virtual translation of the point P and the other involving a virtual rotation around that same point. However, since a virtual translation leaves the vector triad unchanged, only the virtual rotation alters the vectors of the triad. Consequently, to determine the new triad, we need only consider the virtual rotation $\delta\boldsymbol{\vartheta}$. Under an infinitesimal rotation $\delta\boldsymbol{\vartheta}$, an arbitrary vector $\mathbf{p}$ transforms to the vector $\mathbf{p} + \delta\boldsymbol{\vartheta} \times \mathbf{p}$. For example, the vector $\mathbf{t}$ becomes the vector $\mathbf{t} + \delta\boldsymbol{\vartheta} \times \mathbf{t}$. However, following the kinematic definition of a band in Sect. 1, the vector triad of the midline of a band is *material*, that is upon imposing the virtual rotation, the original $\mathbf{t}$-vector translates into a new $\mathbf{t}$-vector, the original $\mathbf{b}$-vector translates into a new $\mathbf{b}$-vector, and so on. We now use this principle and write the following expression for the triad at the point Q:

$$\left.\begin{aligned} &\mathbf{t} + \delta\boldsymbol{\vartheta} \times \mathbf{t}, \\ &\mathbf{n} + \delta\boldsymbol{\vartheta} \times \mathbf{n}, \\ &\mathbf{b} + \delta\boldsymbol{\vartheta} \times \mathbf{b}. \end{aligned}\right\} \tag{6}$$

The point Q' may be reached in two ways. First, consider doing so from P'. This involves a transition corresponding to the application of a virtual rotation. In principle, conditions identical to those involved in the transition from P to Q hold; one just needs to be careful to use the vectors corresponding to the triad at Q'. These vectors are given through the formulas (5). Further, the virtual rotation at P' is no longer $\delta\boldsymbol{\vartheta}$, but rather is $\delta\boldsymbol{\vartheta} + d\delta\boldsymbol{\vartheta}$, since the point under consideration is advanced by a distance ds along the midline. On the basis of these considerations, the accompanying vector triad at the point P' becomes

$$\left.\begin{aligned} &\mathbf{t} + K\mathbf{n}ds + (\delta\boldsymbol{\vartheta} + d\delta\boldsymbol{\vartheta}) \times (\mathbf{t} + K\mathbf{n}ds), \\ &\mathbf{n} + (-K\mathbf{t} + W\mathbf{b})ds + (\delta\boldsymbol{\vartheta} + d\delta\boldsymbol{\vartheta}) \times \big(\mathbf{n} + (-K\mathbf{t} + W\mathbf{b})ds\big), \\ &\mathbf{b} - W\mathbf{n}ds + (\delta\boldsymbol{\vartheta} + d\delta\boldsymbol{\vartheta}) \times (\mathbf{b} - W\mathbf{n}ds). \end{aligned}\right\} \tag{7}$$

The other way to reach Q' goes through Q. This is a displacement along the fixed (virtually displaced) midline. Thus, the Frenet–Serret formulas (2) hold for this transition. One only needs to keep in mind that the vector triad at Q, as given in (6), corresponds to the starting point and that the curvature and twist of the midline now take the values

$$K + \delta K \quad \text{and} \quad W + \delta W$$

provided in (3). The computation defines a vector triad,

$$\left.\begin{aligned}&\mathbf{t}+\delta\boldsymbol{\vartheta}\times\mathbf{t}+(K+\delta K)(\mathbf{n}+\delta\boldsymbol{\vartheta}\times\mathbf{n})ds,\\&\mathbf{n}+\delta\boldsymbol{\vartheta}\times\mathbf{n}+\left[-(K+\delta K)\mathbf{t}+\delta\vartheta\times t+(W+\delta W)(\mathbf{b}+\delta\boldsymbol{\vartheta}\times\mathbf{b})\right]ds,\\&\mathbf{b}+\delta\boldsymbol{\vartheta}\times\mathbf{b}-(W+\delta W)(\mathbf{n}+\delta\boldsymbol{\vartheta}\times\mathbf{n})ds,\end{aligned}\right\}\tag{8}$$

for the point Q'. We now have two representations for the triad in the point Q', namely (7) and (8). Consequently, the two sets of equations must describe the same triads of vectors. Through pairwise comparison of the corresponding vectors in (7) and (8), one arrives, after neglecting all terms of third order in infinitesimally small quantities, at three relations between infinitesimally small quantities of second order. These relations are

$$\left.\begin{aligned}&\mathbf{n}\delta K ds=d\delta\boldsymbol{\vartheta}\times\mathbf{t},\\&-\mathbf{t}\delta K ds+\mathbf{b}\delta W ds=d\delta\boldsymbol{\vartheta}\times\mathbf{n},\\&-\mathbf{n}\delta W ds=d\delta\boldsymbol{\vartheta}\times\mathbf{b}.\end{aligned}\right\}\tag{9}$$

The variable $d\delta\boldsymbol{\vartheta}$ appearing in (9) is the virtual rotation of the vector triad in P' relative to the vector triad in P. The system of equations (9) allows for the determination of the components of $d\delta\boldsymbol{\vartheta}$. The relationship

$$d\delta\boldsymbol{\vartheta}=(\mathbf{t}\delta W+\mathbf{b}\delta K)ds\tag{10}$$

is a general solution to (9). *Together with the constraint* $\delta(ds)=0$, (10) *can be recognized to provide a foundation for the theory of bands.*

3 The Equations of Static Equilibrium and the Differential Equations of a MÖBIUS Band

The considerations of the two preceding paragraphs hold for bands of arbitrary (that is, not necessarily infinitesimally narrow) width which have a planar shape in the absence of stress and a midline representing a geodesic line on the band surface. From now on, discussion is restricted to *infinitesimally narrow* bands.

Consider a planar cut through the point P on the midline of such a band. The sectional plane shall be the normal plane of the band at the point P. The stresses appearing in the sectional plane, which act from the part of the band with larger s to the part of the band with smaller s, shall be represented by a force and a moment, both acting at the point P. Using the accompanying vector triad at P as a basis for decomposing the force and moment, we write

$$\mathfrak{K}=T\mathbf{t}+N\mathbf{n}+B\mathbf{b}\tag{11}$$

for the force and

$$\mathfrak{M}=\mathfrak{T}\mathbf{t}+\mathfrak{N}\mathbf{n}+\mathfrak{B}\mathbf{b}\tag{12}$$

for the moment.

With the requirement that no external forces act on the band, the equilibrium conditions for the band are

$$\frac{d\mathfrak{K}}{ds}=\mathbf{0}\tag{13}$$

and

$$\frac{d\mathfrak{M}}{ds} + \mathbf{t} \times \mathfrak{K} = \mathbf{0}. \tag{14}$$

Using the decomposition (11) and (12), differentiating in accord with the Frenet–Serret formulas (2), and decomposing the resulting relations into components yields six scalar equations for the six scalar components T, N, B, $\mathfrak{T}$, $\mathfrak{N}$, and $\mathfrak{B}$. The midline curvature K as well as the midline twist W appear in these equations, which read

$$\left.\begin{aligned}
&\frac{dT}{ds} - KN = 0,\\
&KT + \frac{dN}{ds} - WB = 0,\\
&WN + \frac{dB}{ds} = 0,\\
&\frac{d\mathfrak{T}}{ds} - K\mathfrak{N} = 0,\\
&K\mathfrak{T} + \frac{d\mathfrak{N}}{ds} - W\mathfrak{B} - B = 0,\\
&W\mathfrak{N} + \frac{d\mathfrak{B}}{ds} + N = 0.
\end{aligned}\right\} \tag{15}$$

Notice that (15) do not include a connection between deformation and stress. Consequently, they hold for arbitrary bodies with pronounced midline (rigid or elastic ropes, wires, bands, etc.). To attack our problem, which relates to the equations of the band, from this perspective, we must provide a relationship between deformation and stress. To achieve this, we invoke and follow the principle of virtual displacements.

The theory of virtual work of the internal forces in a one-dimensional continuum with material midline can be found in the literature [2–5]. In general, the shear force is always a reaction force, that is, the virtual work of the shear force vanishes identically. Further, for inextensible ropes—and the band is imagined to be of such a nature—the tensile force is a reaction force as well and the corresponding virtual work thus also vanishes identically. The remaining contribution to the internal virtual work δA_i is the virtual work of the moment $\mathfrak{M}$, which is

$$\delta A_i = -\int_{s_1}^{s_2} \mathfrak{M} \cdot d\delta\boldsymbol{\vartheta}, \tag{16}$$

where $\delta\boldsymbol{\vartheta}$ is the virtual rotation of the accompanying vector triad of the midline of the band. However, in (10) of the preceding paragraph, we found the most general form for $d\delta\boldsymbol{\vartheta}$ that is compatible with the properties of a band. If we now use (10) together with the component representation (12) of $\mathfrak{M}$, and plug it into (16), we arrive at an expression,

$$\delta A_i = -\int_{s_1}^{s_2} (\mathfrak{T}\delta W + \mathfrak{B}\delta K)ds, \tag{17}$$

for the virtual work of the internal forces. *We thus see that, for a band, the moment* $\mathfrak{M}$ *is also a reaction moment.*

The relationship (17) holds for every object that may be called *band* in the sense of the definition of bands in Sect. 1. The result (17) is a general relation that holds for arbitrary bands. Now, an arbitrary band is a rather complex object. In the current work, we do not seek to develop a general theory of bands, but rather to explicitly treat only MÖBIUS bands. A MÖBIUS band is characterized energetically through its elastic potential

$$U = A\frac{(K^2 + W^2)^2}{K^2}, \tag{18}$$

where A is a positive constant (see Sadowsky [1], in which a justification for this elastic potential is provided). The elastic energy E of a band segment from $s = s_1$ to $s = s_2$ is then given by

$$E = \int_{s_1}^{s_2} U ds = A\int_{s_1}^{s_2} \frac{(K^2 + W^2)^2}{K^2} ds, \tag{19}$$

and the variation of the energy, caused by the virtual deformation of the band, is given as

$$\delta E = \int_{s_1}^{s_2} \left(\frac{\partial U}{\partial W}\delta W + \frac{\partial U}{\partial K}\delta K\right) ds. \tag{20}$$

However, since $\delta A_i = -\delta E$, and since δK and δW are independent variations, comparison of (17) and (20) leads to the conclusion that

$$\mathfrak{T} = A\frac{4W(K^2 + W^2)}{K^2}, \tag{21}$$

$$\mathfrak{B} = A\frac{2(K^4 - W^4)}{K^3}, \tag{22}$$

with $A > 0$ constant.

The six Eqs. (15) and the two Eqs. (21) and (22) form a system of eight equations for eight unknowns

$$T,\ N,\ B,\ \mathfrak{T},\ \mathfrak{N},\ \mathfrak{B},\ K,\quad \text{and}\quad W.$$

We shall refer to these eight equations as the differential equations of a MÖBIUS band, since through the integration of these equations one may determine the curvature K and the twist W as a function of the arc length s. This then is what may be called a natural description of the midline of the band, and this description determines the unique space curve of the midline of the band up to arbitrary rigid motions. Naturally, these are differential equations that need to be supplemented with appropriate boundary conditions at $s = 0$ and $s = l$ (where l is the length of the MÖBIUS band); these boundary conditions shall account for the correct closure of the strip to a MÖBIUS band.

Of course, one could eliminate the six coefficients from (11) and (12) from the eight equations and in the end arrive at two equations for the two geometric variables K and W (this elimination is possible without further difficulties), however the result becomes more complicated as a result, and compromises clarity to such an extent that this elimination is not conducted here.

The coefficients in (11) and (12) may be expressed rather clearly through K, W, and their derivatives with respect to arc length.

The expressions for the coefficients are

$$T = AC - U = A\left(C - \frac{(K^2 + W^2)^2}{K^2}\right),$$

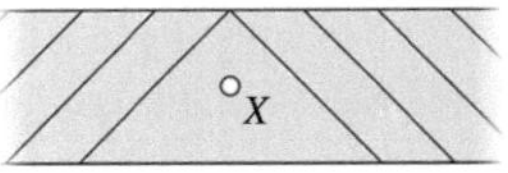

Fig. 2 Adaptation of Fig. 2 from the original version of the paper

where C is an arbitrary integration constant,

$$N = -\frac{A}{K}\frac{d}{ds}\frac{(K^2+W^2)^2}{K^2}, \tag{23}$$

$$B = \frac{2AW}{K^3}\left(K^2+W^2\right)^2 + 4A\frac{d}{ds}\left(\frac{1}{K}\frac{d}{ds}\frac{W(K^2+W^2)}{K^2}\right), \tag{24}$$

$$\mathfrak{N} = \frac{4A}{K}\frac{d}{ds}\frac{W(K^2+W^2)}{K^2}, \tag{25}$$

($\mathfrak{T}$ and $\mathfrak{B}$, cf. (21) and (22)).

These six equations further need to satisfy the conditions

$$KT + \frac{dN}{ds} - WB = 0 \tag{26}$$

and

$$WN + \frac{dB}{ds} = 0. \tag{27}$$

Plugging (25) into (26) and (27) would give the two differential equations for the curvature K and the twist W.

4 A Peculiar Implication for the Shape of a Möbius Band

Due to a lack of space, this last section can only be presented in the shortest possible form. Following the considerations with the help of a paper model and scissors is therefore recommended, as these considerations might otherwise remain incomprehensible.

The midline of the band possesses a singular point X that can be found in the following way: a Möbius band possesses a symmetry axis defined such that it is congruent with itself after a rotation of 180° about that axis. The symmetry axis intersects the band at two points such that the axis coincides with the binormal **b** at one of these points. In Fig. 2, this is the point X.

Let φ denote the angle between the rectilinear generators of the band through the point P and **b**; then

$$\tan\varphi = \frac{W}{K}. \tag{28}$$

For the point X,

$$\mathfrak{B} = 0 \tag{29}$$

due to symmetry.

To arrive at an additional conclusion, consider an *experiment with a band model*: this shows that

$$\lim_{P\to X}\varphi \neq 0, \tag{30}$$

$$\lim_{P\to X}\mathfrak{T} \neq 0. \tag{31}$$

The experiment corresponding to (30) consists of observing a band; for (31), one needs to cut the boundaries of a band at X to infer the moment $\mathfrak{T}$ from the twist of the band in the weakened cross section. By (29), (30), (31), and the governing equations (21) and (22) of the band, it transpires that

$$\lim_{P\to X} K \neq 0, \qquad \lim_{P\to X} W \neq 0, \quad \text{and} \quad \lim_{P\to X} \varphi = 45^\circ. \tag{32}$$

The last limit in (32) points to the following peculiar fact:

A MÖBIUS *band consists of a planar, right triangle. The curved, analytic portion of the band connects continuously with the two legs of the right triangle; it connects with continuous tangential plane, but with discontinuous curvature.*

Acknowledgements The authors thank Roger L. Fosdick, Russel E. Todres, and David M. Kleiman for enlightening discussions during the translation of Sadowsky's paper, discussions which led to the inclusion of the Appendix for clarification of Sadowsky's original considerations.

Appendix: Interpretation and Explanation of Sect. 4 of Sadowsky's Paper

Unfortunately, Sect. 4 of Sadowsky's paper is too brief to contain illustrative explanations as to why his "peculiar implication" is so remarkable and important. In this appendix, we attempt to provide an additional illustrative explanation of Sadowsky's early observation. To this end, we discuss the last section in detail and use recent simulation results by Kleiman et al. [6].

The curvature of the midline K represents the curvature along the tangent direction of the midline, which cuts through the rectilinear generators of the band surface at an angle. This angle is $90^\circ \pm \varphi$, where φ is the angle used by Sadowsky in (28). This geometric consideration (or a similar geometric argument) leads to (28).

Further, the (rotational) symmetry axis of a Möbius band can indeed easily be found with a paper model. In the simulation results, the symmetry axis coincides with the z-axis of the coordinate system, as shown in Fig. 3. Due to symmetry, the bending moment $\mathfrak{B}$ in (22) vanishes at X (see (29)). Using the condition (29) in (22) yields

$$K^4 - W^4 = 0 \tag{33}$$

and, thus,

$$K = \pm W \tag{34}$$

at X, which includes the possibility that $K = 0$ and $W = 0$ at X.

Now, suppose that $K \neq 0$ at X. In view of (34),

$$\frac{W}{K} = \pm 1 \tag{35}$$

and, thus,

$$\tan\varphi = \pm 1. \tag{36}$$

The angle φ would accordingly be

$$\varphi = \pm\frac{1}{4}\pi = \pm 45^\circ. \tag{37}$$

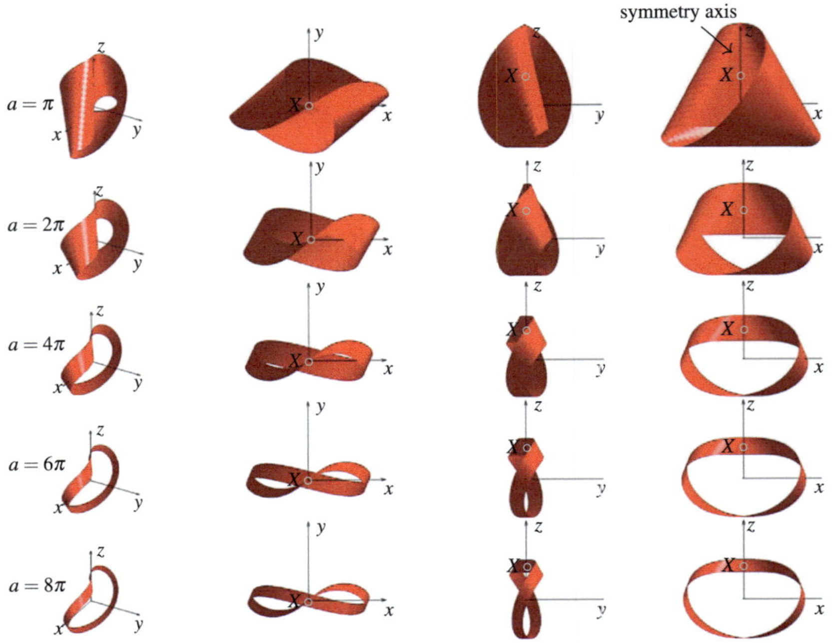

Fig. 3 Location of the singular point X and the (rotational) symmetry axis: Equilibrium shapes of approximately developable Möbius bands for different aspect ratios a obtained from simulations with a lattice model by Kleiman et al. [6]. The band is rotated into its main axes and the approximate location of the point X is indicated. With the current choice of coordinate system, the (rotational) symmetry axis coincides with the z-axis

Equation (37) implies that there must be two perpendicular rectilinear generators of the bent surface at X, which means that the bent surface must be flat at X. However, a flat surface at X contradicts the supposition that $K \neq 0$ at X. Consequently,

$$K = 0 \quad \text{and} \quad W = 0 \tag{38}$$

at X. Note that the rectilinear generator at X coincides with $\mathbf{b}$ at X by symmetry. The result (38) is confirmed by plots generated by our numerical simulations shown in Fig. 4.

Now, consider the point P on the midline close to X. The curvature K cannot be zero at this point. This becomes clear from the following observation. If K was zero at P, the surface would have zero curvature along the rectilinear generator through P as well as along the midline tangent direction through P. Since these two tangent directions are not equal, the surface would have to be flat at P, and not bent, as is presumed. Consequently,

$$K \neq 0 \tag{39}$$

at P. With (39) holding at P, W may or may not vanish. If W vanishes at P, then the rectilinear generator through P would have to coincide with $\mathbf{b}$ at P, since $\varphi = 0$ by (28). That means that, for $W = 0$, the rectilinear generator at P is perpendicular to the tangent direction of the midline at P, in which case K represents the single non-zero principal curvature of the bent surface at P. In addition, for $W = 0$, in view of (21), there would be no twisting moment around the axis tangent to the midline at P. However, this condition of $K \neq 0$ and $W = 0$ cannot persist for all P on the midline, since such a configuration would

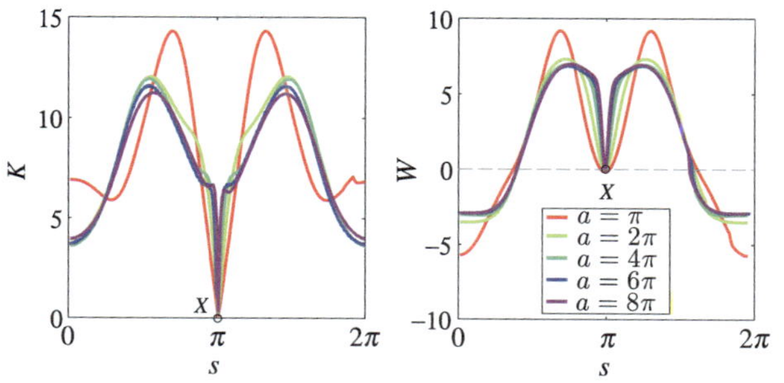

Fig. 4 The curvature and twist (in arbitrary units) of developable Möbius bands show the singular point X with $K = 0$ and $W = 0$: Curvature and twist of the centerline of the developable Möbius bands for different aspect ratios a. The singular nature of this point becomes more prominent for increasing aspect ratio a. Results are obtained from simulations with a lattice model by Kleiman et al. [6] where complete results, validation, and discussion can be found

not be consistent with an isometric mapping to a bent (and twisted) Möbius band. In other words, for K to vanish, W must vanish as well (corresponding to the singular point X), whereas W may vanish without K vanishing, but not for all points. This is confirmed by our simulation results (Fig. 4), which consistently show three zeros for W and non-zero W on the rest of the band.

In view of the preceding discussion, we conclude that what Sadowsky was trying to communicate with (30), (31), and (32) is that any point P on the midline in a neighborhood of X has $K \neq 0$ and must also have $W \neq 0$, which, according to (21), ensures that there is a non-zero twisting moment about the tangent axis through P and, according to (28), that $\varphi \neq 0$ at P. Moreover, the condition $W \neq 0$ persists up to the point X, where both K and W are zero. In other words, since $K = 0$ and $W = 0$ at X and $K \neq 0$ and $W \neq 0$ at P, the generators connect to the flat point X at $\pm 45°$. This corresponds to the "planar, right triangle" to which Sadowsky refers. In fact, such planar, right triangles are well observed in the simulation results shown in Fig. 3, especially for low aspect ratios. For high aspect ratios, the singular nature of the curvature and twist at X becomes more prominent, as shown in Fig. 4. Considering that Sadowsky did not have access to simulation results, his early observation is quite remarkable and important.

References

1. Sadowsky, M.: Ein elementarer Beweis für die Existenz eines abwickelbaren Möbiusschen Bandes und die Zurückführung des geometrischen Problems auf ein Variationsproblem. Sitz.ber. Preuss. Akad. Wiss. Berl. Philos.-Hist. Kl. **22**, 412–415 (1930)
2. Hamel, G.: Über die Mechanik der Drähte und Seile. Sitzungsber. Berl. Math. Ges. **25**, 3–8 (1926)
3. Hamel, G.: Die Axiome der Mechanik. In: von Geiger, H., Scheel, K. (eds.) Handbuch der Physik, vol. 5, pp. 1–42. Springer, Berlin (1927)
4. Hamel, G.: Über Seilsteifigkeit. Z. Angew. Math. Mech. **7**, 114–119 (1927)
5. Hamel, G.: Lehrbuch der elementaren Mechanik, 1st edn., vol. 36. Teubner, Leipzig (1912), § 36
6. Kleiman, D.M., Hinz, D.F., Fried, E.: Stretchable Möbius bands. Proceedings of the National Academy of Sciences of the United States of America (submitted)

DOI 10.1007/978-94-017-7300-3_4
Reprinted from *Journal of Elasticity* Journal, DOI 10.1007/s10659-014-9491-4

Translation of Michael Sadowsky's Paper "The Differential Equations of the MÖBIUS Band"

Denis F. Hinz · Eliot Fried

Received: 8 August 2014 / Published online: 21 August 2014

Abstract This article is a translation of Michael Sadowsky's original paper "Die Differentialgleichungen des MÖBIUSschen Bandes" in Jahresbericht der Deutschen Mathematiker-Vereinigung **39** (2. Abt. Heft 5/8, Jahresversammlung vom 16. bis 23. September), 49–51 (1929), which is a short version of his paper "Theorie der elastisch biegsamen undehnbaren Bänder mit Anwendungen auf das MÖBIUS'sche Band" in 3. internationaler Kongress für technische Mechanik, Stockholm, 1930.

Keywords Möbius band · Energy functional · Bending energy

Mathematics Subject Classification 74K20 · 74K10 · 53A04 · 74G55 · 01A75

Translation of the Original Paper

In a previously completed work, Sadowsky [1] provided a proof for the *existence* of a developable MÖBIUS band and provided a variational context for the underlying geometric problem. In the present work, the problem of determining the equilibrium shape of a MÖBIUS band formed from an elastic strip is treated as a static mechanical problem.

Let P denote an arbitrary point on the midline of the MÖBIUS band, let s denote the arclength of the midline, let $\mathbf{t}$, $\mathbf{n}$, and $\mathbf{b}$ denote the accompanying vector triad at P, and let K and W denote the curvature and twist of the midline at P. Considering a cut perpendicular to the midline at P, the stresses along the line of the cut may be reduced to a force $\mathfrak{K}$ and a moment $\mathfrak{M}$ both acting at P.

Citations of this translation should refer also to Sadowsky's original paper, as cited in the Abstract.

D.F. Hinz · E. Fried (✉)
Mathematical Soft Matter Unit, Okinawa Institute of Science and Technology, Okinawa 904-0495, Japan
e-mail: eliot.fried@oist.jp

D.F. Hinz
e-mail: dfhinz@gmail.com

The representations of $\mathfrak{K}$ and $\mathfrak{M}$ relative to the triad **t**, **n**, and **b** shall be denoted as

$$\mathfrak{K} = T\mathbf{t} + N\mathbf{n} + B\mathbf{b}, \tag{1}$$

and

$$\mathfrak{M} = \mathfrak{T}\mathbf{t} + \mathfrak{N}\mathbf{n} + \mathfrak{B}\mathbf{b}. \tag{2}$$

From a mechanical perspective, the band is defined through the following three requirements:

1. The band is an object with distinguished midline.
2. The vector triad **t**, **n**, and **b** accompanying the midline is material.
3. The midline is inextensible.

(The second requirement results from the band being part of the rectifying surface of its midline.)

Applying a virtual deformation to the band and invoking the Frenet–Serret formulas results in

$$d\delta\boldsymbol{\vartheta} = (\mathbf{t}\delta W + \mathbf{b}\delta K)ds, \tag{3}$$

and in

$$\delta ds = 0, \tag{4}$$

where $\delta\boldsymbol{\vartheta}$ is the virtual rotation of the vector triad at P.

The following considerations hold only for infinitesimally narrow bands.

The virtual work of the internal forces is[1]

$$\delta A_i = -\int_{s_1}^{s_2} (\mathfrak{T}\delta W + \mathfrak{B}\delta K)ds. \tag{5}$$

The mean curvature H of the band surface at P is[1]

$$H = \frac{K^2 + W^2}{2K}. \tag{6}$$

The elastic potential E of the band is proportional to the square of the mean curvature (granted the presumption that the band surface is a developable surface)

$$E = A\frac{(K^2 + W^2)^2}{K^2}. \tag{7}$$

Since $\delta A_i = -\int_{s_1}^{s_2} \delta E ds$ and, in view of (5) and (7), it follows that

$$\mathfrak{T} = \frac{\partial E}{\partial W} = A\frac{4W(K^2 + W^2)}{K^2} \tag{8}$$

and

$$\mathfrak{B} = \frac{\partial E}{\partial K} = A\frac{2(K^4 - W^4)}{K^3}, \tag{9}$$

[1] HAMEL, Über die Mechanik der Drähte und Seile, Sitzungsberichte der Berliner Mathematischen Gesellschaft, XXV, 1925/26. Also HAMEL Die Axiome der Mechanik, Handbuch der Physik, herausgegeben von H. GEIGER und KARL SCHEEL, Band V. In the latter work further references may be found.

where A is a positive material constant. The six components appearing in (1) and (2) satisfy six known *equilibrium equations*[1]. Using (8) and (9), those equations read

$$\begin{cases} T = A\Big(C - \dfrac{(K^2+W^2)^2}{K^2}\Big), \\ N = -\dfrac{A}{K}\dfrac{d}{ds}\dfrac{(K^2+W^2)^2}{K^2}, \\ B = \dfrac{2AW}{K^3}\left(K^2+W^2\right)^2 + 4A\dfrac{d}{ds}\left(\dfrac{1}{K}\dfrac{d}{ds}\dfrac{W(K^2+W^2)}{K^2}\right), \\ \mathfrak{N} = \dfrac{4A}{K}\dfrac{d}{ds}\dfrac{W(K^2+W^2)}{K^2}, \end{cases} \tag{10}$$

and

$$\begin{cases} KT + \dfrac{dN}{ds} - WB = 0, \\ WN + \dfrac{dB}{ds} = 0, \end{cases} \tag{11}$$

where C is an arbitrary constant of integration.

If the expressions (10) were to be used in (11), two differential equations for K and W, respectively, would result. The integration of these equations would yield the natural midline of the band. Equations (8), (9), (10), and (11) shall be referred to as the equations of the MÖBIUS band.

The midline of the band possesses a singular point X that can be found in the following way: a MÖBIUS band possesses a symmetry axis defined such that it is congruent with itself after a rotation of 180° about that axis. The symmetry axis intersects the band at two points such that the axis coincides with the binormal **b** at one of these points; this is the point X.

Let φ denote the angle between the rectilinear generators of the band through the point P and **b**; then[1]

$$\tan\varphi = \frac{W}{K}. \tag{12}$$

For the point X,

$$\mathfrak{B} = 0 \tag{13}$$

due to symmetry.

To arrive at an additional conclusion, consider an *experiment with a band model*:[2] this shows that

$$\lim_{P\to X} \varphi \neq 0, \tag{14}$$

$$\lim_{P\to X} \mathfrak{T} \neq 0. \tag{15}$$

The experiment corresponding to (14) consists of observing a band; for (15), one needs to cut the boundaries of a band at X to infer the moment $\mathfrak{T}$ from the twist of the band in the

[2]For an interpretation of the ensuing discussion and result, including figures generated from recent numerical simulations, refer to D.F. Hinz and E. Fried, Translation and interpretation of Michael Sadowsky's paper "Theory of elastically bendable inextensible bands with applications to the MÖBIUS band", appearing in the present volume of the *Journal of Elasticity*.

weakened cross-section. From (13), (14), (15), and the governing equations (8) and (9) of the band it transpires that

$$\lim_{P\to X} K \neq 0, \qquad \lim_{P\to X} W \neq 0, \quad \text{and} \quad \lim_{P\to X} \varphi = 45^\circ. \tag{16}$$

The last limit in (16) points to the following peculiar fact:

A MÖBIUS *band consists of a planar, right triangle. The curved, analytic portion of the band connects continuously with the two legs of the right triangle; it connects with continuous tangential plane, but with discontinuous curvature.*

References

1. Sadowsky, M.: Ein elementarer Beweis für die Existenz eines abwickelbaren Möbiusschen Bandes und die Zurückführung des geometrischen Problems auf ein Variationsproblem. Sitz.ber. Preuss. Akad. Wiss. Phys.-Math. Kl. **22**, 412–415 (1930)

DOI 10.1007/978-94-017-7300-3_5
Reprinted from *Journal of Elasticity* Journal, DOI 10.1007/s10659-014-9489-y

Translation of W. Wunderlich's "On a Developable Möbius Band"

Russell E. Todres

Received: 23 July 2014 / Published online: 26 August 2014

Abstract The following is a translation of Walter Wunderlich's article "Über ein abwickelbares Möbiusband", which appeared in the *Monatshefte für Mathematik* **66** (1962), 276–289 and was dedicated to Prof. Dr. Paul Funk on the occasion of his 75th birthday. Wunderlich summarizes Sadowsky's work (Sitzber. Preuss. Akad. Wiss. **22**:412–415, 1930; Verhandlungen des 3. Internationalen Kongresses für Technische Mechanik, II (Stockholm, 1930), pp. 444–451, Sveriges Litografiska Tryckerier, Stockholm, 1931) on developable Möbius bands and improves Sadowsky's upper bound of the dimensionally-reduced variational description for determining the configuration of a Möbius band whose width is small in comparison to its length. Attempting to reproduce the equilibrium depiction of a band of finite width, using a rational-algebraic developable, Wunderlich then extends Sadowsky's results by presenting perhaps the first successful model of a closed, analytic, developable Möbius band with associated thinness bounds. This translation makes Wunderlich's work accessible to the broader research community at a time of growing interest in and relevance of thin-walled structural elements.

Keywords Möbius bands · Differential geometry · Developable surfaces

Mathematics Subject Classification 53A04 · 74G10 · 74G55 · 74K10 · 74K20 · 01A75

1 Introduction

To illustrate a one-sided surface, *F.A. Möbius* described a band model, which has since become famous, and which arises by connecting the narrow edges of a long, rectangular strip

Superscripted Arabic numbers refer to footnotes in the original, while superscripted lowercase Roman numerals are used for translator's notes appearing at the end. References are those which appear in the original paper, available at: https://eudml.org/doc/177173.

R.E. Todres (✉)
Mathematical Soft Matter Unit, Okinawa Institute of Science and Technology, Okinawa 904-0495, Japan
e-mail: russell.todres@oist.jp

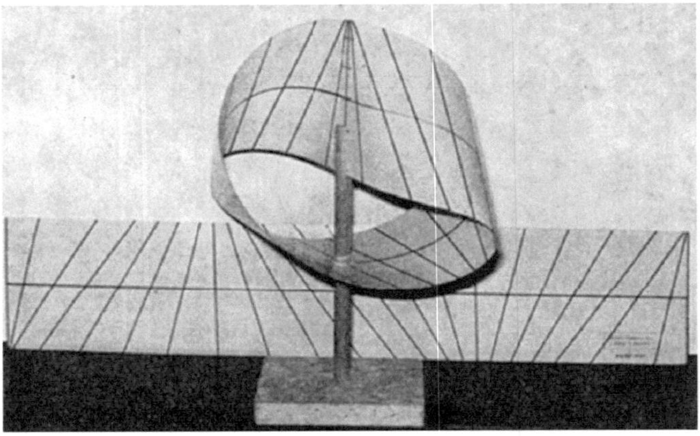

Fig. 1 Equilibrium form of the Möbius band

of paper so that all diagonally opposed vertices meet [1]. Since the material is essentially inextensible, a *developable surface* results, on which the long, originally straight centerline of the rectangle depicts a *closed, geodesic line m*. Conversely, the band surface can thus be considered as the *rectifying developable*[i] of m, namely, the envelope of tangent planes normal to the osculating planes of m.

With regard to the pure topological meaning of the model, the developability of the band is basically irrelevant, and so simple, analytical representations of *non-developable Möbius bands* are not lacking. A natural, elementary, geometrical way to generate such a Möbius band is as follows: Take a circle, which rotates about an axis lying in its plane but not intersecting the circle; a diameter, which simultaneously turns in the circle's surface with half the angular velocity, then generates a zone consisting of a skew, metric, special *cubic ruled surface*, which runs within a ring torus, is bounded by a sixth-order space curve, and has been considered by several authors [2–6]. In the projective generalization, it is therefore valid that a one-sided band with the character of a Möbius band exists along each ellipse situated on a general third-order ruled surface.

The thought of perhaps attaining a *developable model* of a Möbius band by means of considering a third-order developable on a conic section k and associated with a cubic ruled surface leads to an unsatisfactory result due to the following reasons: The generators of the developable are the conjugate surface tangents to the tangents of k; at the point T of k where the plane containing the conic section touches the surface, the tangent of k is however osculant, i.e., the tangent coincides with its conjugate, and the edge of regression[ii] of the developable approaches k tangentially. This disruptive singularity cannot be eliminated no matter how narrow the band, and every other plane curve k is ruled out for the same reason, even if its geodesic character is not considered. Since the plane containing k must lie in the tangent plane of the developable, k is a tangential section of the developable and is therefore touched by the edge of regression.

These circumstances may have contributed to occasional doubts about the theoretical existence of developable Möbius bands, as no less than H.A. Schwarz, among others, is supposed to have expressed. To remedy such reservations, *M. Sadowsky* presented a simple, elementary band model composed of three cylindrical and three planar sections [7].

Such doubts seem strange, however, when the Möbius band model is actually made, since it succeeds easily with a sufficient thinness ratio of the rectangular strip—approximately $<1/5$. The exact upper-limit of the ratio is not yet known. Incidentally, with a sufficiently stiff material, a well-defined, *stable equilibrium form* appears (Fig. 1), which only seems weakly dependent on the band width. *M. Sadowsky* [8] also strove to determine this equilibrium shape and the internally occurring stresses but did not however arrive at a final solution of this extremely difficult problem. Proceeding from the fact that the work of deformation necessary to bend a thin, flat, elastic, small plate over a cylinder is proportional to the area F and the square of the normal curvature $\bar{\varkappa}$,[1] it follows from the principle of *Minimum Deformation Energy* that the equilibrium form under discussion is a solution of the *variational problem*

$$\int \bar{\varkappa}^2 \cdot \mathrm{d}F = \mathrm{Min}, \tag{1}$$

where $\bar{\varkappa}$ is the principal normal curvature perpendicular to the rectilinear generator (bending line) of the developable portrayed by means of the Möbius band.

With s denoting the measured arc length along the centerline of the band, $\vartheta = \vartheta(s)$ the angle subtending the bending line and the centerline, and, imagining oneself in the projected rectangular strip—length L, width $2b$—with Cartesian coordinates u, v introduced in the length and transverse directions $(0 \le u \le L, -b \le v \le b)$, the *point of regression* V of the generator

$$u = s + v \cdot \cot\vartheta \tag{2}$$

belonging to the arc length s is then determined by[2]

$$V = \frac{\mathrm{d}s}{\mathrm{d}\vartheta} \cdot \sin^2\vartheta. \tag{3}$$

If the variables s and v are used as (surface) coordinates, then the surface element is

$$\mathrm{d}F = \mathrm{d}u\,\mathrm{d}v = \frac{V - v}{V}\,\mathrm{d}s\,\mathrm{d}v. \tag{4}$$

Since the principal normal curvature $\bar{\varkappa}$ along the generator (2) is inversely proportional to the distance from the point of regression, it is related to the principal normal curvature $\bar{\varkappa}_0 = \bar{\varkappa}_0(s)$ along the centerline $v = 0$ through[iii]

$$\bar{\varkappa} : \bar{\varkappa}_0 = V : (V - v), \tag{5}$$

and the integral (1) takes the following form:

$$\int \bar{\varkappa}^2\,\mathrm{d}F = \iint \frac{\bar{\varkappa}_0^2 V}{V - v}\,\mathrm{d}s\,\mathrm{d}v = \int_0^L \bar{\varkappa}_0^2 V \log\frac{V + b}{V - b}\,\mathrm{d}s. \tag{6}$$

For an "*infinitesimally thin*" *band* $(b \ll L)$, the integral (6) reduces to $2b\int \bar{\varkappa}_0^2\,\mathrm{d}s$. Considering that, according to the Euler formula for parabolic surface points, the common *curvature* $\varkappa = \varkappa(s)$ *of the centerline* is linked to the principal normal curvature $\bar{\varkappa}_0$ by $\varkappa = \bar{\varkappa}_0(s)\sin^2(s)$, and drawing on the familiar relation $\cot\vartheta = \tau/\varkappa$ from the theory of space

[1] The proportionality factor has the value $Eh^3/24$, where E is the modulus of elasticity and h the thickness.

[2] In the full interval $0 \le s \le L$, $|V(s)| > b$ must be maintained to keep the singularities of depicted points of regression outside of the rectangular region.

curves, where $\tau = \tau(s)$ is the *torsion of the centerline*, agreement is reached with *Sadowsky* for the description of the centerline by

$$M = \int_0^L \bar{\varkappa}_0^2 \, \mathrm{d}s = \int_0^L \frac{\varkappa^2}{\sin^4 \vartheta} \, \mathrm{d}s = \int_0^L \frac{(\varkappa^2 + \tau^2)^2}{\varkappa^2} \, \mathrm{d}s = \mathrm{Min}. \tag{7}$$

In *dimensionless form*, the centerline of the infinitesimally thin Möbius band therefore emerges from among all closed space curves, by which the osculating plane makes half a rotation after a full circuit, and is characterized by the minimal value of the quantity LM. The cylindrical strip model of *Sadowsky* [7] mentioned above delivers the bound[iv]

$$LM < 15\pi^2 = 148.04. \tag{8}$$

By an appropriate change of the angle formed by the three cylinder axes—by Sadowsky, all 60°—the bound can actually be reduced to

$$LM < 13.527\pi^2 = 133.51. \tag{9}$$

2 Reproduction of the Equilibrium Model

Although the present article provides no solution to the variational problem, an *approximate portrayal* of the equilibrium model depicted in Fig. 1 ought to be attempted by means of a *rational-algebraic developable*, which may simultaneously yield the first example ever of a *closed, analytic, developable Möbius band.*

Initially, on closer examination of the equilibrium form, the existence of a *symmetry axis* becomes evident, as represented in the model of Fig. 1 by the supporting rod. This axis intersects the centerline m of the Möbius band at two points A and C: At A, it is the principal normal of m and the surface normal of the band, while at C, on the other hand, it is the binormal of m and the generator of the band surface. We take the symmetry axis as the z axis of the Cartesian coordinate system to be used, whose origin O should lie in the middle of the section AC; we place the x axis parallel to the tangent of m in C.

In the normal projection onto the xy plane ("plan view"), the midline m appears as a figure eight with a turning node at the center of symmetry (see Fig. 2(a)), while the normal projection on the yz plane ("elevation") takes the form of a pure oval (see Fig. 2(b)), which can be approximated by the *ellipse*

$$y^2 + \frac{z^2}{c^2} = 1, \quad \text{with } c = \frac{4}{5}. \tag{10}$$

The power series expansion of m as a function of x in a neighborhood of the vertex C will begin as follows:

$$y = \lambda x^3 + \cdots, \qquad z = c\left(1 - \frac{\lambda^2}{2}x^6 + \cdots\right). \tag{11}$$

We now raise the *central projection* of the midline m from its singular point $C(0, 0, c)$ onto the xy plane—the "stereographic projection" of the elliptical cylinder (10), if you will:

$$\xi = \frac{cx}{c - z}, \qquad \eta = \frac{cy}{c - z}. \tag{12}$$

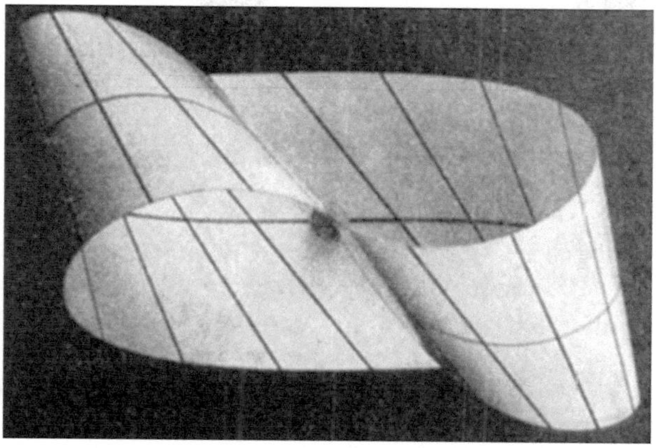

(a)

(b)

Fig. 2 (**a**) Top view. (**b**) Front view

The curved branch emanating from the center of projection is thereby changed into a parabolic one, which, according to (11), will have the asymptotic development

$$\xi \approx \frac{2}{\lambda^2 x^5}, \qquad \eta \approx \frac{2}{\lambda x^3}. \tag{13}$$

Correspondingly (and considering the central symmetry relative to the origin), the attempt may be made to specify the centrally projected midline by means of a parameter t in the form

$$\xi = \alpha t + \beta t^3 + \gamma t^5, \qquad \eta = \delta t + \varepsilon t^3. \tag{14}$$

Measurement of the photograph, Fig. 2(a), leads to the following guides for an appropriate determination of the coefficients: (a) at the point $A(0, 0, -c)$, belonging to the parameter value $t = 0$, the direction of travel is approximately $x : y = 3 : 4$, which determines the relation $\alpha : \delta = 3 : 4$; (b) on assigning the parameter value $t = 1$ to the available curve point $B(1, 1, 0)$, one obtains the requirements $\alpha + \beta + \gamma = \delta + \varepsilon = 1$; (c) so that the direction $x : y = \xi : \eta$ grows monotonically from A with t, as this apparently the case is, we must have $\beta\delta - \alpha\varepsilon \geq 0$. Together, this results in the following conditions:

$$\alpha = \frac{3}{4}(1 - \varepsilon) > 0, \qquad \beta \geq \frac{3}{4}\varepsilon > 0, \qquad 0 < \gamma \leq \frac{1}{4}, \qquad \delta = 1 - \varepsilon > 0, \quad 0 < \varepsilon < 1. \tag{15}$$

A viable assumption, obtained from a series of trials and which nicely approximates the figure eight of the plan view is:

$$\alpha = \frac{1}{2}, \qquad \beta = \frac{1}{3}, \qquad \gamma = \frac{1}{6}, \qquad \delta = \frac{2}{3}, \qquad \varepsilon = \frac{1}{3}. \tag{16}$$

For the following, it proves advantageous to transform to *homogeneous coordinates* with

$$x_0 : x_1 : x_2 : x_3 = 1 : x : y : z - c = \frac{1}{2}\left(1 + \eta^2\right) : \xi : \eta : -c. \tag{17}$$

The *midline m* described in this way, viz.,

$$\begin{aligned} x_0 &= \frac{1}{2}\left(1 + \delta^2 t^2 + 2\delta\varepsilon t^4 + \varepsilon^2 t^6\right), \\ x_1 &= \alpha t + \beta t^3 + \gamma t^5, \\ x_2 &= \delta t + \varepsilon t^3, \\ x_3 &= -c, \end{aligned} \tag{18}$$

can be recognized as a *rational space curve of sixth-order*. The curve is depicted in plan, elevation, and side view in Fig. 3 using (16). Ultimately, a curve of lower order appears to be impossible.

Now, the *band surface* itself as the *rectifying developable of the midline m* is to be determined. For this, one needs the first two derivatives

$$\begin{aligned} \dot{x}_0 &= \delta^2 t + 4\delta\varepsilon t^3 + 3\varepsilon^2 t^5, \\ \dot{x}_1 &= \alpha + 3\beta t^2 + 5\gamma t^4, \\ \dot{x}_2 &= \delta + 3\varepsilon t^2, \\ \dot{x}_3 &= 0, \end{aligned} \tag{19}$$

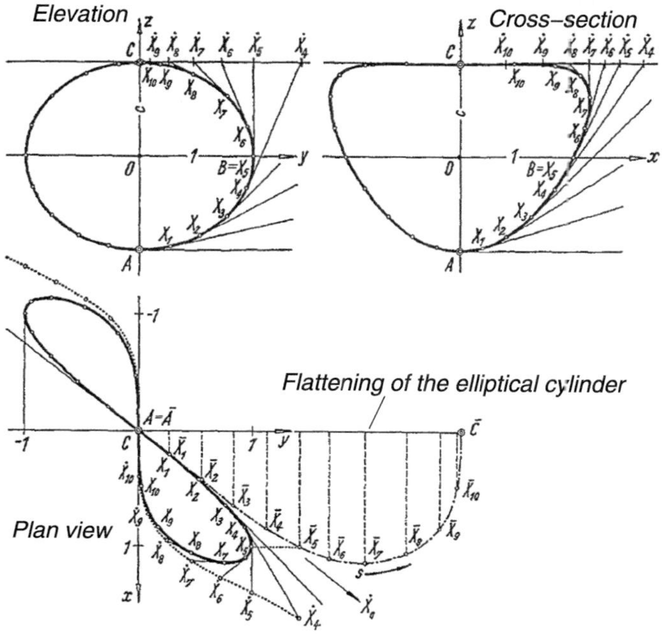

Fig. 3 Space curve of order 6 as a suitable centerline of a developable Möbius band

and

$$\begin{aligned}\ddot{x}_0 &= \delta^2 + 12\delta\varepsilon t^2 + 15\varepsilon^2 t^4,\\ \ddot{x}_1 &= 6\beta t + 20\gamma t^3,\\ \ddot{x}_2 &= 6\varepsilon t,\\ \ddot{x}_3 &= 0.\end{aligned} \tag{20}$$

For the point $X(x_i)$ on the curve, (19) and (20) respectively constitute a point $\dot{X}(\dot{x}_i)$ of the *tangent* of m and a point $\ddot{X}(\ddot{x}_i)$ of the *osculating plane* of m. From that, we first determine the *infinite point*[v] $U(u_i)$ *of the tangent* according to

$$u_i = x_0\dot{x}_i - \dot{x}_0 x_i \quad (i = 0, 1, 2, 3) \tag{21}$$

as

$$\begin{aligned}u_0 &= 0,\\ u_1 &= \frac{1}{2}\big[\alpha + (3\beta - \alpha\delta^2)t^2 + \cdots - \gamma\varepsilon^2 t^{10}\big],\\ u_2 &= \frac{1}{2}\big[\delta + (3\varepsilon - \delta^3)t^2 + \cdots - 3\varepsilon^3 t^8\big],\\ u_3 &= c\big(\delta^2 t + 4\delta\varepsilon t^3 + 3\varepsilon^2 t^5\big).\end{aligned} \tag{22}$$

Further, from the equation of the *osculating curve* $X\dot{X}\ddot{X}$ (with running coordinate y_i)

$$\begin{vmatrix} y_0 & x_0 & \dot{x}_0 & \ddot{x}_0 \\ y_1 & x_1 & \dot{x}_1 & \ddot{x}_1 \\ y_2 & x_2 & \dot{x}_2 & \ddot{x}_2 \\ y_3 & -c & 0 & 0 \end{vmatrix} \equiv v_0 y_0 + v_1 y_1 + v_2 y_2 + v_3 y_3 = 0, \tag{23}$$

we determine the *infinite point* $V(0 : v_1 : v_2 : v_3)$ *of the binormal* with

$$\begin{aligned} v_1 &= -c\big(\delta^3 + 9\delta^2\varepsilon t^2 + 27\delta\varepsilon^2 t^4 + 27\varepsilon^3 t^6\big), \\ v_2 &= c\big[\alpha\delta^2 + 3\delta(4\alpha\varepsilon - \beta\delta)t^2 + \cdots + 15\gamma\varepsilon^2 t^8\big], \\ v_3 &= 3(\beta\delta - \alpha\varepsilon)t + \cdots + 3\gamma\varepsilon^3 t^{11}. \end{aligned} \tag{24}$$

The *rectifying plane* $X\dot{X}V$ is then represented by

$$\begin{vmatrix} y_0 & x_0 & \dot{x}_0 & 0 \\ y_1 & x_1 & \dot{x}_1 & v_1 \\ y_2 & x_2 & \dot{x}_2 & v_2 \\ y_3 & -c & 0 & v_3 \end{vmatrix} \equiv w_0 y_0 + w_1 y_1 + w_2 y_2 + w_3 y_3 = 0, \tag{25}$$

with

$$\begin{aligned} w_0 &= -c^2\alpha^2\big(\alpha^2 + \delta^2\big) - \cdots - 6\gamma^2\varepsilon^4 t^{18}, \\ w_1 &= v_2 u_3 - v_3 u_2 = \left[c^2\alpha\delta^4 - \frac{3\delta}{2}(\beta\delta - \alpha\varepsilon)\right]t + \cdots + \frac{9}{2}\gamma\varepsilon^6 t^{19}, \\ w_2 &= v_3 u_1 - v_1 u_3 = \left[c^2\delta^5 + \frac{3\alpha}{2}(\beta\delta - \alpha\varepsilon)\right]t + \cdots - \frac{3}{2}\gamma^2\varepsilon^5 t^{21}, \\ w_3 &= v_1 u_2 - v_2 u_1 = -\frac{c\delta^2}{2}\big(\alpha^2 + \delta^2\big) - \cdots + \frac{15c}{2}\gamma^2\varepsilon^4 t^{18}. \end{aligned} \tag{26}$$

Thereby, we have also obtained the *infinite point* $W(0 : w_1 : w_2 : w_3)$ *of the principal normal* of m.

We can already tell from the rudimentary representation (26) that the developable surface representing our Möbius band is a *rational developable of class 21*. To determine its order, we still need the plane defined by differentiating with respect to t as

$$\dot{w}_0 y_0 + \dot{w}_1 y_1 + \dot{w}_2 y_2 + \dot{w}_3 y_3 = 0, \tag{27}$$

with the coefficients

$$\begin{aligned} \dot{w}_0 &= \cdots - 108\gamma^2\varepsilon^4 t^{17}, \\ \dot{w}_1 &= \left[c^2\alpha\delta^4 - \frac{3\delta}{2}(\beta\delta - \alpha\varepsilon)\right] + \cdots + \frac{171}{2}\gamma\varepsilon^6 t^{18}, \\ \dot{w}_2 &= \left[c^2\delta^5 + \frac{3\alpha}{2}(\beta\delta - \alpha\varepsilon)\right] + \cdots - \frac{63}{2}\gamma^2\varepsilon^5 t^{20}, \\ \dot{w}_3 &= \cdots + 135c\gamma^2\varepsilon^4 t^{17}. \end{aligned} \tag{28}$$

The line of intersection of the planes (25) and (27) is the *generator* passing through X of the rectifying developable. Calculating the Plücker coordinates $q_{ik} = w_i \dot{w}_k - w_k \dot{w}_i$, we recognize from q_{12} that our developable is of *order 39*.

A partially numerical, partially graphical procedure was used for the *graphical representation* of the Möbius band in Fig. 5.[vi] The coordinates x_i of X, $\dot{x}_i$ of $\dot{X}$, $\ddot{x}_i$ of $\ddot{X}$,[vii] u_i of U, v_i of V, and w_i of W were computed for the parameter values $t = 0, 0.2, 0.4, 0.6, 0.8, 1, 1.2, 1.5, 2, 3, 6$, and ∞. Based on these values, the *central projection of the ideal triangle* UVW with vertices at infinity was drawn from the origin O onto the picture plane $x = 1$ (Fig. 4(a)); likewise, as a supplement and check, the analogous, and in this case, co-linear projection on the plane $z = 1$ was made (Fig. 4(b)). The image of UVW is a self-polar triangle of the imaginary circle[viii] $y^2 + z^2 + 1 = 0$ or $x^2 + y^2 + 1 = 0$, whose centroid is therefore its orthocenter. The circle mentioned above may be interpreted as a reference conic section of an *elliptic metric* according to the Cayley–Klein model. The loci of the points of U and V ("pursuit curves" of the tangent surface and polar developable of m) have the illustrated triangle sides UW and VW as tangents and the sketch of the side UV as common elliptic normal. They form a pair of (polar reciprocal) *elliptic parallel curves* $\pi/2$ apart. Their common elliptic evolute—the envelope of the face UV—depicts the *pursuit curve of the developable band* and is polar reciprocal to the locus of points of W (pursuit curve of the principal normal surface).[3] The contact point E of UV with the envelope—verifiable in the figure as the pole of the tangent of the last-mentioned pursuit curve—denotes the infinite point of the developable generator belonging to the point X and, upon location, can be added in the determined direction OE in the main image of Fig. 5.

Since $w_1 = w_2 = \dot{w}_0 = \dot{w}_3 = 0$, (25) and (27) yield the direction of the developable generator at the starting point $A(t = 0)$ as

$$\begin{aligned} x : y : z &= y_1 : y_2 : y_3 = -\dot{w}_2 : \dot{w}_3 : 0 \\ &= 3\alpha(\beta\delta - \alpha\varepsilon) + 2c^2\delta^5 : 3\delta(\beta\delta - \alpha\varepsilon) - 2c^2\alpha\delta^4 : 0. \end{aligned} \tag{29}$$

With the assumption (16), this leads to $y/x = -372/6121 = -0.0608$. This represents a certain deviation from the proportions in the study model, where according to Fig. 2, by contrast, a weak, positive direction can be identified.

As mentioned at the outset, at the endpoint $C(t = \infty)$, the developable generator coincides with the z axis and, due to symmetry considerations, represents an apex tangent of the edge of regression. For the apex itself, from (25) with $y_1 = y_2 = 0$, we determine the coordinates

$$z - c = \frac{y_3}{y_0} = \lim_{t\to\infty} \frac{w_0}{w_3} = \frac{4}{5c}. \tag{30}$$

This limits the possible *band width* $2b$ in all cases to $b \leq 4/5c$, which, with the present assumption, is $b \leq 1$.

To actually delimit the band, the *development* of the developable was initially constructed in Fig. 6. The arc length s of the centerline m, which here becomes a straight line, was derived from the flattening of the elliptic support cylinder (10) in Fig. 3, while the direction angle $\vartheta = \sphericalangle UOE$ was taken from a simple folding over in Fig. 4. The parallel lines in Fig. 6, lying a distance of $\pm b$ away from the straightened midline, form the flattened *border*

[3] According to (22), (24), and (26), the loci of U, V, W are—also in the central elevation—rational curves of respective order 10, 11, and 21.

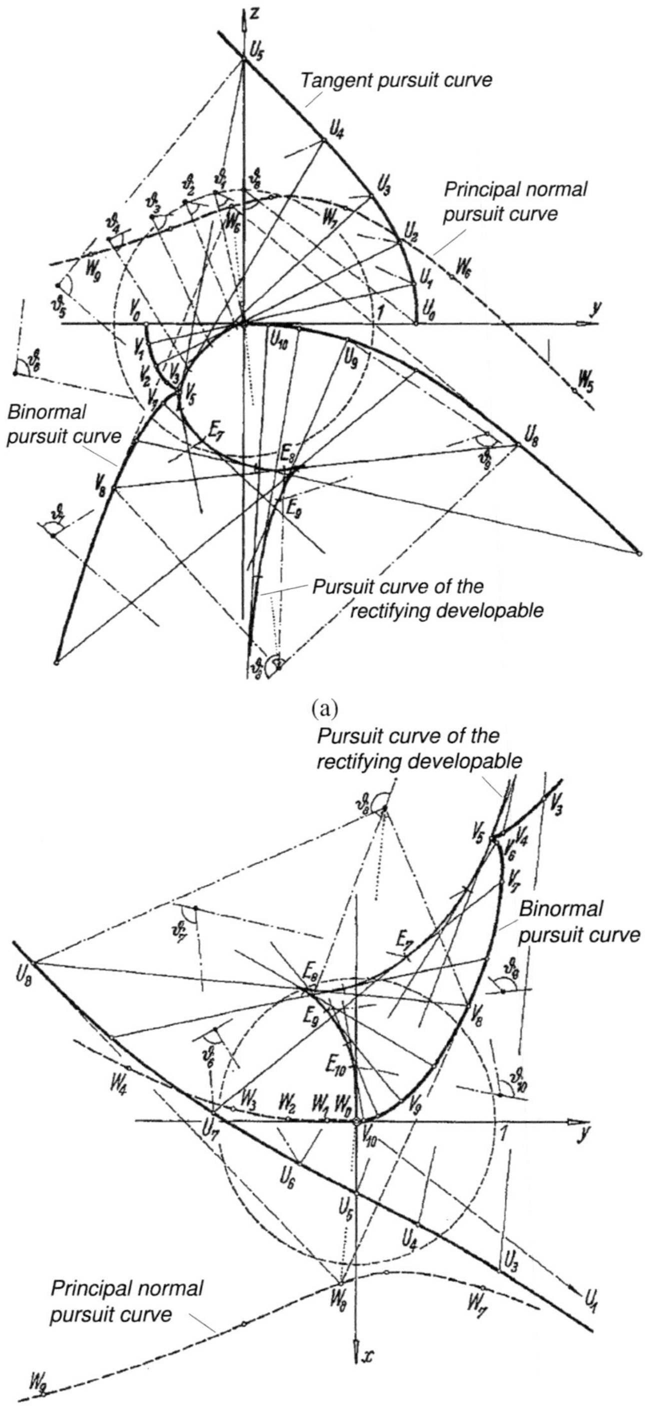

(a)

(b)

Fig. 4 (**a**) Central projection of the infinite plane from O onto the plane $x = 1$. (**b**) Central projection of the infinite plane from O onto the plane $z = 1$

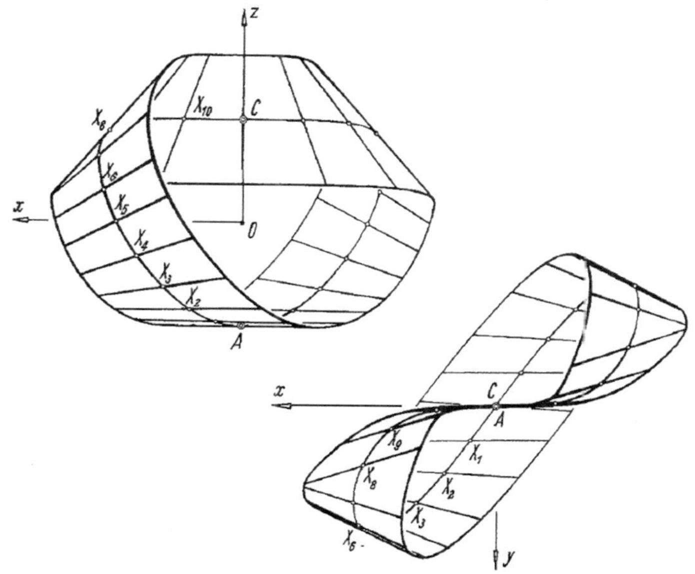

Fig. 5 Algebraic Möbius band in elevation and side view

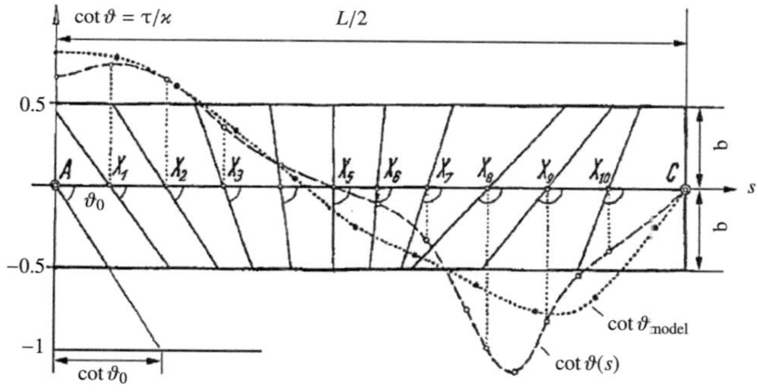

Fig. 6 Development of the band

of the Möbius band and intersect the generating lines, which can be drawn in the picture according to the rules of descriptive geometry. The boundary curve so obtained is certainly a *closed, algebraic geodesic* of the developable band. In the present setting, the obvious, permissible strip width $2b = 1$ was used as a basis.

The plot of the function $\vartheta(s)$ shown in Fig. 6 for the algebraic developable (dashed line) exhibits, in contrast to its prevailing dependence on s in the mock-up (dotted line),[4] a surprisingly irregular character, which further restricts the permissible band width in addition to the limit (30). In this regard, the reproduction carried out here does not yet seem completely satisfactory.

[4]The highlighted generators of the developable on the cardboard model were determined as the shadow boundaries in the cone of light of a projection device.

The computed identification of the dependence $\vartheta(s)$ would have required considerable effort, which is why it must be left undone for the time being. The numerical implementation of the algebraic model would however be of interest to achieve, for example, an improvement of the bounds (9). The assumption (16) certainly does not directly correspond to an "infinitesimally thin" Möbius band, which could be better represented by a sufficiently narrow steel band and which would definitely require somewhat modified coefficients.

Acknowledgements I thank Eliot Fried for his suggestion to translate this important work as well as his considerable editing and technical help with the manuscript. Michael Ban and Denis Hinz also provided valuable linguistic clarifications.

References

1. Möbius, F.A.: Über die Bestimmung des Inhaltes eines Polyeders (On the determination of the volume of a polyhedron). Ber. Verh. Sächs. Ges. Wiss. **17**, 31–68 (1865); Gesammelte Werke, Band II (Collected Works, vol. II), p. 484. Hirzel, Leipzig (1886)
2. Maschke, H.: Note on the unilateral surface of Moebius. Trans. Am. Math. Soc. **1**, 39 (1900)
3. Scheffers, G.: Einführung in die Theorie der Flächen (2. Aufl.) (Introduction to the Theory of Surfaces, 2nd edn.), pp. 41–43. Veit, Leipzig (1913)
4. Weyl, H.: Die Idee der Riemannschen Fläche (The Concept of the Riemann Surface), p. 26. Teubner, Leipzig (1913)
5. Krames, J.: Die Regelfläche dritter Ordnung, deren Striktionslinie eine Ellipse ist (The ruled surface of order 3 whose line of striction is an ellipse). Sitzber. Akad. Wiss. Wien **127**, 563–568 (1918) (Wunderlich comment: The strange and distinguishing feature determined therein is that the line of striction, which in general is of eighth order for cubic ruled surfaces, reduces to an ellipse in the presented special case by means of splitting of minimal generators)
6. Klíma, J.: O zborcené ploše, jejíž část je topologicky ekvivalentní s Möbiovým listem (On a skew surface, part of which is topologically equivalent to the Möbius band). Čas. Pěst. Mat. Fys. **65**(4), 211–216 (1936)
7. Sadowsky, M.: Ein elementarer Beweis für die Existenz eines abwickelbaren Möbiusschen Bandes und Zurückfürhing des geometrischen Problems auf ein Variationsproblem (An elementary proof for the existence of a developable Möbius band and the attribution of the geometric problem to a variational problem). Sitzber. Preuss. Akad. Wiss. **22**, 412–415 (1930). See English translation by Hinz and Fried in this issue
8. Sadowsky, M.: Theorie der elastisch biegsamen undehnbaren Bänder mit Anwendungen auf das Möbius'sche Band (Theory of elastically bendable inextensible bands with applications to the Möbius band). In: Oseen, C.W., Weibull, W. (Hrsg.) Verhandlungen des 3. Internationalen Kongresses für Technische Mechanik, Teil II: Elastizität, Plastizität, Festigkeit, Ballistik und rationelle Mechanik, Stockholm, 24–29 August 1930, S. 444–451 (Oseen, C.W., Weibull, W. (Eds.), Proceedings of the 3rd International Congress for Applied Mechanics, Part II: Elasticity, Plasticity, Strength, Ballistics, and Rational Mechanics, Stockholm, 24–29 August 1930, pp. 444–451. Sveriges Litografiska Tryckerier, Stockholm, 1931). See English translation by Hinz and Fried in this issue

Translator's Notes

i. Wunderlich uses the term *torse*, which has herein been replaced by the more common and synonymous *developable*.
ii. *Edge of regression* (cuspidal line) is used for the German *Gratlinie*.
iii. The use of $\varkappa_0$ in (5) in the original instead of $\bar{\varkappa}_0$ appears to be an error.
iv. For clarity, periods are used here instead of the commas Wunderlich used to denote decimals.
v. Synonymous with *ideal point*.
vi. The order of Figs. 5 and 6 is herein reversed from the original for clarity.
vii. The use of $\ddot{X}_i$ in the original instead of $\ddot{X}$ appears to be an error.
viii. *Imaginary circle* was chosen for the German *nullteiliger Kreis*.

DOI 10.1007/978-94-017-7300-3_6
Reprinted from *Journal of Elasticity* Journal, DOI 10.1007/s10659-014-9475-4

Gamma-Limit of a Model for the Elastic Energy of an Inextensible Ribbon

Nicholas O. Kirby · Eliot Fried

Received: 7 November 2013 / Published online: 20 March 2014

Abstract A Γ-convergence result involving the elastic bending energy of a narrow inextensible ribbon is established. As a consequence of the result, the energy is reduced to a one-dimensional integral, over the centerline of the ribbon, in which the aspect ratio of the ribbon appears as a small parameter. That integral is observed to increase monotonically with the aspect ratio. The Γ-limit of the family of energies is taken in a Sobolev space of centerlines with nonvanishing curvature. In that space, it is shown that the Γ-limit is a functional first proposed by Sadowsky in the context of narrow ribbons that form Möbius bands. The results obtained here do not apply to such ribbons, since the centerline of a Möbius band must have at least one inflection point. As a first step toward dealing with such inflection points, a result concerning the lower semicontinuity of the Sadowsky functional with inflection points comprising a set of measure zero within the domain of an arclength parameterization is presented.

Keywords Low-dimensional media · Dimensional reduction · Curvature elasticity · Sadowsky functional · Torsion · Sequential lower semicontinuity · Weak convergence

Mathematics Subject Classification 49Q10 · 49S05 · 82B21

1 Introduction

An inextensible ribbon is modeled as a two-dimensional surface that is geometrically constrained to be isometric to a rectangle of given length ℓ and width $2w$. The dimensionless

N.O. Kirby
Department of Mechanical Engineering, University of Washington, Seattle, WA 98195, USA
e-mail: nokirby@uw.edu

E. Fried (✉)
Mathematical Soft Matter Unit, Okinawa Institute of Science and Technology, 1919-1 Tancha, Onna-son, Kunigami-gun, Okinawa, 904-0495 Japan
e-mail: eliot.fried@oist.jp

parameter $\varepsilon = 2w/\ell$ is referred to as the aspect ratio of the ribbon. As Giomi and Mahadevan [10] note, certain biopolymers, such as DNA, and graphene and silicene nanoribbons are examples of ribbons with very small aspect ratios. Granted that the curvature $\tilde{\kappa}$ of the centerline $\mathcal{C}$ of the ribbon is nonvanishing, the geometric constraint yields a parametrization of the ribbon in terms of $\mathcal{C}$.

To determine the equilibrium shape of an elastic, inextensible ribbon subject to imposed end conditions, it suffices to minimize its net potential energy. Here, it is assumed that the elastic energy density ϕ of the ribbon is an isotropic, quadratic function of the Weingarten map, and thereby a symmetric, quadratic function of the principle curvatures of the ribbon. Upon completing the square, ϕ admits a representation in terms of the mean and Gaussian curvatures H and K of the ribbon of the form

$$\phi = \frac{D}{2}(H - H_0)^2 + CK, \tag{1}$$

where D and C are constant moduli and H_0 is the spontaneous mean curvature. The expression (1) was proposed by Germain [9]. The particular version of (1) considered here, in which H_0 is taken to be zero, was considered by Poisson [12].

In the limit $\varepsilon \to 0$ of vanishing aspect ratio, Sadowsky [15] argued that the energy of a ribbon forming a Möbius band should be proportional to

$$\mathcal{F} = \int_{\mathcal{C}} \tilde{\kappa}^2 \big(1 + \eta^2\big)^2 \, \mathrm{d}\xi, \tag{2}$$

where η is the ratio $\tilde{\tau}/\tilde{\kappa}$, with $\tilde{\tau}$ being the torsion of the centerline $\mathcal{C}$, and where ξ denotes arclength along $\mathcal{C}$. The properties of the functional (2) were studied in some detail by Wunderlich [19]. Recently, Starostin and van der Heijden [16] used the variational bicomplex formalism to investigate the equilibrium equations for the problem associated with minimizing the functional

$$\mathcal{F}_\varepsilon = \int_{\mathcal{C}} \tilde{\kappa}^2 \big(1 + \eta^2\big)^2 \frac{1}{\varepsilon \ell \dot{\eta}} \ln\!\left(\frac{2 + \varepsilon \ell \dot{\eta}}{2 - \varepsilon \ell \dot{\eta}}\right) \mathrm{d}\xi, \tag{3}$$

for $\varepsilon > 0$, where a superposed dot indicates differentiation with respect to the arclength parameter ξ. Upon inspection, it is evident that the Sadowsky functional $\mathcal{F}$ is the pointwise limit of the elastic energy $\mathcal{F}_\varepsilon$ as $\varepsilon \to 0$. However, the question of whether the Sadowsky functional (2) is the proper variational limit (that is, the Γ-limit) of the elastic energy (3) of a ribbon with a given centerline remains unanswered. This question is settled herein for curves with nonvanishing curvature that are parametrized by arclength and are elements of certain Sobolev spaces.

As Randrup and Røgen [13] remark, the centerline of a nonorientable developable, like a developable Möbius band, must have at least one point at which the curvature vanishes. The problem of establishing the Sadowsky functional as the Γ-limit for a space of centerlines containing those corresponding to a Möbius band is left for future work. However, a result in this direction is provided.

The first steps in the analysis are identical to those appearing in the papers of Wunderlich [19] and Starostin and van der Heijden [16] and also in the thesis of Yong [20]. These steps deliver an expression for the elastic energy of the ribbon in terms of the shape of its centerline and depending parametrically on the aspect ratio ε of the ribbon. In particular, the energy is given by

$$E = \frac{\varepsilon \ell D}{2} \int_0^\ell \tilde{\kappa}^2(\xi)\big(1 + \eta^2(\xi)\big)^2 g\big(\varepsilon \ell \dot{\eta}(\xi)\big) \, \mathrm{d}\xi, \tag{4}$$

where $g\colon \mathbb{R} \to [1, +\infty]$ is defined by

$$g(x) = \begin{cases} 1, & \text{if } x = 0, \\ \frac{1}{x}\ln(\frac{2+x}{2-x}), & \text{if } |x| < 2 \text{ and } x \neq 0, \\ +\infty, & \text{if } |x| \geq 2. \end{cases} \tag{5}$$

Thereafter, it is shown that, upon taking ε to zero, the Sadowsky functional is not merely the pointwise limit of the one-parameter family $2E/\varepsilon D$ of functionals but is also the Γ-limit in a naturally chosen space of curves with nonvanishing curvature. The keys to the argument are the monotonic dependence of the family of energy functionals on ε and judicious applications of some known facts about functionals with convex integrands.

The results contained here should be viewed as first steps toward a full understanding of the dimensional reduction problem considered by Sadowsky, and undoubtedly the problem will be revisited in the future. Aside from approximation results regarding convergence of minimizers of (3) to minimizers of (2), important open questions include the possibility of centerlines with inflection points. An anticipated difficulty to be overcome is the lack of a polynomial growth bound, from below, for the integrand of the Sadowsky functional (2).

There may be some temptation to view the results gathered here as being relevant to the work of Friesecke, James and Müller [7]. In that work, the elastic energy of a deformed rectangle is defined on the space of surfaces parameterized by functions $u \in W^{2,2}([0,\ell]\times[-w,w];\mathbb{R}^3)$ satisfying $u_x \cdot u_y = |u_x|^2 - 1 = |u_y|^2 - 1 = 0$. Here, however, the surface energy is defined in the smooth setting, reduced to a one-dimensional integral formulation, and only then is the space of functions to which centerlines are stipulated to belong introduced. Another future direction would involve working within the two-dimensional setting of Friesecke, James and Müller [7] and deriving the reduced energy.

2 The Energy of an Elastic Band

Consider a surface isometric to a rectangle with base ℓ and height $2w$. Let

$$\varepsilon = \frac{2w}{\ell}. \tag{6}$$

By the Theorema Egregium of Gauss [8], this surface must have Gaussian curvature equal to zero; that is, the surface must be developable. Struik [17, Sect. 2.8] proves that any developable surface in $\mathbb{R}^3$ is a ruled surface. Given a space curve with length ℓ, arclength parametrization $\mathbf{r}$, Frenet frame $\{\mathbf{t}, \mathbf{n}, \mathbf{b}\}$, curvature $\tilde{\kappa}$ with isolated zeroes, and torsion $\tilde{\tau}$, define the ratio

$$\eta = \frac{\tilde{\tau}}{\tilde{\kappa}}. \tag{7}$$

Graustein [11, Sect. 52] shows that a curve on a developable surface is a geodesic if and only if the surface is the rectifying developable of the curve. Graustein [11, Sect. 28] also shows that the unique rectifying developable of a space curve with centerline having arclength parameterization $\mathbf{r}$ and width $2w$ can be expressed in the form

$$\mathbf{x}(\xi,\zeta) = \mathbf{r}(\xi) + \zeta\big[\mathbf{b}(\xi) + \eta(\xi)\mathbf{t}(\xi)\big], \quad (\xi,\zeta) \in [0,\ell]\times[-w,w]. \tag{8}$$

Given the centerline $\mathbf{r}$, consider the energy of an elastic ribbon $\mathcal{S}$ of width $2w$ parameterized by $\mathbf{x}$. Wunderlich [19] shows that the nonvanishing principal curvature κ_1 of $\mathcal{S}$ is given by

$$\kappa_1 = \frac{\tilde{\kappa}(1+\eta^2)}{|1+\zeta\dot{\eta}|}, \tag{9}$$

where the dot denotes differentiation with respect to arclength. The bending energy E of $\mathcal{S}$ therefore takes the form

$$E = \frac{D}{2}\int_{\mathcal{S}} \kappa_1^2 \,\mathrm{d}A = \frac{D}{2}\int_0^{\ell}\int_{-w}^{w} \frac{\tilde{\kappa}^2(\xi)(1+\eta^2(\xi))^2}{|1+\zeta\dot{\eta}(\xi)|}\,\mathrm{d}\zeta\,\mathrm{d}\xi, \tag{10}$$

where D is a measure of flexural rigidity. Evaluating the integral on the far right-hand side of (10) over the width $2w$ of the ribbon yields

$$E = Dw\int_0^{\ell} \tilde{\kappa}^2(\xi)\big(1+\eta^2(\xi)\big)^2 g\big(2w\dot{\eta}(\xi)\big)\,\mathrm{d}\xi, \tag{11}$$

where $g\colon \mathbb{R}\to[1,+\infty]$ is defined by

$$g(x) = \begin{cases} 1, & \text{if } x = 0,\\ \frac{1}{x}\ln\!\left(\frac{2+x}{2-x}\right), & \text{if } |x| < 2 \text{ and } x \neq 0,\\ +\infty, & \text{if } |x| \geq 2. \end{cases} \tag{12}$$

2.1 Nondimensionalization

To nondimensionalize the problem, introduce a unit speed parametrization $\mathbf{u}\colon [0,1]\to\mathbb{R}^3$ of the centerline defined such that $\mathbf{u}(s) = \mathbf{r}(\ell s)/\ell$ for each $s\in[0,1]$. Let I denote the open interval $(0,1)$. For a measurable set $V\subset[0,1]$, consider the family of functionals $\mathcal{F}_\varepsilon(\cdot,V)\colon W^{3,p}(I;\mathbb{R}^3)\to[0,+\infty]$ defined such that

$$\mathcal{F}_\varepsilon(\mathbf{u},V) = \begin{cases} \int_V \kappa^2(1+\eta^2)^2 g(\varepsilon\eta')\,\mathrm{d}s, & \text{if } \mathbf{u}\in W^{3,p}(I;\mathbb{R}^3) \text{ such that } \eta\in W^{1,1}(I),\\ +\infty, & \text{otherwise.} \end{cases} \tag{13}$$

In (13) and hereafter, a prime indicates differentiation with respect to $s = \xi/\ell$ and the symbols κ and τ denote the dimensionless counterparts of the curvature $\tilde{\kappa}$ and torsion $\tilde{\tau}$, given in terms of $\mathbf{u}$ by

$$\kappa = \left|\mathbf{u}''\right| \quad\text{and}\quad \tau = \frac{\mathbf{u}'\cdot(\mathbf{u}''\times\mathbf{u}''')}{|\mathbf{u}''|^2}. \tag{14}$$

By the chain rule, $\kappa = \ell\tilde{\kappa}$ and $\tau = \ell\tilde{\tau}$, whereby η as defined in (7) admits the alternative representation $\eta = \tau/\kappa$. It then follows that η and η' may be expressed in terms of $\mathbf{u}$ by

$$\begin{aligned} \eta &= \frac{\mathbf{u}'\cdot(\mathbf{u}''\times\mathbf{u}''')}{|\mathbf{u}''|^3},\\ \eta' &= \frac{\mathbf{u}'\cdot(\mathbf{u}''\times\mathbf{u}'''')}{|\mathbf{u}''|^3} - \frac{3(\mathbf{u}''\cdot\mathbf{u}''')[\mathbf{u}'\cdot(\mathbf{u}''\times\mathbf{u}''')]}{|\mathbf{u}''|^5}. \end{aligned} \tag{15}$$

3 Notation

For a function $\mathbf{u} : I \to \mathbb{R}^3$ with $k \geq 1$ weak derivatives $\mathbf{u}', \mathbf{u}'', \mathbf{u}^3, \ldots, \mathbf{u}^k$ and the k-tuple $D^{[k]}\mathbf{u}$ be defined by

$$D^{[k]}\mathbf{u} = \left(\mathbf{u}, \mathbf{u}', \ldots, \mathbf{u}^k\right). \tag{16}$$

Define function spaces Y, X^p, Y^{κ_m}, and $X^{\kappa_m,p}$ by

$$\begin{aligned} Y &:= \left\{\mathbf{u} \in C^\infty\left([0,1]; \mathbb{R}^3\right) : \left|\mathbf{u}'(s)\right| = 1 \text{ for all } s \in I\right\}, \\ X^p &:= \mathrm{cl}_{W^{3,p}(I;\mathbb{R}^3)}(Y), \\ Y^{\kappa_m} &:= \left\{\mathbf{u} \in Y : \left|\mathbf{u}''(s)\right| \geq \kappa_m \text{ for all } s \in I\right\}, \\ X^{\kappa_m,p} &:= \mathrm{cl}_{W^{3,p}(I;\mathbb{R}^3)}\left(Y^{\kappa_m}\right), \end{aligned} \tag{17}$$

where for a Banach space $\mathcal{B}$ and $\mathcal{A} \subset \mathcal{B}$ a subset of that space, $\mathrm{cl}_{\mathcal{B}}(\mathcal{A})$ denotes the closure of $\mathcal{A}$ with respect to the norm of $\mathcal{B}$.

4 Γ-Limit

The goal of the ensuing analysis is to determine a value of p such that the sequence $\{\mathcal{F}_\varepsilon(\cdot, I)\}$ of functionals defined in (13) has Γ-limit

$$\mathcal{F}(\mathbf{u}, I) = \int_I \kappa^2\left(1 + \eta^2\right)^2 \mathrm{d}s, \tag{18}$$

with respect to weak convergence in $X^{\kappa_m,p}$ and with respect to strong convergence in $W^{3,p}(I; \mathbb{R}^3)$. The functional $\mathcal{F}(\cdot, I)$ defined by (18) is called the Sadowsky functional.

4.1 Existence of Γ-Limit

In particular, following De Giorgi [6], given any sequence $\{\varepsilon_j\}$ with $\varepsilon_j > 0$ and $\varepsilon_j \to 0$ and any element $\mathbf{u} \in X$:

1. for every sequence $\{\mathbf{u}_j\}$ with $\mathbf{u}_j \in X$ such that $\mathbf{u}_j \to \mathbf{u}$ in X, $\mathcal{F}(\mathbf{u}, I)$ is bounded above in accord with

$$\mathcal{F}(\mathbf{u}, I) \leq \liminf_{j\to\infty} \mathcal{F}_{\varepsilon_j}(\mathbf{u}_j, I); \tag{19}$$

2. there exists a sequence $\{\mathbf{u}_j\}$ converging to $\mathbf{u}$ such that $\mathcal{F}(\mathbf{u}, I)$ is bounded below in accord with

$$\mathcal{F}(\mathbf{u}, I) \geq \limsup_{j\to\infty} \mathcal{F}_{\varepsilon_j}(\mathbf{u}_j, I). \tag{20}$$

For any such sequence $\{\varepsilon_j\}$, it is possible to extract a decreasing subsequence $\{\varepsilon_{j_k}\}$. Since the integrand of $\mathcal{F}_\varepsilon(\cdot, I)$ increases with ε (regardless of the sign of η'), it can be deduced that the Γ-limit of the sequence of functionals $\mathcal{F}_{\varepsilon_{j_k}}(\cdot, I)$ exists and is given by

$$\Gamma\text{-}\lim_{k\to\infty} \mathcal{F}_{\varepsilon_{j_k}}(\cdot, I) = \mathrm{sc}\left(\inf_{k\in\mathbb{N}} \mathcal{F}_{\varepsilon_{j_k}}(\cdot, I)\right) = \mathrm{sc}\left(\inf_{\varepsilon>0} \mathcal{F}_\varepsilon(\cdot, I)\right), \tag{21}$$

where $\mathrm{sc}(F)$ is the lower semicontinuous envelope of F; that is, for any $\mathbf{u} \in X$,

$$\mathrm{sc}(F)(\mathbf{u}) = \sup\{G(\mathbf{u}) : G \text{ is lower semicontinuous}, G \leq F\}. \tag{22}$$

The relevant fact concerning monotone sequences of functionals is verified by Braides [2, Sect. 1.8].

As a first step toward establishing the Sadowsky functional as the Γ-limit of the sequence $\{\mathcal{F}_\varepsilon(\cdot, I)\}$, it is useful to compute the functional $\tilde{\mathcal{F}}$ defined for each $\mathbf{u} \in X^p$ by

$$\tilde{\mathcal{F}}(\mathbf{u}) = \inf_{\varepsilon>0} \mathcal{F}_\varepsilon(\mathbf{u}, I). \tag{23}$$

Lemma 1 *Let $X = X^p$ or $X^{\kappa_m,p}$. Given $\mathbf{u} \in X$, $\tilde{\mathcal{F}}$ defined in accord with* (23) *is given by*

$$\tilde{\mathcal{F}}(\mathbf{u}) = \begin{cases} \mathcal{F}(\mathbf{u}, I), & \text{if } \mathbf{u} \in X \cap \{\mathbf{u} : \eta' \in L^\infty(I)\}, \\ +\infty, & \text{otherwise.} \end{cases} \tag{24}$$

Proof By the Monotone Convergence Theorem (see, for instance, Wheeden and Zygmund [18]), if $\mathcal{F}_\varepsilon(\mathbf{u}, I)$ is finite for some $\varepsilon > 0$, then the limits involved in (23) may be exchanged to give

$$\begin{aligned} \inf_{\varepsilon>0} \mathcal{F}_\varepsilon(\mathbf{u}, I) &= \lim_{\varepsilon\searrow 0} \int_I \kappa^2(1+\eta^2)^2 g(\varepsilon\eta')\,\mathrm{d}s \\ &= \int_I \lim_{\varepsilon\searrow 0} \kappa^2(1+\eta^2)^2 g(\varepsilon\eta')\,\mathrm{d}s \\ &= \mathcal{F}(\mathbf{u}). \end{aligned} \tag{25}$$

Notice that $\tilde{\mathcal{F}}(\mathbf{u}, I) \neq \mathcal{F}(\mathbf{u}, I)$ if and only if $\mathcal{F}_\varepsilon(\mathbf{u}, I) = +\infty$ for all $\varepsilon > 0$ and $\mathcal{F}(\mathbf{u}, I) < +\infty$.

Suppose that $\mathcal{F}(\mathbf{u}, I) < +\infty$. Let $A = \{s \in I : \eta'(s) = 0\}$, $B_\varepsilon = \{s \in I : 0 < |\eta'(s)| < 2/\varepsilon\}$, and $C_\varepsilon = I \setminus (A \cup B_\varepsilon) = \{s \in I : |\eta'(s)| \geq 2/\varepsilon\}$. Then

$$\mathcal{F}_\varepsilon(\mathbf{u}, I) = \mathcal{F}(\mathbf{u}, A) + \mathcal{F}_\varepsilon(\mathbf{u}, B_\varepsilon) + \infty(C_\varepsilon), \tag{26}$$

where ∞ is the set function defined such that, given any measurable set V,

$$\infty(V) = \begin{cases} +\infty & \text{if } \mu(V) > 0, \\ 0 & \text{otherwise.} \end{cases} \tag{27}$$

However, $\mathcal{F}_\varepsilon(\mathbf{u}, I) = +\infty$ for all $\mathbf{u}$ such that $\|\eta'\|_\infty = +\infty$ since, in that case $\infty(C_\varepsilon) = +\infty$ for all $\varepsilon > 0$. On the other hand, if $\|\eta'\|_\infty < +\infty$, then for $\varepsilon < 2/\|\eta'\|_\infty$ it follows that $\infty(C_\varepsilon) = 0$ and, by Hölder's inequality, that

$$\mathcal{F}_\varepsilon(\mathbf{u}, B_\varepsilon) \leq \mathcal{F}(\mathbf{u}, B_\varepsilon)\big\|g(\varepsilon\eta'(\cdot))\big\|_\infty \leq g(\varepsilon\|\eta'\|_\infty)\mathcal{F}(\mathbf{u}, B_\varepsilon) < +\infty. \tag{28}$$

Hence, $\tilde{\mathcal{F}}(\mathbf{u}, I) \neq \mathcal{F}(\mathbf{u}, I)$ if and only if $\mathcal{F}(\mathbf{u}, I) < +\infty$ and $\|\eta'\|_\infty = +\infty$. □

Lemma 1 and (21) lead to the conclusion that

$$\Gamma\text{-}\lim_{\varepsilon\to 0^+} \mathcal{F}_\varepsilon(\cdot, I) = \mathrm{sc}(\tilde{\mathcal{F}}), \tag{29}$$

where $\tilde{\mathcal{F}}$ is defined as in (24).

4.2 Curves with Curvature Bounded from Below

Consider now the problem in which the space curve parameterized by $\mathbf{u}$ has (dimensionless) curvature κ greater than some constant. In particular, for $p > 1$ and $\kappa_m > 0$, take $X^{\kappa_m,p}$ as defined in (17). By the compact embedding $W^{3,p}(I;\mathbb{R}^3) \hookrightarrow C^{2,1-1/p}([0,1];\mathbb{R}^3)$, any function $\mathbf{u} \in X^{\kappa_m,p}$ satisfies the pointwise constraints $|\mathbf{u}'(s)| = 1$ and $|\mathbf{u}''(s)| \geq \kappa_m$ for almost every $s \in I$. For information on the salient embedding results, see Adams and Fournier [1].

4.3 The Sadowsky Functional is Lower Semicontinuous

Let $f : \mathbb{R}^9 \times \mathbb{R}^9 \to \mathbb{R}$ be defined at $(\mathbf{U},\mathbf{P}) = ((\mathbf{u}_0,\mathbf{u}_1,\mathbf{u}_2),(\mathbf{p}_1,\mathbf{p}_2,\mathbf{p}_3))$ by

$$f(\mathbf{U},\mathbf{P}) = \left(\frac{\alpha(\kappa_m^{-1}|\mathbf{u}_2|)}{\kappa_m}\right)^2 \left(|\mathbf{p}_2|^2 + \alpha(|\mathbf{u}_1|)^2 \left(\frac{\alpha(\kappa_m^{-1}|\mathbf{u}_2|)}{\kappa_m}\right)^4 \left[\mathbf{u}_1 \cdot (\mathbf{u}_2 \times \mathbf{p}_3)\right]^2\right)^2, \tag{30}$$

where $\alpha\colon [0,\infty) \to \mathbb{R}$ are defined by

$$\alpha(t) = \begin{cases} \frac{1}{t}, & \text{if } t \geq 1, \\ 1, & \text{otherwise.} \end{cases} \tag{31}$$

Consider the functional $\widehat{\mathcal{F}}$ defined by

$$\widehat{\mathcal{F}}(\mathbf{U}) = \int_I f\big(\mathbf{U}(s), \mathbf{U}'(s)\big)\, \mathrm{d}s. \tag{32}$$

Notice that on the set $X^{\kappa_m,p}$, the functional $\mathcal{F}$ may be evaluated via

$$\mathcal{F}(\mathbf{u}; I) = \widehat{\mathcal{F}}\big(D^{[2]}\mathbf{u}\big). \tag{33}$$

The map f has the following properties: f and the derivative $f_{\mathbf{P}}$ of f with respect to its second argument are continuous; f is nonnegative; and f is convex in its second argument. Tonelli's semicontinuity theorem (see, for instance, Buttazzo, Giaquinta and Hildebrandt [3, Sect. 3.1]) implies that $\widehat{\mathcal{F}}$ is sequentially weakly lower semicontinuous in $W^{1,p}(I;\mathbb{R}^9)$ for all $p \geq 1$. To present the theorem, it is useful to recall the following terminology.

Definition 1 A functional $\mathcal{G}\colon \mathcal{B} \to \mathbb{R}$ is *sequentially weakly lower semicontinuous* in a Banach space $\mathcal{B}$, if for every $\mathbf{x} \in \mathcal{B}$ and every sequence $\{\mathbf{x}_k\} \in \mathcal{B}$ that converges weakly in $\mathcal{B}$ to $\mathbf{x}$, the condition

$$\mathcal{G}(\mathbf{x}) \leq \liminf_{k\to\infty} \mathcal{G}(\mathbf{x}_k) \tag{34}$$

is satisfied.

Theorem 1 (Tonelli's semicontinuity theorem) *Let $I \subset \mathbb{R}$ be a bounded open interval, and for $n \geq 1$ let $g\colon \mathbb{R} \times \mathbb{R}^n \times \mathbb{R}^n \to \mathbb{R}$ be a function with the following properties*:

1. *g and $g_{\mathbf{P}}$ are continuous in $(s,\mathbf{U},\mathbf{P})$*;
2. *g is nonnegative or bounded from below by an L^1-function*;
3. *g is convex in $\mathbf{P}$.*

The functional $\mathcal{G}$ defined by

$$\mathcal{G}(\mathbf{U}) = \int_I g\big(\mathbf{U}(s), \mathbf{U}'(s)\big)\, \mathrm{d}s \tag{35}$$

is then sequentially weakly lower semicontinuous in $W^{1,p}(I;\mathbb{R}^n)$ for all $p \geq 1$.

A corollary to the observation that $\widehat{\mathcal{F}}$ satisfies the hypotheses of the preceding theorem is stated in the following lemma.

Lemma 2 *The Sadowsky functional $\mathcal{F}$ defined by* (33) *is sequentially weakly lower semicontinuous in $W^{3,p}(I;\mathbb{R}^3)$.*

Proof Let a sequence $\{\mathbf{u}_k\} \in W^{3,p}(I;\mathbb{R}^3)$ that weakly converges in $W^{3,p}(I;\mathbb{R}^3)$ to some $\mathbf{u}$. Then the associated sequences $\{\mathbf{u}_k\}$, $\{\mathbf{u}_k'\}$, $\{\mathbf{u}_k''\}$, and $\{\mathbf{u}_k'''\}$ converge weakly in $L^p(I;\mathbb{R}^3)$ to $\mathbf{u}$, $\mathbf{u}'$, $\mathbf{u}''$, and $\mathbf{u}'''$, respectively. Hence, the sequences $\{D^{[2]}\mathbf{u}_k\}$ and $\{D^{[2]}\mathbf{u}_k'\}$ converge weakly in $L^p(I;\mathbb{R}^9)$ to $D^{[2]}\mathbf{u}$ and $D^{[2]}\mathbf{u}'$, respectively. This implies that the sequence $\{D^{[2]}\mathbf{u}_k\}$ converges weakly in $W^{1,p}(I;\mathbb{R}^9)$. By Tonelli's semicontinuity theorem, the functional $\widehat{\mathcal{F}}$ is sequentially weakly lower semicontinuous, and therefore

$$\mathcal{F}(\mathbf{u}; I) = \widehat{\mathcal{F}}\big(D^{[2]}\mathbf{u}\big) \leq \liminf_{k\to\infty} \widehat{\mathcal{F}}\big(D^{[2]}\mathbf{u}_k\big) = \liminf_{k\to\infty} \mathcal{F}(\mathbf{u}_k; I). \tag{36}$$

This proves the lemma. □

Recall that $g\colon \mathbb{R}\times\mathbb{R}^n\times\mathbb{R}^n \to \mathbb{R}$ is a Carathéodory [4, 5] function if $s \mapsto g(s,\mathbf{U},\mathbf{P})$ is measurable for all $(\mathbf{U},\mathbf{P}) \in \mathbb{R}^n\times\mathbb{R}^n$ and $(\mathbf{U},\mathbf{P}) \mapsto g(s,\mathbf{U},\mathbf{P})$ is continuous for almost every $s \in I$. The following is another proposition proven by Buttazzo, Giaquinta and Hildebrandt [3, Sect. 3.2].

Proposition 1 *Let $g: \mathbb{R}\times\mathbb{R}^n\times\mathbb{R}^n \to \mathbb{R}$ be a Carathéodory function that is convex in its third argument and satisfies*

$$\big|g(s,\mathbf{U},\mathbf{P})\big| \leq c_1|\mathbf{P}|^p + c_2 \tag{37}$$

for some $p > 1$, and let $\{\mathbf{U}_k\}$ be a sequence $W^{1,p}(I;\mathbb{R}^n)$. If $\{\mathbf{U}_k\}$ converges strongly in $W^{1,p}(I;\mathbb{R}^n)$, then $\mathcal{G}(\mathbf{U}_k)$ converges to $\mathcal{G}(\mathbf{U})$.

The choice of f in (30) allows makes it possible to apply this proposition to the functional $\widehat{\mathcal{F}}$. This in turn leads to the following useful lemma, which might be of independent interest.

Lemma 3 *For all $\kappa_m > 0$ and $p \geq 4$, the Sadowsky functional $\mathcal{F}(\cdot; I)$ is continuous on $X^{\kappa_m,p}$ with respect to strong convergence in $W^{3,p}(I;\mathbb{R}^3)$.*

Proof It is clear that $f: \mathbb{R}^9\mathbb{R}^9 \to \mathbb{R}$ as defined in (30) is a Carathéodory function. Furthermore, from the definition (31) of α, it follows that

$$t\alpha(t) \leq 1 \quad \text{and} \quad \alpha(t) \leq 1. \tag{38}$$

Hence, for any $(\mathbf{U}, \mathbf{Q}) \in \mathbb{R}^9 \times \mathbb{R}^9$ and any $p \geq 4$, f obeys the inequality

$$\left|f(\mathbf{U}, \mathbf{P})\right| \leq \frac{1}{\kappa_m^2}\left(|\mathbf{p}_2|^2 + \frac{|\mathbf{p}_3|^2}{\kappa_m^2}\right)^2 \leq c_1 |\mathbf{P}|^4 \leq c_1 |\mathbf{P}|^p + c_1, \tag{39}$$

where $c_1 = \max\{1, 1/\kappa_m^2\}/\kappa_m^2$. By Proposition 1, it follows that $\widehat{\mathcal{F}}$ is continuous on $W^{1,p}(I;\mathbb{R}^9)$.

Since strong convergence of $\{\mathbf{u}_k\}$ in $W^{3,p}(I;\mathbb{R}^3)$ implies strong convergence of $\{D^{[2]}\mathbf{u}_k\}$ in $W^{1,p}(I;\mathbb{R}^9)$, this proves the lemma. □

Theorem 2 *If $p \geq 4$, then any sequentially weakly lower semicontinuous function $\mathcal{G}\colon X^{\kappa_m,p} \to \mathbb{R}$ that obeys $\mathcal{G}(\mathbf{u}) \leq \mathcal{F}(\mathbf{u}, I)$ on the dense subset Y^{κ_m} also satisfies the inequality*

$$\mathcal{G}(\mathbf{u}) \leq \mathcal{F}(\mathbf{u}, I) \tag{40}$$

for all $\mathbf{u} \in X^{\kappa_m,p}$. It follows that for $\mathbf{u} \in X^{\kappa_m,p}$ and $p \geq 4$

$$\Gamma\text{-}\lim_{\varepsilon \searrow 0} \mathcal{F}_\varepsilon(\mathbf{u}, I) = \mathcal{F}(\mathbf{u}, I) \tag{41}$$

with respect to the sequential weak topology on $W^{3,p}(I;\mathbb{R}^3)$.

Proof Let $\mathbf{u} \in X^{\kappa_m,p}$ be given. By hypothesis, $\mathcal{G}$ is weakly lower semicontinuous, and, therefore, if there exists a sequence $\{\mathbf{u}_n\} \subset Y^{\kappa_m}$ such that $\mathbf{u}_n \rightharpoonup \mathbf{u}$ in $W^{3,p}(I;\mathbb{R}^3)$, then $\mathcal{G}$ is bounded above in accord with

$$\mathcal{G}(\mathbf{u}) \leq \liminf_{n\to\infty} \mathcal{G}(\mathbf{u}_n) \leq \liminf_{n\to\infty} \mathcal{F}(\mathbf{u}_n, I). \tag{42}$$

Clearly, any curve $\mathbf{u} \in Y^{\kappa_m}$ satisfies $\|\eta'\|_\infty < +\infty$. Therefore, it suffices to find a sequence $\{\mathbf{u}_n\} \subset Y^{\kappa_m}$ such that $\mathbf{u}_n \rightharpoonup \mathbf{u}$ in $W^{3,p}(I;\mathbb{R}^3)$ and the limit

$$\lim_{n\to\infty} \mathcal{F}(\mathbf{u}_n, I) = \mathcal{F}(\mathbf{u}, I)$$

holds. By definition of $X^{\kappa_m,p}$, it is possible to choose a sequence $\{\mathbf{u}_n\} \subset Y$ such that $\mathbf{u}_n \to \mathbf{u}$ strongly (and, therefore weakly, as well) in $W^{3,p}(I;\mathbb{R}^3)$. By Lemma 3, $\mathcal{F}(\mathbf{u}, I) = \lim_{n\to\infty} \mathcal{F}(\mathbf{u}_n, I)$, and the conclusion follows. □

Theorem 3 *If $p \geq 4$, any lower semicontinuous function $\mathcal{G}\colon X^{\kappa_m,p} \to \mathbb{R}$ that obeys $\mathcal{G}(\mathbf{u}) \leq \mathcal{F}(\mathbf{u}, I)$ on the dense subset Y^{κ_m} also satisfies the inequality*

$$\mathcal{G}(\mathbf{u}) \leq \mathcal{F}(\mathbf{u}, I) \tag{43}$$

for all $\mathbf{u} \in X^{\kappa_m,p}$. It follows that for $\mathbf{u} \in X^{\kappa_m,p}$

$$\Gamma\text{-}\lim_{\varepsilon \searrow 0} \mathcal{F}_\varepsilon(\mathbf{u}, I) = \mathcal{F}(\mathbf{u}, I) \tag{44}$$

with respect to the strong topology on $W^{3,p}(I;\mathbb{R}^3)$ for $p \geq 4$.

Proof Let $\mathbf{u} \in X^{\kappa_m,p}$ be given. As before, it suffices to find a sequence $\{\mathbf{u}_n\} \subset Y^{\kappa_m}$ such that $\mathbf{u}_n \to \mathbf{u}$ in $W^{3,p}(I; \mathbb{R}^3)$ and the limit

$$\lim_{n\to\infty} \mathcal{F}(\mathbf{u}_n, I) = \mathcal{F}(\mathbf{u}, I)$$

holds. By Lemma 3, $\mathcal{F}(\mathbf{u}, I) = \lim_{n\to\infty} \mathcal{F}(\mathbf{u}_n, I)$ and the conclusion follows. □

The theorem has the following important corollary.

Corollary 1 *On the set $X^{+,p} = \bigcup_{\kappa_m>0} X^{\kappa_m,p}$, the Sadowsky functional is the Γ-limit of the elastic energy $\mathcal{F}_\varepsilon(\cdot, I)$ with respect to the sequential weak topology and with respect to the strong topology on $W^{3,p}(I; \mathbb{R}^3)$.*

Proof Let $\mathbf{u} \in X^{+,p}$. Then $\mathbf{u} \in X^{\kappa_m,p}$ for some $\kappa_m > 0$. That the limsup condition (20) is satisfied follows from the Γ convergence of $\mathcal{F}_\varepsilon(\cdot, I)$ to $\mathcal{F}(\cdot, I)$ in $X^{\kappa_m,p}$.

Let $\{\mathbf{u}_j\}$ be a sequence in $X^{+,p}$ such that $\mathbf{u}_j \to \mathbf{u}$ either strongly or weakly. Then $\mathbf{u} \in X^{\kappa_m,p}$ for some $\kappa_m > 0$ and, for either mode of convergence, $\mathbf{u}_j'' \to \mathbf{u}$ uniformly on $[0, 1]$. Hence, $\mathbf{u}_j \in X^{\kappa_m/2,p}$ for $j \geq J$ for some sufficiently large $J > 0$. Thus, (19) holds since $\mathcal{F}_{\varepsilon_j}(\cdot, I)$ Γ-converges to $\mathcal{F}(\cdot, I)$ in $X^{\kappa_m/2,p}$. □

The set $X^{+,p}$ consists of those arclength parameterized curves in $W^{3,p}(I; \mathbb{R}^3)$ well-approximated by a sequence of smooth curves for which the infima of their curvatures are uniformly bounded away from zero. The space $X^{+,p}$ is the natural space of curves with nonvanishing curvature described in the Introduction.

5 Lower Semicontinuity at Curves with Isolated Inflection Points

This analysis is limited in the sense that nonorientable ribbons do not have centerlines in the space $X^{+,p}$, since all such centerlines have strictly positive curvature. Bearing in mind that the Sadowsky functional was proposed in the context of work concerning the shape of a Möbius band, it is of interest to determine whether the Γ-convergence result extends to a space of centerlines which allows, at least, for isolated inflection points. The following lemma provides a step in this direction.

Lemma 4 *The Sadowsky functional $\mathcal{F}$ is sequentially weakly lower semicontinuous in $W^{3,p}(I; \mathbb{R}^3)$ at functions $\mathbf{u} \in X^p$ such that $\{t \in [0, 1] : \mathbf{u}''(t) = 0\}$ has measure zero for $p > 1$.*

Proof Given an element $\mathbf{u} \in X^p$ for $p > 1$, choose a $C^{2,1-1/p}([0, 1]; \mathbb{R}^3)$ representative of $\mathbf{u}$—also denoted by $\mathbf{u}$, and define $Z = \{s \in [0, 1] : \mathbf{u}''(s) = 0\}$. Clearly Z is closed, and $\{I_j\}_{j\in\mathcal{J}}$ be the family of nonoverlapping open intervals $I_j = (a_j, b_j)$ on which $\mathbf{u}''(s) \neq 0$. Since there are at most countably many such intervals, take $\mathcal{J} \subset \mathbb{N}$. Since Z has measure zero, it is clear that

$$\mathcal{F}(\mathbf{w}, I) = \mathcal{F}\left(\mathbf{w}, \bigcup_{j\in\mathcal{J}} I_j\right), \tag{45}$$

for any $\mathbf{w} \in X^p$.

Let $\{\mathbf{u}_n\}$ be a sequence in X^p such that $\mathbf{u}_n \rightharpoonup \mathbf{u}$ weakly in $W^{3,p}(I;\mathbb{R}^3)$, and let $j \in \mathcal{J}$ be given. Choose $K_j \in \mathbb{N}$ so large that $K_j > 2/(b_j - a_j)$. For $k \geq K_j$, then define the intervals $I_{j,k}$ by

$$I_{j,k} = \left(a_j + \frac{1}{k}, b_j - \frac{1}{k}\right), \tag{46}$$

with the remainder denoted by $R_{j,k} = I_j \setminus I_{j,k}$. By the continuity of $\mathbf{u}''$, the curvature $|\mathbf{u}''|$ is bounded below on each interval $I_{j,k}$ such that

$$m_{j,k} = \inf_{t \in I_{j,k}} \left|\mathbf{u}''(t)\right| > 0. \tag{47}$$

By the uniform convergence $\mathbf{u}_n'' \to \mathbf{u}''$ on $[0, 1]$, granted that N_j is sufficiently large it can be concluded that $|\mathbf{u}_n''(t)| > m_{j,k}/2$ for all $t \in I_{j,k}$ and $n \geq N_j$. Since for any $\varepsilon > 0$ it is feasible to choose $K_j > 0$ such that if $k_j > K_j$, then $\mathcal{F}(\mathbf{u}, I_{j,k_j}) > \mathcal{F}(\mathbf{u}, I_j) - \varepsilon 2^{-j}$ if $\mathcal{F}(\mathbf{u}, I_j)$ is finite and $\mathcal{F}(\mathbf{u}, I_{j,k_j}) > 1/\varepsilon$ otherwise.

Suppose that $\mathcal{F}(\mathbf{u}, I_j)$ is finite for all $j \in \mathcal{J}$. By Fatou's lemma applied to the counting measure, the countable additivity of $\mathcal{F}(\mathbf{w}; \cdot)$ as a set function, and the positivity of $\mathcal{F}(\cdot, A)$ for any measurable set $A \subset I$

$$\begin{aligned} \liminf_{n\to\infty} \mathcal{F}(\mathbf{u}_n, I) &= \liminf_{n\to\infty} \sum_{j\in\mathcal{J}} \mathcal{F}(\mathbf{u}_n, I_j) \\ &\geq \sum_{j\in\mathcal{J}} \liminf_{n\to\infty} \mathcal{F}(\mathbf{u}_n, I_{j,k_j}). \end{aligned}$$

Since $\mathcal{F}(\cdot, I_{j,k_j})$ is sequentially weakly lower semicontinuous at functions such that $|\mathbf{u}''| \geq m$ for some $m > 0$, it follows that

$$\begin{aligned} \liminf_{n\to\infty} \mathcal{F}(\mathbf{u}_n, I) &\geq \sum_{j\in\mathcal{J}} \mathcal{F}(\mathbf{u}, I_{j,k_j}) \\ &= \mathcal{F}\left(\mathbf{u}, \bigcup_{j\in\mathcal{J}} I_j\right) - \varepsilon, \end{aligned}$$

wherein the countable additivity of $\mathcal{F}(\mathbf{u}, \cdot)$ as a set function is used.

If there is at least one interval I_J for which $\mathcal{F}(\mathbf{u}, I_J)$ is infinite, then $\mathcal{F}(\mathbf{u}, I)$ is infinite. Moreover, it follows that

$$\begin{aligned} \liminf_{n\to\infty} \mathcal{F}(\mathbf{u}_n, I) &\geq \liminf_{n\to\infty} \mathcal{F}(\mathbf{u}_n, I_{J,k_J}) \\ &\geq \frac{1}{\varepsilon} \end{aligned}$$

and, since $\varepsilon > 0$ is arbitrary, the conclusion holds. □

6 Discussion

It has been shown that the Γ-convergence in a space $X^{+,p}$ of ribbons with centerlines having nonvanishing curvature with respect to weak and strong convergence of the centerlines in

$W^{3,p}(I;\mathbb{R}^3)$. The elastic energy $\mathcal{F}_\varepsilon$ of such ribbons may be written as a single integral along their centerline depending parametrically on the aspect ratio ε of the ribbon. For a fixed centerline $\mathbf{u}$, the elastic energy $\mathcal{F}_\varepsilon(\mathbf{u}, I)$ is monotonically increasing in the aspect ratio, and the Γ-limit result follows upon showing that the lower semicontinuous envelope of the point-wise limit $\tilde{\mathcal{F}}$ is the Sadowsky functional in $X^{+,p}$ for $p \geq 4$ with respect to weak convergence in $W^{3,p}(I;\mathbb{R}^3)$.

It has also been shown that the Sadowsky functional is sequentially weakly lower semicontinuous at functions $\mathbf{u}$ in X^p with isolated inflection points. Since Γ-limits are lower semicontinuous, this is a necessary but not sufficient condition for the Sadowsky functional to be the Γ-limit in a space containing such functions. To establish that $\mathcal{F}$ is the Γ-limit of $\mathcal{F}_\varepsilon$ with respect to the sequential weak (strong) topology on $W^{3,p}(I;\mathbb{R}^3)$, it remains to construct a sequence of functions $\{\mathbf{u}_n\}$ in X^p with η_n satisfying $\eta_n' \in L^\infty(I)$, $\mathbf{u}_n \to \mathbf{u}$ in weakly (strongly, respectively) $W^{3,p}(I;\mathbb{R}^3)$, and $\lim_{n\to\infty} \mathcal{F}(\mathbf{u}_n, I) = \mathcal{F}(\mathbf{u}, I)$. Such a construction would firmly establish Sadowsky's functional as the variational limit in a space including centerlines of developables that form Möbius bands—the context in which that functional was originally derived. A possible obstruction is due to the use of the Frenet frame, and it might therefore be worth following the lead of Rappaport and Rabin [14] and developing an analogue to the Γ-convergence result using the Darboux frame for the centerline.

From the perspective of direct methods in the calculus of variations, it is of interest to prove the following natural conjecture.

Conjecture 1 *For each $\kappa_m > 0$ and $p \geq 4$, smooth functions are strongly dense in the space*

$$\Xi^{\kappa_m,p} = \left\{\mathbf{u} \in W^{3,p}\left(I;\mathbb{R}^3\right) : \left|\mathbf{u}'(s)\right| = 1 \text{ and } \left|\mathbf{u}''(s)\right| \geq \kappa_m \text{ for a.e. } s \in I\right\}. \quad (48)$$

In particular, $X^{\kappa_m,p} = \Xi^{\kappa_m,p}$ and consequently $X^{\kappa_m,p}$ is sequentially weakly closed.

Although this direction is not pursued here, it is also of interest to show that there exist minima of the functionals $\mathcal{F}_\varepsilon(\cdot, I)$ and $\mathcal{F}(\cdot, I)$ on the set of $\mathbf{u} \in X^{\kappa_m,p}$ subject to boundary conditions for each $\kappa_m > 0$, $\varepsilon > 0$, and $p \geq 4$, and that the minima of $\mathcal{F}_\varepsilon(\cdot, I)$ converge to the minimum of $\mathcal{F}(\cdot, I)$. Establishing the above conjecture is likely to be an important step toward establishing the existence of minimizers in $X^{\kappa_m,p}$.

References

1. Adams, R.A., Fournier, J.J.F.: Sobolev Spaces. Academic Press, San Diego (2003)
2. Braides, A.: Γ-Convergence for Beginners. Oxford University Press, London (2002)
3. Buttazzo, G., Giaquinta, M., Hildebrandt, S.: One-Dimensional Variational Problems. Oxford University Press, London (1998)
4. Carathéodory, C.: Über den Variabilitätsbereich der Koeffizienten von Potenzreihen, die gegebene Werte nicht annehmen. Math. Ann. **64**(1), 95–115 (1907)
5. Carathéodory, C.: Über den Variabilitatsbereich der Fourier'schen Konstanten von positiven harmonischen Funktionen. Rend. Circ. Mat. Palermo **32**(1), 197–217 (1911)
6. De Giorgi, E., Franzoni, T.: Su un tipo di convergenza variazionale. Atti Accad. Naz. Lincei, Rend. Cl. Sci. Fis. Mat. Nat. **58**(8), 842–850 (1975)
7. Friesecke, G., James, R.D., Müller, S.: A theorem on geometric rigidity and the derivation of nonlinear plate theory from three-dimensional elasticity. Commun. Pure Appl. Math. **55**(11), 1461–1506 (2002)
8. Gauss, C.F.: Disquisitiones generales circa superficies curvas. Comm. Soc. Gottingen **6**, 1823–1827 (1827)
9. Germain, S.: Recherches sur la théorie des surfaces élastiques. Huzard-Courcier (1821)
10. Giomi, L., Mahadevan, L.: Statistical mechanics of developable ribbons. Phys. Rev. Lett. **104**(23), 238104 (2010)

11. Graustein, W.C.: Differential Geometry. Macmillan Co., New York (1935)
12. Poisson, S.D.: Mémoire sur les surfaces élastiques. Mém. Cl. Sci. Mathém. Phys. Inst. de Fr. **2**, 167–226 (1812)
13. Randrup, T., Røgen, P.: Sides of the Möbius strip. Arch. Math. **66**(6), 511–521 (1996)
14. Rappaport, S.M., Rabin, Y.: Differential geometry of polymer models: worm-like chains, ribbons and Fourier knots. J. Phys. A, Math. Theor. **40**(17), 4455 (2007)
15. Sadowsky, M.: Ein elementarer Beweis für die Existenz eines abwickelbaren Möbiusschen Bandes und Zurückfügrung des geometrischen Problems auf ein Variationsproblem. Sitzber. Preussischen Akad. der Wiss. Philos.-hist. Kl. **22**, 412–415 (1930)
16. Starostin, E.L., van der Heijden, G.H.M.: The equilibrium shape of an elastic developable Möbius strip. Proc. Appl. Math. Mech. **7**(1), 2020115–2020116 (2007)
17. Struik, D.J.: Lectures on Classical Differential Geometry. Dover, New York (1961)
18. Wheeden, R.L., Zygmund, A.: Measure and Integral. Dekker, New York (1977)
19. Wunderlich, W.: Über ein abwickelbares Möbiusband. Monatshefte Math. **66**(3), 276–289 (1962)
20. Yong, E.H.: Problems at the nexus of geometry and soft matter: rings, ribbons and shells. Ph.D. thesis, Harvard University (2012)

DOI 10.1007/978-94-017-7300-3_7
Reprinted from *Journal of Elasticity* Journal, DOI 10.1007/s10659-014-9487-0

"Wunderlich, Meet Kirchhoff": A General and Unified Description of Elastic Ribbons and Thin Rods

Marcelo A. Dias · Basile Audoly

Received: 30 January 2014 / Published online: 26 August 2014

Abstract The equations for the equilibrium of a thin elastic ribbon are derived by adapting the classical theory of thin elastic rods. Previously established ribbon models are extended to handle geodesic curvature, natural out-of-plane curvature, and a variable width. Both the case of a finite width (Wunderlich's model) and the limit of small width (Sadowksky's model) are recovered. The ribbon is assumed to remain developable as it deforms, and the direction of the generatrices is used as an internal variable. Internal constraints expressing inextensibility are identified. The equilibrium of the ribbon is found to be governed by an equation of equilibrium for the internal variable involving its second-gradient, by the classical Kirchhoff equations for thin rods, and by specific, thin-rod-like constitutive laws; this extends the results of Starostin and van der Heijden (Nat. Mater. 6(8):563–567, 2007) to a general ribbon model. Our equations are applicable in particular to ribbons having geodesic curvature, such as an annulus cut out in a piece of paper. Other examples of application are discussed. By making use of a material frame rather than the Frenet–Serret frame, the present work unifies the description of thin ribbons and thin rods.

Keywords Elastic plates · Elastic rods · Energy minimization

Mathematics Subject Classification 74K20 · 74K10 · 74G65

1 Introduction

A ribbon is an elastic body whose dimensions (typical length L, width w and thickness h) are all very different, $L \gg w \gg h$. While previous work has been focussed on the case

M.A. Dias
School of Engineering, Brown University, Providence, RI 02912, USA
e-mail: marcelo_dias@brown.edu

B. Audoly (✉)
Sorbonne Universités, UPMC Univ Paris 06, CNRS, UMR 7190 Institut Jean Le Rond d'Alembert, 75005 Paris, France
e-mail: audoly@lmm.jussieu.fr

of rectangular ribbons, we consider the general case of ribbons having non-zero natural curvatures, both in the out-of-plane and the in-plane directions. This extension includes ribbon geometries such as those obtained by cutting a piece of paper along two arbitrary curves.

From a mechanical perspective, elastic ribbons lie halfway between the 1D case of thin rods (for which $w \sim h$), and the 2D case of thin elastic plates or shells (for which $w \sim L$). On one hand, their elastic energy is given by the theory of thin elastic plates or shells. On the other hand, ribbons look like 1D structures (thin rods) when observed from the large scale L: this suggests that they can be described by the classical equations for thin elastic rods. This article is concerned with the following problem of dimensional reduction: starting from a thin, developable shell model, can one recover the 1D equations of equilibrium applicable to thin rods?

This work builds up on a few seminal articles. The dimensional reduction has already been carried out at the energy level and in the particular case of rectangular, naturally flat ribbons: Sadowsky [21] derived a 1D energy functional for a narrow ribbon (small w), and his work was later generalized by Wunderlich [28] to a finite width w. Their dimensional reduction was made possible by focussing on developable configurations of the ribbon, which are preferred energetically in the thin limit, $h \ll w$. Developable surfaces are special cases of ruled surfaces, i.e., they are spanned by a set of straight lines called generatrices or rulings: the 1D elastic energy of Wunderlich is based on a reconstruction of the surface of the ribbon in terms of its center-line and of the angle between the generatrices and the center-line tangent. We use a similar parameterization here and derive the 1D energy functional for a developable, but not necessarily rectangular, ribbon.

Next comes the question of minimizing this 1D energy to solve the equilibrium problem. Upper bounds for the energy have been obtained by inserting trial forms of the ribbon into the 1D energy, as was done in the context of the elastic Möbius strip [19, 28]. Finding equilibrium solutions, however, requires one to derive the equations of equilibrium by a variational method. This has been done in a beautiful article by Starostin and van der Heijden [25] for naturally flat and rectangular ribbons. They found equilibrium equations that bear a striking resemblance with the Kirchhoff equations governing the equilibrium of thin rods. Their result was later extended to helical ribbons [24], another case where geodesic curvature is absent. Here, we want to revisit and extend their work in the following ways.

First, the derivation of Starostin and van der Heijden, based on the variational bicomplex formalism, uses a different approach than the classical theory of thin rods. The final equations, however, look similar to the Kirchhoff equations for the equilibrium of thin rods. In fact, previous work on thin ribbons has developed as a field largely independent from the vast literature on thin rods. This is unfortunate in view of their deep similarities. Here, we advocate the viewpoint that a ribbon is just a special kind of a thin rod, having an internal parameter and being subjected to kinematical constraints—this is quite similar to the way the incompressibility constraint is handled in 3D elasticity. These specificities can be incorporated naturally into the classical theory of thin rods, as we show. Doing this allows one to recycle much of the existing knowledge on thin rods. In particular, the equations of equilibrium for ribbons are derived in close analogy with those for rods, and in a straightforward way.

Second, we make use of directors, as in the classical theory of rods. By contrast, Wunderlich has introduced a parameterization of the mid-surface of the ribbon based on the Frenet–Serret frame associated with the center-line. Wunderlich's energy, in particular, is defined in terms of the Frenet–Serret notions of torsion and curvature. This parameterization has a drawback: it is specific to the case where the center-line is a geodesic, as we show.

By working instead with directors, we can naturally extend Wunderlich's model to ribbons that have geodesic curvature, i.e. to ribbons curved in their own plane such as an annulus cut out from a piece of paper.

In the present work we make use of several ideas introduced in a recent article [10], where we have shown that the buckling of a curved strip cut out from a piece of paper and folded along its central circle [11] can be analyzed using the language of thin rods. This was done by identifying the relevant geometrical constraints and constitutive laws. Here, we do not consider any fold but allow for more general geometries (non-uniform width and geodesic curvature).

This article is organized as follows. In Sect. 2, we extend the parameterization of developable surfaces introduced by Wunderlich: making use of the frame of directors, we account for the geodesic curvature of the center-line and a variable width. In Sect. 3, Wunderlich's energy functional is extended. In Sect. 4, the equilibrium equations of a general ribbon are derived by a variational method adapted from the theory of thin rods. In Sect. 5, we recover known ribbon models in the special case of a geodesic center-line ($\kappa_g = 0$) and constant width w. In Sect. 6, we present some equilibrium problems for ribbons having geodesic curvature as possible illustrations of our theory.

2 Geometry of a Developable Ribbon

2.1 Developable Transformation from Reference to Current Configuration

As we consider developable ribbons, we can assume that the reference configuration is planar.[1] This planar reference configuration is not necessarily stress-free (we shall address the case of naturally curved ribbons). In the reference configuration, a material line $\mathbf{X}(S)$, called the *center-line*, is traced out on the ribbon, see Fig. 1a. Here, S is the arc-length along the center-line as measured in reference configuration, $|\mathbf{X}'(S) = 1|$. Primes denote derivation with respect to arc-length, and boldface characters denote vectors. The surface of the ribbon is oriented by prescribing a constant unit vector $\mathbf{D}_2$, perpendicular to the plane of the ribbon. Let $\mathbf{D}_3(S) = \mathbf{X}'(S)$ be the unit tangent to the center-line. The vectors $\mathbf{D}_1(S) = \mathbf{D}_2 \times \mathbf{D}_3(S)$, $\mathbf{D}_2$ and $\mathbf{D}_3(S)$ then form an orthonormal frame.

A deformed, developable configuration of the ribbon is specified by the functions

$$\big(\mathbf{x}(S), \mathbf{d}_1(S), \mathbf{d}_2(S), \mathbf{d}_3(S), \eta(S)\big), \tag{1}$$

which are subjected to geometrical constraints derived later. Here $\mathbf{x}(S)$ is the deformed center-line, $\mathbf{d}_i(S)$ (for $i = 1, 2, 3$) define the *frame of directors* (also called the *material frame*) and $\eta(S)$, defined below, encodes the definition of the generatrices.

The third director is chosen to be the tangent to the deformed center-line,

$$\mathbf{x}'(S) = \mathbf{d}_3(S), \tag{2*}$$

and the second director $\mathbf{d}_2(S)$ is defined to be normal to the ribbon at $\mathbf{x}(S)$ as in the reference configuration, see Fig. 1b. The directors are defined to be orthonormal,

$$\mathbf{d}_i(S) \cdot \mathbf{d}_j(S) = \delta_{ij}, \tag{3*}$$

[1] For a closed developable ribbon, there may not exist any *global* planar configurations—see the example in Sect. 6.2. In that case, we introduce an arbitrary cut in the planar configuration of reference.

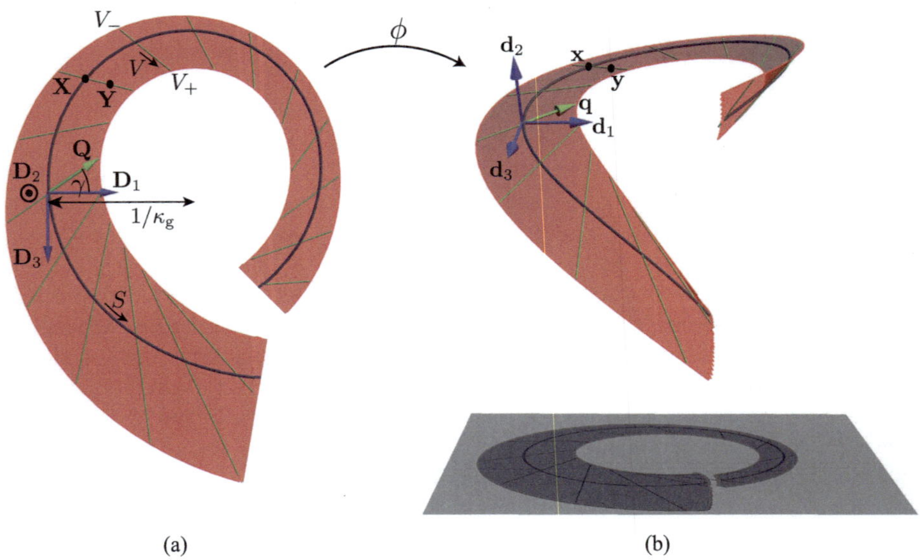

Fig. 1 Geometry of a developable ribbon (**a**) in the planar, undeformed configuration, and (**b**) in the actual configuration. The direction of the generatrices is measured by the parameter η in the model, which is the tangent of the angle γ between the director $\mathbf{d}_1$ and the generatrix direction $\mathbf{q}$ (the angle γ shown in the figure is negative, and $\eta < 0$ here)

where δ_{ij} stands for Kronecker's symbol. This implies $|\mathbf{x}'| = |\mathbf{d}_3| = 1 = |\mathbf{X}'(S)|$. By construction, the directors $\mathbf{d}_i$ are *material* vectors: contrary to the Frenet–Serret frame associated with the center-line, they follow the rotation of the ribbon.

As the ribbon is inextensible, it remains developable by Gauss' *Theorema egregium*. Smooth, developable surfaces are ruled [23]: there exists a one-parameter family of straight lines, called *generatrices*, that sweeps out over the entire surface. As in previous work [25, 28], we define $\eta(S)$ to be the tangent of the angle γ between $\mathbf{d}_1$ and the generatrix. Then, the vector

$$\mathbf{q}(\eta, S) = \eta(S)\mathbf{d}_3(S) + \mathbf{d}_1(S) \tag{4}$$

spans the generatrix.[2] Therefore, the transformation from the reference to the deformed configuration can be expressed as the mapping ϕ:

$$\phi\colon \quad \mathbf{Y} = \mathbf{X}(S) + V\mathbf{Q}(\eta, S) \quad \mapsto \quad \mathbf{y} = \mathbf{x}(S) + V\mathbf{q}(\eta, S). \tag{5}$$

Here, V is a coordinate along the generatrix, $\mathbf{Y}$ and $\mathbf{y}$ denote a current point along the ribbon in reference and actual configurations, respectively. The vector $\mathbf{Q}$ is defined as $\mathbf{Q}(\eta, S) = \eta(S)\mathbf{D}_3(S) + \mathbf{D}_1(S)$: it defines the direction of the generatrix brought back in the reference configuration.

We use the longitudinal and transverse coordinates (S, V) to parameterize the ribbon's surface. S varies in the interval $0 \le S \le L$, where L is the curvilinear length of the center-line. The transverse coordinate V varies in a domain $V_-(\eta, S) \le V \le V_+(\eta, S)$. The end-

[2]The vector $\mathbf{q}$ depends both on the unknown *function* $\eta(\cdot)$ and on the arc-length parameter S; hence the arguments shown in the left-hand side of Eq. (4).

points $V_\pm(\eta, S)$ of the interval are such that the points $\mathbf{Y}_\pm(S) = \mathbf{X}(S) + V_\pm(\eta, S)\mathbf{Q}(\eta, S)$ lie on the edges of the ribbon. The functions $V_\pm(\eta, S)$ capture the relative position of the edges and of the center-line, and are called the *edge functions*. Explicit expressions are derived in Sect. 2.2 for some ribbon geometries.

From Eq. (3), the directors define an orthonormal frame for any value of the arc-length parameter S. Therefore, there exists a vector $\boldsymbol{\omega}(S)$ called the Darboux vector or the rotation gradient, such that

$$\mathbf{d}_i'(S) = \boldsymbol{\omega}(S) \times \mathbf{d}_i(S) \tag{6}$$

for $i = 1, 2, 3$. The operation $\times$ denotes the cross product in the Euclidean space. The components of the rotation gradient in the basis of directors, $\omega_i(S) = \boldsymbol{\omega}(S) \cdot \mathbf{d}_i(S)$ measure the amount of bending ($i = 1, 2$) and twisting ($i = 3$) of the center-line. An explicit expression is

$$\omega_i(S) = \frac{1}{2} \sum_{j=1}^{3} \sum_{k=1}^{3} \epsilon_{ijk}\, \mathbf{d}_j'(S) \cdot \mathbf{d}_k(S), \tag{7*}$$

where ϵ_{ijk} represents the permutation symbol: $\epsilon_{ijk} = 1$ when (i, j, k) is an even permutation of the indices, $\epsilon_{ijk} = -1$ when it is an odd permutation, and $\epsilon_{ijk} = 0$ otherwise. In the language of the geometry of surfaces, the directors frame $(\mathbf{d}_1, \mathbf{d}_2, \mathbf{d}_3)$ is called the Darboux frame associated with the center-line curve, and the strains ω_1, ω_2, and ω_3 are respectively the normal curvature, the geodesic curvature, and the geodesic torsion. Note that the strain measures, ω_i, are based on the frame of directors (a material frame), while previous work used the Frenet–Serret frame associated with the center-line. Working with a frame of directors offers many advantages: it extends naturally to the case of non-geodesic center-lines, allows one to use the same language as in the theory of rods, and to remove the artificial singularities displayed by the Frenet–Serret frame near inflection points or straight segments.

2.2 Edge Functions

The edge functions $V_\pm(\eta, S)$ encode the relative positions of the edges of the ribbon with respect to the center-line. Expressions for $V_\pm$ are derived below for the cases of a rectangular and an annular ribbon.

The case of a rectangular ribbon is quite simple. We use the central axis of the ribbon as the center-line. In reference configuration, the equation of the edges is $(\mathbf{Y} - \mathbf{X}) \cdot \mathbf{D}_1 = \pm w/2$, where w is the width of the ribbon. Inserting the parameterization of $\mathbf{Y}$ from Eq. (5), this yields

$$V_\pm(\eta, S) = \pm\frac{w}{2} \quad \text{(rectangular ribbon)}, \tag{8}$$

where the $\pm$ labels the edges.

The case of an annular ribbon is studied next. We use an arc of a circle as the center-line, having its radius given by the inverse of the geodesic curvature, κ_g^{-1}, and a constant width w of the ribbon. In reference configuration, the center $\mathbf{C}$ of the circular center-line is $\mathbf{C} = \mathbf{X}(S) + \kappa_g^{-1}\mathbf{D}_1(S)$. Therefore, the equation for the edges is $(\mathbf{Y}(S, V) - \mathbf{C})^2 = (\kappa_g^{-1} \mp \frac{w}{2})^2$. Using Eq. (5) and the definition of $\mathbf{Q}$, this shows that the edge functions $V_\pm$ are the roots V of the second-order polynomial

$$\left(V - \frac{1}{\kappa_g}\right)^2 + (\eta V)^2 = \left(\frac{1}{\kappa_g} \mp \frac{w}{2}\right)^2.$$

Solving for V, we find

$$V_{\pm}(\eta, S) = \frac{1}{\kappa_g} \frac{1 - \sqrt{1 \mp (1+\eta^2) w \kappa_g (1 \mp \frac{w\kappa_g}{4})}}{1+\eta^2} \quad \text{(annular ribbon).} \tag{9}$$

Letting the curvature of the annulus go to zero, $\kappa_g \to 0$, we recover $V_{\pm} \to \pm w/2$ which is consistent with Eq. (8).

In the general case of a ribbon having a variable width, or when the center-line has non-constant curvature κ_g, $V_{\pm}$ may be available through an implicit equation and not necessarily in closed form.

2.3 Constraints Expressing Developability

The condition of inextensibility of the ribbon imposes some kinematical constraints on the unknowns listed in (1), and on the curvature and twisting strains ω_i calculated by Eqs. (6)–(7). These constraints are derived as follows.

The center-line is a curve drawn on the surface of the ribbon. Its geodesic curvature is defined by $\kappa_g = \mathbf{x}'' \cdot \mathbf{d}_1 = \mathbf{d}_3' \cdot \mathbf{d}_1 = \omega_2$. It is a classical result of the differential geometry of surfaces [23] that the geodesic curvature is conserved upon isometric deformations of a surface. Therefore, κ_g is prescribed by the reference configuration: $\kappa_g(S) = \mathbf{D}_3'(S) \cdot \mathbf{D}_1(S)$. We write this geodesic constraint as

$$\mathcal{C}_g(\omega_2, S) = 0 \tag{10a*}$$

where

$$\mathcal{C}_g(\omega_2, S) = \kappa_g(S) - \omega_2. \tag{10b*}$$

Note that $\omega_2 = 0$ when $\kappa_g = 0$, i.e. when the center-line is a geodesic. In that case, the derivative of the tangent is $\mathbf{d}_3' = \boldsymbol{\omega} \times \mathbf{d}_3 = \omega_1(-\mathbf{d}_2)$ and $(-\mathbf{d}_2)' \cdot \mathbf{d}_1 = \omega_3$. Here, we recognize the definition of the Frenet–Serret frame $(\mathbf{d}_3, -\mathbf{d}_2, \mathbf{d}_1)$ associated with the center-line: the Frenet–Serret curvature and torsion are ω_1 and ω_3, respectively. Therefore, in the particular case when the center-line is a geodesic, our directors coincide with the Frenet–Serret frame. It is much more convenient to work with the directors in general.

The second constraint expresses the developability of the ruled surface spanned by the generatrices. It is found in classical textbooks of differential geometry [23], and is rederived in Appendix:

$$\mathcal{C}_d(\omega_1, \omega_3, \eta) = 0, \tag{11a*}$$

where

$$\mathcal{C}_d(\omega_1, \omega_3, \eta) = \eta\omega_1 - \omega_3. \tag{11b*}$$

2.4 Area Element

To integrate the elastic energy along the surface of the ribbon, we will need the expression for the area element da. It is calculated in Appendix from the Jacobian of the transformation ϕ:

$$da = |\partial_S \mathbf{y} \times \partial_V \mathbf{y}| \, dS \, dV = \left(1 - \frac{V}{V_c(\eta, S)}\right) dS \, dV, \tag{12}$$

where the auxiliary quantity V_c is defined as

$$V_c(\eta, \eta', S) = \frac{1}{(1+\eta^2)\kappa_g(S) - \eta'}. \tag{13*}$$

Since $da = 0$ at $V = V_c$, the transformation ϕ is singular there. The quantity V_c can be interpreted as the value of the transverse coordinate V where the generatrix intersects neighboring generatrices, i.e., intersects its own caustic, called the *striction curve* [23]. We assume that the striction curve stays outside of the physical domain, so that the curvature tensor is nowhere singular on the ribbon: either $V_c < V_- \leq V \leq V_+$ or $V_c > V_+ \geq V \geq V_-$. This implies that

$$\frac{1 - V_+/V_c}{1 - V_-/V_c} > 0 \tag{14a}$$

and

$$\left|\frac{V}{V_c}\right| < 1 \tag{14b}$$

2.5 Curvature Tensor of the Deformed Ribbon

In order to write the elastic energy of the ribbon, we need the curvature tensor $\mathbf{K}$ at the arbitrary point (S, V) of the ribbon surface. This geometrical calculation, at the heart of Wunderlich's energy, is given in Appendix. The dependence of the curvature tensor, $\mathbf{K}(\eta, \eta', \omega_1, S, V)$, on the transverse coordinate is imposed by the developability condition to be

$$\mathbf{K}(\eta, \eta', \omega_1, S, V) = \frac{\mathbf{K}_0(\eta, \omega_1)}{1 - \frac{V}{V_c(\eta, \eta', S)}}, \tag{15}$$

where $\mathbf{K}_0(\eta, \omega_1) = \mathbf{K}(\eta, \eta', \omega_1, S, V = 0)$ denotes the curvature tensor evaluated along the center-line, defined by

$$\mathbf{K}_0(\eta, \omega_1) = -\omega_1\big(\mathbf{d}_3 \otimes \mathbf{d}_3 - \eta(\mathbf{d}_3 \otimes \mathbf{d}_1 + \mathbf{d}_1 \otimes \mathbf{d}_3) + \eta^2 \mathbf{d}_1 \otimes \mathbf{d}_1\big). \tag{16*}$$

Here, the curvature tensor is expressed by its components in the orthonormal basis $(\mathbf{d}_3(S), \mathbf{d}_1(S))$ spanning the plane tangent to the ribbon at (S, V). Inserting Eq. (16) into Eq. (15) yields $\det \mathbf{K} = \det \mathbf{K}_0 = 0$ which is consistent with Gauss' *theorema egregium* (conservation of Gauss curvature by isometric deformations of a surface).

3 Elastic Energy

Being modeled as an inextensible plate, the deformed ribbon is isometrically mapped to its planar underformed configuration. The constraint of isometric mapping is enforced by Eqs. (10a), (10b), (11a), (11b). As a result, the stretching energy cancels. The only contribution to the elastic energy of the ribbon is the bending energy E. For a plate made up of a homogeneous isotropic Hookean solid,

$$E = \frac{D}{2}\iint \big[(1-\nu)\,\mathrm{tr}\big((\mathbf{K} - \mathbf{K}_r)^2\big) + \nu\,\mathrm{tr}^2(\mathbf{K} - \mathbf{K}_r) - (1-\nu)\,\mathrm{tr}\big(\mathbf{K}_r^2\big) - \nu\,\mathrm{tr}^2\,\mathbf{K}_r\big]\,da,$$

where $D = Yh^3/(12(1-\nu^2))$ is the bending modulus of the plate, ν the Poisson's ratio, Y the Young's modulus, h the thickness of the ribbon. This expression is based on the classical formula for plates [18], and has been modified to take natural (or reference) curvature into account through the natural curvature tensor $\mathbf{K}_\mathrm{r}$. The last two constant terms appearing in the second line are included to make the energy density zero when the ribbon is flat ($\mathbf{K} = \mathbf{0}$), which is a convenient convention.

The energy can be simplified by expanding the squared matrices, using the identity $\mathrm{tr}^2\,\mathbf{K} = \mathrm{tr}(\mathbf{K}^2) + 2\det\mathbf{K}$ valid for 2×2 matrices, and the developability condition $\det\mathbf{K} = 0$ that follows from Eq. (15):

$$E = \frac{D}{2}\iint \left(\mathrm{tr}\left(\mathbf{K}^2\right) - 2\mathbf{Q}_\mathrm{r} : \mathbf{K}\right)\mathrm{d}a, \tag{17}$$

where the symmetric tensor $\mathbf{Q}_\mathrm{r}$ capturing the effect of the reference curvature is defined by

$$\mathbf{Q}_\mathrm{r} = \mathbf{K}_\mathrm{r} + \nu\,\mathrm{cof}\,\mathbf{K}_\mathrm{r}, \tag{18}$$

and $\mathrm{cof}\,\mathbf{K}_\mathrm{r}$ denotes the cofactor matrix:

$$\mathrm{cof}\,\mathbf{K}_\mathrm{r} = \left(K^\mathrm{r}_{11}\mathbf{d}_3 \otimes \mathbf{d}_3 - K^\mathrm{r}_{13}\left(\mathbf{d}_1 \otimes \mathbf{d}_3 + \mathbf{d}_1 \otimes \mathbf{d}_3\right) + K^\mathrm{r}_{33}\mathbf{d}_1 \otimes \mathbf{d}_1\right). \tag{19}$$

The symbols K^r_{ij} denote the components of the natural curvature $\mathbf{K}_\mathrm{r}$ in the tangent basis $(\mathbf{d}_3, \mathbf{d}_1)$, with $i, j = 3, 1$.

Inserting the expression of the area element $\mathrm{d}a$ from Eq. (12) and the known dependence of $\mathbf{K}$ on the transverse coordinate from Eq. (15) into Eq. (17) we get

$$E = \frac{D}{2}\int_0^L \left[\mathrm{tr}\left({\mathbf{K}_0}^2\right)\int_{V_-}^{V_+}\frac{\mathrm{d}V}{1 - V/V_\mathrm{c}} - 2\mathbf{Q}_\mathrm{r} : \mathbf{K}_0 \int_{V_-}^{V_+}\mathrm{d}V\right]\mathrm{d}S.$$

Here, by factoring the natural curvature tensor $\mathbf{Q}_\mathrm{r}$ out of the integral along the generatrices, we have assumed that $\mathbf{Q}_\mathrm{r}$ varies on the typical length-scale L (length of the ribbon) and thus can be considered constant on the much smaller length-scale w (width of the ribbon). Our model thus handles non-uniform geometries $\mathbf{Q}_\mathrm{r}(S)$, even though we omit the argument S in $\mathbf{Q}_\mathrm{r}$ for the sake of legibility.

Integrating along V, we obtain the one-dimensional energy functional

$$E\left(\eta, \eta', \omega_1\right) = \frac{D}{2}\int_0^L \left[\left(-V_\mathrm{c}\ln\frac{1 - V_+/V_\mathrm{c}}{1 - V_-/V_\mathrm{c}}\right)\mathrm{tr}\left({\mathbf{K}_0}^2\right) - 2(V_+ - V_-)\mathbf{Q}_\mathrm{r} : \mathbf{K}_0\right]\mathrm{d}S. \tag{20*}$$

This functional extends Wunderlich's result to the case of a ribbon with natural out-of-plane curvature (through the term depending on $\mathbf{Q}_\mathrm{r}$), a variable width (through the functions $V_\pm$), and geodesic curvature (through the dependence of $V_\pm$ and V_c on κ_g). The argument of the logarithm is always positive, as is noted in the inequality (14a).

4 Equations of Equilibrium

In this section, we use the calculus of variations to derive the equations of equilibrium for a general ribbon model. Because of our parameterization based on directors, we do this simply by extending the classical derivation of the equations of equilibrium for thin elastic rods, using the principle of virtual work. We have used a similar approach in a recent paper [10, Eqs. (9b) and (A.2b)] to derive the equations of equilibrium of a thin annular strip folded along its central circle.

4.1 Principle of Virtual Work for a Ribbon

If the external load is conservative, the equilibrium of the ribbon is found by minimizing the total potential energy, which is the sum of the potential energy associated with the external load and of the elastic energy $E(\eta, \eta', \omega_1)$ in Eq. (20). This is a constrained minimization problem, and we need to extend the classical variational derivation [2, 6, 7, 9, 13, 26] equilibrium for thin rods to take into account the presence of the *kinematical constraints* $\mathcal{C}_\mathrm{d}$ and $\mathcal{C}_\mathrm{g}$, and of an *internal variable* $\eta(S)$.

To handle the case of non-conservative loads, we derive the equations of equilibrium using the more general framework of the principle of virtual work. A virtual motion of the ribbon is specified by a virtual displacement $\hat{\mathbf{x}}(S)$ of the center-line, a virtual rotation $\hat{\boldsymbol{\psi}}(S)$ of the orthonormal frame of directors, a virtual variation $\hat{\mathbf{d}}_i(S)$ of the directors, a virtual variation $\hat{\eta}(S)$ of the parameter defining the direction of the generatrices, and the virtual changes of material curvature and twisting strains $\hat{\omega}_i(S)$.

We start by noting that Eqs. (2), (3) and (7) define what is known as an inextensible Euler–Bernoulli rod. When written in incremental form, they yield relations between the virtual quantities $\hat{\mathbf{x}}$, $\hat{\boldsymbol{\psi}}$, $\hat{\mathbf{d}}_i$ and $\hat{\omega}_i$. In the principle of virtual work, these relations are viewed as constraints, and are handled by a constraint term $\mathcal{W}_\mathrm{cEB}$, which we call the virtual work of the Euler–Bernoulli constraints. Our treatment of these constraints does not differ from the standard theory of inextensible Euler–Bernoulli rods, and we shall therefore omit the details.

The virtual work of a ribbon includes the following contributions. First, the virtual internal work is equal to minus the first variation of the elastic energy $\mathcal{W}_\mathrm{i} = -\hat{E}$, as usual for elasticity problems. Second, the external virtual work reads $\mathcal{W}_e = \int_0^L (\mathbf{p}\cdot\hat{\mathbf{x}} + \mathbf{c}\cdot\hat{\boldsymbol{\psi}})\,\mathrm{d}S$ where $\mathbf{p}$ and $\mathbf{c}$ are the density of applied force and moment onto the center-line, per unit arc-length. For static problems, the virtual work of acceleration is zero. The constraints are treated by constraint terms involving Lagrange multipliers: the constraints applicable to an inextensible Euler–Bernoulli model are included in a constraint term $\mathcal{W}_\mathrm{cEB}$ as explained above; the constraints $\mathcal{C}_\mathrm{g}$ and $\mathcal{C}_\mathrm{d}$, which are specific to a ribbon, are treated by two contributions $\mathcal{W}_\mathrm{cg} = \int_0^L \lambda_\mathrm{g}\hat{\mathcal{C}}_\mathrm{g}\,\mathrm{d}S$ and $\mathcal{W}_\mathrm{cd} = \int_0^L \lambda_\mathrm{d}\hat{\mathcal{C}}_\mathrm{d}\,\mathrm{d}S$, which we call, respectively, the virtual work of the geodesic and developability constraints. Here, $\lambda_\mathrm{g}(S)$ and $\lambda_\mathrm{d}(S)$ denote two Lagrange multipliers, and $\hat{\mathcal{C}}_\mathrm{i}$ is the first variation of any of the constraints, $i \in \{\mathrm{d}, \mathrm{g}\}$. Combining all the contributions, we write the principle of virtual work as

$$
\begin{aligned}
&-\int_0^L \left(\frac{\partial \mathcal{E}}{\partial \eta'}\hat{\eta}' + \frac{\partial \mathcal{E}}{\partial \eta}\hat{\eta} + \frac{\partial \mathcal{E}}{\partial \omega_1}\hat{\omega}_1\right)\mathrm{d}S + \mathcal{W}_\mathrm{e} + \mathcal{W}_\mathrm{cEB}\cdots \\
&+\int_0^L \lambda_\mathrm{g}\frac{\partial \mathcal{C}_\mathrm{g}}{\partial \omega_2}\hat{\omega}_2\,\mathrm{d}S + \int_0^L \lambda_\mathrm{d}\left(\sum_{i=1}^{3}\frac{\partial \mathcal{C}_\mathrm{d}}{\partial \omega_i}\hat{\omega}_i + \frac{\partial \mathcal{C}_\mathrm{d}}{\partial \eta}\hat{\eta}\right)\mathrm{d}S = 0,
\end{aligned}
$$

where $\mathcal{E}$ is the elastic energy density defined by $E = \int_0^L \mathcal{E}(\eta, \eta', \omega_1)\,\mathrm{d}S$.

Rearranging the terms, we have

$$
\mathcal{W}_\eta + \mathcal{W}_\mathrm{i}^\mathrm{rod} + \mathcal{W}_\mathrm{e} + \mathcal{W}_\mathrm{cEB} = 0, \tag{21}
$$

where we have grouped the terms depending on the internal variable η,

$$\mathcal{W}_\eta = -\int_0^L \left[\frac{\partial \mathcal{E}}{\partial \eta'}\hat{\eta}' + \left(\frac{\partial \mathcal{E}}{\partial \eta} - \lambda_{\mathrm{d}}\frac{\partial \mathcal{C}_{\mathrm{d}}}{\partial \eta}\right)\hat{\eta}\right] \mathrm{d}S, \tag{22}$$

and the terms depending on the virtual changes of strain,

$$\mathcal{W}_{\mathrm{i}}^{\mathrm{rod}} = -\int_0^L \sum_{i=1}^{3}\left(\frac{\partial \mathcal{E}}{\partial \omega_i} - \sum_{j\in\{\mathrm{g},\mathrm{d}\}} \lambda_j \frac{\partial \mathcal{C}_j}{\partial \omega_i}\right)\hat{\omega}_i \, \mathrm{d}S. \tag{23}$$

Equation (21) expresses the principle of virtual work for a ribbon: equilibrium configurations are such that this equation is satisfied for any kinematically admissible virtual displacement. Making use of the strong similarity with the principle of virtual work for a thin rod, we can now derive the equations of equilibrium for ribbons easily.

4.2 Equations of Equilibrium

The equation of equilibrium with respect to η comes from the term $\mathcal{W}_\eta$. Integrating by parts and canceling the coefficient of $\hat{\eta}$, we find, as in [25],

$$-\frac{\mathrm{d}}{\mathrm{d}S}\left(\frac{\partial \mathcal{E}}{\partial \eta'}\right) + \frac{\partial \mathcal{E}}{\partial \eta} - \lambda_{\mathrm{d}}\frac{\partial \mathcal{C}_{\mathrm{d}}}{\partial \eta} = 0. \tag{24}$$

Through its first term, this equation depends on the second arc-length derivative of η. Boundary terms have been omitted: the derivation of the boundary condition for η is irrelevant to the case of a closed ribbon, and is left to the reader.

We proceed to the second term $\mathcal{W}_{\mathrm{i}}^{\mathrm{rod}}$ in Eq. (21). We observe that it can be put into the usual form of the internal virtual work of a thin rod [2, 6, 7, 26], namely

$$\mathcal{W}_{\mathrm{i}}^{\mathrm{rod}} = -\int_0^L \mathbf{M}\cdot\sum_{i=1}^{3}(\hat{\omega}_i\,\mathbf{d}_i)\,\mathrm{d}S, \tag{25}$$

when we identify the internal moment $\mathbf{M}(S)$ based on Eq. (23):

$$\begin{aligned}\mathbf{M} &= \sum_{i=1}^{3}\left(\frac{\partial \mathcal{E}}{\partial \omega_i} - \lambda_{\mathrm{d}}\frac{\partial \mathcal{C}_{\mathrm{d}}}{\partial \omega_i} - \lambda_{\mathrm{g}}\frac{\partial \mathcal{C}_{\mathrm{g}}}{\partial \omega_i}\right)\mathbf{d}_i \\ &= \left(\frac{\partial \mathcal{E}}{\partial \omega_1} - \eta\lambda_{\mathrm{d}}\right)\mathbf{d}_1 + \lambda_{\mathrm{g}}\,\mathbf{d}_2 + \lambda_{\mathrm{d}}\,\mathbf{d}_3.\end{aligned} \tag{26}$$

This equation is one of the main results of our paper. It expresses the constitutive law of the ribbon in the language of thin rods. The case of a folded annular strip has been considered recently by the same authors, and similar constitutive laws have been obtained [10, Eqs. (9b) and (A.2b)]. In Eq. (26), the first term in the parenthesis yields the usual constitutive law $\mathbf{M} = \sum_i \frac{\partial \mathcal{E}}{\partial \omega_i}\,\mathrm{d}_i$ for an unconstrained, non-linearly elastic thin rod. The two other terms are constraint terms (the isotropic pressure term entering in the constitutive law of an incompressible elastic solid is a constraint term of the same kind).

Let us now consider the contributions to the principle of virtual work that involve the virtual motion of the center-line and directors, i.e., those depending on $(\hat{\mathbf{x}}, \hat{\boldsymbol{\psi}}, \hat{\mathbf{d}}_i, \hat{\omega}_i)$. These contributions are $\mathcal{W}_{\mathrm{i}}^{\mathrm{rod}} + \mathcal{W}_{\mathrm{e}} + \mathcal{W}_{\mathrm{cEB}}$, as the other contribution $\mathcal{W}_\eta$ concerns the internal degree of freedom only. These are exactly the same three terms as those entering in the principle of virtual work for an inextensible Euler–Bernoulli rod. It is well known [2, 6, 7, 26] that the corresponding equations of equilibrium are the Kirchhoff equations expressing the balance of forces and moments on a small chunk of the center-line,

$$\mathbf{R}'(S) + \mathbf{p}(S) = \mathbf{0}, \tag{27a*}$$

$$\mathbf{M}'(S) + \mathbf{x}'(S) \times \mathbf{R}(S) + \mathbf{c}(S) = \mathbf{0}. \tag{27b*}$$

This remark saves us the effort of rederiving these equations of equilibrium. Note that there is no constitutive law associated with the internal force $\mathbf{R}(S)$, as $\mathbf{R}(S)$ is the Lagrange multiplier associated with the Euler–Bernoulli constraint in Eq. (2).

To sum up, an elastic ribbon is governed by the same equations as an inextensible Euler–Bernoulli rod, up to two small changes: (i) the presence of the internal degree of freedom η yields the new equation of equilibrium (24), and (ii) the presence of constraint terms in the constitutive law; see Eq. (26).

4.3 Complete Set of Equations for an Elastic Ribbon

The complete set of equations for the equilibrium of a ribbon are summarized as follows. The geometry and the natural shape of the ribbon are prescribed by the geodesic curvature $\kappa_{\mathrm{g}}(S)$, by the edge functions $V_{\pm}(\eta, S)$, and by the tensor $\mathbf{Q}_{\mathrm{r}}$ defined in Eq. (18). The external loading is prescribed by the functions $\mathbf{p}(S)$ and $\mathbf{c}(S)$, subject to the global balance of forces and moments. The equilibrium configuration of the ribbon is sought in terms of the following *unknowns*: the center-line $\mathbf{x}(S)$, the directors $\mathbf{d}_i(S)$, the parameter $\eta(S)$ capturing the direction of the generatrices, and the Lagrange multipliers $M_2(S) = \lambda_{\mathrm{g}}(S)$ and $M_3(S) = \lambda_{\mathrm{d}}(S)$.

The *geometrical* equations are the compatibility of the tangent and the third director in Eq. (2), the orthonormality of the directors in Eq. (3), and the developability constraints in Eqs. (10a), (10b)–(11a), (11b). The rotation gradient $\omega_i(S)$ is found using Eq. (7).

In terms of the unknowns, one expresses the coordinate V_{c} of the striction curve using Eq. (13), and the curvature tensor $\mathbf{K}_0$ along the center-line using Eq. (16). Based on the definition of the energy functional in Eq. (20), one can then calculate the internal moment by the *constitutive law*, which we rewrite from Eq. (26) as

$$\mathbf{M}(S) = \left(\frac{\partial \mathcal{E}}{\partial \omega_1} - \eta M_3\right)\mathbf{d}_1(S) + M_2(S)\,\mathbf{d}_2(S) + M_3(S)\,\mathbf{d}_3(S). \tag{28*}$$

The *equations of equilibrium* are the standard Kirchhoff equations (27), together with Eq. (24) for the equilibrium of η, which we rewrite as

$$-\frac{\mathrm{d}}{\mathrm{d}S}\left(\frac{\partial \mathcal{E}}{\partial \eta'}\right) + \frac{\partial \mathcal{E}}{\partial \eta} - M_3\omega_1 = 0. \tag{29*}$$

If the ribbon is closed, these equations must be complemented by periodic boundary conditions.[3] If it is open, boundary conditions corresponding to the type of support must be enforced—their derivation has been left to the reader.

[3]Finding the conditions for a continuous curve in space to be closed is not a trivial problem and it may not have a close solution. This problem was posed by N.V. Efimov [12] and W. Frenchel [14]. Frenchel posed the

The equations tagged by a star, such as Eqs. (28) and (29) above, form the complete set of equations governing the equilibrium of a ribbon.

5 Special Cases

We review ribbon models known in the literature, that are specific to the case $\kappa_g = 0$, in the light of our general model.

5.1 Naturally Straight, Rectangular Ribbons

This model was first introduced by Wunderlich in his study of the shape of a Möbius band [28]. The Möbius band is a naturally flat and straight ribbon having a constant width w: we set $\mathbf{Q}_r = \mathbf{0}$, $\kappa_g = 0$. Then, by Eq. (13), the coordinate of the striction curve is $V_c = -1/\eta'$. The relevant edge functions are given by Eq. (8) as $V_\pm = \pm w/2$. Inserting into the energy functional in Eq. (20), we have

$$E_W = \frac{Dw}{2} \int_0^L \omega_1{}^2 (1+\eta^2)^2 \frac{1}{\eta' w} \ln\left(\frac{1+\eta' w/2}{1-\eta' w/2}\right) \mathrm{d}S, \tag{30}$$

where we have used $\mathrm{tr}(\mathbf{K}_0{}^2) = \omega_1{}^2(1 + \eta^2)^2$. Wunderlich's energy functional is recovered [28]. It has been analyzed recently and is relevant to the elastic Möbius strip [19, 25], and to open developable ribbons [17].

Starostin and van der Heijden [25] have worked out the corresponding equations of equilibrium (for the naturally flat, rectangular ribbon). Here, we have recovered their results as a special case: their equations [4] are identical[4] to the Kirchhoff equations (27) derived above; their equations [5] are the constitutive law (28) and the equilibrium for the internal variable η.

5.2 Sadowsky's Limit: Narrow Rectangular Ribbons

The limiting case of Wunderlich's energy in Eq. (30) corresponding to $w \to 0$ was derived by Sadowsky [21] much before Wunderlich. It reads

$$E_S = \frac{Dw}{2} \int_0^L \omega_1{}^2 (1+\eta^2)^2 \,\mathrm{d}S. \tag{31}$$

This *narrow strip* model applies to a rectangular, naturally flat ribbon, when the deformation is sufficiently small for the striction curve to remain far from the physical edge of the ribbon, $|V_c| \gg w$. This model captures the geometrically non-linear coupling between bending and

problem asking what are the necessary and sufficient conditions of closure given the curvature and torsion of a space curve. By integrating the Frenet–Serret equations, the result yields an infinite series of integrals with no closed form [4].

[4]The quantities $(2w, \mathbf{t}, \mathbf{n}, \mathbf{b}, \tau, \kappa, \eta, g, \mathbf{M}, \mathbf{F}, M_t, M_b)$ in their notation must be identified with the quantities $(w, \mathbf{d}_3, -\mathbf{d}_2, \mathbf{d}_1, \omega_3, \omega_1, \eta, E, -\mathbf{M}, -\mathbf{R}, -M_3, -M_1)$ in our notation, respectively. The last four minus signs introduced here arise because they use a non-standard convention for the sign of the internal force $\mathbf{R}$ and moment $\mathbf{M}$—by contrast, we use the usual convention that $\mathbf{R}$ and $\mathbf{M}$ measure the force and moment applied across an imaginary cut by the downstream part of the ribbon onto the upstream part, where 'downstream' and 'upstream' refer to the direction of increasing arc-length coordinate S.

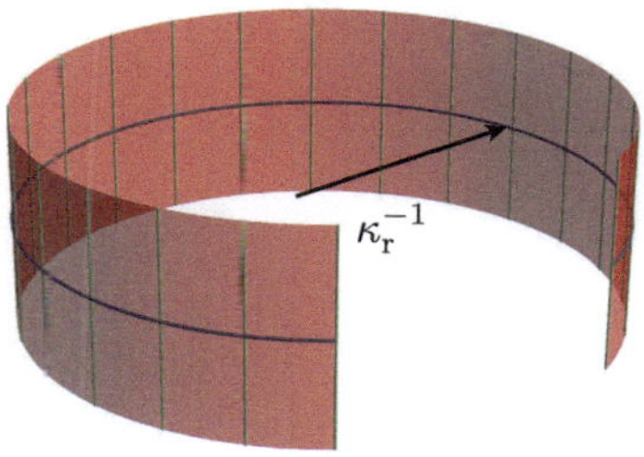

Fig. 2 A ribbon cut out in a cylindrical shell having natural curvature κ_r (Sect. 6.1)

twisting modes. It has been applied to the statistical mechanics of developable ribbons [15] and to the analysis of elastic strips comprising a central fold [10].

The equations governing the equilibrium of Wunderlich's strip model, derived in reference [25], apply as a special case to the equilibrium of Sadowsky's narrow strip model. An alternative derivation appears in reference [10, Sect. 2].

5.3 Helical Ribbons

The case of a naturally helical ribbon has been studied recently [5, 24]. Its geodesic curvature is zero, $\kappa_g = 0$. Assuming that the helical shape is stress-free, the natural curvature tensor reads, from Eq. (16),

$$\mathbf{K}_r = -\omega_1^r\,\mathbf{d}_3 \otimes \mathbf{d}_3 + \omega_3^r(\mathbf{d}_3 \otimes \mathbf{d}_1 + \mathbf{d}_1 \otimes \mathbf{d}_3) - \frac{(\omega_3^r)^2}{\omega_1^r}\,\mathbf{d}_1 \otimes \mathbf{d}_1,$$

where ω_1^r and ω_3^r are the two parameters defining the radius and the step of the helix. The equations of equilibrium, derived in [24] by the same method as in [25], can be recovered from the general model derived above.

When subjected to moderate forces and moments compatible with the helical symmetries, these ribbons remain helical—such helical configurations are relevant to the opening of chiral seed pods [1], a phenomenon driven by residual internal stresses [27]. In the presence of larger or less symmetric loads, non-helical solutions are possible [5, 24].

6 Illustrations

With the aim to illustrate our results and motivate further studies, we present a few problems that could be solved using the equations derived in this work.

6.1 Buckling of a Cylindrical Ribbon

We consider the cylindrical ribbon shown in Fig. 2. Its natural curvature is denoted by κ_r. When laid flat, the ribbon is a rectangle of size $L \times w$. In its natural configuration, the direction supported by its width w is aligned with the axis of the cylinder: its natural curvature is an out-of-plane curvature $\kappa_r = \omega_1^r$, its geodesic curvature being zero, $\kappa_g = 0$. The tensor of natural curvature reads $\mathbf{K}_r = -\kappa_r\,\mathbf{d}_3 \otimes \mathbf{d}_3$, as shown by inserting $\omega_1 = \kappa_r$ and $\eta = \omega_3/\omega_1 = 0$ into Eq. (16). Using the definition of the internal moment $\mathbf{Q}_r$ in Eq. (18), we have

$$\mathbf{Q}_r = -\kappa_r(\mathbf{d}_3 \otimes \mathbf{d}_3 + \nu\,\mathbf{d}_1 \otimes \mathbf{d}_1). \tag{32}$$

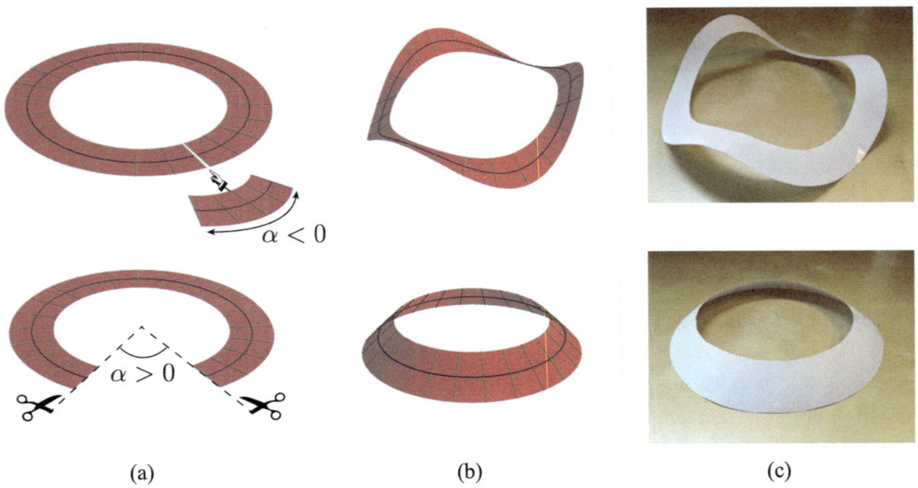

Fig. 3 Undercurved (*top row*) and overcurved (*bottom row*) annuli (Sect. 6.2): (**a**) preparation of the two states, (**b**) sketch of typical equilibrium shapes, (**c**) experiments using paper models

By Eq. (20), the energy functional governing this strip model reads

$$E = \frac{Dw}{2} \int_0^L \left[{\omega_1}^2 \left(1+\eta^2\right)^2 \frac{1}{w\eta'} \ln\left(\frac{1+\eta' w/2}{1-\eta' w/2}\right) - 2\kappa_r \omega_1 \left(1+\nu\eta^2\right) \right] \mathrm{d}S, \tag{33}$$

which is simply Wunderlich's energy, complemented by a term depending on natural curvature. For a narrow strip (Sadowsky's limit, $\eta' w \to 0$), this becomes

$$E = \frac{Dw}{2} \int_0^L \left[{\omega_1}^2 \left(1+\eta^2\right)^2 - 2\kappa_r \omega_1 \left(1+\nu\eta^2\right) \right] \mathrm{d}S. \tag{34}$$

The buckling of a closed ribbon of this type, caused by a mismatch of the natural curvature and the curvilinear length of the ribbon ($\kappa_r \neq 2\pi/L$), is studied in another article [22] based on the equations derived here.

6.2 Overcurved and Undercurved Annular Ribbons

The undercurved and overcurved annuli shown in Fig. 3 are simple examples of ribbons having non-zero curvature. They can be obtained as a paper model by cutting out an annulus from a sheet of paper. Assume the reference line to be middle line of the annular strip: its curvature defines the geodesic curvature $\kappa_g \neq 0$. If a sector with angle α is removed from the annulus and the two newly formed ends are pasted together, the annulus becomes overcurved: its curvilinear length $L = (2\pi - \alpha)/\kappa_g$ is less than that of a circle with curvature κ_g, namely $2\pi/\kappa_g$. The undercurved annuli can be achieved by pasting together two identical annuli, each one being such that $\alpha < \pi$—the effective value of α after pasting is then negative, $\alpha < 0$. In the presence of undercurvature or overcurvature, the annulus will buckle out-of-plane. This problem is a variant of the problem of the *folded* annular strip, which we have studied recently [10, 11]. The buckling of the annular ribbon without a fold has not yet

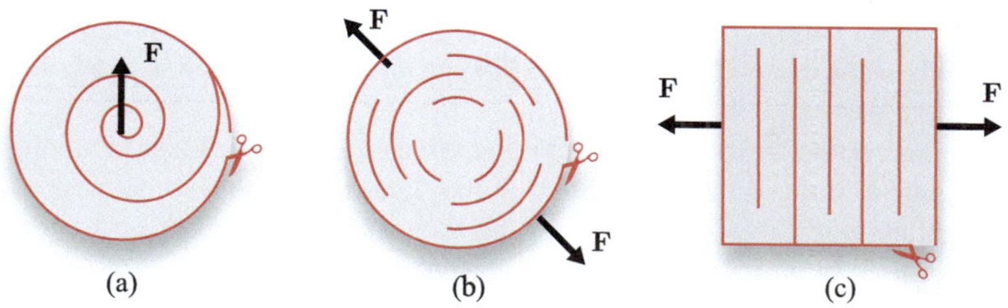

Fig. 4 Patterns cut out in a piece of paper can produce various 3D shapes when pulled (Sect. 6.3): (**a**) spiral cut pattern, (**b**) concentric, circular cut patterns, (**c**) straight, alternating cut patterns

been analyzed to the best of our knowledge. This can be done using the equations derived in this article.

6.3 3D Kirigami from 2D Cut-Out Patterns

Complex 3D shapes can be obtained by pulling on a thin sheet of paper that has been cut along arbitrary lines; see Fig. 4. This special form of kirigami (the art of *cutting paper*) involves tuning the geometry of the cuts to produce various 3D shapes. For instance, the alternated concentric cuts shown in Fig. 4b are the basis of a commercial paper model that produces a bowl-like shape. In all these examples, the center-line is not a geodesic. The example shown in Fig. 4c, for instance, is made by patching rectangular strips together, and it can be viewed as a single strip having a variable width and a zigzagging center-line. The unfolding of these 3D kirigami could be analyzed using the general ribbon model derived in this article.

7 Conclusion

We have presented a general theory of ribbons that fits into the well-established framework of thin elastic rods. By working with a frame of directors (material frame) instead of the theory of Frenet–Serret frame, we could extend the energy functional and the equilibrium equations to a general ribbon geometry. In particular, geodesic curvature, out-of-plane curvature, and a variable width have been taken into account. Instead of using the inextensibility conditions to eliminate degrees of freedom, as in previous work, we have treated them as constraints in the sense of the calculus of variations. This allowed us to view an elastic ribbon as a special kind of a thin elastic rod—namely, a kinematically constrained, hyperelastic rod possessing an internal degree of freedom. We could also explain the deep similarities between the theories of ribbons and thin rods.

This unifying view bridges the gap between classical rod theory and the theory of elastic ribbons—the latter has been developed largely as an independent subject so far. This makes it possible to reuse the large body of numerical and analytical methods available for thin rods. Numerical methods for simulating elastic ribbons, for instance, are currently limited to the calculation of non-linear equilibria, using numerical continuation, see, e.g., [25]. Numerical continuation is a powerful tool but its use can be quite impractical. Even though this is standard for thin rods [8], we are not aware of any simulation method that can predict the dynamics of ribbons, or account for self-contact. When extended to the dynamical case by including the virtual power of acceleration, the constrained variational formulation presented here offers a natural way of porting existing numerical methods for the dynamics of thin rods, such as the finite-element method [29] or methods based on discrete differential geometry [3], to ribbons.

Here we have focused on developable configurations. Under large loads involving a combination of tension and twisting, ribbons can adopt non-developable configurations [5, 16, 17, 20]. In future work, it would be interesting to extend the parameterization of ribbons presented here to account for deviations from the constraint of local area preservation.

Acknowledgements We would like to thank E. Fried for his detailed comments and suggestions on the manuscript. MAD thankfully acknowledges support from NSF Grant No. CBET-0854108.

Appendix: Curvature Tensor of a Developable Surface

Here, we use the notation introduced in Sect. 2.1 to prove the geometrical identities announced in Sects. 2.3 to 2.5, relevant to developable surfaces. We consider a general ruled surface, enforce the condition of developability, and derive the expression of the curvature tensor at an arbitrary point on the surface. By doing so, we extend the expressions obtained by Wunderlich [28] and by Starostin and van der Heijden [25] to account for the geodesic curvature κ_{g} of the center-line.

Let us first calculate the tangent vectors at an arbitrary point $\mathbf{y}(S, V)$ of the surface:

$$\mathbf{y}_{,S}(S, V) = \mathbf{d}_3(S) + V\mathbf{q}'(S), \tag{35a}$$

$$\mathbf{y}_{,V}(S, V) = \mathbf{q}(S), \tag{35b}$$

where $\mathbf{q}'(S)$ denotes the total derivative of $\mathbf{q}$ defined in Eq. (4) as $\mathbf{q} = \eta\mathbf{d}_3 + \mathbf{d}_1$.

Using Eq. (6), we write the following equation

$$\mathbf{q}' = \eta\omega_2\mathbf{d}_1 + (\omega_3 - \eta\omega_1)\mathbf{d}_2 + (\eta' - \omega_2)\mathbf{d}_3, \tag{36}$$

which implies

$$\mathbf{q} \times \mathbf{q}' = (\eta\omega_1 - \omega_3)\mathbf{q}^\perp + \frac{1}{V_{\mathrm{c}}}\mathbf{d}_2, \tag{37}$$

where V_{c} is the quantity defined by Eq. (13), and $\mathbf{q}^\perp$ is the vector

$$\mathbf{q}^\perp = \mathbf{d}_2 \times \mathbf{q} = -\mathbf{d}_3 + \eta\mathbf{d}_1. \tag{38}$$

Later on, we shall show that $\mathbf{q}^\perp$ is a vector perpendicular to $\mathbf{q}$ lying in the plane tangent to the surface; hence the notation.

The classical condition for a ruled surface to be developable [23, Sect. 3.II] is that the following three vectors are linearly dependent: the tangent $\mathbf{d}_3$ to the center-line (called the directrix in the context of the geometry of surfaces), the vector $\mathbf{q}$ spanning the generatrices, and its derivative with respect to the arc-length along the center-line. This is expressed by $(\mathbf{q} \times \mathbf{q}') \cdot \mathbf{d}_3 = 0$. In view of Eq. (37), this yields $\eta\omega_1 = \omega_3$, which is the constraint of developability announced in Eq. (11a).

As a result, $\mathbf{q}' \cdot \mathbf{d}_2 = 0$, and so $\mathbf{y}_{,S} \cdot \mathbf{d}_2 = 0$. On the other hand, Eq. (35b) shows that $\mathbf{y}_{,V} \cdot \mathbf{d}_2 = 0$. The director $\mathbf{d}_2(S)$ is orthogonal to both tangents: $\mathbf{d}_2$ is a unit normal at any point of the developable surface.

The element of area on the surface reads

$$da = |\mathbf{y}_{,S} \times \mathbf{y}_{,V}|\,dS\,dV = \left|\mathbf{d}_3 \times \mathbf{q} + V\mathbf{q}' \times \mathbf{q}\right| dS\,dV = \left|\left(1 - \frac{V}{V_c}\right)\mathbf{d}_2\right| dS\,dV. \tag{39}$$

Noting that $1 - V/V_c > 0$ by the inequality (14b), we arrive at the result announced in Eq. (12).

To compute the curvature tensor $\mathbf{K}(S, V)$, we note that the direction of the generatrix is a principal direction of zero curvature, since the surface is developable. Therefore, there exists some scalar field $k(S, V)$ such that

$$\mathbf{K} = k\mathbf{q}^{\perp} \otimes \mathbf{q}^{\perp}. \tag{40}$$

The quantity k in equation above can be found by contracting with $\mathbf{y}_{,S}$ on both sides of the equation to give:

$$\mathbf{y}_{,S} \cdot \mathbf{K} \cdot \mathbf{y}_{,S} = k\left(\mathbf{q}^{\perp} \cdot \left(\mathbf{d}_3 + V\mathbf{q}'\right)\right)^2 = k\left(-1 + \frac{V}{V_c}\right)^2. \tag{41}$$

By the definition of the curvature tensor (second fundamental form) [23], the left-hand side of the resulting identity is the normal projection of the second derivative $\mathbf{y}_{,SS}$:

$$\mathbf{y}_{,S} \cdot \mathbf{K} \cdot \mathbf{y}_{,S} = \mathbf{y}_{,SS} \cdot \mathbf{d}_2 = \left(\mathbf{d}_3' + V\mathbf{q}''\right) \cdot \mathbf{d}_2 = -\omega_1 + V\left(\frac{d(\mathbf{q}' \cdot \mathbf{d}_2)}{dS} - \mathbf{q}' \cdot \mathbf{d}_2'\right). \tag{42}$$

In this equation, $\mathbf{q}' \cdot \mathbf{d}_2 = \omega_3 - \eta\omega_1 = 0$ by the developability condition, and $\mathbf{q}' \cdot \mathbf{d}_2' = \mathbf{q}' \cdot (\boldsymbol{\omega} \times \mathbf{d}_2) = \mathbf{q}' \cdot (\omega_1 \mathbf{q} \times \mathbf{d}_2) = -\omega_1 \mathbf{d}_2 \cdot (\mathbf{q} \times \mathbf{q}') = -\omega_1 / V_c$. Therefore,

$$\mathbf{y}_{,S} \cdot \mathbf{K} \cdot \mathbf{y}_{,S} = -\omega_1\left(1 - \frac{V}{V_c}\right). \tag{43}$$

From Eqs. (41) and (43), we can solve for k, giving $k = -\omega_1/(1 - V/V_c)$. Inserting this result into Eq. (40) yields the expression of curvature tensor announced in Eqs. (15) and (16). The curvature tensor keeps the same form as in the case of zero geodesic curvature [25, 28] provided that the proper definition of V_c in terms of κ_g is used; see Eq. (13).

References

1. Armon, S., Efrati, E., Kupferman, R., Sharon, E.: Geometry and mechanics in the opening of chiral seed pods. Science (New York) **333**(6050), 1726–1730 (2011). http://www.ncbi.nlm.nih.gov/pubmed/21940888. doi:10.1126/science.1203874
2. Audoly, B., Pomeau, Y.: Elasticity and Geometry: From Hair Curls to the Nonlinear Response of Shells. Oxford University Press, London (2010)
3. Bergou, M., Wardetzky, M., Robinson, S., Audoly, B., Grinspun, E.: Discrete elastic rods. ACM Trans. Graph. **27**(3), 63:1–63:12 (2008)
4. Cheng-Chung, H.: A Differential-Geometric Criterion for a Space Curve to be Closed. Proceedings of the American Mathematical Society **83**(2), 357–361 (1981). http://www.jstor.org/stable/2043528. doi:10.2307/2043528
5. Chopin, J., Kudrolli, A.: Helicoids, wrinkles, and loops in twisted ribbons. Phys. Rev. Lett. **111**(17), 174302 (2013)
6. Chouaïeb, N.: Kirchhoff's problem of helical solutions of uniform rods and stability properties. Ph.D. thesis, École polytechnique fédérale de Lausanne, Lausanne, Switzerland (2003)

7. Cohen, H.: A non-linear theory of elastic directed curves. International Journal of Engineering Science **4**(5), 511–524 (1966). http://www.sciencedirect.com/science/article/pii/0020722566900139. doi:10.1016/0020-7225(66)90013-9
8. Coleman, B., Swigon, D.: Theory of supercoiled elastic rings with self-contact and its application to DNA plasmids. Journal of Elasticity **60**(3), 173–221 (2000). doi:10.1023/A:1010911113919
9. Cosserat, E., Cosserat, F.: Théorie des Corps déformables. A. Hermann et Fils, Paris (1909)
10. Dias, M.A., Audoly, B.: A non-linear rod model for folded elastic strips. J. Mech. Phys. Solids **62**, 57–80 (2014). http://linkinghub.elsevier.com/retrieve/pii/S0022509613001658. doi:10.1016/j.jmps.2013.08.012
11. Dias, M.A., Dudte, L.H., Mahadevan, L., Santangelo, C.D.: Geometric Mechanics of Curved Crease Origami. Phys. Rev. Lett. **109**(11), 1–5 (2012). http://link.aps.org/doi/10.1103/PhysRevLett.109.114301. doi:10.1103/PhysRevLett.109.114301
12. Efimov, N.V.: Some problems in the theory of space curves. Uspekhi Mat. Nauk **2**(3), 193–194 (1947). http://mi.mathnet.ru/umn6961
13. Ericksen, J.L.: Simpler static problems in nonlinear theories of rods. International Journal of Solids and Structures **6**(3), 371–377 (1970). http://www.sciencedirect.com/science/article/pii/0020768370900454. doi:10.1016/0020-7683(70)90045-4
14. Frenchel, W.: On the differential geometry of closed space curves. Bulletin of the American Mathematical Society **57**(1), 44–54 (1951). http://projecteuclid.org/euclid.bams/1183515801
15. Giomi, L., Mahadevan, L.: Statistical mechanics of developable ribbons. Phys. Rev. Lett. **104**, 238104 (2010). http://link.aps.org/doi/10.1103/PhysRevLett.104.238104. doi:10.1103/PhysRevLett.104.238104
16. Green, A.E.: The elastic stability of a thin twisted strip. II. Proc. R. Soc. Lond. A **161**, 197–220 (1937)
17. Korte, A.P., Starostin, E.L., van der Heijden, G.H.M.: Triangular buckling patterns of twisted inextensible strips. Proc. R. Soc. A, Math. Phys. Eng. Sci. **467**(2125), 285–303 (2010). http://rspa.royalsocietypublishing.org/cgi/doi/10.1098/rspa.2010.0200. doi:10.1098/rspa.2010.0200
18. Love, A.E.H.: A Treatise on the Mathematical Theory of Elasticity. Dover, New York (1944)
19. Mahadevan, L., Keller, J.B.: The shape of a Möbius band. Proc. R. Soc. A, Math. Phys. Eng. Sci. **440**, 149–162 (1993)
20. Mockensturm, E.M.: The elastic stability of twisted plates. J. Appl. Mech. **68**(4), 561–567 (2001)
21. Sadowsky, M.: Ein elementarer Beweis für die Existenz eines abwickelbares Möbiusschen Bandes und Zurückfürung des geometrischen Problems auf ein Variationsproblem. Sitzungsber. Preuss. Akad. Wiss. **22**, 412–415 (1930)
22. Seffen, K.A., Audoly, B.: Buckling of a closed, naturally curved ribbon (2014, in preparation)
23. Spivak, M.: A Comprehensive Introduction to Differential Geometry, vol. 3, 3rd edn. Publish or Perish, Inc., Houston (1999)
24. Starostin, E., van der Heijden, G.: Tension-induced multistability in inextensible helical ribbons. Phys. Rev. Lett. **101**(8), 084301 (2008). http://link.aps.org/doi/10.1103/PhysRevLett.101.084301. doi:10.1103/PhysRevLett.101.084301
25. Starostin, E.L., van der Heijden, G.H.M.: The shape of a Möbius strip. Nat. Mater. **6**(8), 563–567 (2007). http://www.ncbi.nlm.nih.gov/pubmed/17632519. doi:10.1038/nmat1929
26. Steigmann, D.J., Faulkner, M.G.: Variational theory for spatial rods. J. Elast. **33**(1), 1–26 (1993)
27. Wu, Z.L., Moshe, M., Greener, J., Therien-Aubin, H., Nie, Z., Sharon, E., Kumacheva, E.: Three-dimensional shape transformations of hydrogel sheets induced by small-scale modulation of internal stresses. Nat. Commun. **4**, 1586 (2013). http://www.ncbi.nlm.nih.gov/pubmed/23481394. doi:10.1038/ncomms2549
28. Wunderlich, W.: Über ein abwickelbares Möbiusband. Monatshefte Math. **66**(3), 276–289 (1962). http://link.springer.com/10.1007/BF01299052. doi:10.1007/BF01299052
29. Yang, Y., Tobias, I., Olson, W.K.: Finite element analysis of DNA supercoiling. J. Chem. Phys. **98**(2), 1673–1686 (1993)

DOI 10.1007/978-94-017-7300-3_8
Reprinted from *Journal of Elasticity* Journal, DOI 10.1007/s10659-014-9495-0

Equilibrium Shapes with Stress Localisation for Inextensible Elastic Möbius and Other Strips

E.L. Starostin · G.H.M. van der Heijden

Received: 27 February 2014 / Published online: 27 August 2014

Abstract We formulate the problem of finding equilibrium shapes of a thin inextensible elastic strip, developing further our previous work on the Möbius strip. By using the isometric nature of the deformation we reduce the variational problem to a second-order one-dimensional problem posed on the centreline of the strip. We derive Euler–Lagrange equations for this problem in Euler–Poincaré form and formulate boundary-value problems for closed symmetric one- and two-sided strips. Numerical solutions for the Möbius strip show a singular point of stress localisation on the edge of the strip, a generic response of inextensible elastic sheets under torsional strain. By cutting and pasting operations on the Möbius strip solution, followed by parameter continuation, we construct equilibrium solutions for strips with different linking numbers and with multiple points of stress localisation. Solutions reveal how strips fold into planar or self-contacting shapes as the length-to-width ratio of the strip is decreased. Our results may be relevant for curvature effects on physical properties of extremely thin two-dimensional structures as for instance produced in nanostructured origami.

Keywords Möbius strip · Inextensible ribbon · Developable surface · Switching point · Equilibrium · Invariant variational formulation · Stress localisation · Conical surface

Mathematics Subject Classification (2000) 74K20 · 34B15 · 53A05

1 Introduction

We do not know who was the first to take a thin flexible strip of paper (papyrus, parchment, birch bark, animal skin, palm leaf or whatever material), to join its ends in space so that

E.L. Starostin · G.H.M. van der Heijden (✉)
Department of Civil, Environmental & Geomatic Engineering, University College London,
Gower Street, London WC1E 6BT, UK
e-mail: g.heijden@ucl.ac.uk

E.L. Starostin
e-mail: ucesest@ucl.ac.uk

the long edges of the strip make a single closed curve and to admire the resulting shape. Certainly it happened a long time ago. However, the mathematical description of such an object was given only in 1858, independently by August Ferdinand Möbius and Johann Benedict Listing. The Möbius strip is not only interesting from an aesthetic or mathematical point of view but also as a generic example of a strip whose equilibrium shape demonstrates principal characteristic features of other elastic ribbons.

Since Möbius the eponymous strip has had a significant impact on human culture. Its intriguing beauty has inspired many artists including M.C. Escher [11]. In engineering, pulley belts are often used in the form of Möbius strips in order to wear 'both' sides equally. At a much smaller scale, Möbius strips have recently been formed in ribbon-shaped $NbSe_3$ crystals under certain growth conditions involving a large temperature gradient [55, 56]. The mechanism proposed by Tanda et al. to explain this behaviour is a combination of Se surface tension, which makes the crystal bend, and twisting as a result of bend-twist coupling due to the crystal nature of the ribbon. Gravesen & Willatzen [17] computed quantum eigenstates of a particle confined to the surface of a developable Möbius strip and compared their results with earlier calculations by Yakubo et al. [62]. They found curvature effects in the form of a splitting of the otherwise doubly degenerate ground state wave function (see also [27]). Thus qualitative changes in the physical properties of Möbius strip structures (for instance nanostrips) may be anticipated and it is of physical interest to know the exact shape of a free-standing strip. It has also been theoretically predicted that a novel state appears in a superconducting Möbius strip placed in a magnetic field [23]. Möbius strip geometries have furthermore been proposed to create optical fibres with tunable polarisation [46] and the Möbius strip topology has been suggested for compact resonators that have a resonance frequency that is half that of a cylindrical loop of the same size [41].

There exist infinitely many realisations of the Möbius strip and attempts have been made to define a unique, 'canonical', shape that is in some sense the simplest [42, 44, 47, 48, 61]. One way to approach this problem is to ask for the shape adopted by a free-standing Möbius strip made of a thin material (like paper); see Fig. 1.[1] In the idealised case when we can neglect the material's stretching, the shape may be found as that minimising the elastic energy associated with bending. A strip that deforms without stretching has constant Gaussian curvature under deformation. If such a strip is intrinsically flat (i.e., has zero Gaussian curvature) then it deforms into a developable surface: it can be unrolled onto a plane [13, 49]. Sadowsky was probably the first who formulated the problem in this way [44, 45]: find the shape of a Möbius strip by minimisation of the integral of the squared mean curvature over a deformed rectangular domain under the constraint of developability. Sadowsky went further and also derived the equations that describe the equilibrium shape of a thin inextensible elastic strip in the limit of vanishing width [45]. For a strip of finite width, Wunderlich showed how to reduce the variational problem to a one-dimensional one: he carried out analytical integration over the width of the strip so that the variational problem reduces to minimisation of the remaining one-dimensional integral over the centreline of the strip [61]. He didn't derive the Euler–Lagrange equilibrium equations for this reduced variational problem. It was

[1] Manufacturing an accurate model of an elastic Möbius strip is not entirely straightforward. If we glue the ends together with a small region of overlap then the resulting model will have non-uniform thickness, which affects the bending stiffness locally and hence the equilibrium shape in space. Welding the ends of a metal or plastic strip would have similar effects. To overcome these problems the model shown in Fig. 1 was created by printing the computed solution on a strip of double the length and then closing it after winding it twice, thus creating a strip of double the thickness. The two layers were then glued together without the need for an overlap and the thickness of the resulting strip is constant everywhere. The colouring was carefully phased in accordance with the actual shape.

Fig. 1 Photo of a paper Möbius strip of aspect ratio $L/(2w) = 2\pi$. Inextensibility of the material causes the surface to adopt a characteristic developable shape indicated by the straight generators. The colouring varies according to the bending energy density, from *violet* for regions of low bending to *red* for regions of high bending

proven recently that the Sadowsky functional is the Γ-limit of the Wunderlich functional for centrelines with non-vanishing curvature [26].

An alternative approach has been to model a narrow elastic Möbius strip as a thin inextensible elastic rod with non-circular cross-section taken in the limit where one of the bending stiffnesses tends to infinity [33]. Although the centreline shapes obtained using this approach are superficially in good agreement with those of a real narrow Möbius strip, such a model fails to reproduce the generic characteristic features of the surface of twisted inextensible strips.

Approximate equations for deformations of wide strips were derived in the mid-1950s by Mansfield [34, 35]. These equations predict the distribution of generators of the developable surface while ignoring the actual three-dimensional geometry. This work was followed up in [2], where localisation of stresses at two diagonally opposite corners was found for a strip in its first buckling mode. The actual shape of the strip was not computed.

The Euler–Lagrange equations for the Sadowsky functional describing equilibria of a narrow developable strip were derived in [21]. Some explicit solutions of the same variational problem were presented in [8, 22]. Folded narrow annular strips were studied in [9].

The geometrically exact equilibrium equations for an elastic developable strip were presented in [50] together with their numerical solution for Möbius strips of various aspect ratios. Here we give a more detailed discussion of the theory of developable strips including a more self-contained analysis of the variational problem and a complete formulation, and solution, of the boundary-value problem for the Möbius strip. We then extend the work to closed one- and two-sided strips of different topology.

2 Geometry of a Developable Strip

Consider an inextensible ribbon that, when developed onto a plane, makes a strip that is bounded by two parallel straight lines. By the developability property, the strip, however deformed, can be reconstructed from its centreline, i.e., the line that is equidistant from both

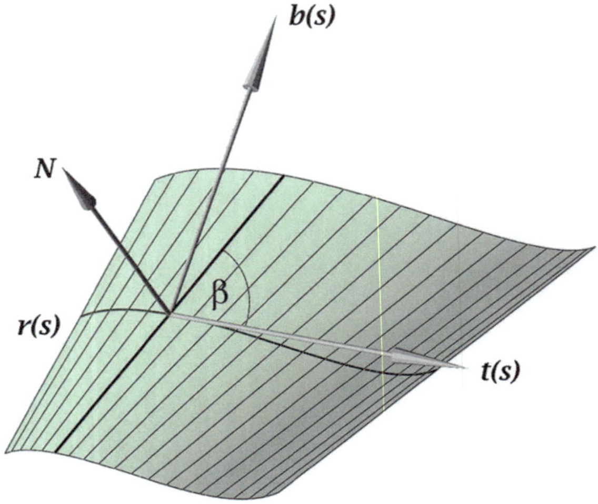

Fig. 2 A developable strip is made up of straight generators in the rectifying plane of tangent, $\boldsymbol{t}$, and binormal, $\boldsymbol{b}$, to the centreline, $\boldsymbol{r}$. The generators make an angle β with the tangent. $\boldsymbol{N}$ is the unit normal to the surface of the strip

parallel lines in the intrinsic geometry of the strip. We denote this line by $\boldsymbol{r}(s) \in \mathbb{R}^3$, where $s \in [0, L]$ is arclength along the curve and L is its length. We assume that $\boldsymbol{r}(s)$ is a regular curve of differentiability class C^3.

Let $\boldsymbol{t} = \boldsymbol{r}'$ be the unit tangent vector (here and in what follows a prime denotes the derivative with respect to s). In points where the curvature $\kappa = |\boldsymbol{t}'| \neq 0$ we define the unit principal normal $\boldsymbol{n} = \boldsymbol{t}'/\kappa$ and the unit binormal $\boldsymbol{b} = \boldsymbol{t} \times \boldsymbol{n}$. Like any regular curve in $\mathbb{R}^3$, $\boldsymbol{r}(s)$ is determined, up to Euclidean motion, by its curvature $\kappa(s)$ and torsion $\tau(s) = -\boldsymbol{b}'(s) \cdot \boldsymbol{n}(s)$.

A non-planar developable surface is either (part of) a cylinder, a cone or a so-called rectifying developable [49]. An analytical flat Möbius strip can only be of the latter type [42] (analyticity rules out Sadowsky's [44] construction of a Möbius strip from planar and cylindrical pieces). Given a curve with non-vanishing curvature there is a unique rectifying developable on which this curve is a geodesic curve. The unit normal $\boldsymbol{N}$ to the surface at the curve is then aligned with the principal normal to the curve and the surface is the envelope of rectifying planes orthogonal to the principal normal to the curve. Then

$$\begin{aligned} \boldsymbol{x}(s,t) &= \boldsymbol{r}(s) + t\big[\boldsymbol{b}(s) + \eta(s)\boldsymbol{t}(s)\big], \\ \tau(s) &= \eta(s)\kappa(s), \quad s = [0, L],\ t = [-w, w], \end{aligned} \tag{1}$$

is a parametrisation of a rectifying developable strip with centreline $\boldsymbol{r}$ and width $2w$ [39]. The straight lines $s = \text{const.}$ are the generators of the surface. They make an angle $\beta = \arctan(1/\eta)$ with the positive tangent direction of the curve $\boldsymbol{r}(s)$ (see Fig. 2). The short edges $s = 0$ and $s = L$ of the strip in particular are generators and for closure of the strip we require $|\eta(0)| = |\eta(L)|$ ($\eta(L) = \eta(0)$ [parallelogram] for a two-sided surface and $\eta(L) = -\eta(0)$ [isosceles trapezium] for a one-sided surface).

We compute the first fundamental form of the surface coordinate patch given by Eq. (1):

$$\mathbb{I} = \mathrm{d}\boldsymbol{x} \cdot \mathrm{d}\boldsymbol{x} = E\,\mathrm{d}s^2 + 2F\,\mathrm{d}s\,\mathrm{d}t + G\,\mathrm{d}t^2,$$

where

$$\begin{aligned} E &= \boldsymbol{x}_s \cdot \boldsymbol{x}_s = \left(1 + t\eta'\right)^2, \\ F &= \boldsymbol{x}_s \cdot \boldsymbol{x}_t = \eta\left(1 + t\eta'\right), \\ G &= \boldsymbol{x}_t \cdot \boldsymbol{x}_t = 1 + \eta^2. \end{aligned}$$

The area element is $\mathrm{d}\sigma = \|\boldsymbol{x}_s \times \boldsymbol{x}_t\|\, \mathrm{d}t\, \mathrm{d}s = \sqrt{EG - F^2}\, \mathrm{d}t\, \mathrm{d}s = (1 + t\eta')\, \mathrm{d}t\, \mathrm{d}s$. The unit normal to the surface $\boldsymbol{N}(s,t) = \frac{\boldsymbol{x}_s \times \boldsymbol{x}_t}{\|\boldsymbol{x}_s \times \boldsymbol{x}_t\|} = -\boldsymbol{n}(s)$ is constant along any generator. We also need the second fundamental form defined by

$$\mathrm{I\!I} = -\mathrm{d}\boldsymbol{x} \cdot \mathrm{d}\boldsymbol{N} = e\, \mathrm{d}s^2 + 2f\, \mathrm{d}s\, \mathrm{d}t + g\, \mathrm{d}t^2,$$

where

$$\begin{aligned} e &= -\boldsymbol{N}_s \cdot \boldsymbol{x}_s = -\kappa\left(1 + t\eta'\right), \\ f &= -\boldsymbol{N}_s \cdot \boldsymbol{x}_t = 0, \\ g &= -\boldsymbol{N}_t \cdot \boldsymbol{x}_t = 0. \end{aligned}$$

We can now compute the shape operator (or Weingarten map) S, i.e., the linear operator on the tangent plane defined by $S(\boldsymbol{T}) = -\partial_{\boldsymbol{T}} \boldsymbol{N}$, where $\boldsymbol{T}$ is a unit tangent vector to the surface [49]. $S(\boldsymbol{T})$ is the gradient of the unit normal to the surface in the tangent direction $\boldsymbol{T}$. It therefore encodes information about the curvature of the surface. In fact, the eigenvalues of the shape operator at each point are the principal curvatures at this point and the eigenvectors are the principal directions. We have $S(\boldsymbol{x}_s) = -\boldsymbol{N}_s$, $S(\boldsymbol{x}_t) = -\boldsymbol{N}_t$, and hence in the basis $(\boldsymbol{x}_s, \boldsymbol{x}_t)$,

$$S = \begin{pmatrix} E & F \\ F & G \end{pmatrix}^{-1} \begin{pmatrix} e & f \\ f & g \end{pmatrix} = \frac{1}{EG - F^2} \begin{pmatrix} eG - fF & fG - gF \\ fE - eF & gE - fF \end{pmatrix} = \begin{pmatrix} \frac{-\kappa(1+\eta^2)}{1+t\eta'} & -\kappa\eta \\ 0 & 0 \end{pmatrix}.$$

We then compute the Gaussian curvature as

$$K := \det S = \kappa_1 \kappa_2 = \frac{eg - f^2}{EG - F^2} = 0,$$

confirming that it vanishes identically, while for the mean curvature we find

$$H := \frac{1}{2} \operatorname{tr} S = \frac{1}{2}(\kappa_1 + \kappa_2) = \frac{eG + gE - 2fF}{2(EG - F^2)} = -\frac{\kappa(1 + \eta^2)}{2(1 + t\eta')}, \tag{2}$$

where κ_1 and κ_2 are the two principal curvatures of the surface.

As to the space-curve geometry of the centreline $\boldsymbol{r}$ we recall the following. After choosing a coordinate system we may identify the orientations of the Frenet–Serret frame $\{\boldsymbol{t}, \boldsymbol{n}, \boldsymbol{b}\}$ attached to $\boldsymbol{r}$ with elements of the group of orthogonal 3×3 matrices:

$$R(s) := \big(\boldsymbol{t}(s), \boldsymbol{n}(s), \boldsymbol{b}(s)\big) \in SO(3).$$

This defines a skew-symmetric 3×3 matrix in the Lie algebra $\mathfrak{so}(3)$ as follows:

$$\widehat{\omega} = R^{\mathsf{T}} R', \tag{3}$$

where we have introduced the 'hat' isomorphism between skew-symmetric matrices $\widehat{\mathsf{w}} = \begin{pmatrix} 0 & -w_3 & w_2 \\ w_3 & 0 & -w_1 \\ -w_2 & w_1 & 0 \end{pmatrix}$ in $\mathfrak{so}(3)$ and axial (or rotation) vectors $\mathsf{w} = (w_1, w_2, w_3)^\mathsf{T}$ in $\mathbb{R}^3$.[2] By definition of the Frenet–Serret frame, we have $\omega_1 = \tau$, $\omega_2 = 0$, $\omega_3 = \kappa$. For a curve on a surface we define its Darboux frame $\{\boldsymbol{t}, \boldsymbol{N}, \boldsymbol{t} \times \boldsymbol{N}\}$. The rotation (Darboux) vector has components $(\tau_g, \kappa_g, \kappa_N)$, which have their own names, geodesic torsion, geodesic curvature and normal curvature, respectively. They relate to the curvature and torsion of the curve as $\tau_g = \tau + \chi'$, $\kappa_g = \kappa \sin \chi$, $\kappa_N = \kappa \cos \chi$, where χ measures the angle from $\boldsymbol{n}$ to $\boldsymbol{N}$ [49]. Here we consider only the case of a geodesic centreline and we thus have $\boldsymbol{n} = -\boldsymbol{N}$, hence $\chi = \pi$ and $\tau_g = \tau = \omega_1$, $\kappa_g = 0 = \omega_2$, $\kappa_N = -\kappa = -\omega_3$.

The Frenet–Serret frame is discontinuous at an inflection point where the normal $\boldsymbol{n}$ (and hence the binormal $\boldsymbol{b}$) flips ($\boldsymbol{n} \to -\boldsymbol{n}$). For a continuous description through an inflection point we can however use a generalised Frenet–Serret frame (whose normal and binormal are plus or minus the normal and binormal of the Frenet–Serret frame and such that the frame is continuous through an inflection point) and let κ be the signed curvature [39, 42].

2.1 Edge of Regression

The asymptotic completion of a developable strip is defined as the surface obtained by extending all generators to infinity in both directions, i.e., taking t in all of $\mathbb{R}$ in Eq. (1). From Eq. (2) we see that the mean curvature H becomes singular if the parameter t in Eq. (1) equals $-1/\eta'$. If $\eta' = 0$ then there is no singularity on the extended generator in the asymptotic completion. We call points on the centreline where $\eta' = 0$, 'cylindrical'. At such points the mean curvature H is independent of t, so the principal curvatures are constant along the local generator. Away from cylindrical points we can define the curve $\boldsymbol{x}_e(s) = \boldsymbol{r}(s) - \frac{1}{\eta'(s)}[\boldsymbol{b}(s) + \eta(s)\boldsymbol{t}(s)]$, which is called the edge of regression. When η' changes sign the edge of regression jumps within the asymptotic completion from one side of the centreline to the other. (For a graphical representation the reader may wish to look ahead to Fig. 10 for the case of a Möbius strip.) The strip cannot be wider than the critical value of t, i.e., we require that $w|\eta'| \leq 1$. The developable surface is then the envelope of the tangents to the edge of regression (i.e., of the generators of the surface) and is therefore also the tangent developable. This envelope meets the edge of regression in two sheets that form cusps in the normal plane to $\boldsymbol{x}_e$ [37].

The edge of regression may have its own singularities. By differentiation we find $\boldsymbol{x}_e'(s) = \frac{\eta''(s)}{\eta'^2(s)}[\boldsymbol{b}(s) + \eta(s)\boldsymbol{t}(s)]$. If $\eta'' = 0$ the tangent to the edge of regression is discontinuous. This corresponds to a cusp point on the curve and a swallow tail singularity of the asymptotic completion of the strip. We call isolated points of the centreline where $\eta'' = 0$, 'conical'. Clearly, for a smooth centreline, there must be a conical point between any pair of cylindrical points.

3 The Energy Functional

The bending energy for an arbitrary Kirchhoff–Love shell of thickness $2d$ can be written as the following integral over the surface of the strip [12, 38, 63]:

$$U = \frac{D}{2} \iint_{\Omega} \left[\nu \big(\mathrm{tr}(\Delta S)\big)^2 + (1-\nu)\,\mathrm{tr}\big((\Delta S)^2\big)\right] \mathrm{d}\sigma, \tag{4}$$

[2]Throughout we adopt the notation that for any vector $\boldsymbol{v} \in \mathbb{R}^3$ the sans-serif symbol v denotes the triple of components $(v_1, v_2, v_3)^\mathsf{T} = (\boldsymbol{v} \cdot \boldsymbol{t}, \boldsymbol{v} \cdot \boldsymbol{n}, \boldsymbol{v} \cdot \boldsymbol{b})^\mathsf{T}$ in the Frenet–Serret frame.

where $D = 2Yd^3/[3(1-\nu^2)]$ is the flexural rigidity, ν is Poisson's ratio, Y is Young's modulus and $\Delta S = S - S_0$; here and later the subscript 0 refers to the undeformed reference configuration. By expressing both S and S_0 in principal axes of S_0, we may write

$$S = \begin{pmatrix} \cos\Delta\beta & -\sin\Delta\beta \\ \sin\Delta\beta & \cos\Delta\beta \end{pmatrix} \begin{pmatrix} \kappa_1 & 0 \\ 0 & \kappa_2 \end{pmatrix} \begin{pmatrix} \cos\Delta\beta & \sin\Delta\beta \\ -\sin\Delta\beta & \cos\Delta\beta \end{pmatrix}, \quad S_0 = \begin{pmatrix} \kappa_{1,0} & 0 \\ 0 & \kappa_{2,0} \end{pmatrix},$$

where $\Delta\beta = \beta - \beta_0$ is the angle between the principal curvature axes in the deformed and undeformed states. Equation (4) then becomes [63]

$$U = D\iint_\Omega \left[(1+\nu)(\Delta H)^2 + (1-\nu)\left((\Delta A)^2 + 4AA_0\sin^2\Delta\beta\right)\right] d\sigma, \tag{5}$$

where $\Delta H = H - H_0$ and $A = \sqrt{H^2 - K}$.

In our case of a developable surface obtained by isometric deformation from its unstressed state we have zero Gaussian curvature K (the undeformed surface must be flat, although not necessarily planar). This means that one of the principal curvatures is zero (say $\kappa_2 = 0$) and we may write

$$U = \frac{D}{2}\iint_\Omega \left[(\kappa_1 - \kappa_{1,0})^2 + 2(1-\nu)\kappa_1\kappa_{1,0}\sin^2\Delta\beta\right] d\sigma. \tag{6}$$

The fact that the strip is fully determined by its centreline suggests reduction of this double integral to a single integral over the centreline. However, this reduction looks intractable for an arbitrary undeformed shape given by $\kappa_{1,0}$. We develop the general theory a little further, and consider two special cases, in Appendix A, but here we proceed by assuming that the strip is planar when relaxed. Thus we set $\kappa_{1,0} = 0$ and have, with $\kappa_1 = 2H$,

$$U = \frac{D}{2}\int_0^L \int_{-w}^{w} \frac{\kappa^2(1+\eta^2)^2}{1+t\eta'}\, dt\, ds. \tag{7}$$

The t-integration can be carried out analytically [61], and we arrive at

$$U = Dw\int_0^L h(\kappa, \eta, \eta')\, ds \tag{8}$$

with

$$\begin{aligned} h(\kappa, \eta, \eta') &= \kappa^2(1+\eta^2)^2 V(w\eta'), \\ V(w\eta') &= \frac{1}{2w\eta'}\log\left(\frac{1+w\eta'}{1-w\eta'}\right). \end{aligned} \tag{9}$$

Note that for strips with no intrinsic curvature, equilibrium shapes do not depend on the material properties: Young's modulus is a simple factor and Poisson's ratio does not enter the energy expression. Also note that in the limit of narrow strips, $w\eta' \to 0$, we have $V(w\eta') \to 1$ and no derivative enters the integrand in Eq. (8) (cf. [44, 45]). The integrand h reduces to κ^2 of the planar Euler elastica if $\eta \equiv 0$, the surface then being cylindrical with the generator everywhere perpendicular to the centreline $\boldsymbol{r}$.

For solutions of closed strips or strips with fixed end points we impose a constant end-to-end distance constraint by adding the following integral expression to the bending energy:

$$W = -\boldsymbol{F}\cdot\left[\boldsymbol{r}(L) - \boldsymbol{r}(0)\right] = -\int_0^L \boldsymbol{F}\cdot\boldsymbol{t}\, ds,$$

where we have used inextensibility of the centreline, i.e., $\boldsymbol{t} = \boldsymbol{r}'$, and $\boldsymbol{F}$ is a (constant) Lagrange multiplier (with the physical meaning of an internal force). Finally, we have to impose the constraint that the centreline is geodesic (straight in the intrinsic geometry), i.e., $\omega_2 = \kappa_g \equiv 0$. We enforce this constraint by adding the integral $T = \int_0^L M_2\omega_2 \,\mathrm{d}s$, where $M_2 = M_2(s)$ is another (local) Lagrange multiplier.

In conclusion, equilibrium shapes of a thin, inextensible and intrinsically planar strip are given by stationary points of the functional

$$\mathcal{A} = \bar{U} + W + T = \int_0^L \mathcal{L}\,\mathrm{d}s, \quad \text{with } \mathcal{L} = h + M_2\omega_2 - \boldsymbol{F}\cdot\boldsymbol{t}, \tag{10}$$

where $\mathcal{L}$ is the Lagrangian of the problem and $\bar{U} = U/(Dw)$ (all force and moments in the rest of the paper are thus normalised by Dw). This represents a 1D variational problem on a curve in $\mathbb{R}^3$ cast in Euclidean invariant form.

Although we are here interested in closed-strip solutions, the functional $\mathcal{A}$ can also be used to study open-strip solutions. If for such open strips the ends are free to move relative to each other then W represents the work done by any applied end force $\boldsymbol{F}$. However, since we use the parametrisation of Eq. (1) in the t-integration to obtain the one-dimensional integral (8), the short edges of the strip must be generators and therefore straight. This is quite natural for fixed ends (held by a straight clamp covering the entire edge of the strip) but constitutes a restriction on deformations considered for strips with free ends.

4 Variational Principle and Equilibrium Equations

4.1 Variational Principle

To derive the equilibrium equations for the functional $\mathcal{A}$ we follow the recent higher-order variational approach of [16]. This approach yields equations in a particularly elegant and transparent geometrical form if we view the Lagrangian $\mathcal{L}$ as a function of the rotation matrix R and its derivatives, i.e., as defined on the (higher-order) tangent bundle of the symmetry group of our problem, the Lie group $SO(3)$. The theory in [16] then gives a symmetry-reduced variational problem with Euler–Lagrange equations in Euler–Poincaré form. Similar treatments of higher-order variational problems on curves can be found in [24, 52].

We can transform $\mathcal{L}$ into suitable form by expressing κ, η, η' and ω_2 in terms of the Frenet–Serret frame viewed as a rotation matrix, R (see Sect. 2). The result is the second-order Lagrangian $\mathcal{L} = \mathcal{L}_F(R, R', R'') : T^{(2)}SO(3) \to \mathbb{R}$, where $T^{(2)}SO(3)$ is the second-order tangent bundle and $\boldsymbol{F} \in \mathbb{R}^3$ is regarded as a parameter. The functional $\mathcal{A}$ may then be written as

$$\mathcal{A} = \int_0^L \mathcal{L}_F\big(R, R', R''\big)\,\mathrm{d}s = \int_0^L \bar{\mathcal{L}}\big(R, R', R''\big)\,\mathrm{d}s - \int_0^L \boldsymbol{F}\cdot\boldsymbol{t}\,\mathrm{d}s, \tag{11}$$

where $\boldsymbol{t} = R\mathrm{e}_1$ and $\bar{\mathcal{L}}$ collects all terms not depending on $\boldsymbol{F}$. The parameter-dependent Lagrangian $\mathcal{L}_F$ is not invariant under $SO(3)$ because of the W term, which is only invariant under rotations S^1 about $\boldsymbol{F}$. Thus $\boldsymbol{F}$ breaks $SO(3)$ symmetry of $\mathcal{L}_F$. However, if we view $\mathcal{L} = \mathcal{L}_F$ as a function defined on $T^{(2)}SO(3) \times \mathbb{R}^3$ with $SO(3)$ acting by left multiplication on both $T^{(2)}SO(3)$ and the parameter manifold $\mathbb{R}^3$ then this $\mathcal{L}$ is left-invariant under $SO(3)$.

We are then in the situation of Sect. 3.3 of [16] and can apply Euler–Poincaré reduction for Lie groups acting on manifolds to obtain a reduced variational principle.

We consider curves in $SO(3)$ with given end point conditions $R(0)$, $R'(0)$, $R(L)$, $R'(L)$ (these fix the end generators in space). According to Hamilton's principle such a curve is an extremal of $\mathcal{A}$ ($\delta\mathcal{A}=0$) under variations of the curve that vanish at the end points, i.e., $\delta R(0)=\delta R'(0)=\delta R(L)=\delta R'(L)=0$, while $\boldsymbol{F}$ is held fixed, if and only if it is a solution of the Euler–Lagrange equations. Here we have used the usual notation for infinitesimal variations of a field variable $v(s)$: for a smooth one-parameter family of curves $v_\epsilon(s)$ we write $\delta v(s)=\frac{\mathrm{d}}{\mathrm{d}\epsilon}v_\epsilon(s)|_{\epsilon=0}$.

The reduced variational principle is

$$\delta\int_0^L l\big(\omega,\omega',\mathsf{F}\big)\,\mathrm{d}s=\delta\left(\int_0^L \bar{l}\big(\omega,\omega'\big)\,\mathrm{d}s-\int_0^L \mathsf{F}\cdot\mathsf{t}\,\mathrm{d}s\right)=0 \tag{12}$$

for $l:\ 2\mathfrak{so}(3)\times\mathbb{R}^3\to\mathbb{R}$ given by $l(\omega,\omega',\mathsf{F})=\mathcal{L}_{\boldsymbol{F}}(R,R',R'')$, i.e., l is just $\mathcal{L}_{\boldsymbol{F}}$ expressed in different variables, and the constrained variations $\delta\omega$ of ω and $\delta\omega'=(\delta\omega)'$ of ω' of the form $\delta\omega=\varpi'+\omega\times\varpi$ with ϖ another axial vector and $\omega\times\varpi:=[\widehat{\omega},\widehat{\varpi}]$, the Lie bracket of $\mathfrak{so}(3)$, $\widehat{\varpi}=R^{\mathsf{T}}(\delta R)\in\mathfrak{so}(3)$ and $\delta\mathsf{F}=-\widehat{\varpi}\mathsf{F}$. The vanishing variation conditions for the non-reduced functional Eq. (11) imply $\varpi(0)=\varpi'(0)=\varpi(L)=\varpi'(L)=\mathsf{0}$ and therefore $\delta\omega(0)=\delta\omega(L)=\mathsf{0}$. The reduced variables are $(\widehat{\omega},\mathsf{F})=(R^{\mathsf{T}}R',R^{\mathsf{T}}\boldsymbol{F})\in\mathfrak{so}(3)\times\mathbb{R}^3$. Note that under this reduction the parameter $\boldsymbol{F}$ acquires field status.

F satisfies the equation

$$(\partial_s+\widehat{\omega})\mathsf{F}=0, \tag{13}$$

with initial condition $\mathsf{F}(0)=R^{\mathsf{T}}(0)\boldsymbol{F}$. The Euler–Lagrange equations for the reduced functional take the Euler–Poincaré form

$$(\partial_s+\widehat{\omega})\mathsf{M}=\frac{\partial l}{\partial\mathsf{F}}\times\mathsf{F}, \tag{14}$$

with M defined by

$$\mathsf{M}=\mathcal{E}_\omega(l), \tag{15}$$

where $\mathcal{E}_\zeta(k):=\frac{\partial k}{\partial\zeta}-\frac{\mathrm{d}}{\mathrm{d}s}(\frac{\partial k}{\partial\zeta'})$ is the Euler–Lagrange operator for the variable ζ.

For further analysis we rewrite Eqs. (13), (14), (15) as the following system: (a) balance equations for the components of the internal force $\mathsf{F}=(F_1,F_2,F_3)^{\mathsf{T}}$ and moment $\mathsf{M}=(M_1,M_2,M_3)^{\mathsf{T}}$ expressed in the Frenet–Serret frame [24, 52]

$$\mathsf{F}'+\omega\times\mathsf{F}=0, \tag{16}$$

$$\mathsf{M}'+\omega\times\mathsf{M}+\mathsf{t}\times\mathsf{F}=0, \tag{17}$$

and (b) the 'constitutive' equations

$$M_j=\frac{\partial l}{\partial\omega_j}-\frac{\mathrm{d}}{\mathrm{d}s}\frac{\partial l}{\partial\omega_j'},\qquad j=1,2,3. \tag{18}$$

Here we have used that $\frac{\partial l}{\partial\mathsf{F}}=-\mathsf{t}=-\mathsf{e}_1$, $\mathsf{e}_1\equiv(1,0,0)^{\mathsf{T}}$.

The equations have $|\mathsf{F}|^2$ and $\mathsf{F}\cdot\mathsf{M}$ as first integrals.

4.2 Equations in the Original Variables κ, η

We note that Eq. (18) for $j = 2$ is trivial. For $j = 1$ and 3 we use a contact transformation $(\omega_1, \omega_3) = (\tau, \kappa) \to (\eta, \kappa)$ [52] to obtain

$$M_1 = \frac{\partial \eta}{\partial \omega_1} \mathcal{E}_\eta(h) = \frac{1}{\kappa} \mathcal{E}_\eta(h), \tag{19}$$

$$M_3 = \frac{\partial h}{\partial \omega_3} + \frac{\partial \eta}{\partial \omega_3} \mathcal{E}_\eta(h) = \frac{\partial h}{\partial \kappa} - \frac{\eta}{\kappa} \mathcal{E}_\eta(h), \tag{20}$$

or, in more compact form,

$$\partial_\kappa h - \eta M_1 - M_3 = 0, \tag{21}$$

$$\mathcal{E}_\eta(h) - \kappa M_1 = 0 \tag{22}$$

(cf. [50] where the signs of moments and forces are taken opposite). We see that Eq. (21) contains only the first derivative of η (coming from the energy density h), while Eq. (22) also has a term with the second derivative of η. Note that for the infinitesimally narrow strip both Eqs. (21), (22) are algebraic. To use Eqs. (16), (17) we represent the Darboux vector as $\omega = (\kappa\eta, 0, \kappa)^\top$.

To bring Eqs. (21), (22) into a convenient form for numerical solution, we first differentiate Eq. (21) with respect to s. On substitution of the moment derivatives from Eq. (17) this gives $(\eta M_1 + M_3)' = \eta' M_1 - F_2$. Combining this equation with Eq. (22) we obtain the following third-order system, linear in terms of the highest derivatives, i.e., κ' and η'':

$$4a_1\left(1+\eta^2\right)^2\kappa' + 8b_2\kappa\eta\left(1+\eta^2\right) = -b_3 M_1 + b_4 F_2, \tag{23}$$

$$4a_1\kappa\left(1+\eta^2\right)^2\eta'' + 4a_3\kappa\eta\left(1+\eta^2\right) = a_2 M_1 - a_4 F_2, \tag{24}$$

where

$$\begin{aligned}
a_1(\eta') &= \frac{1}{4}\left(2V_{\eta'}^2 - V V_{\eta'\eta'}\right), \\
a_2(\eta') &= V + \eta' V_{\eta'}, \\
a_3(\eta') &= V a_2, \\
a_4(\eta') &= V_{\eta'}, \\
b_2(\eta') &= \frac{1}{2}\left[\eta'\left(V_{\eta'}^2 - V V_{\eta'\eta'}\right) - V V_{\eta'}\right], \\
b_3(\eta') &= \frac{1}{2}\left(2V_{\eta'} + \eta' V_{\eta'\eta'}\right), \\
b_4(\eta') &= \frac{1}{2} V_{\eta'\eta'}.
\end{aligned} \tag{25}$$

Note that the above equations become singular if the curvature vanishes or if $a_1 = 0$.

4.3 Hamiltonian

Legendre transformation of the second-order reduced Lagrangian l gives the symmetry-reduced Hamiltonian

$$\mathcal{H}(\pi_1, \omega, \pi_2, \mathsf{F}) = \pi_1 \cdot \omega + \pi_2 \cdot \omega' - l(\omega, \omega', \mathsf{F})$$

where $\pi_1 = \frac{\partial l}{\partial \omega} - \frac{\mathrm{d}}{\mathrm{d}s}\left(\frac{\partial l}{\partial \omega'}\right) = \mathsf{M}$, $\pi_2 = \frac{\partial l}{\partial \omega'}$ are the reduced Ostrogradsky momenta [16]. The Euler–Lagrange equations for the reduced Lagrangian l derived in Sect. 4 are equivalent to Hamilton's equations for the reduced Hamiltonian $\mathcal{H}$ with respect to a non-canonical Poisson bracket. It is easy to show that $\sum_{j=1}^{3} \frac{\partial l}{\partial \omega'_j} \omega'_j = \frac{\partial h}{\partial \eta'} \eta'$. So in terms of η and our other variables the Hamiltonian can be written as

$$\mathcal{H} = M_1 \kappa \eta + M_3 \kappa + \frac{\partial h}{\partial \eta'} \eta' - h(\kappa, \eta, \eta') + F_1. \tag{26}$$

For a uniform strip with h not explicitly depending on arclength s, $\mathcal{H}$ is a conserved quantity, as can be verified directly by differentiating the right-hand side of Eq. (26) with respect to s and using Eqs. (16), (17), (21) and (22) to show that $\mathcal{H}' = 0$.

After substitution of h from Eq. (9) we can rewrite Eq. (26) in the explicit form:

$$\mathcal{H} = \kappa(M_1 \eta + M_3) + \kappa^2 (1 + \eta^2)^2 \left[\frac{1}{1 - (w\eta')^2} - \frac{1}{w\eta'} \log\left(\frac{1 + w\eta'}{1 - w\eta'} \right) \right] + F_1.$$

4.4 Symmetries

The equilibrium equations (Eqs. (16), (17), (23), (24)) are invariant under the reversing involutions R_1 and R_2:

R_1:

$$F_1 \to F_1, \quad F_2 \to -F_2, \quad F_3 \to F_3, \quad M_1 \to M_1, \quad M_2 \to -M_2, \quad M_3 \to M_3,$$
$$\kappa \to \kappa, \quad \eta \to \eta, \quad s \to -s,$$

R_2:

$$F_1 \to F_1, \quad F_2 \to F_2, \quad F_3 \to -F_3, \quad M_1 \to M_1, \quad M_2 \to M_2, \quad M_3 \to -M_3,$$
$$\kappa \to -\kappa, \quad \eta \to -\eta, \quad s \to -s,$$

and the non-reversing involution

S:

$$F_1 \to F_1, \quad F_2 \to F_2, \quad F_3 \to -F_3, \quad M_1 \to -M_1, \quad M_2 \to -M_2, \quad M_3 \to M_3,$$
$$\kappa \to \kappa, \quad \eta \to -\eta, \quad s \to s.$$

Further involutions exist but will not be required. Note that R_2 requires κ to be interpreted as the signed curvature.

4.5 Kinematics Equations

Reconstruction of the centreline of the strip requires solving for the tangent $\boldsymbol{t}$ and integrating this to get $\boldsymbol{r}$. We choose a parametrisation of the Frenet–Serret frame $\{\boldsymbol{t}, \boldsymbol{n}, \boldsymbol{b}\}$ in terms of three Euler angles ψ, ϑ and φ [32]:

$$\boldsymbol{t} = \begin{pmatrix} \sin\vartheta\cos\psi \\ \sin\vartheta\sin\psi \\ \cos\vartheta \end{pmatrix}, \qquad \boldsymbol{n} = \begin{pmatrix} -\sin\psi\sin\varphi + \cos\vartheta\cos\psi\cos\varphi \\ \cos\psi\sin\varphi + \cos\vartheta\sin\psi\cos\varphi \\ -\sin\vartheta\cos\varphi \end{pmatrix},$$

$$\boldsymbol{b} = \begin{pmatrix} -\sin\psi\cos\varphi - \cos\vartheta\cos\psi\sin\varphi \\ \cos\psi\cos\varphi - \cos\vartheta\sin\psi\sin\varphi \\ \sin\vartheta\sin\varphi \end{pmatrix}.$$

The Euler angles are related to the Darboux vector by the kinematics equations

$$\begin{aligned} \psi' &= \kappa\sin\varphi\csc\vartheta, \\ \vartheta' &= \kappa\cos\varphi, \\ \varphi' &= \kappa(\eta - \sin\varphi\cot\vartheta). \end{aligned} \tag{27}$$

Note that the angle ϑ should not approach 0 or π. To guarantee this we have to choose the third axis z of the laboratory reference frame such that the tangent $\boldsymbol{t}$ would never align with $\pm z$.

To find the centreline $\boldsymbol{r}$ we solve Eq. (27) in conjunction with the equation $\boldsymbol{r}' = \boldsymbol{t}$, or, writing $\boldsymbol{r} = (x, y, z)^{\mathsf{T}}$,

$$\begin{aligned} x' &= \sin\vartheta\cos\psi, \\ y' &= \sin\vartheta\sin\psi, \\ z' &= \cos\vartheta. \end{aligned} \tag{28}$$

4.6 Full System of Equations

For ease of reference we collect here together all the equations derived (i.e., Eqs. (16), (17), (23), (24), (27), (28)):

$$\begin{aligned} F_1' - \kappa F_2 &= 0, \\ F_2' + \kappa F_1 - \kappa\eta F_3 &= 0, \\ F_3' + \kappa\eta F_2 &= 0, \\ M_1' - \kappa M_2 &= 0, \\ M_2' + \kappa M_1 - \kappa\eta M_3 - F_3 &= 0, \\ M_3' + \kappa\eta M_2 + F_2 &= 0, \\ 4a_1\left(1+\eta^2\right)^2\kappa' + 8b_2\kappa\eta\left(1+\eta^2\right) &= -b_3 M_1 + b_4 F_2, \\ 4a_1\kappa\left(1+\eta^2\right)^2\eta'' + 4a_3\kappa\eta\left(1+\eta^2\right) &= a_2 M_1 - a_4 F_2, \\ \psi' &= \kappa\sin\varphi\csc\vartheta, \end{aligned} \tag{29}$$

$$\vartheta' = \kappa \cos\varphi,$$
$$\varphi' = \kappa(\eta - \sin\varphi \cot\vartheta),$$
$$x' = \sin\vartheta \cos\psi,$$
$$y' = \sin\vartheta \sin\psi,$$
$$z' = \cos\vartheta.$$

This system is of 15th order with variables F_1, F_2, F_3, M_1, M_2, M_3, κ, η, ψ, ϑ, φ, x, y, z. In the following sections we formulate boundary-value problems for these equations that exploit the symmetries identified in Sect. 4.4.

5 The Möbius Strip

5.1 Properties of a One-Sided Strip

Randrup & Røgen [39] have shown that along the centreline of a one-sided rectifying developable strip an odd number of switching points must occur where $\kappa = \tau = 0$ and the principal normal to the centreline flips (i.e., makes a 180° turn). Moreover, when approaching a switching point, τ goes to zero at least as fast as κ, meaning that η is bounded. From Eq. (2) it follows that through the inflection point goes an umbilic line, i.e., a line on which both principal curvatures are equal, namely zero. (Incidentally, developable Möbius strips without switching points may exist if the surface does not contain a closed geodesic [7].) To make the twisted nature of the Möbius strip precise we note that a closed centreline with a continuous and periodic twist rate (here $\tau(s)$) defines a closed cord [15], for which one can define a linking number Lk [15]. Any closed ribbon of a cord of half-integer Lk is one-sided. The simplest example, with $Lk = \pm\frac{1}{2}$, gives the classical Möbius strip with half a turn of twist (the opposite values of Lk corresponding to mirror images of each other). The linking number Lk so defined is identical to the Möbius twisting number of a closed curve in an embedded surface in $\mathbb{R}^3$ introduced in [40]. In [40] it was shown that within the set of flat surfaces, flat strips with the same Möbius twisting number, and whose centrelines are of the same knot type, belong to the same isotopy class. It is known that for every isotopy class there exists a developable Möbius strip with a closed geodesic centreline (each such class for instance contains a strip that is obtained by isometric deformation from a planar rectangular domain) [29].

5.2 Boundary-Value Problem

Plastic or paper models of a Möbius strip suggest that equilibrium shapes possess an axis of π-rotational symmetry. It seems unlikely that nonsymmetric solutions exist. We formulate a boundary-value problem for such a C_2-symmetric Möbius strip by imposing boundary conditions at $s = 0$ and $s = L/2$. Involution R_1 is then used to obtain a solution on the full interval $[-L/2, L/2]$ by π rotation about the axis through both end points. Thus we specify the following boundary conditions for the system of Eqs. (29) over half a strip (see Fig. 3):

$$F_2(0) = 0, \tag{30}$$
$$M_2(0) = 0, \tag{31}$$

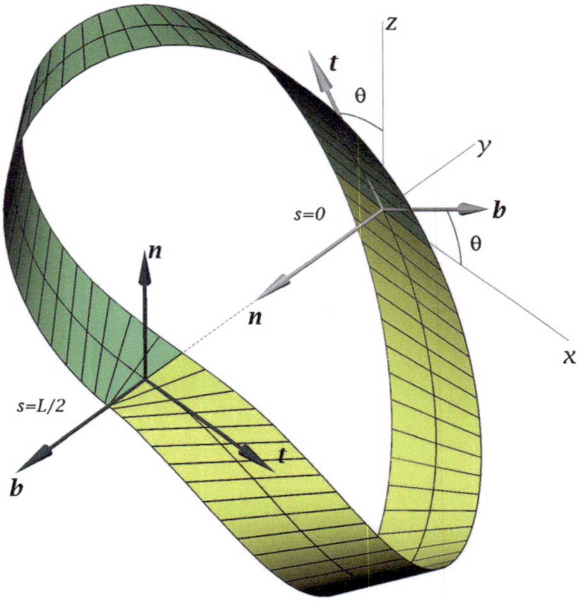

Fig. 3 Möbius strip made of two congruent pieces (*green* and *yellow*). The y-axis is the axis of C_2-symmetry and is (negatively) aligned both with the principal normal at the cylindrical point at $s = 0$ and with the binormal at the inflection point at $s = L/2$. The Frenet–Serret frame $\{\boldsymbol{t}, \boldsymbol{n}, \boldsymbol{b}\}$ is shown at the beginning (*grey*) and end (*black*) of the arclength interval $[0, L/2]$. The shape shown is an actual solution for aspect ratio $L/(2w) = 5\pi$

$$\kappa(L/2) = 0, \tag{32}$$

$$\eta'(0) = 0, \tag{33}$$

$$\psi(0) = \pi, \qquad \psi(L/2) = 2\pi, \tag{34}$$

$$\vartheta(L/2) = \frac{\pi}{2}, \tag{35}$$

$$\varphi(0) = \frac{\pi}{2}, \qquad \varphi(L/2) = \pi, \tag{36}$$

$$x(0) = 0, \qquad x(L/2) = 0, \tag{37}$$

$$y(0) = 0, \tag{38}$$

$$z(0) = 0, \qquad z(L/2) = 0. \tag{39}$$

The conditions Eqs. (37)–(39) fix one end of the half strip at the origin and align the rotational symmetry axis with the y-axis. The conditions Eqs. (34)–(36) orient the Frenet–Serret frames at both ends as follows:

$$\boldsymbol{t}(0) = \begin{pmatrix} -\sin\vartheta_0 \\ 0 \\ \cos\vartheta_0 \end{pmatrix}, \qquad \boldsymbol{n}(0) = \begin{pmatrix} 0 \\ -1 \\ 0 \end{pmatrix}, \qquad \boldsymbol{b}(0) = \begin{pmatrix} \cos\vartheta_0 \\ 0 \\ \sin\vartheta_0 \end{pmatrix},$$

$$\boldsymbol{t}(L/2) = \begin{pmatrix} 1 \\ 0 \\ 0 \end{pmatrix}, \qquad \boldsymbol{n}(L/2) = \begin{pmatrix} 0 \\ 0 \\ 1 \end{pmatrix}, \qquad \boldsymbol{b}(L/2) = \begin{pmatrix} 0 \\ -1 \\ 0 \end{pmatrix},$$

where $\vartheta_0 = \vartheta(0)$ is as yet unknown. It turns out that this choice of angles avoids the Euler-angle singularity at $\theta = 0$ in all our subsequent computations. Taken together these position and angle conditions eliminate all rigid-body degrees of freedom of the strip. We see that the tangent plane to the strip is orthogonal to the axis of symmetry at $s = 0$ and contains this axis at $s = L/2$. Moreover, the principal normal $\boldsymbol{n}(0)$ points in the negative y-direction, as does the binormal $\boldsymbol{b}(L/2)$.

The left end condition Eq. (33) specifies the end point $s = 0$ as a cylindrical point, while the right end condition Eq. (32) specifies the end point $s = L/2$ as a point with vanishing curvature (which together with the symmetry property means that it is an inflection point). Having the inflection point at the end of the interval prevents us from having to integrate through the inflection point. Note that the κ thus obtained is non-negative over the entire arclength interval. To maintain force and moment balance when assembling the entire strip we require that the components along the π-rotation symmetry axis vanish at $s = 0$, hence Eqs. (30) and (31).

The final boundary condition comes from Eq. (21). Since we differentiated this equation to derive our system of ODEs, we need to impose Eq. (21) as a boundary condition to fix the integration constant. Using that $\eta'(0) = 0$ and taking the limit $\lim_{s\to 0} \partial_\kappa g = 2\kappa(0)(1 + \eta^2(0))^2$, we may impose this condition in the form

$$M_3(0) = -\eta(0)M_1(0) - 2\kappa(0)\big(1 + \eta^2(0)\big)^2.$$

Since the force vector is constant in space, Eq. (30) implies $F_3(L/2) = 0$ because $\boldsymbol{n}(0) = \boldsymbol{b}(L/2)$. Vanishing curvature and torsion at $s = L/2$ together with Eq. (21) entails $M_3(L/2) = 0$. Note that $F_3(L/2) = 0$ and $M_3(L/2) = 0$ are in the fixed-point set of R_2. One can similarly verify that the boundary conditions cause an R_1 symmetric solution also to be R_2 symmetric. We could therefore also have used the R_2 involution about the inflection point at $s = L/2$ to obtain the full Möbius strip from the computed half strip.

Note that the boundary conditions do not include the condition $\eta(L/2) = 0$. This condition is not required because it is enforced by symmetry: the boundary conditions at $s = L/2$ enforce a switching point at $s = L/2$ for the symmetric full strip (i.e., $\boldsymbol{n}(-L/2) = -\boldsymbol{n}(L/2)$ and $\boldsymbol{b}(-L/2) = -\boldsymbol{b}(L/2)$) and according to the Randrup & Røgen properties of Sect. 5.1, at such points $\eta = 0$, in addition to $\kappa = 0$. For a nonsymmetric solution, η in fact need not be zero at switching points. The fact that we always find η tending to zero at $s = L/2$ (see below) is therefore evidence for the non-existence of nonsymmetric solutions. For the infinitely narrow strip, η in fact does approach a nonzero value, namely 1, at the switching point [45], which shows that there is no symmetric solution in that case.

The boundary-value problem is solved numerically by the continuation code AUTO [10]. There are significant numerical difficulties solving this problem, as each end of the integration interval has a singularity. At the switching point enforced at $s = L/2$, $|\eta'|$ is always found to tend to $1/w$, giving a logarithmic singularity in Eq. (9) (this singular behaviour is consistent with analytical asymptotic results in [25]). In practice, to compute a starting solution, we first compute an approximate solution with $\kappa(L/2) \approx 0.1$ to stay away from the singularity at $s = L/2$. When all other boundary conditions are satisfied we 'pull' the solution into the singularity by continuing $\kappa(L/2)$ to zero as far as possible, typically reaching values of 0.001. At this point, we typically have $\eta(L/2)$ of the same order of magnitude while $1/w - |\eta'(L/2)|$, the distance from the singularity, is typically as small as 10^{-6}. At the other end ($s = 0$), numerical convergence requires Taylor expansions of the coefficients a_i, b_i in Eqs. (25) about $\eta' = 0$ to be used for a small interval around $s = 0$ to eliminate the removable singularity of V in Eq. (9) (we use expansions up to fourth order). As a check on

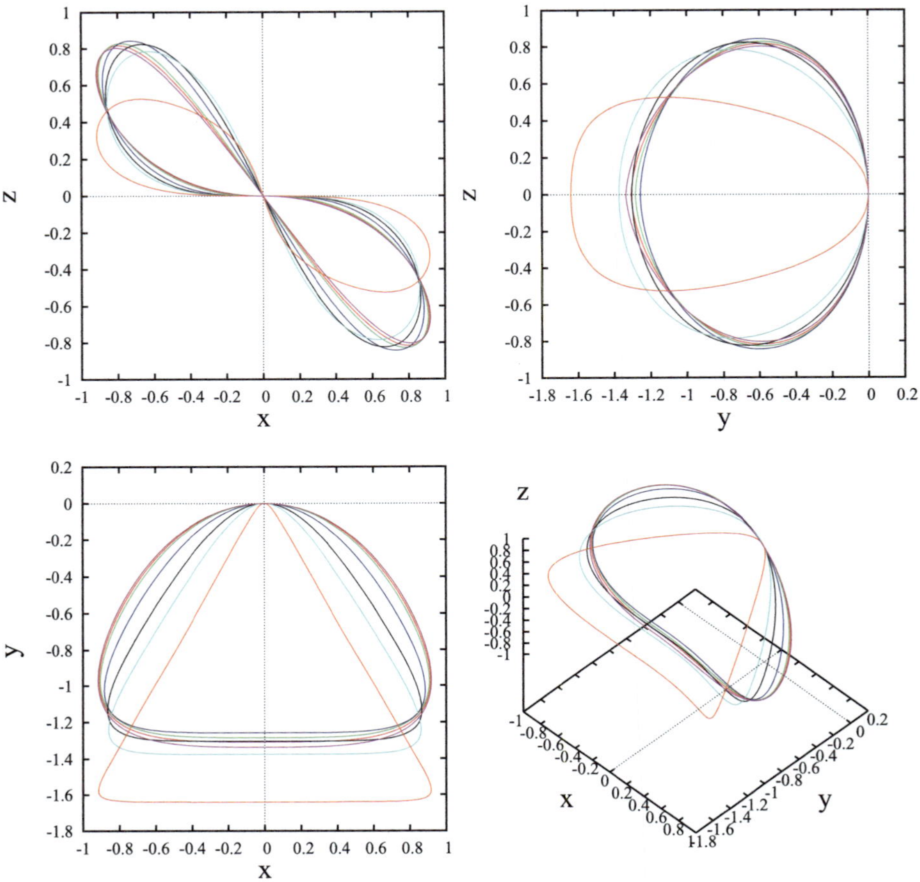

Fig. 4 Projections and 3D shape of the centreline of Möbius strips for $w = 0$ (*magenta*), 0.1 (*red*), 0.2 (*green*), 0.5 (*blue*), 0.8 (*black*), 1.0 (*cyan*) and 1.5 (*orange*). ($L = 2\pi$.)

the numerical results, the first integrals $|\mathsf{F}|$ and $\mathsf{F} \cdot \mathsf{M}$ are typically found to be constant to within 10^{-7}.

Figures 4, 5 and 6 show numerically obtained solutions with $Lk = \frac{1}{2}$ (mirror images, having a link $Lk = -\frac{1}{2}$, are obtained by applying the reflection S). There is only one physical parameter in the problem, namely the aspect ratio $L/(2w)$ of the strip. In the computations we have fixed $L = 2\pi$ and varied w. Also shown in the figures is the evolution along the strip of the straight generator. We note the points where the generators start to accumulate. At these points $|w\eta'| \to 1$ and the integrand in Eq. (8) (the energy density) diverges. Where this happens the generator rapidly sweeps through a nearly flat (violet) triangular region, a phenomenon readily observed in a paper Möbius strip (Fig. 1). We also observe two additional (milder) accumulations where no inflection occurs and the energy density remains finite. It can be shown that the energy density is monotonic along a generator. This implies that the (red) regions of high curvature cannot be connected by a generator, as a careful inspection confirms. Bounding the (violet) triangular (more precisely, trapezoidal) regions are two cylindrical generators of constant curvature (and hence constant colour in the figure) that realise local minima for the angle β.

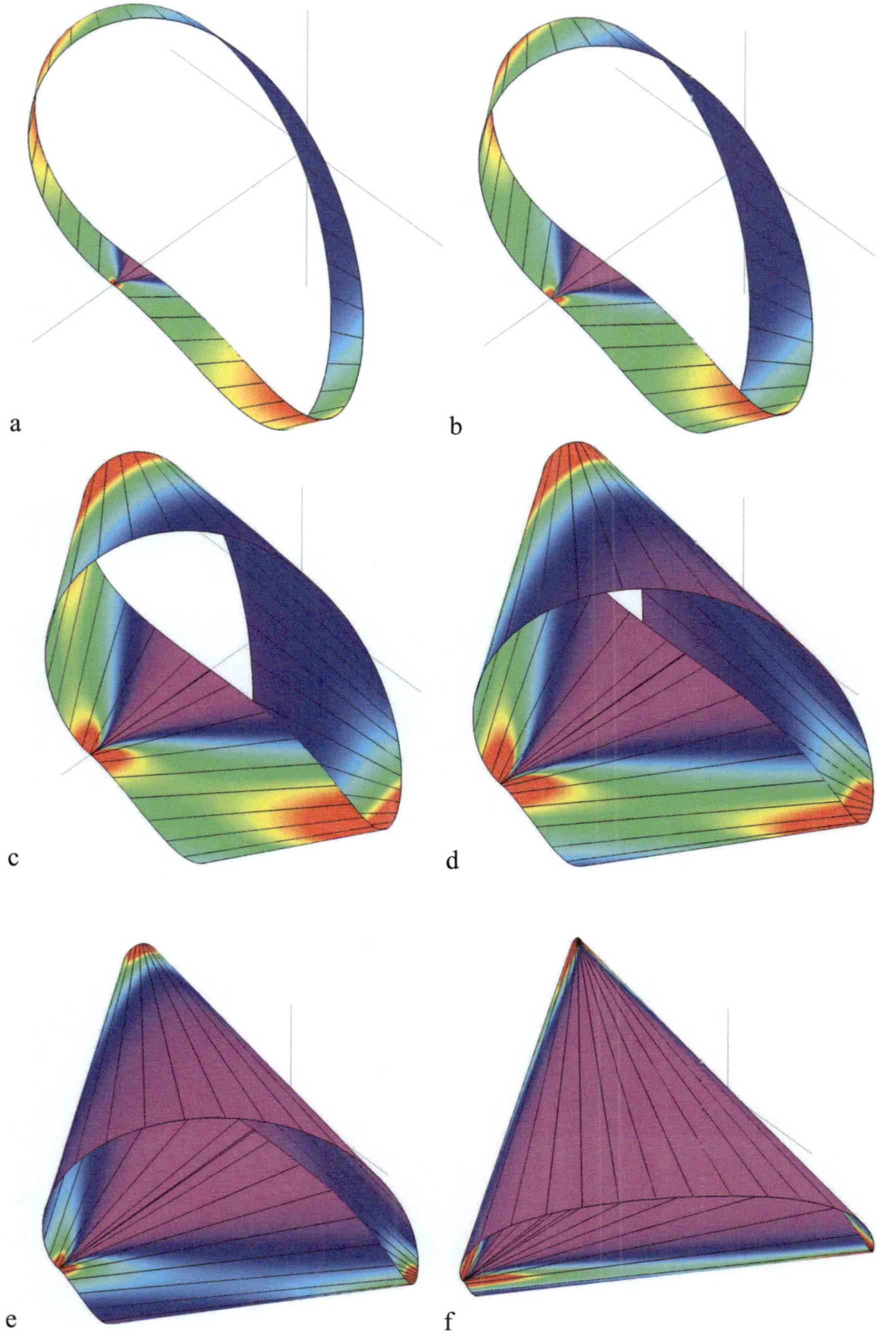

Fig. 5 Computed 3D shapes of the Möbius strip for $w = 0.1$ (**a**), 0.2 (**b**), 0.5 (**c**), 0.8 (**d**), 1.0 (**e**) and 1.5 (**f**). The colouring changes according to the local bending energy density, from *violet* for regions of low bending to *red* for regions of high bending (scales are individually adjusted). Solution (**c**) may be compared with the paper model in Fig. 1 on which the generator field and density colouring have been printed. ($L = 2\pi$.)

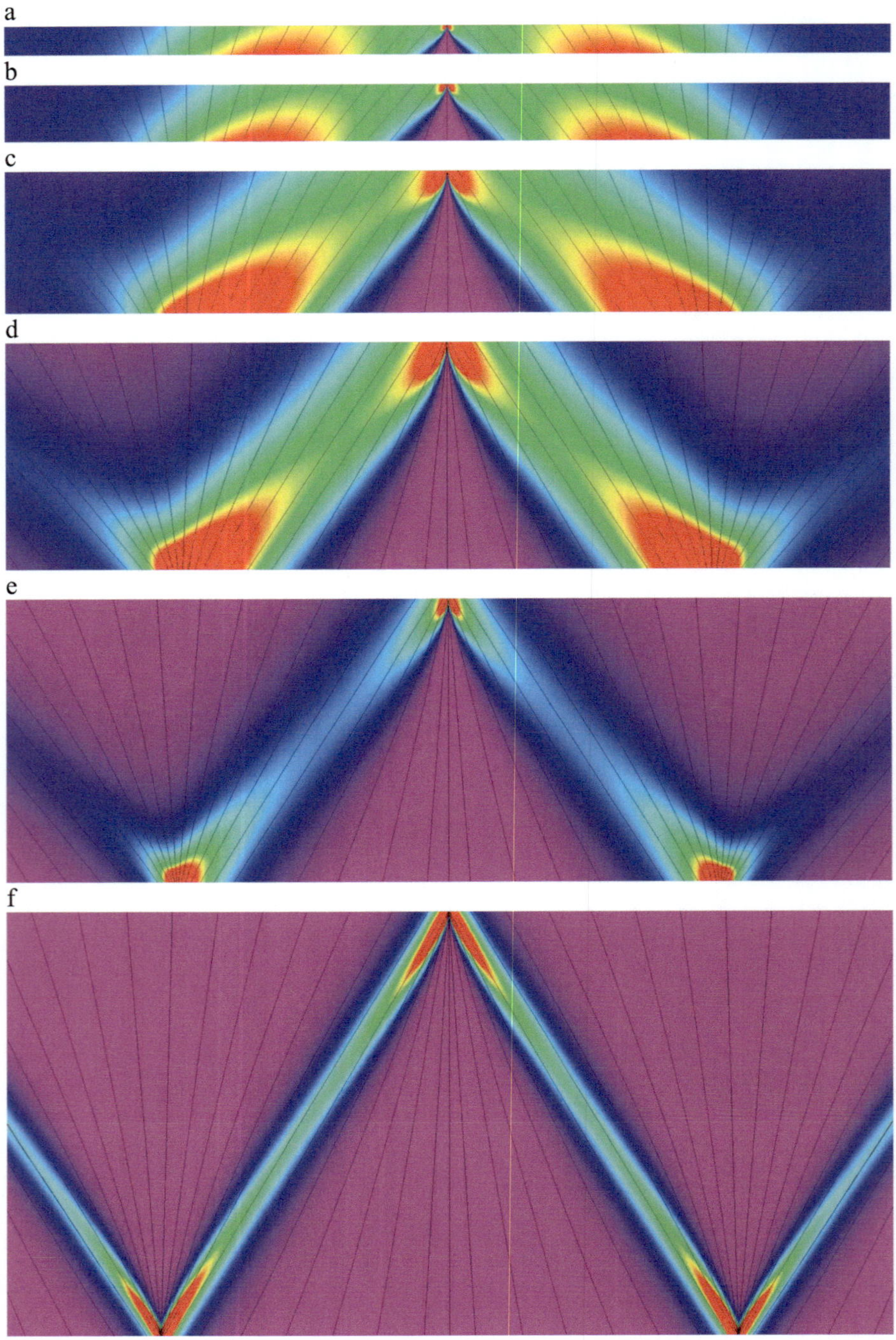

Fig. 6 Developments on the plane of the solutions in Fig. 5: $w = 0.1$ (**a**), 0.2 (**b**), 0.5 (**c**), 0.8 (**d**), 1.0 (**e**) and 1.5 (**f**). The colouring changes according to the local bending energy density, from *violet* for regions of low bending to *red* for regions of high bending (scales are individually adjusted). ($L = 2\pi$.)

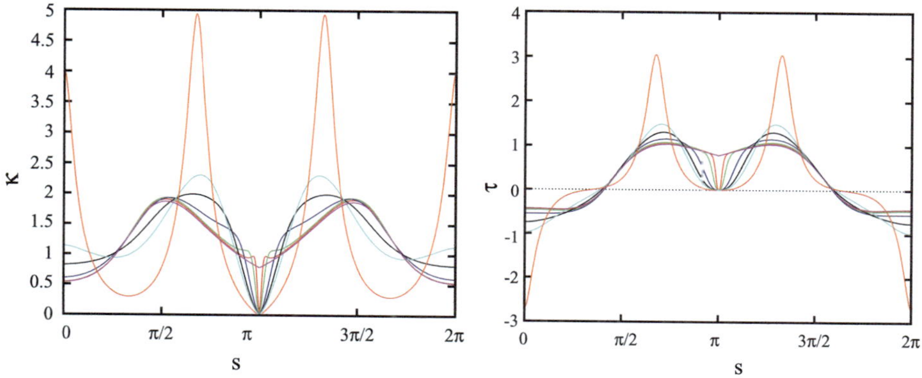

Fig. 7 Curvature and torsion of a Möbius strip. Curvature κ (*left*) and torsion τ (*right*) are shown for $w = 0$ (*magenta*), 0.1 (*red*), 0.2 (*green*), 0.5 (*blue*), 0.8 (*black*), 1.0 (*cyan*) and 1.5 (*orange*). At $s = \pi$ the principal normal changes direction to its opposite. ($L = 2\pi$.)

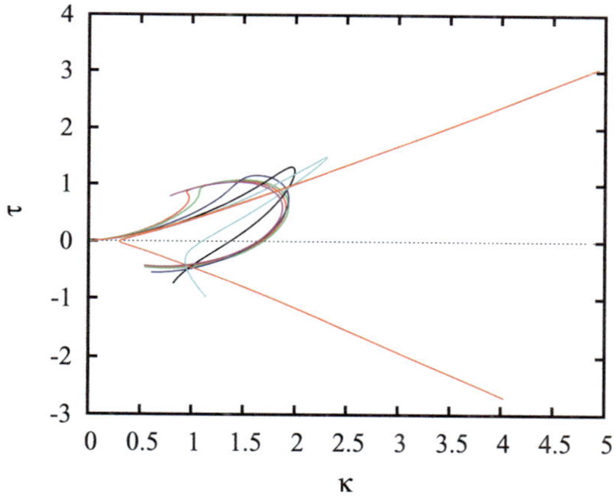

Fig. 8 Diagram of torsion against curvature of the strip's centreline. Colours as in Fig. 7

As w is increased the accumulations and associated triangular regions become more pronounced. At the critical value given by $w = \pi/\sqrt{3}$ the strip collapses into a triple-covered equilateral triangle [3, 48]. The folding process as w is increased towards this flat triangular limit resembles the tightening of tubular knots as they approach the ideal shape of minimum length to diameter ratio [54]. In the flat limit the generators are divided into three groups, intersecting each other in three vertices. The bounding generators of constant curvature become the creases. It has been conjectured that a smooth developable Möbius strip can be isometrically embedded in $\mathbb{R}^3$ only if $w < \pi/\sqrt{3}$ [20], while it has been proven that a smooth developable Möbius strip can be immersed in $\mathbb{R}^3$ only if $w < 2$ [20]. Interestingly, smaller (in fact, arbitrarily small) values for the aspect ratio $L/(2w)$ can be obtained if one allows for additional folding [3, 13].

Figures 7 and 8 give plots of curvature and torsion. The Randrup & Røgen property that $\kappa = \eta = 0$ at an odd number of points is confirmed in Fig. 7 and can also be seen in Fig. 6 at the centre of the images where the generator makes an angle of 90° with the centreline. As the maximum value $w_c = \pi/\sqrt{3} = 1.8138\ldots$ is approached, both curvature and torsion become increasingly peaked about $s = 0, 2\pi/3$ and $4\pi/3$. In the limit all bending and torsion

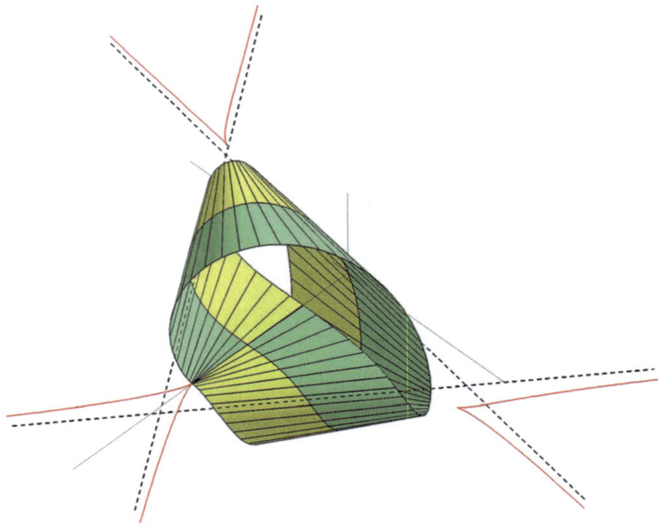

Fig. 9 Equilibrium shape of the Möbius strip for $w = 0.7$. *Red curves* show the edge of regression. *Dashed lines* mark the asymptotic directions. ($L = 2\pi$.)

is concentrated at the creases of the flat triangular shape. Going to the other extreme, we find that the solution in the limit of zero width has non-vanishing curvature, so that the Randrup & Røgen conditions are not satisfied. Given that the Frenet–Serret frame flips at $s = \pi$ this means that the curvature is discontinuous. In addition, η tends to 1, giving a limiting generator angle $\beta = 45°$. Both these properties were anticipated in [45]. This shows that the zero-width limit is singular and suggests that the Sadowsky problem has only a solution with discontinuous curvature.

5.3 Edge of Regression

The edge of regression for an equilibrium shape consists of three components (the red curves in Fig. 9). Each component has a cusp, corresponding to a swallow tail singularity of the asymptotic completion of the strip. This is the minimum number of singularities that a rectifying developable Möbius strip can have [37]. Note that precisely one cusp lies on the edge of the strip at the umbilic generator (this is always observed, for all widths). The edge of regression goes to infinity when approaching a cylindrical point (having $\eta' = 0$) and asymptotically tends to the direction of the generator at this point (see Fig. 10). Thus three cusps (corresponding to 'conical' points, having $\eta'' = 0$) alternate with three 'cylindrical' points.

5.4 Energy, Twist and Writhe of the Strip's Centreline

The normalised energy $\bar{U} = U/(Dw)$ of the strip is shown in Fig. 11. The only meaningful energy estimate available in the literature, at the particular value of $w = \pi/30$, is that of Gravesen & Willatzen [17], who minimise $\bar{U}$ for the parametrised family of developable Möbius strips (not in equilibrium) constructed in [39]. Their solution gives $\bar{U} = 9.22$; at the same width we find the much lower value $\bar{U} = 5.93$.

Also shown in Fig. 11 is the twist $Tw = \frac{1}{2\pi}\int_0^L \tau \, ds$ of the strip as a function of w. Tw may be written as $Tw = Lk - Wr$, where Lk is the linking number of the Möbius cord

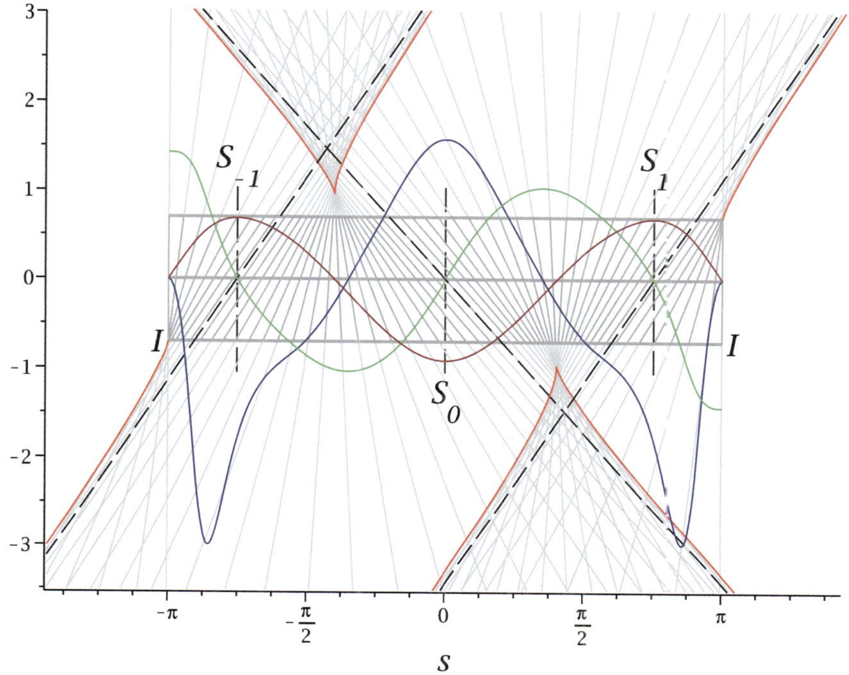

Fig. 10 Development of the Möbius strip for $w = 0.7$ with graphs of $\eta(s)$ (*brown*), $\eta'(s)$ (*green*) and $\eta''(s)$ (*blue*). I marks the inflection point while the S_i mark cylindrical points. Generators are drawn in *grey* with extensions outside the strip illustrating the asymptotic completion in projection. *Red curves* show the edge of regression where extensions of the generators intersect. Inclined *dashed lines* mark the asymptotic directions. ($L = 2\pi$.)

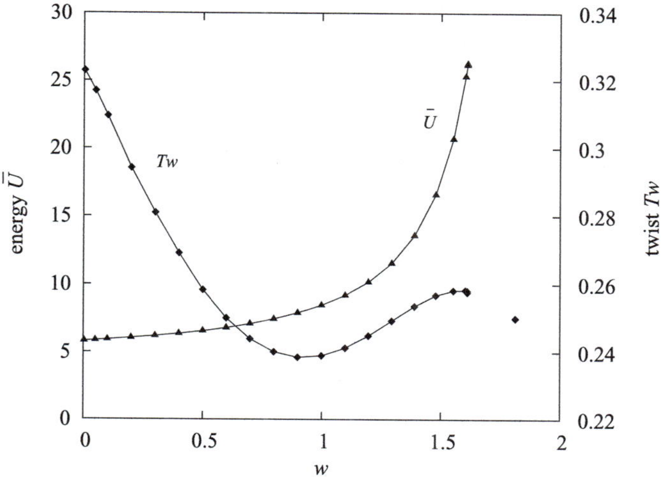

Fig. 11 Normalised energy $\bar{U}$ and total torsion (twist) Tw as functions of half-width w. The results suggest that the energy diverges as the critical value $w_c = \pi/\sqrt{3} = 1.8138\ldots$ is approached. In the same (flat) limit both Tw and Wr can be analytically shown to be $1/4$. ($L = 2\pi$.)

and Wr is the writhe of the centreline. Wr is commonly used as a measure for the spatial deformation of a 3D curve. It can be defined as the crossing number of the curve averaged over all viewing directions [14].

6 Developable Strips of Higher Linking Number—Möbius Surgery

6.1 Modified Boundary-Value Problem for D_n-Symmetric Solutions

The Möbius strip defines only one example of a boundary-value problem for twisted sheets. A natural generalisation is to strips with linking numbers other than $\pm\frac{1}{2}$. The Randrup & Røgen properties of Sect. 5.1 also hold for these generalised Möbius strips and our techniques, exploiting symmetry, can readily be applied to such problems.

We have seen that the Möbius strip has π rotational symmetry about $\boldsymbol{n}$ (described by R_1) at the cylindrical point at $s=0$ (labelled S_0 in Fig. 10), where $\eta'=0$, and π rotational symmetry about $\boldsymbol{b}$ (described by R_2) at the inflection point at $s=L/2$ (labelled I in Fig. 10), where $\kappa=0$. Continuity of forces and moments across these points is ensured because the corresponding normal and binormal components vanish: $F_2(0)=0=M_2(0)$ and $F_3(L/2)=0=M_3(L/2)$. (Note that the rotational symmetry is also confirmed in Figs. 5 and 6 by the generators at these two points having constant colour, corresponding to the curvature being constant along these generators.)

We can therefore use involution R_1 at cylindrical points and involution R_2 at inflection points to construct solutions to the equilibrium equations with an arbitrary number of such points. In fact, the Möbius half strip has a second, intermediate, cylindrical point (S_1 in Fig. 10) with constant-colour generator. This point is not a point of full symmetry as F_2 and M_2 are nonzero. However, it turns out that by a slight change of orientation of the Frenet–Serret frame we can deform the strip such as to make these components zero. We then have a choice of taking either the (long) segment S_0I or the (short) segment S_1I as elementary piece in constructing more complicated equilibrium shapes. Note from Fig. 10 that one difference between these two pieces is that S_1I has positive η (and hence positive torsion τ), while on S_0I η (and hence τ) changes sign. The shapes constructed by combining these pieces will in general not be closed. However, by imposing symmetric closure conditions we can obtain D_n-symmetric solutions for $n=2,3,4,5,\ldots$.

We thus reformulate the boundary-value problem as follows. We drop the two frame orientation conditions at $s=0$ in the previous set of boundary conditions in Sect. 5.2 and specify, on the interval $[0, L/(2n)]$ (see Fig. 12),

$$F_2(0)=0, \qquad F_3\big(L/(2n)\big)=0, \tag{40}$$

$$M_2(0)=0, \qquad M_3\big(L/(2n)\big)=0, \tag{41}$$

$$\kappa\big(L/(2n)\big)=0, \tag{42}$$

$$\eta'(0)=0, \tag{43}$$

$$\psi\big(L/(2n)\big)=2\pi, \tag{44}$$

$$\vartheta\big(L/(2n)\big)=\frac{\pi}{2}, \tag{45}$$

$$\phi\big(L/(2n)\big)=\pi, \tag{46}$$

$$x\big(L/(2n)\big)=0, \tag{47}$$

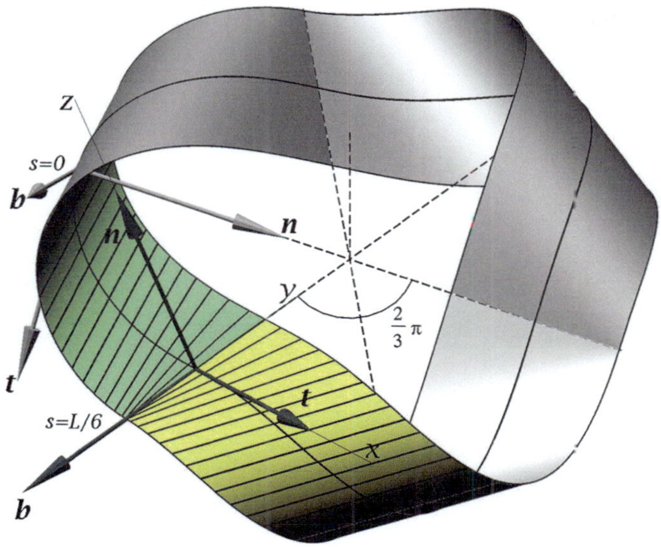

Fig. 12 A D_3-symmetric solution for $(m,n)=(2,3)$. The D_3-symmetry axis is vertical (*dashed*). The three C_2-symmetry axes are also shown *dashed*; each is aligned with the principal normal at one intersection with the strip and with the binormal at the opposite intersection. The Frenet–Serret frame $\{\boldsymbol{t},\boldsymbol{n},\boldsymbol{b}\}$ is shown at the beginning (*grey*) and end (*dark*) of the arclength interval $[0, L/6]$. Two of the six congruent pieces are coloured (*green* and *yellow*); one is obtained from the other by rotation through π about the binormal $\boldsymbol{b}$ at the inflection point $s=L/6$. The angle between the principal normal at $s=0$ and the binormal at $s=L/6$ equals $2\pi/3$. The shape shown is an actual solution for aspect ratio $L/(2w)=9.87$

$$y\big(L/(2n)\big)=0, \tag{48}$$

$$z\big(L/(2n)\big)=0. \tag{49}$$

These boundary conditions 'cut out' a piece between cylindrical and inflection points (either S_0I or S_1I) of the Möbius strip solution.

We then add two symmetric closure conditions for a solution consisting of a sequence of $2n$ congruent pieces satisfying the above boundary conditions. The first condition enforces coplanarity of $\boldsymbol{n}(0)$, $\boldsymbol{b}(L/(2n))$ and $\boldsymbol{r}(L/(2n))-\boldsymbol{r}(0)$:

$$\big(\boldsymbol{n}(0)\times\boldsymbol{b}\big(L/(2n)\big)\big)\cdot\big(\boldsymbol{r}\big(L/(2n)\big)-\boldsymbol{r}(0)\big)=z(0)n_x(0)-x(0)n_z(0)=0. \tag{50}$$

In explicit form, after using the boundary conditions (40)–(49), this reads

$$z(0)\big(-\sin\psi(0)\sin\varphi(0)+\cos\vartheta(0)\cos\psi(0)\cos\varphi(0)\big)+x(0)\sin\vartheta(0)\cos\varphi(0)=0. \tag{51}$$

The second condition enforces an angle $\frac{m}{n}\pi$, $\frac{m}{n}\in\mathbb{Q}\setminus\mathbb{Z}$, $m<n$, between $\boldsymbol{n}(0)$ and $\boldsymbol{b}(L/(2n))$:

$$\boldsymbol{n}(0)\cdot\boldsymbol{b}\big(L/(2n)\big)=\cos\left(m\frac{\pi}{n}\right). \tag{52}$$

This condition guarantees that concatenation of $2n$ copies of the elementary piece produces a closed strip (with n inflection points). In our variables the condition reads

$$-\cos\psi(0)\sin\varphi(0)-\cos\vartheta(0)\sin\psi(0)\cos\varphi(0)=\cos\left(m\frac{\pi}{n}\right). \tag{53}$$

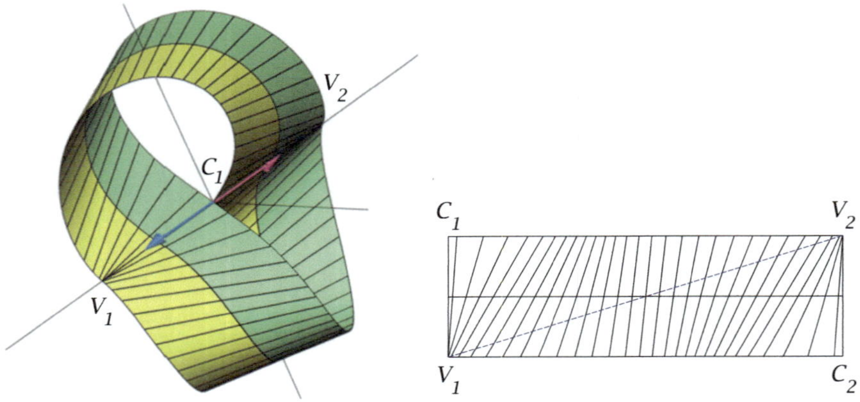

Fig. 13 Two-sided closed strip for $(m,n)=(1,2)$ with $Lk=1$ and aspect ratio $L/(2w)=6.589$ (strip (b) in Fig. 14), $|C_1V_1|=|C_2V_2|=2w$. *Arrows* show the contact forces. *Left*: Equilibrium shape in 3D. C_i is a point of self-contact. *Right*: Development of one half of the strip with straight generators shown. The *blue diagonal line* is a geodesic

The D_n-symmetry axis is directed along the vector $(z(0),0,-x(0))^\intercal$ and intersects the y-axis at

$$\begin{aligned} y_c &= \left[\left(\boldsymbol{b}\big(L/(2n)\big)-\boldsymbol{n}(0)\cos\left(m\frac{\pi}{n}\right)\right)\cdot\big(\boldsymbol{r}\big(L/(2n)\big)-\boldsymbol{r}(0)\big)\right]\csc^2\left(m\frac{\pi}{n}\right)\\ &= \big(\boldsymbol{r}\big(L/(2n)\big)-\boldsymbol{r}(0)\big)\cdot\boldsymbol{b}\big(L/(2n)\big)\\ &\quad -\sqrt{\big(\boldsymbol{r}\big(L/(2n)\big)-\boldsymbol{r}(0)\big)^2-\big(\big(\boldsymbol{r}\big(L/(2n)\big)-\boldsymbol{r}(0)\big)\cdot\boldsymbol{b}\big(L/(2n)\big)\big)^2}\cot\left(m\frac{\pi}{n}\right)\\ &= y(0)-\sqrt{x^2(0)+z^2(0)}\cot\left(m\frac{\pi}{n}\right). \end{aligned}$$

Note that the C_2-symmetric Möbius strip solution of Sect. 5, having $(m,n)=(0,1)$, is degenerate for these boundary conditions as the first symmetry condition, Eq. (51), is identically satisfied since $x(0)=z(0)=0$.

6.2 Two-Sided Strip with $(m,n)=(1,2)$—Figure-of-Eight ($Lk=1$)

For $(m,n)=(1,2)$ we have a two-sided solution with $Lk=1$ in the shape of a figure-of-eight. A simple instance of this solution is a cylindrical shape with planar centreline and with all generators orthogonal to this centreline ($\eta\equiv 0$). This is the figure-of-eight elastica solution with self-intersection [32]. A non-selfintersecting solution may be obtained by considering a slightly displaced solution with non-planar centreline and allowing for self-contact (see Fig. 13). For a D_2-symmetric solution we assume, prompted by paper models, that this self-contact occurs at the inflection point. If we then also assume that the contact at the edge of the strip is frictionless then the contact force will only have a component in the binormal direction, i.e., along the umbilical generator through the inflection point. We can compute such solutions by dropping the second boundary condition in Eq. (40) and imposing the geometrical contact condition $y_c=w$, which now becomes $y(0)=w$, instead. The end

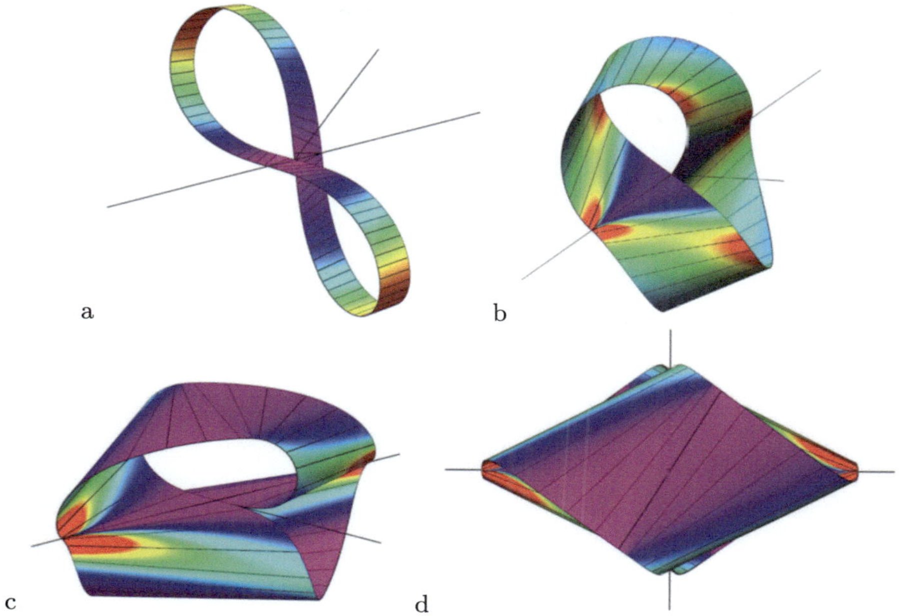

Fig. 14 D_2-symmetric shapes of a two-sided developable closed strip for $(m,n)=(1,2)$ with $Lk=1$ and aspect ratio $L/(2w)=24.420$ (**a**), 6.589 (**b**) and 4.218 (**c**, **d**) (two views). This last value is the lowest value without self-intersections of the strip. Rotational symmetry axes are shown

force $F_3(L/4)$ is then allowed to become nonzero. For a physically realisable solution we require $F_3(L/4)<0$ so that it is balanced by an equal and opposite contact force in the (positive) binormal direction coming from the other half of the strip (see Fig. 13).

Three examples of strips with self-contacts, constructed from (short) $S_1 I$ pieces of the Möbius strip, are shown in Fig. 14. The solutions have three cylindrical points between the two inflection points, which means that in deforming the $S_1 I$ piece to satisfy the symmetric closure conditions an extra intermediate cylindrical point has appeared. The solutions in Fig. 14(a, b) have, however, retained the property of $S_1 I$ that they have positive torsion, while those in (c) and (d) have slightly negative torsion (in the top and bottom violet regions in the figure).

Dependence of the contact force $2F_3(L/4)$ on the aspect ratio of the strip is plotted in Fig. 15. The contact force vanishes only in the zero-width limit $w=0$. This seems to suggest that no finite-width solutions exist that are free of self-contact. However, there may be different types of $(m,n)=(1,2)$ solutions with no self-contact and indeed Fig. 16 below gives an example of exactly such a solution.

Figure 14 shows that as the width w increases the figure-of-eight increasingly flattens. The aspect ratio of the strip is bounded from below. To see this note from Fig. 13 (*left*) that the points V_1 and V_2 of the umbilic generators lying on the outer edges opposite to the contact points are separated by twice the width of the strip: $|V_1 V_2| = 4w$. This means that in the planar development of the strip the straight line connecting V_1 and V_2 cannot be shorter than $4w$ (Fig. 13, *right*), so we conclude that the length of the centreline cannot be less than $4w\sqrt{3}$ and the aspect ratio $L/(2w)$ cannot go below $2\sqrt{3}$ (consistent with the asymptotic behaviour displayed in Fig. 15).

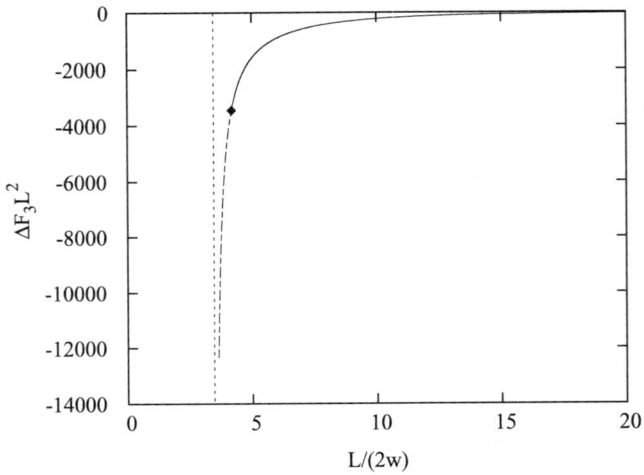

Fig. 15 Contact force $\Delta F_3 = 2F_3(L/4)$ for D_2-symmetric two-sided strip with $(m,n) = (1,2)$, having $Lk = 1$, as a function of aspect ratio $L/(2w)$. The *diamond* marks the point where the tangents at contacting points become aligned. Solutions on the *dashed continuation of the curve* have self-intersections. The *vertical asymptote* corresponds to the limiting aspect ratio $2\sqrt{3}$

If we imagine continuation of the sequence in Fig. 14 to a completely flattened triple-covered rhombus, with diagonals of length $4w$ and $4w/\sqrt{3}$, then this solution would have exactly the limiting aspect ratio. (By allowing other such sharp folds we can in fact reach strips of the same topology with lower aspect ratio.)

However, numerical results show that, for our *equilibrium* solutions, the strip starts to self-intersect near the contact point before this flat state is reached. This indicates that shapes with a single contact do not exist beyond a critical value of w. Shapes with multiple contact points are beyond the scope of the present work, but it appears that the contact point bifurcates giving rise to other, nonsymmetric, solutions. The onset of this self-intersection coincides with the tangents to the edges at the contact point becoming aligned with each other (at the diamond in Fig. 15). This tangent alignment happens when $\boldsymbol{t}(L/2)$ becomes parallel to the rotational symmetry axis, which can be represented as the vector $\boldsymbol{n}(0) \times \boldsymbol{b}(L/2)$. Hence the condition is $\boldsymbol{t}(L/2) \times (\boldsymbol{n}(0) \times \boldsymbol{b}(L/2)) = 0$, which simplifies to $\boldsymbol{n}(0) \cdot \boldsymbol{t}(L/2) = 0$ and further to $x(0) = 0$. In our continuation run this condition is met at aspect ratio $L/(2w) = 4.218$. Figures 14(c, d) show the solution at this value.

We finally show, in Fig. 16, an example of a $(m,n) = (1,2)$ strip with $Lk = 1$ constructed from two (long) S_0I pieces of the Möbius strip. The solution is not self-contacting and has the additional cylindrical points between inflection points from the S_0I piece. The number of cylindrical and conical points is therefore the same as that of the solutions in Fig. 14. At the four conical points η and κ are small and in this sense the conical points are close to being inflection points; they share with the two inflection points the features of radiating sharp ridges that bound flat triangular regions (similar to the Möbius strip solutions in Figs. 5 and 6). In projection the solution has almost perfect D_6 symmetry. Despite the vastly different appearances, the main difference between the solutions in Fig. 14(a, b) and Fig. 16 is that the former have all positive torsion while the latter has a torsion that changes sign (at non-inflection points where the generator is perpendicular to the centreline).

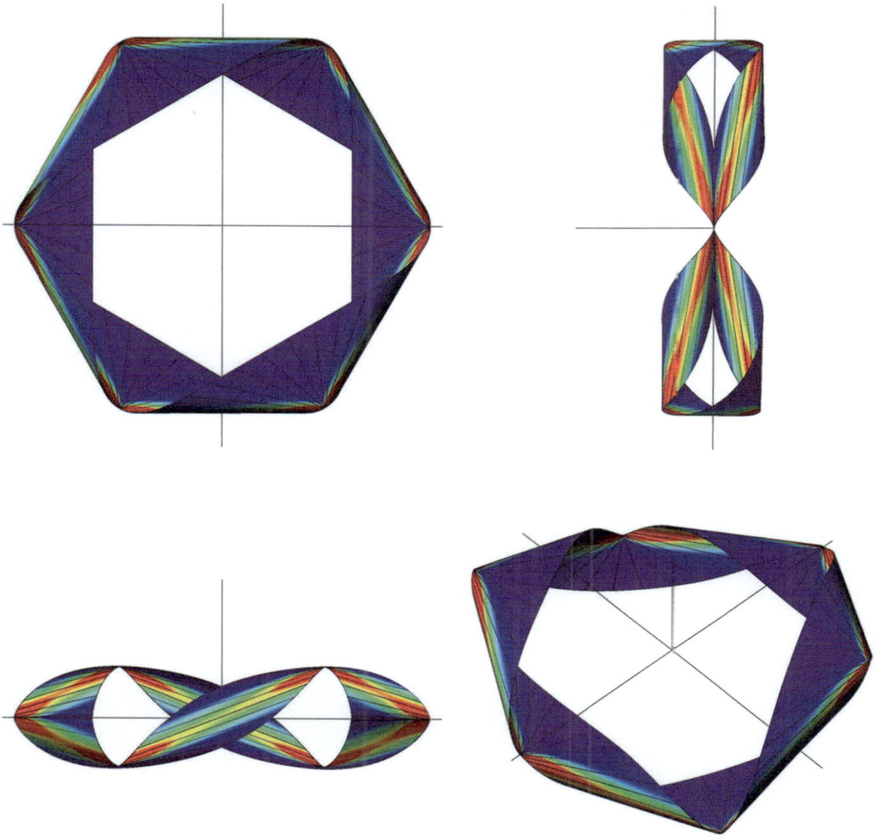

Fig. 16 Different views of a developable closed strip for $(m,n)=(1,2)$ with $Lk=1$ and aspect ratio $L/(2w)=15.789$ and with D_2-symmetry axes indicated

6.3 One-Sided Strip with $(m,n)=(2,3)$—Strip with $Lk=3/2$

For $(m,n)=(2,3)$, taking three copies of a S_1I piece of the Möbius strip, we get a one-sided strip with $Lk=\frac{3}{2}$ (also known from Escher's work [11]). Three examples at different aspect ratios are shown in Fig. 17, with the last (c) showing the shape at initial (triple) self-contact at the central axis, when $y_c=w$. The development of one third of this shape is shown in Fig. 18. The points C_1 and C_2 coincide in space at the point of triple contact. All three solutions have positive torsion and, unlike for the figure-of-eight solutions in Fig. 14, no intermediate cylindrical points have appeared in the deformation of the S_1I piece.

D_3-symmetric solutions for wider strips, without self-intersection, can be obtained by imposing an impenetrability constraint and introducing a contact force. Thus we assume that the single triple contact point is maintained by three contact forces of equal magnitude acting along the respective umbilic binormals (the lines C_iV_i, $i=1,2,3$, in Fig. 18). The contact forces are then coplanar and therefore self-balancing (note that these forces applied at the edge of the strip produce no moments when transferred to the centreline of the strip).

The boundary-value problem is then modified as in the previous section for the figure-of-eight solution by dropping the second condition in Eq. (40) and adding the geometrical

Fig. 17 3D shapes of a one-sided developable closed strip for $(m, n) = (2, 3)$ with $Lk = \frac{3}{2}$ and aspect ratio $L/(2w) = 54.863$ (**a**), 9.369 (**b**) and 4.744 (**c**)

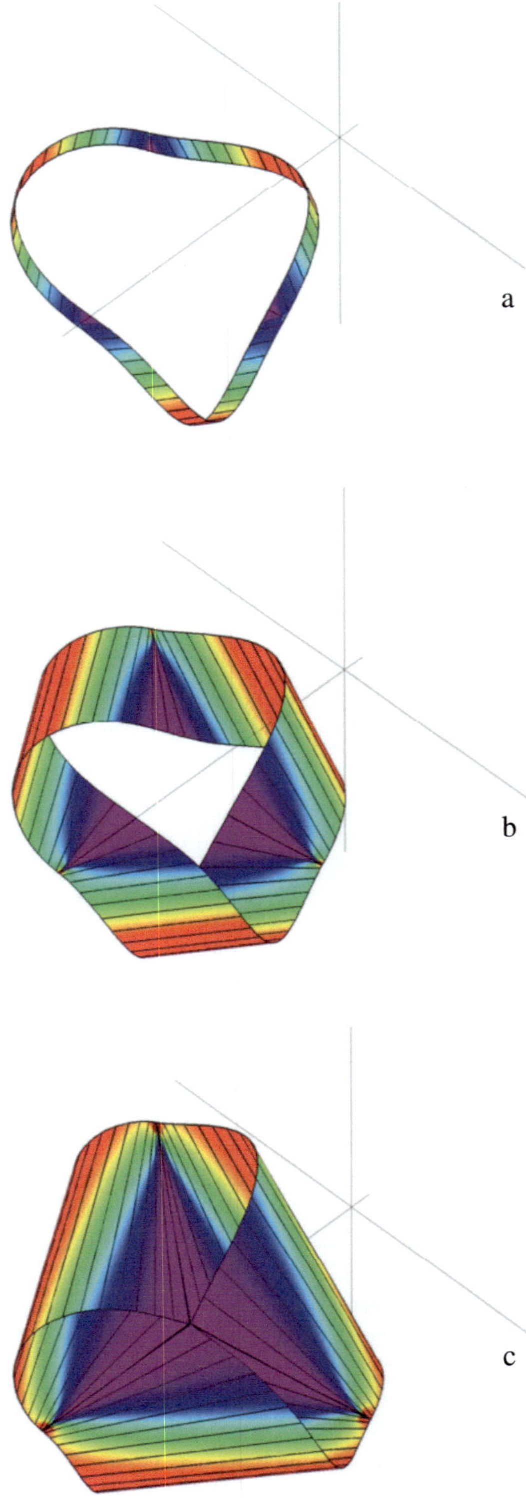

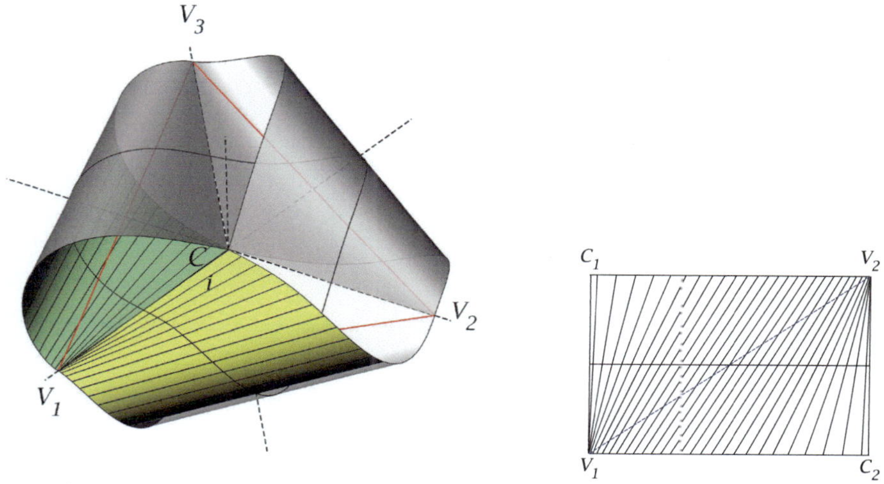

Fig. 18 One-sided closed strip for $(m, n) = (2, 3)$ with $Lk = \frac{3}{2}$ and aspect ratio $L/(2w) = 4.744$ (strip (c) in Fig. 17), $|C_1 V_1| = |C_2 V_2| = 2w$. *Left*: Equilibrium shape in 3D. C_i is a point of triple self-contact. The *straight lines* connecting the points V_i, $i = 1, 2, 3$, do not belong to the surface of the strip. *Right*: Development of one third of the strip with straight generators shown. The *blue diagonal line* is a geodesic

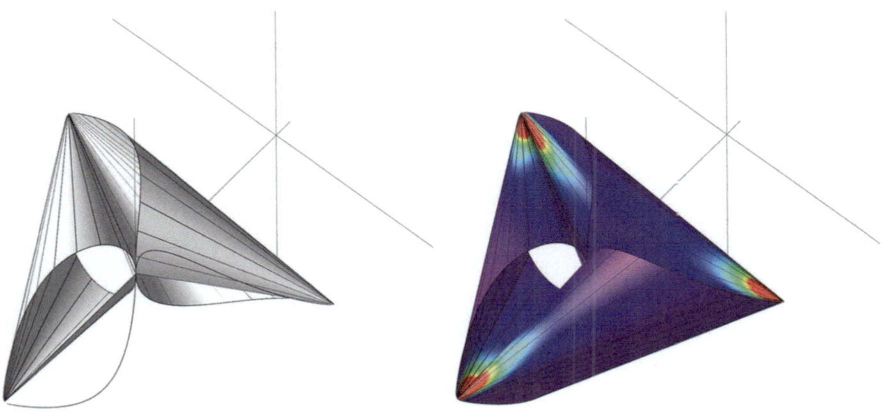

Fig. 19 Self-contacting one-sided closed strip for $(m, n) = (2, 3)$ with $Lk = \frac{3}{2}$ and aspect ratio $L/(2w) = 4.249$. The vertical D_3-symmetry axis is shown. The shape closely approaches an assembly of three conical surfaces. The edge of the strip touches itself in the centre of symmetry in a triple contact point. *Left*: Strip with one third removed to better reveal its shape. *Right*: Strip with colouring according to the local bending energy density, ranging from *violet* for regions of low bending to *red* for regions of high bending

contact condition $y_c = w$ instead. The end force $F_3(L/6)$ is thus allowed to become nonzero. An example of such a self-contacting solution is shown in Fig. 19.

There is again a lower bound to the aspect ratio of the strip. The three coplanar binormals $C_i V_i$ make an equilateral triangle $V_1 V_2 V_3$ with sides $2w\sqrt{3}$ (see Fig. 18) and the distance $V_1 V_2$ measured on the strip's surface cannot be shorter than $2w\sqrt{3}$, this minimum value occurring only if the entire straight line connecting V_1 and V_2 lies in the surface. This mini-

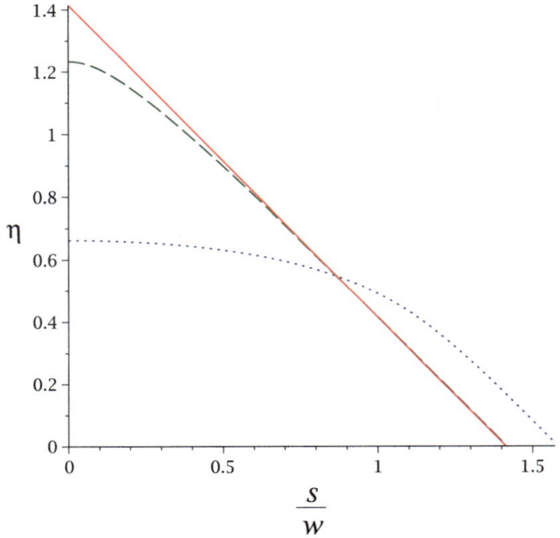

Fig. 20 Dependence of η on normalised arclength for a sixth of the strip with $Lk = \frac{3}{2}$: *blue* (*dotted*) is for the shape in Fig. 17(c), *green* (*dashed*) for the shape in Fig. 19 and *red* (*solid*) for the strip made of cones, in which case $\eta = \sqrt{2} - s/w$

mum value for $|V_1V_2| = 2w\sqrt{3}$ then gives the minimum aspect ratio $L/(2w) = 3\sqrt{2}$. In this limit, V_1V_2 becomes a generator and all other generators pass through the points V_i. The surface is thus conical. One can also show that η becomes a piecewise linear function of arclength: $\eta = \sqrt{2}(|4\{\sqrt{2}s/(8w)\} - 2| - 1)$, $s \in [0, 6\sqrt{2}w]$, where $\{x\}$ denotes the fractional part of x. Figure 20 shows how η approaches this conical limit. Looking at Fig. 19 one can indeed imagine a limiting shape consisting of three conical surfaces with their apices at V_1, V_2 and V_3. The conical patches join each other along the three generators V_1V_2, V_2V_3 and V_3V_1. We study the conical limit of our equations further in Appendix B.

The combined tri-conical surface can be made arbitrarily smooth (by appropriate local change of the curvature). The resulting surface then provides an interesting example of a developable surface that is arbitrarily smooth yet does not have a smooth (just a continuous) field of generators (since $\eta(s)$ lacks differentiability at points where the centreline crosses either of the three generators V_1V_2, V_2V_3 or V_3V_1) [43, 57].

Returning to equilibrium solutions, our computational results suggest that just before the limiting aspect ratio, at $L/(2w) = 4.249$, all three tangents to the edges in the central triple contact point C_i become aligned along the D_3-symmetry axis. This solution is shown in Fig. 19. The strip has no other contacts and does not self-intersect. However, as in the case of the figure-of-eight solution, this alignment of tangents appears to be the onset of self-intersection. To find the limiting shape of a physical strip we would therefore have to account for contact more complicated than single-point contact, a problem we leave for future work.

Experimentation with paper models also suggests that at or shortly after the first contact a pitchfork bifurcation occurs where the D_3-symmetric solution becomes unstable and two stable solutions appear, buckled in opposite directions. For such nonsymmetric solutions the three contacting umbilic generators would no longer be coplanar. They may or may not continue to be perpendicular to the edge of the strip ($\eta = 0$). If they are then there would only be a normal component in addition to the binormal component of the contact force

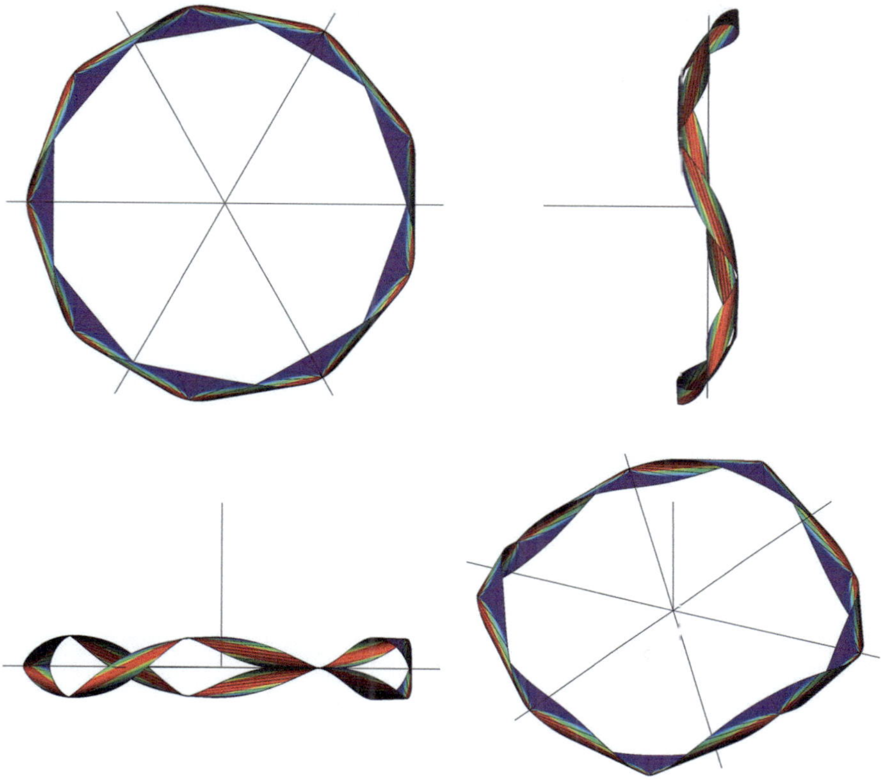

Fig. 21 Different views of a developable closed strip for $(m, n) = (2, 3)$ with $Lk = 3/2$ and aspect ratio $L/(2w) = 47.229$ and with D_3-symmetry axes indicated

at the edge of the strip, and hence at the centreline. This edge component would induce a twisting moment M_1 at the centreline that would also have to be taken into account. If the umbilic generators are no longer perpendicular to the edge then all force and moment components of the contact force would have to be accounted for. In the wide limit we can imagine a folded shape in the form of a tetrahedron with three faces being double-covered right triangles with sides $2w$, $2w$ and $2w\sqrt{2}$, while the fourth face is absent. It is easy to see that such a 'quarter-cube' configuration, with residual C_3-symmetry, has aspect ratio $L/(2w) = 3$.

In Fig. 21 we finally show an example of a $(m, n) = (2, 3)$ strip with $Lk = 3/2$ constructed from three (long) $S_0 I$ pieces of the Möbius strip. The solution is not self-contacting. It follows the pattern of the $Lk = 1$ solution in Fig. 16 and in projection has approximate D_9-symmetry.

6.4 Higher-Order One- and Two-Sided Developable Strips

The series of closed-strip solutions may be continued by increasing the number of congruent pieces. Figure 22 shows strips made of 6, 8, 10 and 12 pieces of type $S_1 I$ for, respectively, $m = 2, 3, 4, 5$ and $n = m + 1$. By construction each strip possesses the corresponding D_n-symmetry. The edges of the strips form torus knots or links of type $(n, 2)$. The centrelines are unknotted. Each of these solutions has n isolated cylindrical points, the same as the number

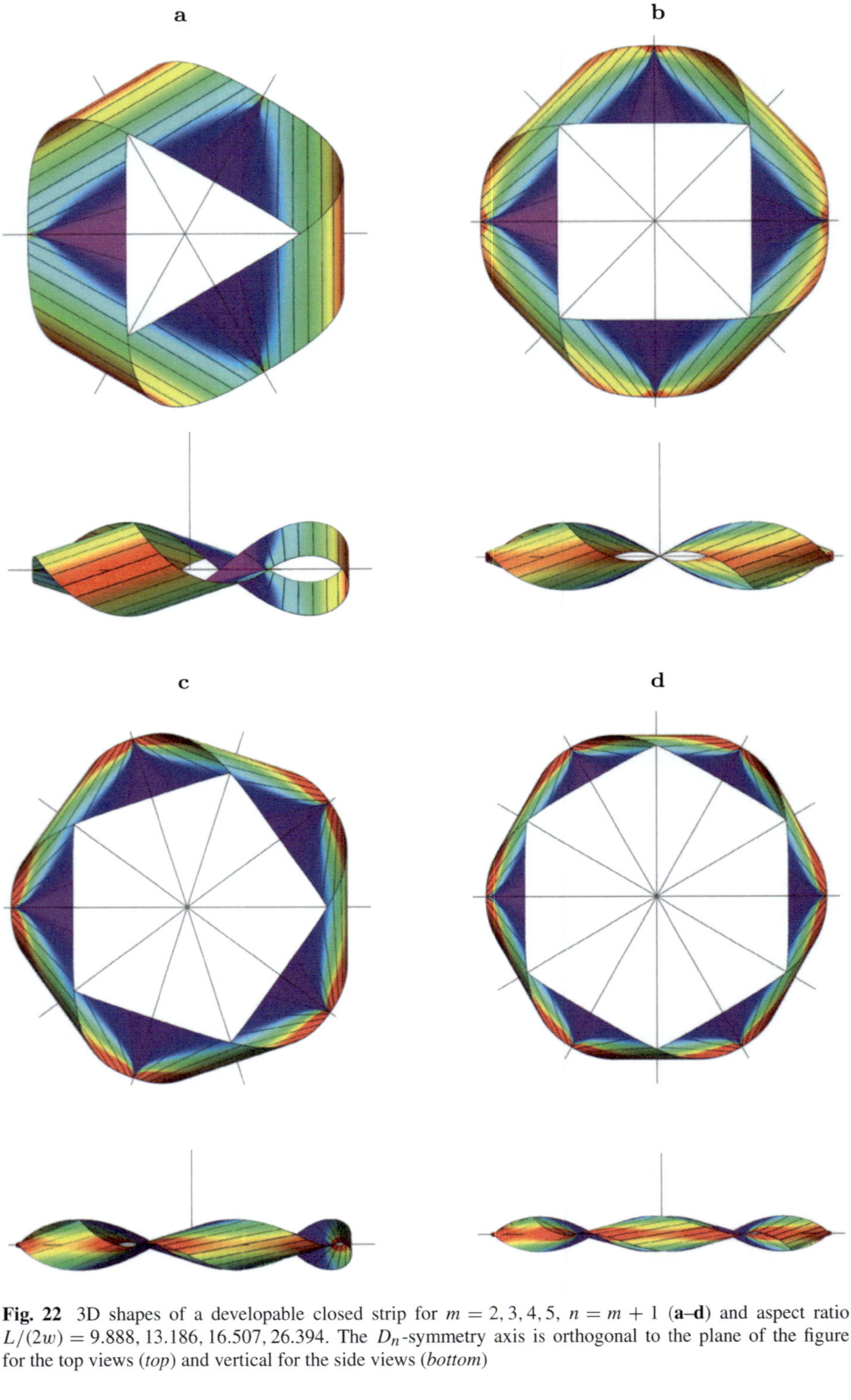

Fig. 22 3D shapes of a developable closed strip for $m = 2, 3, 4, 5$, $n = m + 1$ (**a–d**) and aspect ratio $L/(2w) = 9.888, 13.186, 16.507, 26.394$. The D_n-symmetry axis is orthogonal to the plane of the figure for the top views (*top*) and vertical for the side views (*bottom*)

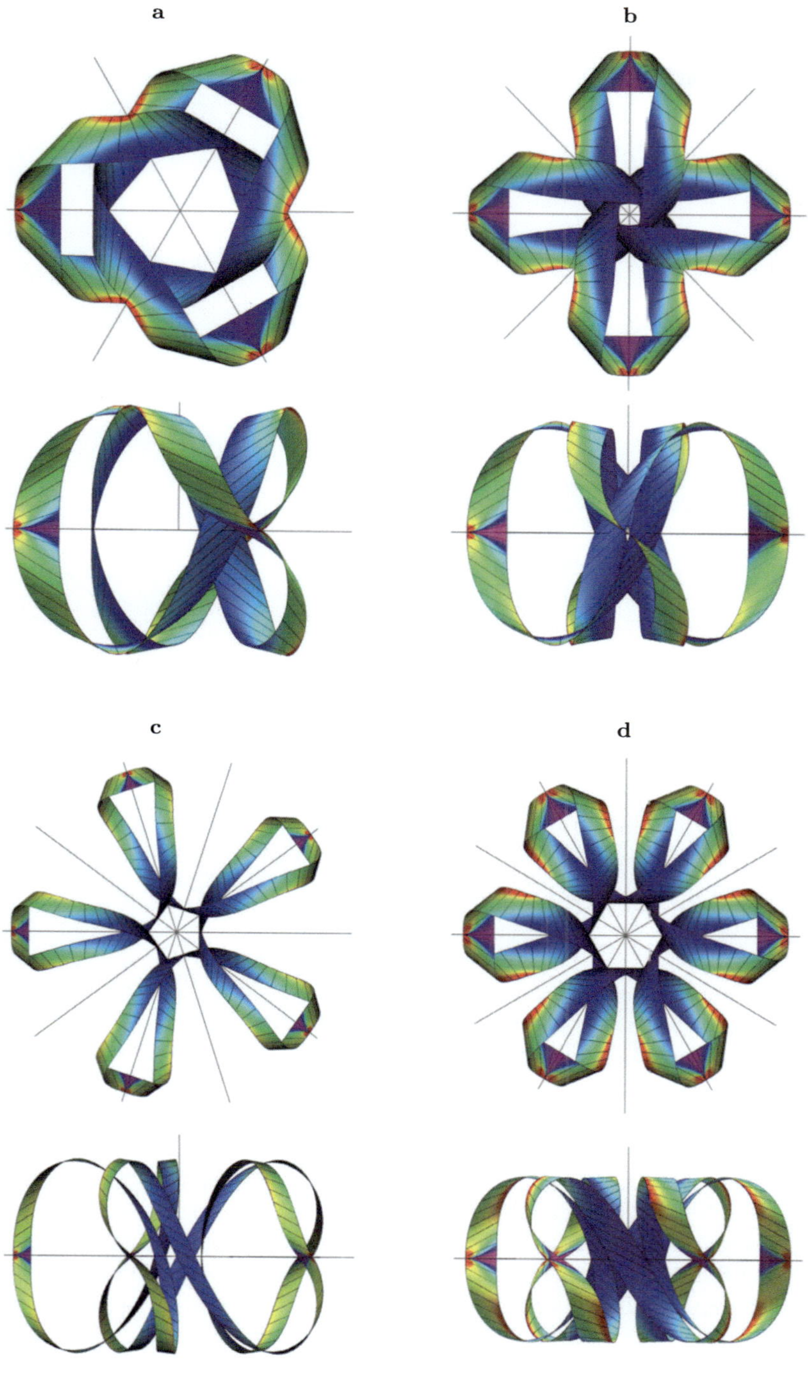

Fig. 23 3D shapes of a developable closed strip for $m = 1$, $n = 3, 4, 5, 6$ (**a**–**d**) and aspect ratio $L/(2w) = 47.134, 62.861, 157.104, 94.272$. The D_n-symmetry axis is orthogonal to the plane of the figure for the top views (*top*) and vertical for the side views (*bottom*). The strips in (**a**) and (**b**) are $(3, 2)$ and $(4, 3)$ torus knots, respectively, while those in (**c**) and (**d**) are unknots

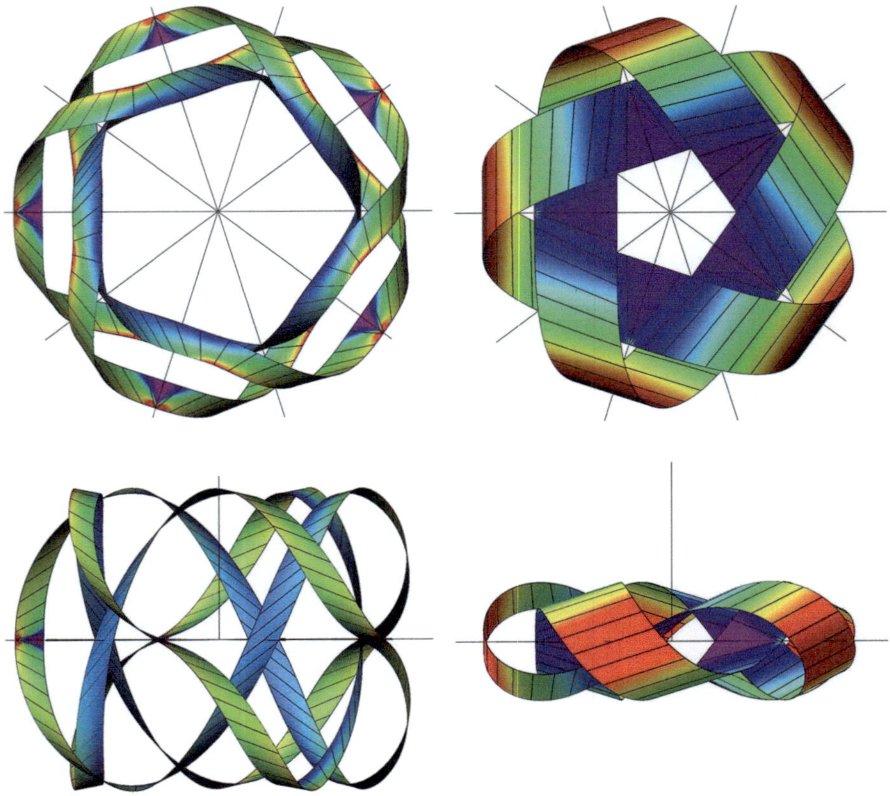

Fig. 24 3D shapes of a developable closed strip for $(m,n) = (2,5)$ (*left*) and $(m,n) = (3,5)$ (*right*) with aspect ratio $L/(2w) = 157.169$ and 26.345, respectively. The D_5-symmetry axis is orthogonal to the plane of the figure for the top views (*top*) and vertical for the side views (*bottom*). The *left* and *right strips* are $(5,3)$ and $(5,2)$ torus knots, respectively

of inflection points. Solutions are alternatingly one- or two-sided. As the width is increased the strips become flatter (as illustrated by the bottom figures in side projection), tending to flat folded shapes with the inflection points at vertices of regular n-gons. One easily verifies that these limiting flat shapes have aspect ratio $L/(2w) = n \cot \frac{\pi}{n}$, $n = 3, 4, 5, \ldots$. The strip for $n = 3$, that we also discussed in the previous section, has self-contact before its width reaches this critical value, while for $n = 4$ self-contact coincides with the final shape, which is a double-covered square of side $2\sqrt{2}w$ (equal to the distance $V_i V_{i+1}$ between singularities on the edge of the strip) and centreline length $L = 8w$. Solutions for $n > 4$ do not have self-contacts. We cannot rule out that D_n-symmetric shapes with multiple central contact exist at higher aspect ratio, but we have not found such elastic strips in equilibrium.

Figure 23 shows a series of D_n-symmetric solutions with $2n$ ($n = 3, 4, 5, 6$) congruent pieces of type $S_0 I$ with an extra cylindrical point in the interior. These shapes therefore have $3n$ cylindrical points $2n$ of which generically do not lie on a C_2-symmetry axis (the number of inflection points remains n). In general, computed solutions may easily have self-intersections (or undergo self-crossings during parameter continuation, which affect the linking number), particularly at smaller aspect ratios, but all presented solutions are non-selfintersecting.

The final two shapes in Fig. 24, for $(m, n) = (2, 5)$ and $(3, 5)$, complement the set of D_5-symmetric solutions made of 10 pieces (the $(m, n) = (4, 5)$ and $(1, 5)$ solutions are shown in Figs. 22 and 23, respectively). Each has 5 inflection points and the left strip has 15 cylindrical points while the right one has only 5.

7 Discussion

We have formulated and numerically solved boundary-value problems for the isometric deformation of flat elastic sheets. Both one- and two-sided topologies have been considered. A common feature in all computed solutions is the existence of singularities of infinite bending energy density on the edge of the strip. Symmetry was used to construct solutions with multiple edge singularities from elementary pieces of the Möbius strip with a single singularity. This technique can be applied to find equilibrium shapes with any number of inflection points. Nonsymmetric solutions could be obtained by solving multiple boundary-value problems with different end conditions but with matching conditions at interior points. Our approach, of course, applies to open strips as well as to closed strips. In [28] we used a technique similar to the Möbius surgery discussed here to construct triangular buckling patterns of twisted strip in good agreement with experimental observations.

We used the developability property to reduce the energy integral to a single integral over the centreline thus obtaining a one-dimensional variational problem. The great advantage of this reduction is that the Euler–Lagrange equations for this problem are ODEs rather than the usual PDEs of plate theory. The relatively small price we pay is that the reduced variational problem is second-order but efficient equilibrium equations, Eqs. (16), (17), (21), (22), were derived in Euler–Poincaré form. A similar approach was taken in [53] to derive equilibrium equations for braided rods.

This reduction generalises to more general strips with prescribed boundaries in the intrinsic geometry, although it may be necessary to choose a non-straight centreline (reference curve) to reduce to $(\kappa_g \neq 0)$. In this case determining the bounds of the t-integration in Eq. (7) may require the solution of algebraic equations, which may render the reduction impractical. For simple geometries, however, for instance for an annular strip, reduction, at least in the narrow limit, is straigthforward [9].

Another generalisation is to strips that are developable but not necessarily planar in their relaxed state. Since we use the parametrisation of a developable surface in Eq. (1), the short edges of the strip must be generators in the deformed state. Subject to this constraint on the deformation, generalisations are possible (see Appendix A for more on this). For instance, for a strip whose relaxed surface belongs to a cylinder (i.e., $\eta = \text{const.}$), the reduction to the centreline can still be carried out (see [51] for an example). Extension to strips with anisotropic elasticity is straightforward.

The Möbius strip is usually treated as a topological or a geometrical object, not a mechanical one. One of the few previous attempts to compute equilibrium shapes of a Möbius strip was by Mahadevan & Keller [33] who employed a thin anisotropic elastic rod model. They obtained asymptotic equations for large values of the aspect ratio of the rod's cross-section. This limit corresponds to perfect alignment of the rod material frame and the Frenet–Serret frame, and the equilibrium equations are therefore the Euler–Lagrange equations for Lagrangian $\bar{l} = \kappa^2(1 + \frac{C}{B}\eta^2)$ in Eq. (12), where B and C are the bending and torsional stiffnesses, respectively. This corresponds to a strip formed by the binormal as it moves along the rod's centreline. The solution to those equations, however, although closed as a rod, does not describe a closed strip, even after the modifications made by the authors.

In particular, it does not satisfy the Randrup & Røgen conditions of Sect. 5.1, and therefore cannot serve as the centreline of a developable Möbius strip, not even a narrow one. It does not capture the essential geometrical properties of such a strip, for instance the fact that the torsion τ is zero in the inflection point (required for the generators to line up).

The geometrical features and accompanying stress localisation of Möbius and other developable strips observed here are seen more widely in problems of elastic sheets such as paper folding or crumpling and fabric draping, and were already noted in [34]. Crumpling of paper is dominated by bending along ridges bounding almost flat regions or facets [31, 58], behaviour that we see back in the nearly flat triangular regions in Fig. 5. In fabric draping, triangular regions are seen to form that radiate out from (approximate) vertices [60]. The formation of these flat triangular regions appears to be a generic feature of nature's response to twisting inextensible sheets. Analytical work on such sheets often *assumes* regions of localisation of bending energy in the form of vertices of conical surfaces [5, 6, 18]. It is known that conical surfaces have infinite elastic energy within the linear elastic theory. A cut-off near the cone apex was therefore introduced in [6] to obtain the equilibrium equations. More recently, local constraints on the surface metric have been incorporated into a variational framework to describe equilibria of generalised conical surfaces made of unstretchable flat sheets [18, 19]. As the examples in our present work show, by using non-conical developable elastic surfaces one can describe bending localisation phenomena without the need for a cut-off.

Importantly, our approach *predicts* the emergence of regions of high bending. Although we prescribe the number of inflection (switching) points of the solution we never impose any a priori constraints on the shape of the strip, neither in the vicinity of the inflection points nor elsewhere. Points of divergence of the bending energy may serve as indicators of positions where out-of-plane tearing (fracture failure mode III) is likely to be initiated. In this respect it is interesting to observe that when one tries to tear a piece of paper one intuitively applies a torsion, thereby creating intersecting creases as in the vertices of the central triangular domains in Fig. 5. A crack originates at the vertex, where the energy density diverges.

Curvature singularities are of interest to studies of twisted graphene-based nanoribbons. All-atom first-principle simulations of zigzag-edged graphene nanoribbons in the form of a Möbius strip show a shape evolution under varying aspect ratio that is in good agreement with our results in Fig. 5 [59]. Other simulation results reveal that the local curvature of nanoribbons increases at defect sites [4]. A seemingly related 'twist localisation' was observed in ab initio calculations of twisted cyclacenes [36].

Recent technological developments, such as nanostructured origami [1] and strain engineering of nanomembranes [30], have demonstrated the possibility of producing extremely thin membrane structures that can be folded in a controlled fashion. These structures have great potential as components of electromechanical devices such as force probes, capacitors, resonators, etc. with potentially unusual physical properties. For instance, silicon nanomembranes (SiNMs), consisting of the same material as bulk Si-based semiconductors, have been shown to become electrical conductors when the membrane is sufficiently thin. There is, therefore, significant interest in the relationship between geometry (and topology) and transport and optical properties such as electrical conductance and photoluminescence. In [27] it was shown that in thin conducting sheets electrons increasingly localise to the high-curvature regions as the sheet folds, with creases forming channels for electron transport. Our methods presented in this paper may be used to find how thin free-standing sheets fold and where creases form.

Appendix A: Energy Functional U for a Non-planar Relaxed Strip

Here we briefly consider the general case of a strip that is developable in its unstressed state but not planar. We rearrange Eq. (6) as

$$U = \frac{D}{2}\iint_{\Omega}\left\{\kappa_1^2 - 2\kappa_1\kappa_{1,0}\left[1+(\nu-1)\sin^2\Delta\beta\right]+\kappa_{1,0}^2\right\}\mathrm{d}\sigma. \tag{54}$$

The relaxed surface of the strip is determined by the centreline $\boldsymbol{r}_0$, which itself is fully specified (up to Euclidean motion) by its curvature $\kappa_0(s)$ and torsion $\tau_0(s)$ so that $\eta_0(s) = \tau_0(s)/\kappa_0(s)$, $\eta_0(s) = \cot\beta_0(s)$. Following the calculation in Sect. 2 we then have for the intrinsic principal curvature $\kappa_{1,0} = -\frac{\kappa_0(1+\eta_0^2)}{1+t\eta_0'}$. Since $\kappa_0(s)$ and torsion $\tau_0(s)$ are given functions, the last term in Eq. (54) integrates to a constant over the fixed domain, which can be dropped. For the remaining two terms we write $U = U_1 + U_2$. The first term, the integral of κ_1^2, gives the single integral of Eq. (8):

$$U_1 = Dw\int_0^L h\left(\kappa,\eta,\eta'\right)\mathrm{d}s.$$

The second term becomes

$$U_2 = -D\int_0^L\int_{-w}^{w}\left\{\frac{\kappa(1+\eta^2)\kappa_0(s_0(s,t))[1+\eta_0^2(s_0(s,t))]}{1+t\eta_0'(s_0(s,t))}\right.$$
$$\left.\times\left[1+(\nu-1)\sin^2\left(\beta-\beta_0\left(s_0(s,t)\right)\right)\right]\right\}\mathrm{d}t\,\mathrm{d}s, \tag{55}$$

where $s_0(s,t)$ solves $s_0 = s + t[\eta(s) - \eta_0(s_0)]$. To interpret this result, note that the strip is parametrised by two families of generators, one for the relaxed strip and one for the deformed strip. Through any point on the centreline of the deformed strip goes exactly one generator of each family. $s_0(s,t)$ is the arclength coordinate of a point on the centreline such that the relaxed generator passing through this point, at angle β_0, crosses the deformed generator passing through centreline point s, at angle β, at distance t from the centreline (see Fig. 25). Because of the property that generators within each family do not intersect inside the surface of the strip, $s_0(s,t)$ is a monotonic function of both s and t. $s_0(s,t)$ may lie outside the interval $[0, L]$, so we will assume that the relaxed state is given for a larger interval if required. Noting that $\sin^2(\beta-\beta_0) = \frac{(\eta-\eta_0)^2}{(1+\eta^2)(1+\eta_0^2)}$ we can write Eq. (55) as

$$U_2 = -D\int_0^L \kappa I_1(s,\eta)\,\mathrm{d}s,$$

where

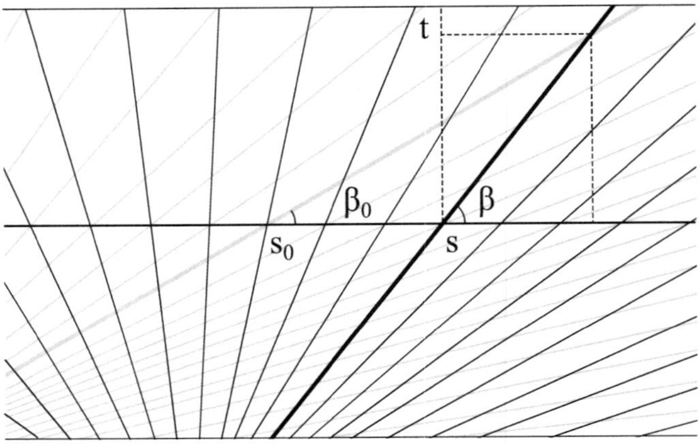

Fig. 25 Coordinates for an intrinsically non-planar strip. The generators for the relaxed state are shown in *grey*, those for the actual, deformed, state in *black*

$$I_1(s,\eta) = \int_{-w}^{w} \frac{\kappa_0(s_0(s,t))}{1+t\eta_0'(s_0(s,t))} \times \left\{\left[1+\eta(s)\eta_0\big(s_0(s,t)\big)\right]^2 + \nu\left[\eta(s)-\eta_0\big(s_0(s,t)\big)\right]^2\right\} \mathrm{d}t.$$

To make further progress we can replace the integration along the generator in I_1 by integration along the centreline, provided that $\eta(s_0(s,t)) \neq \eta(s)$ for all s. Then $(1+t\eta_0')\,\mathrm{d}s_0 = (\eta-\eta_0)\,\mathrm{d}t$ and

$$I_1(s,\eta) = \int_{s_0^-}^{s_0^+} \frac{\kappa_0(s_0)}{\eta(s)-\eta_0(s_0)} \left[\big(1+\eta(s)\eta_0(s_0)\big)^2 + \nu\big(\eta(s)-\eta_0(s_0)\big)^2\right] \mathrm{d}s_0,$$

where $s_0^\pm$ are solutions of the equation $s_0^\pm = s \pm w[\eta(s) - \eta_0(s_0^\pm)]$.

If $\eta = \eta_0$ in an isolated point then I_1 has a finite limit. To see this, let $\eta - \eta_0 = \epsilon$, $|\epsilon| \ll 1$, at, say, $s = 0$. Then we have $s_0 = t(\eta - \eta_0) = t\epsilon$ and $\lim_{\epsilon \to 0} I_1(0,\eta) = 2w\kappa_0(0)(1+\eta^2(0))^2$, which means that the singularity is removable. If $\eta = \eta_0$ on an interval then $s_0 = s$ on that interval and the t-integration in Eq. (55) can be performed immediately, as in U_1.

A.1 Two Special Cases

We consider two special cases where a reduction to a single integral over the centreline of the strip can be obtained explicitly.

- **Cylindrical strip**

  If the relaxed shape of the strip is cylindrical, i.e., $\eta_0 = \text{const}$, then I_1 can be simplified to

$$I_1(s,\eta) = \frac{1}{\eta-\eta_0}\left[(1+\eta_0\eta)^2 + \nu(\eta-\eta_0)^2\right] \int_{s_0^-}^{s_0^+} \kappa_0(s_0)\,\mathrm{d}s_0,$$

  where $s_0^\pm = s \pm w(\eta - \eta_0)$.

For a strip lying on a circular cylinder of constant radius R_0 and having a helical centreline with pitch angle $\pi/2 - \beta_0 = \text{const.}$, we have $\kappa_0 = \sin^2\beta_0/R_0 = 1/[R_0(1 + \eta_0^2)] = \text{const.}$ and

$$U_2 = -\frac{2D}{R_0}\int_0^L \kappa \frac{(1+\eta_0\eta)^2 + \nu(\eta-\eta_0)^2}{1+\eta_0^2}\,\mathrm{d}s$$

(cf. [51]).

- **Corrugated strip**

  Now let κ_0 be a prescribed periodic function, for instance $\kappa_0(s) = A\sin(\upsilon s)$, where the amplitude A and frequency υ are constant. Taking for simplicity $\eta_0 \equiv 0$, then

$$U_2 = -\frac{2DA}{\upsilon}\int_0^L \frac{\kappa\sin(\upsilon w\eta)}{\eta}\left(1+\nu\eta^2\right)\sin(\upsilon s)\,\mathrm{d}s.$$

We note that the integrand now explicitly depends on arclength s.

Appendix B: The Conical Limit

Here we explore the particular (singular) case in which the developable strip adopts a conical shape. The statics of conical surfaces has attracted extensive attention recently [5, 6, 18], especially in models for paper crumpling. Approximately conical pieces appear in our computed solutions for relatively wide strips. Our particular aim here is to describe the limiting tri-conical shape of the D_3-symmetric one-sided strip with $Lk = 3/2$ as the strip's aspect ratio is increased (see Fig. 19).

The limiting strip solution that we are interested in is assembled from six congruent conical pieces. It is sufficient to focus on finding the shape of one such piece for $s \in [0, w\sqrt{2}]$. The edge of regression for this conical piece degenerates into a point, the apex of the cone. The condition for all generators to go through this point, $\eta = \sqrt{2} - s/w$, gives an explicit formula for η (see Fig. 26). This η makes the logarithm in the bending energy density, Eq. (9), divergent. We regularise this energy by cutting off a small triangular domain near the apex of the cone. Thus we compute the bending energy by performing integration in Eq. (7) for $t \in [-w, (1-\epsilon)w]$, $0 < \epsilon \ll 1$ (we can justify this by appealing to the fact that the local curvature becomes too high near the apex for the assumption of an elastic material to remain valid [6]). Then V is a multiplicative constant, $V = -\frac{1}{2}\log\frac{\epsilon}{2}$, in Eq. (9) and therefore does not enter the equilibrium equations. To account for the conical $\eta(s)$ given above we replace the density $h(\kappa, \eta, \eta')$ in Eq. (8) with

$$\tilde{h}(\kappa, \eta) = \kappa^2\left(1+\eta^2\right)^2 + \lambda(s)\left(\eta - \sqrt{2} + s/w\right),$$

where λ is a Lagrange multiplier. Note that this is exactly Sadowsky's density but with prescribed η.

The equations for this density follow from the procedure described in Sects. 4.1 and 4.2. The constitutive Eq. (21) allows us to express

$$\kappa = \frac{\eta M_1 + M_3}{2(1+\eta^2)^2}, \tag{56}$$

which we substitute into the balance equations to obtain a closed system of six non-autonomous equations, Eqs. (16), (17), for the six components $(F_1, F_2, F_3, M_1, M_2, M_3)$.

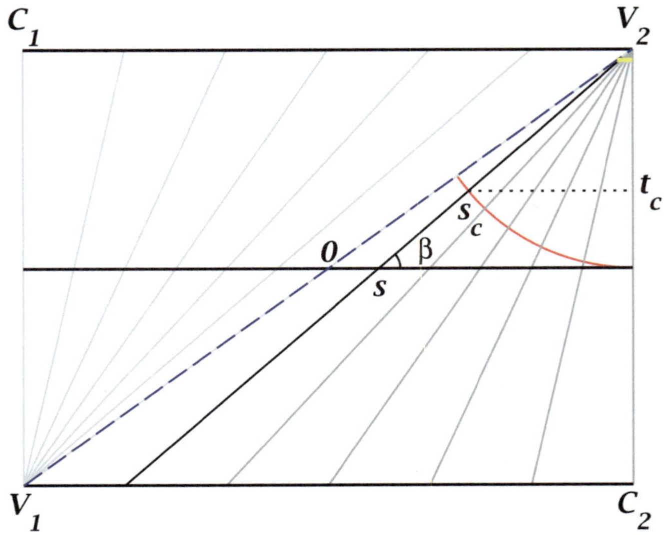

Fig. 26 Development of the conical surface. $|C_1V_1| = |C_2V_2| = 2w$, $|V_1V_2| = 2w\sqrt{3}$. Arclength s is measured along the straight centreline from 0 to the right; s_c is measured along the circular arc of radius w (*red*). The cut-off is indicated by the *yellow line*

We can then formulate a boundary-value problem as in Sect. 6.1 on the interval $[0, w\sqrt{2}]$ to find the forces, moments and the curvature of the centreline and reconstruct the surface of the strip. There is however one difference and that is that for the η as prescribed we cannot satisfy the boundary condition $\eta'(0) = 0$. This means that the resulting assembled strip has only a continuous (and generally not a differentiable) generator at $s = 0$. All other boundary conditions carry over to the present limiting situation. In particular, the condition $\kappa(w\sqrt{2}) = 0$ follows from Eq. (56), the fact that $\eta(w\sqrt{2}) = 0$ and the other condition $M_3(w\sqrt{2}) = 0$. The remaining constitutive Eq. (22) may be used to obtain the Lagrange multiplier $\lambda(s)$. However, instead of pursuing this formulation it is instructive for this conical problem to use a different formulation as we discuss next.

The bending energy of the cone has a particularly simple form if we change to a different reference curve, namely the circular arc of radius w centred at the apex of the cone (see Fig. 26). Let s_c be the arclength coordinate along this arc. It is easy to see that the generators induce a one-to-one mapping between the strip's centreline and the circular arc, $s \to s_c$ for $s \in [0, w\sqrt{2}]$, $s_c \in [0, s_{ce}]$, $s_{ce} = w \arctan\sqrt{2}$, given by $\sqrt{2} - \frac{s}{w} = \tan\frac{s_{ce}-s_c}{w} = \tan(\arctan(\sqrt{2}) - \frac{s_c}{w})$ and $\mathrm{d}s = (1 + \eta^2)\,\mathrm{d}s_c$. Let κ_N be the normal curvature of the new reference curve. It equals the nonzero principal curvature at the same point, which we can compute from Eq. (2) by noting that t is now given by the value t_c at the intersection of the local generator with the circular arc (see Fig. 26). We have $t_c = w(1 - \sin\beta) = w(1 - 1/\sqrt{1+\eta^2})$ and hence $\kappa_N = -\kappa(1 + \eta^2)^{3/2}$. We can then express the bending energy as

$$\int_0^{w\sqrt{2}} \kappa(s)^2 \big(1 + \eta(s)^2\big)^2 \,\mathrm{d}s = \int_0^{s_{ce}} \kappa_N^2(s_c)\,\mathrm{d}s_c.$$

This simple form suggests that the equation governing the normal curvature is that of the Euler elastica, as observed previously in [6]. To demonstrate this we introduce the Darboux frame $\{\boldsymbol{T}, \boldsymbol{N}, \boldsymbol{U}\}$ at the reference arc with the unit tangent $\boldsymbol{T}$, normal to the surface $\boldsymbol{N}$ and

$\boldsymbol{U} = \boldsymbol{T} \times \boldsymbol{N}$. Then we define $R_c(s_c) := (\boldsymbol{T}, \boldsymbol{N}, \boldsymbol{U}) \in SO(3)$ and $\widehat{\omega}_c = R_c^{\mathrm{T}} \frac{\mathrm{d}R_c}{\mathrm{d}s_c} \in \mathfrak{so}(3)$, so that $\omega_{c1} = \tau_g$, $\omega_{c2} = \kappa_g$, $\omega_{c3} = \kappa_N$ (we omit the subindex 'c' in τ_g, κ_g and κ_N). Since our circular arc lies on the conical surface with tangent orthogonal to the generator everywhere, we have $\tau_g \equiv 0$ and $\kappa_g \equiv 1/w$. Then we define the reduced Lagrangian as

$$\mathcal{L}_c = \omega_{c3}^2 + M_{c1}\omega_{c1} + M_{c2}\left(\omega_{c2} - \frac{1}{u}\right) - F_{c1},$$

where $M_{c1}(s_c)$ and $M_{c2}(s_c)$ are Lagrange multipliers. Applying again the procedure described in Sect. 4.1, we obtain the six balance equations for $(F_{c1}, F_{c2}, F_{c3}, M_{c1}, M_{c2}, M_{c3})$ and the constitutive equation

$$M_{c3} = 2\kappa_N. \tag{57}$$

For completeness we give the relationship between the system of forces and moments at the centreline and the equivalent transferred system of forces and moments at the circular arc:

$$\begin{aligned}
F_{c1} &= \frac{1}{\sqrt{1+\eta^2}}(F_1 - \eta F_3),\\
F_{c2} &= -F_2,\\
F_{c3} &= -\frac{1}{\sqrt{1+\eta^2}}(\eta F_1 + F_3),\\
M_{c1} &= \frac{1}{\sqrt{1+\eta^2}}(M_1 - \eta M_3) + w\big(\sqrt{1+\eta^2} - 1\big)F_2,\\
M_{c2} &= -M_2 + w\left(1 - \frac{1}{\sqrt{1+\eta^2}}\right)(F_1 - \eta F_3),\\
M_{c3} &= -\frac{1}{\sqrt{1+\eta^2}}(\eta M_1 + M_3).
\end{aligned} \tag{58}$$

We can also write the Hamiltonian corresponding to the Lagrangian $\mathcal{L}_c$:

$$\mathcal{H}_c = \kappa_N^2 + \frac{M_{c2}}{w} + F_{c1}, \tag{59}$$

which is a conserved quantity. (By using Eqs. (58) this Hamiltonian when expressed in terms of centreline variables becomes $\mathcal{H}_c = \kappa^2(1+\eta^2)^3 - \frac{M_2}{w} + F_1 - \eta F_3$. That this is a conserved quantity, despite the fact that η is a given function of s, can also be verified by differentiation and using the balance equations for the centreline forces and moments.)

From the first and last moment balance equations we find $M_{c2} = \frac{1}{\kappa_N}\frac{\mathrm{d}M_{c1}}{\mathrm{d}s_c} + \frac{2}{w}$ and $F_{c2} = \frac{M_{c1}}{w} - 2\frac{\mathrm{d}\kappa_N}{\mathrm{d}s_c}$. Differentiating the latter and substituting the result together with F_{c1} obtained from Eq. (59) into the second force balance equation we arrive at

$$\frac{\mathrm{d}^2\kappa_N}{\mathrm{d}s_c^2} + \frac{\kappa_N}{2}\left(\kappa_N^2 - H_c + \frac{2}{w^2}\right) = 0, \tag{60}$$

where H_c is the value of the Hamiltonian $\mathcal{H}_c$, as yet undetermined. This is the curvature equation for the planar Euler elastica, which has well-known solutions in terms of Jacobi elliptic functions.

To reconstruct the surface, we first parametrise the vectors with three Euler angles ψ_c, ϑ_c, φ_c:

$$\boldsymbol{T} = \begin{pmatrix} -\sin\psi_c \sin\varphi_c + \cos\vartheta_c \cos\psi_c \cos\varphi_c \\ \cos\psi_c \sin\varphi_c + \cos\vartheta_c \sin\psi_c \cos\varphi_c \\ -\sin\vartheta_c \cos\varphi_c \end{pmatrix},$$

$$\boldsymbol{N} = \begin{pmatrix} -\sin\psi_c \cos\varphi_c - \cos\vartheta_c \cos\psi_c \sin\varphi_c \\ \cos\psi_c \cos\varphi_c - \cos\vartheta_c \sin\psi_c \sin\varphi_c \\ \sin\vartheta_c \sin\varphi_c \end{pmatrix}, \qquad \boldsymbol{U} = \begin{pmatrix} \sin\vartheta_c \cos\psi_c \\ \sin\vartheta_c \sin\psi_c \\ \cos\vartheta_c \end{pmatrix}.$$

The Euler angles are related to the Darboux vector by the kinematics equations

$$\begin{aligned} \frac{\mathrm{d}\psi_c}{\mathrm{d}s_c} &= -\frac{1}{w}\sin\varphi_c \csc\vartheta_c, \\ \frac{\mathrm{d}\vartheta_c}{\mathrm{d}s_c} &= -\frac{1}{w}\cos\varphi_c, \\ \frac{\mathrm{d}\varphi_c}{\mathrm{d}s_c} &= \kappa_N + \frac{1}{w}\sin\varphi_c \cot\vartheta_c. \end{aligned} \tag{61}$$

Note that the angle ϑ_c should not approach 0 or π. Finally, to find the reference arc $\boldsymbol{r}_c = (x, y, z)^\mathsf{T}$ in three-dimensional space we solve Eq. (61) in conjunction with the equation $\frac{\mathrm{d}\boldsymbol{r}_c}{\mathrm{d}s_c} = \boldsymbol{T}$.

To compute limiting tri-conical shapes of our one-sided strip we solve the elastica equation (60) together with Eq. (61) on the interval $[0, s_{ce}]$ subject to appropriate boundary conditions. We choose an orientation such that the tangent $\boldsymbol{T}(0)$ is aligned with the z-axis and the vectors $\boldsymbol{U}(0)$ and $\boldsymbol{U}(s_{ce})$ span the xy-plane, making an angle of $\pi/6$ (cf. Eq. (52)). The xy-plane will then be a symmetry plane for the full strip. We also require the normal curvature to vanish at s_{ce}. All this translates into the boundary conditions

$$\begin{aligned} & & \kappa_N(s_{ce}) &= 0, \\ \psi_c(0) &= -\pi/6, & \psi_c(s_{ce}) &= 0, \\ \vartheta_c(0) &= \pi/2, & \vartheta_c(s_{ce}) &= \pi/2, \\ \varphi_c(0) &= 0, & & \end{aligned}$$

i.e., six boundary conditions for a fifth-order system of equations and one free parameter H_c. Since the parameter w can be trivially scaled out the solution to this boundary-value problem yields a universal tri-conical shape for our strip. Numerical solution gives $H_c w^2 = -1.830652$ and the graphs for the normal curvature κ_N on the circular arc and the curvature κ of the centreline shown in Fig. 27. Figure 28 illustrates the assembly of the entire shape from a single piece in the same manner as described in Sect. 6.3. The resulting shape is barely distinguishable from the shape in Fig. 19 (Fig. 20 compares η for the two solutions). Note that while the normal to the surface is continuous everywhere, the curvature is discontinuous at the generators connecting the apices.

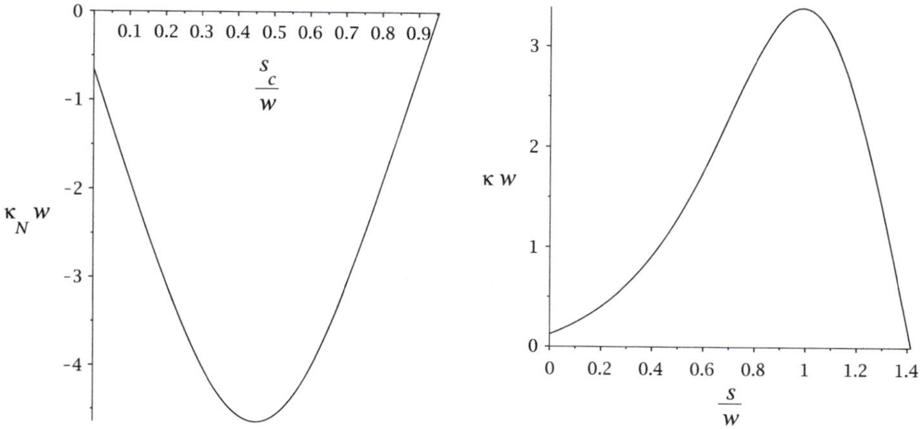

Fig. 27 *Left*: Normal curvature κ_N along the circular reference arc for a sixth of the limiting tri-conical shape of the strip. s_c/w measures normalised length along the arc. *Right*: Curvature κ along the centreline of the conical piece. s/w measures normalised length along the centreline

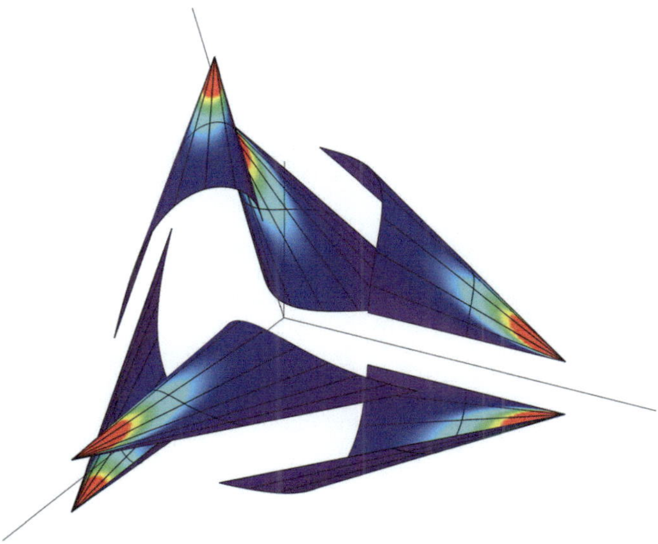

Fig. 28 Assembling the limiting D_3-symmetric one-sided strip from six congruent conical pieces

The shape of the conical strip has been found without simultaneously solving for the forces and moments whose equations are semi-decoupled from the elastica equation (60). Once κ_N has been found the force and moment equations form a system of six linear equations for the components F_{c1}, F_{c2}, F_{c3}, M_{c1}, M_{c2} and M_{c3}. We can solve these equations subject to our standard force and moment conditions, Eqs. (40) and (41), translated into conditions for the above components by using the relationships (58). These four conditions, together with the single constitutive equation (57) and the Hamiltonian (59), both of which are fixed by the geometry, yield a unique solution.

References

1. Arora, W.J., Nichol, A.J., Smith, H.I., Barbastathis, G.: Membrane folding to achieve three-dimensional nanostructures: nanopatterned silicon nitride folded with stressed chromium hinges. Appl. Phys. Lett. **88**(5), 053108 (2006). doi:10.1063/1.2168516
2. Ashwell, D.G.: The inextensional twisting of a rectangular plate. Q. J. Mech. Appl. Math. **15**(1), 91–107 (1962). doi:10.1093/qjmam/15.1.91
3. Barr, S.: Experiments in Topology. Thomas Y. Crowell Company, New York (1964)
4. Caetano, E.W.S., Freire, V.N., dos Santos, S.G., Albuquerque, E.L., Galvão, D.S., Sato, F.: Defects in graphene-based twisted nanoribbons: structural, electronic, and optical properties. Langmuir **25**(8), 4751–4759 (2009). doi:10.1021/la803929f
5. Cerda, E., Chaieb, S., Melo, F., Mahadevan, L.: Conical dislocations in crumpling. Nature **401**, 46–49 (1999). doi:10.1038/43395
6. Cerda, E., Mahadevan, L., Pasini, J.M.: The elements of draping. Proc. Natl. Acad. Sci. USA **101**(7), 1806–1810 (2004). doi:10.1073/pnas.0307160101
7. Chicone, C., Kalton, N.J.: Flat embeddings of the Möbius strip in R^3. Commun. Appl. Nonlinear Anal. **9**, 31–50 (2002)
8. Chubelaschwili, D., Pinkall, U.: Elastic strips. Manuscr. Math. **133**(3–4), 307–326 (2010). doi:10.1007/s00229-010-0369-x
9. Dias, M.A., Dudte, L.H., Mahadevan, L., Santangelo, C.D.: Geometric mechanics of curved crease origami. Phys. Rev. Lett. **109**, 114301 (2012). doi:10.1103/PhysRevLett.109.114301
10. Doedel, E., et al.: AUTO: software for continuation and bifurcation problems in ordinary differential equations. In: User's Manual, Concordia University, Montreal, Canada (2007). http://indy.cs.concordia.ca/auto/
11. Emmer, M.: Visual art and mathematics: the Moebius band. Leonardo **13**(2), 108–111 (1980)
12. Friesecke, G., James, R.D., Mora, M.G., Müller, S.: Derivation of nonlinear bending theory for shells from three-dimensional nonlinear elasticity by gamma-convergence. C. R. Math. **336**, 697–702 (2003). doi:10.1016/S1631-073X(03)00028-1
13. Fuchs, D., Tabachnikov, S.: Mathematical Omnibus: Thirty Lectures on Classic Mathematics. American Mathematical Society, Providence (2007)
14. Fuller, F.B.: The writhing number of a space curve. Proc. Natl. Acad. Sci. USA **68**(4), 815–819 (1971)
15. Fuller, F.B.: Decomposition of the linking number of a closed ribbon: a problem from molecular biology. Proc. Natl. Acad. Sci. USA **75**(8), 3557–3561 (1978)
16. Gay-Balmaz, F., Holm, D.D., Meier, D.M., Ratiu, T.S., Vialard, F.X.: Invariant higher-order variational problems. Commun. Math. Phys. **309**, 413–458 (2012). doi:10.1007/s00220-011-1313-y
17. Gravesen, J., Willatzen, M.: Eigenstates of Möbius nanostructures including curvature effects. Phys. Rev. A **72**, 032108 (2005). doi:10.1103/PhysRevA.72.032108
18. Guven, J., Müller, M.M.: How paper folds: bending with local constraints. J. Phys. A, Math. Theor. **41**(5), 055203 (2008). doi:10.1088/1751-8113/41/5/055203
19. Guven, J., Müller, M.M., Vázquez-Montejo, P.: Conical instabilities on paper. J. Phys. A, Math. Theor. **45**(1), 015203 (2012). doi:10.1088/1751-8113/45/1/015203
20. Halpern, B., Weaver, C.: Inverting a cylinder through isometric immersions and isometric embeddings. Trans. Am. Math. Soc. **230**, 41–70 (1977). http://links.jstor.org/sici?sici=0002-9947%28197706%29230%3C41%3AIACTII%3E2.0.C0%3b2-I
21. Hangan, T.: Elastic strips and differential geometry. Rend. Semin. Mat. (Torino) **63**(2), 179–186 (2005)
22. Hangan, T., Murea, C.: Elastic helices. Rev. Roum. Math. Pures Appl. **50**(5-6), 641–645 (2005)
23. Hayashi, M., Ebisawa, H.: Little-parks oscillation of superconducting Möbius strip. J. Phys. Soc. Jpn. **70**(12), 3495–3498 (2001)
24. Hornung, P.: Euler–Lagrange equations for variational problems on space curves. Phys. Rev. E **81**(6), 066603 (2010). doi:10.1103/PhysRevE.81.066603
25. Hornung, P.: Euler–Lagrange equation and regularity for flat minimizers of the Willmore functional. Commun. Pure Appl. Math. **64**(3), 367–441 (2011). doi:10.1002/cpa.20342
26. Kirby, N., Fried, E.: Gamma-limit of a model for the elastic energy of an inextensible ribbon (2013). arXiv:1307.3540 [math.AP]. 10 p.
27. Korte, A.P., van der Heijden, G.H.M.: Curvature-induced electron localization in developable Möbius-like nanostructures. J. Phys. Condens. Matter **21**(49), 495301 (2009). doi:10.1088/0953-8984/21/49/495301
28. Korte, A.P., Starostin, E.L., van der Heijden, G.H.M.: Triangular buckling patterns of twisted inextensible strips. Proc. R. Soc. Lond. Ser. A **467**(2125), 285–303 (2010). doi:10.1098/rspa.2010.0200
29. Kurono, Y., Umehara, M.: Flat Möbius strips of given isotopy type in $\mathbb{R}^3$ whose centerlines are geodesics or lines of curvature. Geom. Dedic. **134**(1), 109–130 (2008). doi:10.1007/s10711-008-9248-y

30. Lagally, M.G.: Strain engineered silicon nanomembranes. J. Phys. Conf. Ser. **61**(1), 652–657 (2007). doi:10.1088/1742-6596/61/1/131
31. Lobkovsky, A., Gentges, S., Li, H., Morse, D., Witten, T.A.: Scaling properties of stretching ridges in a crumpled elastic sheet. Science **270**(5241), 1482–1485 (1995). doi:10.1126/science.270.5241.1482
32. Love, A.E.H.: A Treatise on the Mathematical Theory of Elasticity Dover, New York (1944), reprinted. Cambridge University Press (1927)
33. Mahadevan, L., Keller, J.B.: The shape of a Möbius band. Proc. R. Soc. Lond. Ser. A **440**, 149–162 (1993)
34. Mansfield, E.H.: The inextensional theory for thin flat plates. Q. J. Mech. Appl. Math. **8**(3), 338–352 (1955). doi:10.1093/qjmam/8.3.338
35. Mansfield, E.H.: The Bending and Stretching of Plates, 2nd edn. Cambridge University Press, Cambridge (1989)
36. Martín-Santamaría, S., Rzepa, H.S.: Twist localisation in single, double and triple twisted Möbius cyclacenes. J. Chem. Soc., Perkin Trans. 2, pp. 2378–2381 (2000). doi:10.1039/B005560N
37. Naokawa, K.: Singularities of the asymptotic completion of developable Möbius strips. Osaka J. Math. **50**(2), 425–437 (2013)
38. Neff, P.: A geometrically exact Cosserat shell-model including size effects, avoiding degeneracy in the thin shell limit. Part I: Formal dimensional reduction for elastic plates and existence of minimizers for positive Cosserat couple modulus. Contin. Mech. Thermodyn. **16**(6), 577–628 (2004). doi:10.1007/s00161-004-0182-4
39. Randrup, T., Røgen, P.: Sides of the Möbius strip. Arch. Math. **66**(6), 511–521 (1996). doi:10.1007/BF01268871
40. Røgen, P.: Embedding and knotting of flat compact surfaces in 3-space. Comment. Math. Helv. **76**, 589–606 (2001)
41. Rohde, U.L., Poddar, A.K., Sundararajan, D.: Printed resonators: Möbius strip theory and applications. Microw. J. **56**(11), 24 (2013)
42. Sabitov, I.K.: Isometric immersions and embeddings of a flat Möbius strip in Euclidean spaces. Izv. Math. **71**(5), 1049–1078 (2007). doi:10.1070/IM2007v071n05ABEH002376
43. Sabitov, I.K.: On the developable ruled surfaces of low smoothness. Sib. Math. J. **50**(5), 919–928 (2009). doi:10.1007/s11202-009-0102-8
44. Sadowsky, M.: Ein elementarer Beweis für die Existenz eines abwickelbaren Möbiusschen Bandes und Zurückführung des geometrischen Problems auf ein Variationsproblem. Sitzungsber. K. Preuss. Akad. Wiss. Berl. **22**, 412–415 (1930)
45. Sadowsky, M.: Theorie der elastisch biegsamen undehnbaren Bänder mit Anwendungen auf das Möbius'sche Band. In: A.C.W. Oseen, W. Weibull (eds.) Verhandl. des 3. Intern. Kongr. f. Techn. Mechanik, 1930, Teil II, pp. 444–451. AB Sveriges Litografiska Tryckerier (1931)
46. Satija, I.I., Balakrishnan, R.: Geometric phases in twisted strips. Phys. Lett. A **373**(39), 3582–3585 (2009). doi:10.1016/j.physleta.2009.07.083
47. Schwarz, G.: A pretender to the title "canonical Moebius strip". Pac. J. Math. **143**(1), 195–200 (1990). http://projecteuclid.org/Dienst/UI/1.0/Summarize/euclid.pjm/1102646207
48. Schwarz, G.E.: The dark side of the Moebius strip. Am. Math. Mon. **97**(10), 890–897 (1990). http://links.jstor.org/sici?sici=0002-9890%28199012%2997%3A10%3C890%3ATDS0TM%3E2.0.C0%3B2-H
49. Spivak, M.: A Comprehensive Introduction to Differential Geometry, vol. 3, 3 edn. (1999). Publish or Perish
50. Starostin, E.L., van der Heijden, G.H.M.: The shape of a Möbius strip. Nat. Mater. **6**(8), 563–567 (2007). doi:10.1038/nmat1929
51. Starostin, E.L., van der Heijden, G.H.M.: Tension-induced multistability in inextensible helical ribbons. Phys. Rev. Lett. **101**(8), 084301 (2008). doi:10.1103/PhysRevLett.101.084301
52. Starostin, E.L., van der Heijden, G.H.M.: Force and moment balance equations for geometric variational problems on curves. Phys. Rev. E **79**(6), 066602 (2009). doi:10.1103/PhysRevE.79.066602
53. Starostin, E.L., van der Heijden, G.H.M.: Theory of equilibria of elastic 2-braids with interstrand interaction. J. Mech. Phys. Solids **64**, 83–132 (2014). doi:10.1016/j.jmps.2013.10.014
54. Stasiak, A., Katritch, V., Kauffman, L.H. (eds.): Ideal Knots, Series on Knots and Everything, vol. 19. World Scientific, Singapore (1998)
55. Tanda, S., Tsuneta, T., Okajima, Y., Inagaki, K., Yamaya, K., Hatakenaka, N.: Crystal topology: a Möbius strip of single crystals. Nature **417**(6887), 397–398 (2002). doi:10.1038/417397a
56. Tanda, S., Tsuneta, T., Toshima, T., Matsuura, T., Tsubota, M.: Topological crystals. J. Phys. IV **131**, 289–294 (2005). doi:10.1051/jp4:2005131074
57. Ushakov, V.: Parameterisation of developable surfaces by asymptotic lines. Bull. Aust. Math. Soc. **54**(3), 411–421 (1996). doi:10.1017/S0004972700021821

58. Vliegenthart, G.A., Gompper, G.: Forced crumpling of self-avoiding elastic sheets. Nat. Mater. **5**(3), 216–221 (2006). doi:10.1038/nmat1581
59. Wang, X., Zheng, X., Ni, M., Zou, L., Zeng, Z.: Theoretical investigation of Möbius strips formed from graphene. Appl. Phys. Lett. **97**(12), 123103 (2010). doi:10.1063/1.3489982
60. Witten, T.A.: Stress focusing in elastic sheets. Rev. Mod. Phys. **79**, 643–675 (2007). doi:10.1103/RevModPhys.79.643
61. Wunderlich, W.: Über ein abwickelbares Möbiusband. Monatshefte Math. **66**(3), 276–289 (1962). doi:10.1007/BF01299052
62. Yakubo, K., Avishai, Y., Cohen, D.: Persistent currents in Möbius strips. Phys. Rev. B **67**, 125319 (2003). doi:10.1103/PhysRevB.67.125319
63. Zorin, D.: Curvature-based energy for simulation and variational modeling. In: Proceedings of the International Conference on Shape Modeling and Applications SMI'05, pp. 198–206. IEEE Comput. Soc., Los Alamitos (2005). doi:10.1109/SMI.2005.14

DOI 10.1007/978-94-017-7300-3_9
Reprinted from *Journal of Elasticity* Journal, DOI 10.1007/s10659-014-9501-6

Bending Paper and the Möbius Strip

Sören Bartels · Peter Hornung

Received: 20 February 2014 / Published online: 18 September 2014

Abstract We present some rigorous results about the bending behaviour of paper. By adapting these results to the Möbius strip, we obtain some qualitative properties of developable Möbius strips which minimize the bending energy. We also provide some numerical simulations which illustrate and strengthen the analytic results.

Keywords Nonlinear elasticity · Plate theory · Mobius strip · Isometric immersions

Mathematics Subject Classification 74K20 · 74B20 · 49Q10 · 65N30 · 53A05

1 Introduction

Consider a flat strip of paper in the plane. It can be modelled by a two dimensional rectangle $S \subset \mathbb{R}^2$ times an interval $(-h/2, h/2)$ of length h, which is the thickness of the sheet of paper. If that sheet is deformed via a deformation $z : S \times (-h/2, h/2) \to \mathbb{R}^3$, then the (scaled) elastic bending energy stored in the deformed configuration $z(S)$ is given by

$$h^{-3} \int_{S\times(-h/2,h/2)} W(\nabla z), \tag{1}$$

where W is some energy density, which for simplicity we assume to be proportional to $\mathrm{dist}^2_{SO(3)}$.

As shown in [10], deformations z^h for which (1) remains asymptotically bounded for $h \ll 1$ have zero membrane energy (asymptotically). More precisely, their asymptotic be-

S. Bartels
Abteilung für Angewandte Mathematik, Albert-Ludwigs-Universität Freiburg, Freiburg, Germany

P. Hornung (✉)
Institut für Angewandte Mathematik, Universität Bonn, Bonn, Germany
e-mail: hornung@iam.uni-bonn.de

haviour is fully captured by the asymptotic midplane deformation $u : S \to \mathbb{R}^3$, which necessarily has zero membrane energy, i.e., it satisfies on S the PDE system

$$\partial_i u \cdot \partial_j u = \delta_{ij}, \tag{2}$$

where

$$\delta_{ij} = \begin{cases} 1 & \text{if } i = j \\ 0 & \text{otherwise.} \end{cases}$$

Maps $u : S \to \mathbb{R}^3$ satisfying (2) on S will be called *bendings* of S. More precisely, the asymptotic midplane deformation u must be a $W^{2,2}$ bending, i.e., a solution to (2) belonging to the Sobolev space $W^{2,2}(S, \mathbb{R}^3)$.

The behaviour of (1) for $h \to 0$ is determined by Kirchhoff's plate functional, which is the following energy functional defined on midplane deformations $u : S \to \mathbb{R}^3$:

$$\mathcal{E}(u) = \begin{cases} \int_S Q(A) & \text{if } u \text{ is a } W^{2,2} \text{ bending of } S, \\ +\infty & \text{otherwise,} \end{cases} \tag{3}$$

where Q is closely related to the quadratic form of linearized elasticity, i.e., the Hessian of W at the identity, and A denotes the second fundamental form of u, i.e., $A_{ij} = n \cdot \partial_j \partial_i u$ where $n = \partial_1 u \wedge \partial_2 u$ is the normal to u. From now on we assume that $Q = |\cdot|^2$, which (up to a constant prefactor) corresponds to the situation when the material is isotropic.

In Sects. 2 and 3 we report some results about $W^{2,2}$ bendings, such as their rigidity properties. Our main focus is on the possibility of approximating a given deformation u with finite energy (3) (i.e. a $W^{2,2}$ bending) by bendings enjoying some nice structural properties; in particular, by smooth ones. Such approximation results are key ingredients in the rigorous derivation (in terms of Γ-convergence) of lower dimensional theories from fully nonlinear three dimensional elasticity, e.g., the derivation of (variants of) (3) from (variants of) (1). They also turn out to be essential in convergence proofs for numerical schemes, cf. Sect. 6 for details. We also present a general way to produce examples of ill-behaved $W^{2,2}$ bendings.

In Sect. 4 we report some rigorous results studying the variational problem (4), i.e., about energy minimizing bendings of arbitrary reference domains. In Sect. 5 we adapt them to the developable Möbius strip of finite width.

In Sect. 6 we present some numerical simulations. We compare our analytical results with these and other numerical simulations and we contrast them with some observations in [28], which were based on an intuitive extrapolation from the infinitesimally narrow strip to the strip with finite width.

Our main motivation for Sects. 5 and 6 is a programme outlined by Sadowsky in [27]. In that article, Sadowsky constructed a developable Möbius strip. This strip consists of pieces of analytic surfaces. As explained in his paper, the ideal shape of the developable Möbius strip is the one which minimizes the energy functional

$$\int H^2 \to \min!, \tag{4}$$

with H the mean curvature and the integral being taken over the whole surface, among all bendings of a fixed reference configuration whose deformed configuration is a Möbius strip. Sadowsky then proceeds to study the infinitesimally narrow Möbius strip. In [28] he derives the corresponding Euler-Lagrange equations.

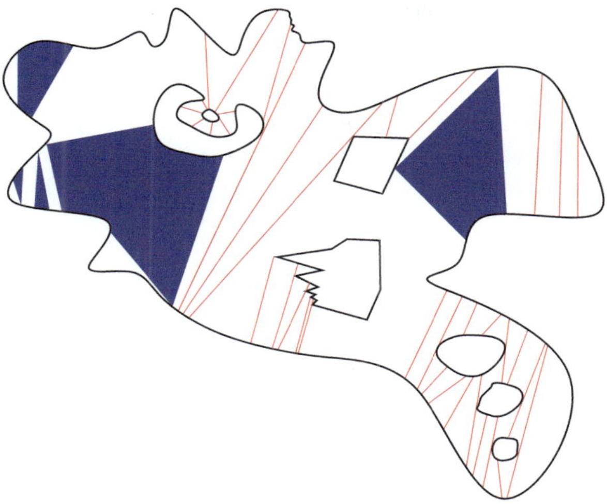

Fig. 1 Illustration of Theorem 2.1: The *blue regions* are the connected components of $C_{\nabla u}$, segments $[x]$ are drawn in *red*. The domain S is nonconvex and is not simply connected

2 Deformation Gradient of a Bending: Countable Developability

The following result from [25] is a variant of a result by Kirchheim [19], who only considered the case of $W^{2,\infty}$ bendings.

Theorem 2.1 *If u is a $W^{2,2}$ bending of S, then $u \in C^1(S, \mathbb{R}^3)$ and the deformation gradient ∇u is 'developable' outside the set*

$$C_{\nabla u} = \{x \in S : \nabla u \text{ is constant in a neighbourhood of } x\}$$

of local constancy of ∇u. This means that for each $x \in S \setminus C_{\nabla u}$ there exists a unique line segment $[x] \subset S$ such that

- *The segment $[x]$ contains the point x, and both endpoints of $[x]$ lie on ∂S.*
- *The deformation gradient ∇u is constant on $[x]$.*

Note that, by uniqueness, two segments $[x]$ and $[y]$ are either identical or disjoint.

Theorem 2.1 can be reduced to the following result by Pogorelov [26, Sect. IX.4, Theorem 1]:

Theorem 2.2 *Let u be a C^1 bending of S such that the spherical image of $u(S)$ (i.e., the image $n(S)$ of S under the normal n to u) has zero area. Then the conclusions of Theorem 2.1 are satisfied.*

The proof in [19] does not make any use of Theorem 2.2, and it is very different from the proof of Theorem 2.2 given in [26]. For simplicity, here we only show how to reduce the proof of Theorem 2.1 to Theorem 2.2.

Proof of Theorem 2.1 using Theorem 2.2 Following the first step in Kirchheim's proof, we will sketch why any $W^{2,2}$ bending u is in fact continuously differentiable in S. (This is false for generic maps in $W^{2,2}(S, \mathbb{R}^3)$ which do not satisfy the isometry constraint.)

The proof is readily reduced to classical facts about maps with positive Jacobian: For any component u_i define $f_\varepsilon : S \to \mathbb{R}^2$ by setting $f_\varepsilon(x) = \nabla u_i(x) + \varepsilon x^\perp$. Then $\det \nabla f_\varepsilon = \varepsilon^2$ because $\nabla^2 u_i$ is symmetric and because $\det \nabla^2 u_i = 0$. This last fact is obvious for smooth bendings, and for $W^{2,2}$ bendings it is proven, e.g., in [11, Proposition 6].

Hence f_ε is in the critical Sobolev space $W^{1,2}$ and has positive Jacobian. By classical results, this implies that f_ε is continuous. Hence ∇u is a uniform limit of continuous maps and therefore it is continuous, too.

As a consequence, any $W^{2,2}$ bending u is a C^1 bending. And the spherical image of u has zero area. The latter follows from the change of variables formula. Hence we can apply Theorem 2.2. □

In Theorem 2.2 one cannot drop the hypothesis on the area of $n(S)$: the well-known convex integration solutions by Nash [24] and Kuiper [20] (cf. also [12]) show that generic C^1 bendings dramatically fail to be rigid. More recent results in this direction seek to find the optimal $\alpha \in (0, 1)$ for which $C^{1,\alpha}$ bendings (of convex shells) are rigid, see [8] and also [6]. For generalizations of Theorem 2.1 in a different direction we refer to [18].

The classical version of Theorem 2.1 deals with C^2 bendings, see [13] and [21]. Their proofs are very different from that of Theorem 2.1 in [19, 25]. However, the conclusion about the level set structure of the deformation gradient is similar for $W^{2,2}$ bendings and for C^2 bendings.

In [23] the authors show that the interior C^1 regularity proven above in fact extends up to the boundary of S when the latter is $C^{1,\alpha}$ for some $\alpha > 0$. This result relies heavily on the developable structure.

We conclude this section by pointing out that the open set $C_{\nabla u}$ consists of countably many connected components U. And each of these components has a very special structure. Namely, its relative boundary $S \cap \partial U$ consists entirely of straight segments $[x]$ as in Theorem 2.1 on which ∇u is constant. We refer to [17, Sect. 3] for precise statements and proofs. In what follows, by a *boundary segment* of a connected component U of $C_{\nabla u}$ we mean a connected component of $S \cap \partial U$. This is always a segment $[x]$ in S as in Theorem 2.1, so it has both endpoints on ∂S and ∇u is constant on $[x]$. We also refer to [13] for some earlier observations regarding the geometry of connected components of $C_{\nabla u}$.

3 Approximation of Bendings with Finite Bending Energy by Smooth Bendings

In this section we review the results of [15, 17], including an overview of the main ideas of the proof. The main result is easy to state. However, for applications one frequently needs some additional facts that are proven in these papers and which do not follow from the main result. The main result is [17, Theorem 1]:

Theorem 3.1 *Let $S \subset \mathbb{R}^2$ be a bounded Lipschitz domain whose boundary is C^1 away from a subset of vanishing one-dimensional Hausdorff measure. Then every $W^{2,2}$ bending of S can be approximated, strongly in $W^{2,2}$, by bendings in $C^\infty(\overline{S}, \mathbb{R}^3)$.*

Remarks

(i) The hypotheses of Theorem 3.1 are satisfied, e.g., when the domain S is Lipschitz and piecewise C^1. This excludes cusps, but it includes most domains which are realistic from a mechanics viewpoint. Moreover, most intermediate steps in the proof of Theorem 3.1 in fact apply to arbitrary Lipschitz domains.

(ii) Note that any straightforward smoothing process, e.g., by convolution with a smoothing kernel, will typically destroy the isometry constraint (2).
(iii) Theorem 3.1 is a 'constructive' result. However, it relies on detailed 'descriptive' results about $W^{2,2}$ bendings which extend the results in [13] also in the classical smooth setting.
(iv) In [25] the analogous result to Theorem 3.1 had been obtained earlier for convex Lipschitz domains $S \subset \mathbb{R}^2$, and with approximating deformations that are C^2 in the interior. The passage from convex to essentially arbitrary Lipschitz domains requires a rather different approach.

In the following sections we give an overview of the proof of Theorem 3.1, including some intermediate steps of independent interest.

3.1 Simple Examples of Ill-Behaved Bendings

The purpose of this section is to construct simple (essentially 1d) examples of $W^{2,2}$ bendings for which the set $C_{\nabla u}$ is ill-behaved: it is the complement of a totally disconnected set of positive measure. These examples show that generic bendings can be very irregular, even if their curvature is uniformly bounded. (A more elaborate, genuinely 2d, example was constructed in [15, Sect. 7]. Its aim was to show that a certain kind of level set segments $[x]$ can be large.)

Let $E \subset (-1, 1)$ be an open and dense set such that $(-1, 1) \setminus E$ has positive length and set $K(t) = \int_0^t (1 - \chi_E(s))ds$. Define the map $u : (-1, 1) \times \mathbb{R} \to \mathbb{R}^3$ by setting

$$u(x_1, x_2) = \left(\int_0^{x_1} \cos K, x_2, \int_0^{x_1} \sin K \right).$$

Let $S \subset (-1, 1) \times \mathbb{R}$ be a bounded domain. Then the restriction of u to S is a $W^{2,\infty}$ bending of S. Moreover,

$$C_{\nabla u} = (E \times \mathbb{R}) \cap S.$$

This set is open, dense in S, and its complement in S has nonzero area because $(-1, 1) \setminus E$ has positive length.

One can easily modify u in a way that converts each connected component U of $C_{\nabla u}$ into a connected component with three boundary segments rather than two. Components with more than two boundary segments will play a role later on.

For example, assume that S is the unit disk and suppose that $r \in (0, 1)$, that E contains the interval $(-r, r)$ (so that $u(x_1, x_2) = (x_1, x_2, 0)$ whenever $|x_1| < r$) and that S is the unit disk centered at the origin. Set $h = (1 - r^2)^{1/2}$ and define the modified bending $\widetilde{u} : S \to \mathbb{R}^3$ by

$$\widetilde{u}(x_1, x_2) = \begin{cases} u(x_1, x_2) & \text{if } x_2 \leq h, \\ (x_1, h + \sin(x_2 - h), \cos(x_2 - h) - 1) & \text{if } x_2 > h. \end{cases}$$

While the original deformation u had $U = S \cap (-r, r) \times \mathbb{R}$ as a connected component of $C_{\nabla u}$, the modified deformation $\widetilde{u}$ has

$$\widetilde{U} = \{(x_1, x_2) \in S : |x_1| < r,\ x_2 < h\},$$

as a connected component. Notice that $\widetilde{U}$ has three boundary segments, namely $\{x_2 = h\} \cap S$ and $\{|x_1| = r\} \cap S$.

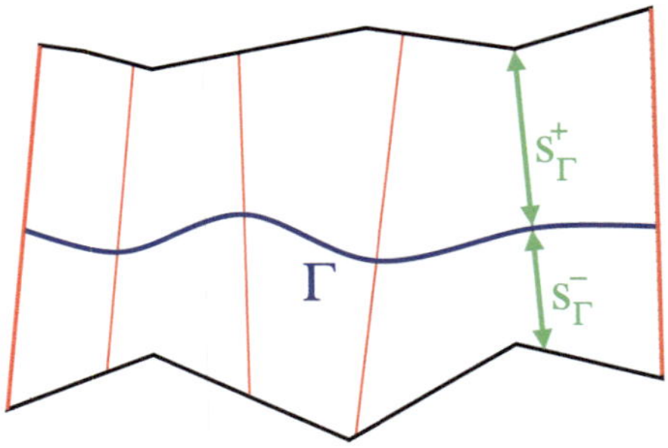

Fig. 2 A line of curvature chart $[\Gamma(0,T)]$. The *upper* and the *lower* boundary belong to ∂S, whereas the *left* boundary is the segment $[\Gamma(0)]$ and the *right* boundary is the segment $[\Gamma(T)]$

3.2 Local Line of Curvature Charts

The natural basic building blocks of a bending are local line of curvature charts. By this we mean subdomains of the reference configuration S that can be fully parametrized by a single line of curvature of the type introduced now.

There is a local correspondence between bendings and pairs of framed curves: Let $\Gamma : [0,T] \to S$ be an arclength parametrized curve in the reference domain S; since $[0,T]$ is compact and Γ is continuous and is contained in S, we are in fact assuming that Γ has a positive distance from ∂S.

The Frenet frame for the arc-length parametrized plane curve Γ is given by

$$R = \begin{pmatrix} \Gamma' \\ N \end{pmatrix},$$

where $N = (\Gamma')^\perp$ is the normal to Γ. The frame satisfies the Frenet equations

$$R' = \begin{pmatrix} 0 & \kappa \\ -\kappa & 0 \end{pmatrix} R, \tag{5}$$

where κ is the curvature of Γ.

With such a curve Γ in S we associate the directed signed distances $s_\Gamma^\pm(t)$ from $\Gamma(t)$ to the 'upper' (resp. 'lower') boundary of S along $N(t)$, see Fig. 2.

Precisely, one defines

$$s_\Gamma^+(t) = \inf\{s > 0 : \Gamma(t) + sN(t) \notin S\},$$
$$s_\Gamma^-(t) = \sup\{s < 0 : \Gamma(t) + sN(t) \notin S\}.$$

We denote by $[\Gamma(t)]$ the segment with endpoints $\Gamma(t) + s_\Gamma^+(t)N(t)$ and $\Gamma(t) + s_\Gamma^-(t)N(t)$. By $[\Gamma(0,T)]$, we denote the union of all $[\Gamma(t)]$ with $t \in (0,T)$. This is the domain covered by the line of curvature chart defined by Γ.

Now in addition to Γ suppose we are given a function $\mu : (0,T) \to \mathbb{R}$. Define the frame $r : [0,T] \to SO(3)$ (with rows denoted by r_i) by solving the ODE system

$$r' = \begin{pmatrix} 0 & \kappa & \mu \\ -\kappa & 0 & 0 \\ -\mu & 0 & 0 \end{pmatrix} r, \tag{6}$$

with suitable initial conditions. Defining $\gamma : [0,T] \to \mathbb{R}^3$ by integrating the first row r_1 of r, we obtain a framed curve (γ, r).

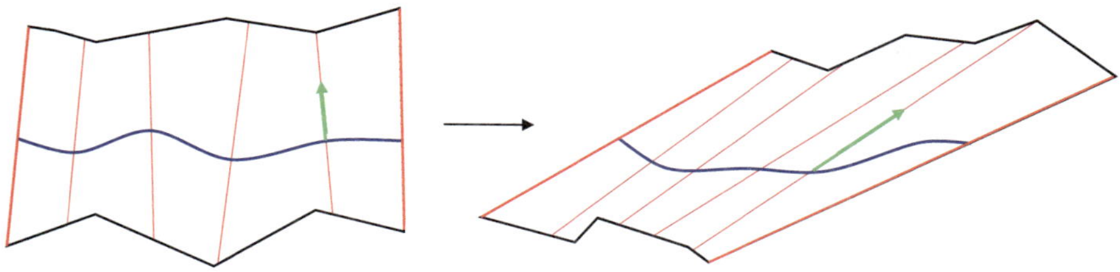

Fig. 3 *Left*: Local reference configuration $[\Gamma(0,T)]$ with curve Γ in *blue*, N in *green* and segments $[\Gamma(t)]$ in *red*. *Right*: Local deformed configuration $u([\Gamma(0,T)])$ with curve γ in *blue* and r_2 in *green*

Now define a map $u : [\Gamma(0,T)] \to \mathbb{R}^3$ by setting

$$u\big(\Gamma(t) + sN(t)\big) = \gamma(t) + sr_2(t) \quad \text{for all } t \in [0,T] \text{ and } s \in \big(s_\Gamma^-(t), s_\Gamma^+(t)\big), \tag{7}$$

cf. Fig. 3. Observe that the normal to u at the points $\Gamma(t) + sN(t)$ agrees with $r_3(t)$ for all $s \in (s_\Gamma^-(t), s_\Gamma^+(t))$.

If Γ is *admissible*, i.e., if

$$\big[\Gamma(t)\big] \text{ does not intersect } \big[\Gamma(t')\big] \quad \text{for any } t' \neq t, \tag{8}$$

then u is well-defined on $[\Gamma(0,T)]$ via (7). And u is automatically a bending. For u defined like this, the curve Γ is a line of curvature of u and the curve γ is a line of curvature lying in $u(S)$, and r is called the Darboux frame of γ, and μ is the normal curvature of γ, while κ is its geodesic curvature. The latter is an intrinsic quantity, hence invariant under bendings, so it agrees with the curvature of the plane curve Γ (which is just the preimage of γ under the deformation u).

This whole procedure can be reversed: given a bending u on S one can (locally on the developable region) find a curve Γ and a function μ such that (7) holds on $[\Gamma(0,T)]$. The curve Γ is automatically admissible.

The plan for the proof of Theorem 3.1 now is to fully cover the developable part of the reference domain by charts of the form $[\Gamma(0,T)]$ and then to smoothen u piecewise on each $[\Gamma(0,T)]$ and glue together these smoothened pieces. We start by outlining the local smoothing procedure on $[\Gamma(0,T)]$.

3.3 Local Smoothing Procedure

Suppose u is a $W^{2,2}$ bending on a subdomain of S of the form $[\Gamma(0,T)]$, on which u is given by (7) for some Γ and some μ, where (γ, r) is determined by (6). Then $\Gamma \in W^{2,\infty}$ and $\mu \in L^2$. The uniform curvature bound on Γ holds because the segments $[\Gamma(t)]$ do not intersect inside S and (by hypothesis) Γ has a positive distance from the boundary of S. Square integrability of μ follows from the fact that

$$\nabla^2 u\big(\Gamma(t) + sN(t)\big) = \frac{\mu(t)}{1 - s\kappa(t)}\, r_3(t) \otimes \Gamma'(t) \otimes \Gamma'(t), \tag{9}$$

which holds for all $s \in (s_\Gamma^-(t), s_\Gamma^+(t))$. We refer to [15] for details.

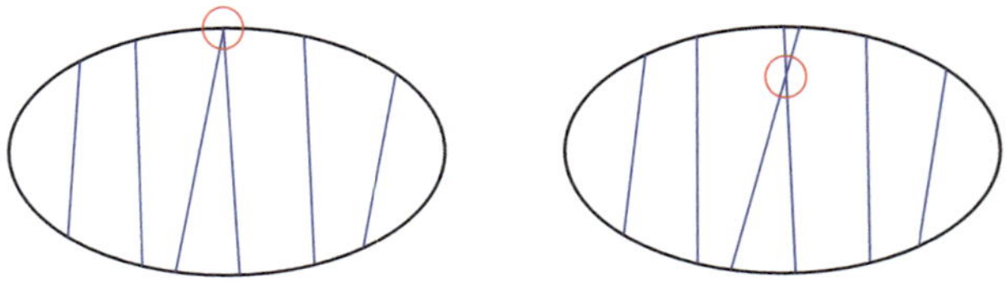

Fig. 4 An unstable level set geometry (*left*) becomes inadmissible under a generic perturbation (*right*)

3.3.1 Naïve Local Smoothing Process

The basic idea of the smoothing process leading to a proof of Theorem 3.1 is very natural:

1. Denote by $\widetilde{\kappa}$ a smooth approximation of κ and by $\widetilde{\mu}$ a smooth approximation of μ.
2. Define $\widetilde{R}$ by solving (5) with $\widetilde{\kappa}$ instead of κ, and then define $\widetilde{\Gamma}$ corresponding to $\widetilde{R}$.
3. Define $\widetilde{r}$ by solving (6) with $\widetilde{\kappa}$, $\widetilde{\mu}$ instead of κ, μ, and then define $\widetilde{\gamma}$ corresponding to $\widetilde{r}$.
4. Define $\widetilde{u}$ by (7) with $\widetilde{\Gamma}$, etc. instead of Γ, etc.

Then $\widetilde{u}$ is automatically a bending of $[\widetilde{\Gamma}(0,T)]$, and if $\widetilde{\kappa}$ and $\widetilde{\mu}$ are close to the original quantities, then we expect $\widetilde{u}$ to be close to u. However, $\widetilde{u}$ is only well-defined if $\widetilde{\Gamma}$ is admissible in the sense of (8).

3.3.2 A More Sophisticated Local Smoothing Process

If the level set structure of the original bending u is 'stable' under perturbations, then $\widetilde{\Gamma}$ will inherit its admissibility from the admissibility of the original curve Γ. Unfortunately, the original level set geometry is unstable in general, as explained in the next paragraph. Therefore, the naïve approach outlined above has to be greatly refined. A more sophisticated approach must take into account the following difficulties:

Unstable Level Set Geometry Naïve modifications of the original bending using the approach outlined in Sect. 3.3.1 can create an inadmissible level set structure. In fact, the level set structure of the original bending u (e.g. Fig. 4 (left)) is related to the curvature κ of some line of curvature Γ for u. Naïve modifications of κ as suggested in Sect. 3.3.1 would give rise to a curve $\widetilde{\Gamma}$ with curvature $\widetilde{\kappa}$, which violates (8). But this means that $\widetilde{\Gamma}$ cannot be the line of curvature of a bending defined on S.

This problem may occur regardless of how close $\widetilde{\kappa}$ is to κ, because two level set segments for u may intersect on ∂S, cf. Fig. 4 (left). The result of a generic small perturbation will typically be as in Fig. 4 (right).

An example in [15] shows that the original level set structure can in fact dramatically fail to be stable.

This problem may also occur on convex reference domains S. In [25] it is circumvented by a preliminary global dilation of the level set structure: replace u by the restriction u_λ of $x \mapsto \lambda u(\frac{x}{\lambda})$ for $\lambda > 1$. Such a dilation procedure only works if λS contains the closure of the original reference configuration S. This is no longer true for generic nonconvex domains S.

Local Boundary Conditions in the Reference Configuration Another key issue is that the smoothened deformation $\widetilde{u}$ defined via (7) should be defined on the same reference subdomain as the original deformation u. But $\widetilde{u}$ is defined on $[\widetilde{\Gamma}(0,\widetilde{T})]$—we write $\widetilde{T}$ from now on in order to avoid confusion, because the actual modifications in [15] also involve a length change of the curve—whereas u is defined in $[\Gamma(0,T)]$. And these sets do not agree for a generic modification $\widetilde{\Gamma}$ of Γ. A necessary condition for them to agree is clearly that the

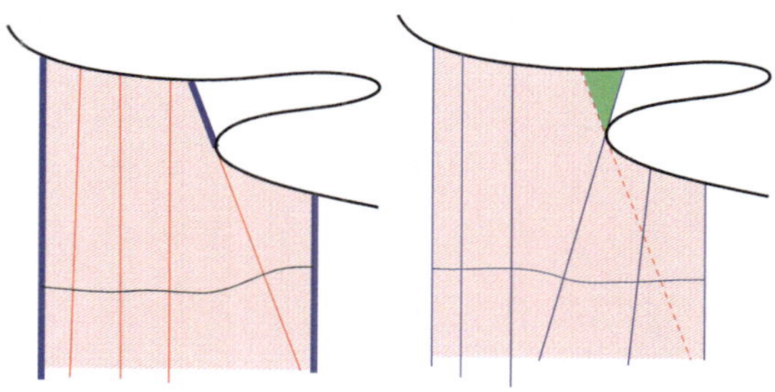

Fig. 5 *Left*: The shaded region is $[\Gamma(0,T)]$. The *bold segments* are $S \cap \partial[\Gamma(0,T)]$. The modified curve (*right*) should parametrize the same reference subdomain as the original curve (*left*). Hence segments intersecting ∂S tangentially should not be moved

'left' boundary segment $[\Gamma(0)]$ and the 'right' boundary segment $[\Gamma(T)]$ remain unchanged by the modification. This is true if

- $\widetilde{\Gamma}(0) = \Gamma(0)$ and $\widetilde{\Gamma}(\widetilde{T}) = \Gamma(T)$ (in fact, the latter condition must be slightly relaxed, cf. [16]).
- $\widetilde{N}(0) = N(0)$ and $\widetilde{N}(\widetilde{T}) = N(T)$.

However, fixing the left and the right segments is not enough to ensure $[\widetilde{\Gamma}(0,\widetilde{T})] = [\Gamma(0,T)]$. This is due to the existence of segments intersecting ∂S tangentially:

Tangential Segments On nonconvex reference domains there will, in general, occur segments $[x]$ which intersect the boundary of the domain S tangentially. So the set $S \cap \partial[\Gamma(0,T)]$ consists not only of $[\Gamma(0)]$ and $[\Gamma(T)]$, but also of (pieces of prolongated) segments $[\Gamma(t)]$ intersecting ∂S tangentially, see Fig. 5. If any of these segments is moved by the perturbation, then typically $[\widetilde{\Gamma}(0,\widetilde{T})] \neq [\Gamma(0,T)]$, cf. Fig. 5. Hence a further requirement on the perturbation $\widetilde{\kappa}$ of κ is that it must not move the segments intersecting ∂S tangentially.

Local Boundary Conditions in the Deformed Configuration The global approximation needed to prove Theorem 3.1 is of course defined on the whole reference domain S. It will be obtained by gluing together local approximations defined on patches of the form $[\Gamma(0,T)]$:

One first fixes a portion $[\Gamma(0,T)]$ of the reference configuration S. To it corresponds the portion $D = u([\Gamma(0,T)])$ of the deformed configuration $u(S)$. One cuts out D leaving the remainder of the deformed configuration $u(S)$ unchanged. Then one replaces D by a smooth approximation $\widetilde{D}$ and one glues $\widetilde{D}$ back into the original deformed configuration. For this to work, the smoothing procedure leading from the original deformed portion D to the approximating deformed portion $\widetilde{D}$ must preserve clamped boundary conditions on $S \cap \partial[\Gamma(0,T)]$. Otherwise, the smoothened portion $\widetilde{D}$ will not fit in a continuously differentiable way into the remaining (unchanged) deformed configuration. So we would like to enforce

$$(\widetilde{u}, \nabla\widetilde{u}) = (u, \nabla u) \quad \text{on } S \cap \partial\big[\Gamma(0,T)\big]. \tag{10}$$

However, this is not possible in general, because $S \cap \partial[\Gamma(0,T)]$ can consist of infinitely many segments, due to segments intersecting ∂S tangentially as in Fig. 5. Since the prescribed data belong to the non-smooth original deformation u, there cannot in general exist a smooth deformation $\widetilde{u}$ agreeing with u on infinitely many segments.

The key observation is that it is enough to preserve boundary conditions only on *long* segments in $S \cap \partial[\Gamma(0,T)]$. (This is essentially because short segments are necessarily very close to the boundary of S. We refer to [17] for details.)

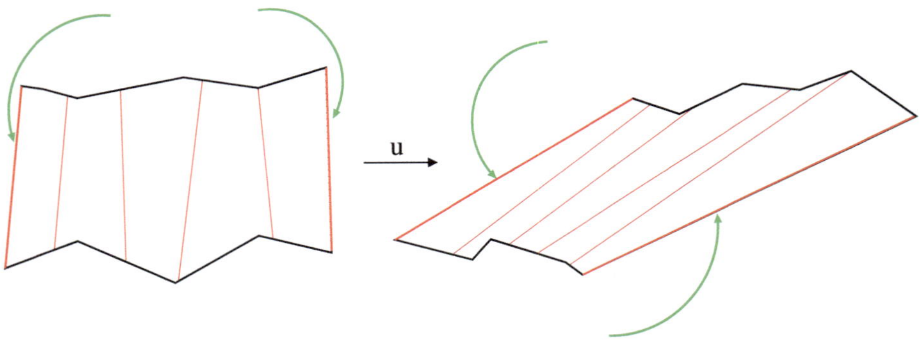

Fig. 6 Reference configuration on the *left*, deformed configuration on the *right*. Clamped boundary conditions on $(u, \nabla u)$ must be imposed on the segments pointed out by the *arrows*

And there is only a finite number of such long boundary segments because the directed distances $s_{\Gamma}^{\pm}$ have bounded variation, see [17, Sect. 5.6.2]. Hence, after possibly subdividing Γ into a finite number of shorter curves, we may assume that $[\Gamma(0)]$ and $[\Gamma(T)]$ are the only long segments in $S \cap \partial[\Gamma(0,T)]$. Hence we only must require that

$$(\widetilde{u}, \nabla\widetilde{u}) = (u, \nabla u) \quad \text{on } \big[\Gamma(0)\big] \cup \big[\Gamma(T)\big]. \tag{11}$$

The translation of (11), cf. Fig. 6 into boundary conditions for the framed curve (γ, r) is this:

$$\tilde{\gamma}(0) = \gamma(0), \qquad \tilde{\gamma}(\widetilde{T}) = \gamma(T), \qquad \tilde{r}(0) = r(0), \qquad \tilde{r}(\widetilde{T}) = r(T). \tag{12}$$

Finally, note that the transition from the approximating deformed portion $\widetilde{D}$ to the remaining deformed configuration must be C^∞. But (10) a priori only ensures to a C^1 transition. A C^∞ transition is be achieved by making the smoothened deformation $\widetilde{u}$ affine in a neighbourhood of $S \cap \partial[\Gamma(0,T)]$.

3.4 The Set $C_{\nabla u}$ of Local Constancy of the Deformation Gradient

Once a local smoothing process on line of curvature charts is available, one has to address the global question: Is it possible to cover the whole reference domain (or its developable part $S \setminus C_{\nabla u}$) by line of curvature patches? It is not possible in general. In fact, on a chart of the form $[\Gamma(0,T)]$, the map ∇u is clearly developable. But Theorem 2.1 only asserts that ∇u is developable away from $C_{\nabla u}$. However the examples in Sect. 3.1 show that the complement of $C_{\nabla u}$ can be totally disconnected. So from Theorem 2.1 alone we cannot conclude that there exists any line of curvature patch at all.

The way out of this dilemma consists of two steps. The first one is to realize that u is developable on a (possibly large) part of $C_{\nabla u}$ as well. The second step is to approximate u by bendings with a nice structure. Each of these well-behaved approximants will then, in turn, be approximated by smooth bendings.

The Developable Part and the Non-developable Part of $C_{\nabla u}$ Theorem 2.1 distinguished $C_{\nabla u}$ from the developable part. This must be sharpened. In fact, some parts of $C_{\nabla u}$ are developable and others are not. The non-developable part of $C_{\nabla u}$ will be denoted by $\widehat{C}_{\nabla u}$. By definition, it consists precisely of those connected components of $C_{\nabla u}$ which have more than two boundary segments, see Fig. 7 (left).

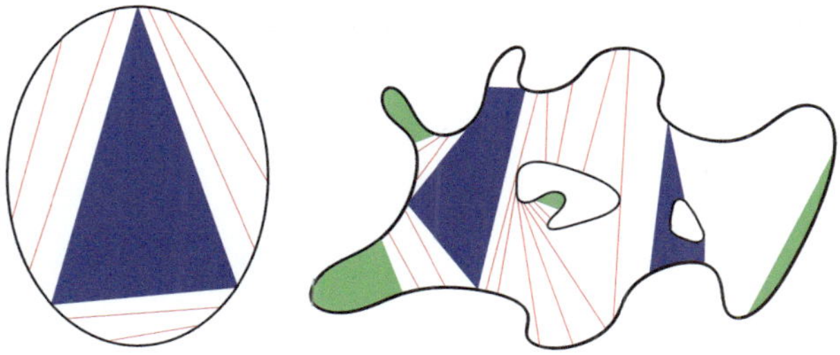

Fig. 7 *Left*: Breakdown of developability on $\widehat{C}_{\nabla u}$: The *shaded triangle* is a connected component of $\widehat{C}_{\nabla u}$ with three boundary segments. *Right*: A typical finitely developable bending as described by Theorem 3.3: The subdomains U_i are *blue* and the subdomains W_i are *green*. And the rest of S can be decomposed into finitely many line of curvature charts

One can prove that ∇u is developable across all components of $C_{\nabla u}$ with at most 2 boundary segments. (If S is convex then the proof is easy; for nonconvex domains it is a bit more subtle, cf. [17, Proposition 9].)

More precisely, through each point $x \in S \setminus \widehat{C}_{\nabla u}$ there exists a segment $[x] \subset S$ (possibly not unique) with endpoints on ∂S and such that ∇u is constant on $[x]$. On $C_{\nabla u} \setminus \widehat{C}_{\nabla u}$ the segments $[x]$ are no longer unique, i.e., there may be several segments through x on which ∇u is constant. Nevertheless, there exists a choice that is globally consistent on $S \setminus \widehat{C}_{\nabla u}$, in the sense that two different segments never intersect inside S. (A precise formulation uses the notion of a ruling, cf. [17] for details.)

Components of $\widehat{C}_{\nabla u}$ Can Behave Poorly The 'developable' connected components of $C_{\nabla u}$ (i.e. those with one or two boundary segments) can be handled in same way as the developable region $S \setminus C_{\nabla u}$. In particular, one can cover it by line of curvature charts.

On the other hand, we saw in Sect. 3.1 that the connected components of $\widehat{C}_{\nabla u}$ can display a very irregular behaviour as well. Firstly, they can accumulate, even in the interior of the domain ($\widehat{C}_{\nabla u}$ can even be locally dense). Secondly, each single connected component of $\widehat{C}_{\nabla u}$ can have infinitely many boundary segments. Hence there is no hope to decompose the developable set $S \setminus \widehat{C}_{\nabla u}$ into nice line of curvature charts, simply because the set $S \setminus \widehat{C}_{\nabla u}$ itself may not be nice at all.

Most Components of $\widehat{C}_{\nabla u}$ are Almost Developable We must clearly modify u, i.e., approximate it by deformations for which the developable set $S \setminus \widehat{C}_{\nabla u}$ is nice. A key observation is that most components of $C_{\nabla u}$ have at most two *long* boundary segments.

More precisely, for any given length $\delta > 0$ there exist at most finitely many components of $C_{\nabla u}$ which have more than two boundary components whose length exceeds δ. We refer to [17, Proposition 8] for details.

3.5 Approximation by Bendings Enjoying a Simple Structure

We now introduce a precise notion of bendings for which the developable set $S \setminus \widehat{C}_{\nabla u}$ is nice: a $W^{2,2}$ bending u is said to be *finitely developable* provided that

- The set $\widehat{C}_{\nabla u}$ consists of finitely many connected components.
- Each connected component of $\widehat{C}_{\nabla u}$ has finitely many boundary segments.

Finitely developable bendings consist of finitely many affine pieces and finitely many developable pieces. The following is a consequence of [17, Theorem 3].

Theorem 3.2 *Every $W^{2,2}$ bending of a bounded Lipschitz domain $S \subset \mathbb{R}^2$ can be approximated by finitely developable bendings.*

More precisely, for any $\delta > 0$ there exists a finitely developable $W^{2,\infty}$ bending $\widetilde{u}$ such that $\|\widetilde{u} - u\|_{W^{2,2}(S)} < \delta$. We even have $\widetilde{u} = u$ outside a strip $B_\delta(\partial S)$ of width δ around ∂S.

The set $\widehat{C}_{\nabla u}$ is nice for any finitely developable bending u. And the complement of its closure consists of finitely many subdomains. (Note that this is in sharp contrast to the examples in Sect. 3.1.) But is it possible to cover each of these subdomains by finitely many line of curvature charts, as seems to be needed in order to implement our smoothing strategy? It is not too hard to see that this is not possible. However, the next key structural result, which is a consequence of [17, Theorem 4], shows that the situation can nevertheless be handled.

In the following statement, by a decomposition of S we mean a collection of mutually disjoint subdomains which cover S up to countably many segments $[x]$. By a line of curvature patch we mean, as usual, a subdomain V of S that is of the form $V = [\Gamma(0, T)]$ for suitable $T > 0$ and a line of curvature $\Gamma \in W^{2,\infty}([0, T], S)$ for u, so that u is of the form (7) for suitable μ.

Theorem 3.3 *If u is a finitely developable bending of a bounded Lipschitz domain S, then S can be decomposed into three kinds of subdomains, which are compatible with the level set structure of the deformation gradient ∇u:*

1. *The finitely many connected components of $\widehat{C}_{\nabla u}$, denoted by $U_1, \ldots, U_M$.*
2. *Finitely many line of curvature patches, denoted by $V_1, \ldots, V_N$.*
3. *The rest, which consists of countably many connected components $W_1, W_2, \ldots$.*

Each of the W_i is contained in an arbitrarily small neighbourhood of ∂S, has precisely one boundary segment $[x_i] = S \cap \partial W_i$, and $\overline{W}_i \cap \overline{W}_j \cap S = \emptyset$ if $i \neq j$. Moreover, $[x_i]$ disconnects S (i.e., W_i can be cut off from the rest of S by cutting along $[x_i]$).

Theorems 3.2 and 3.3 are of independent interest and are key ingredients, e.g., in [29]. Within the proof of Theorem 3.1, the role of Theorem 3.2 is to reduce the problem: We only need to approximate finitely developable bendings, and these enjoy the nice structure provided by Theorem 3.3 cf. Fig. 7 (right). With Theorem 3.3 at hand, the construction plan for smooth approximations to a given finitely developable bending is simply this:

1. Keep $\widehat{C}_{\nabla u}$ as it is: $U_1, \ldots, U_M$.
2. Smoothen u on each subdomain covered by a single line of curvature chart $V_1, \ldots, V_N$ using the procedure outlined in Sect. 3.3.
3. Make u affine on each $W_1, W_2, \ldots$.

4 Minimizers of Kirchhoff's Bending Energy of Plates

If Kirchhoff's plate theory is a good model to describe the bending behaviour of paper, then one expects minimizers of the functional (3) (under suitable boundary conditions or external forces) to attain the shape adopted by a deformed sheet of paper. For this reason one is interested in the qualitative properties of such minimizers. These questions are studied in [16]. Here we summarize the main results and ideas of that paper. As in [27, 28], for simplicity we only consider isotropic materials, for which Q in (3) satisfies $Q(A) = |A|^2$ up to a constant prefactor. So the functional under consideration is

$$\mathcal{E}(u) = \begin{cases} \int_S |A|^2 & \text{if } u \text{ is a } W^{2,2} \text{ bending of } S, \\ +\infty & \text{otherwise.} \end{cases} \tag{13}$$

We will tacitly be using the fact that for $W^{2,2}$ bendings we have $\nabla^2 u = n \otimes A$ almost everywhere. For bendings, the mean curvature is given by $H = \frac{1}{2} \operatorname{Tr} A$.

4.1 Existence of Minimizers

Before studying properties of minimizers, let us verify that they indeed exist. The following result is easy to prove, see [16] for details.

Proposition 4.1 *Let u_0 be a $W^{2,2}$ bending of S and let $\partial_c S \subset \partial S$ be closed. Then there exists a minimizer of $\mathcal{E}$ among all $W^{2,2}$ bendings u of S satisfying the clamped boundary conditions $(u, \nabla u) = (u_0, \nabla u_0)$ on $\partial_c S$.*

The equality $\nabla u = \nabla u_0$ on $\partial_c S$ is understood in the trace sense. The role of the deformation u_0 in Proposition 4.1 is merely to serve as a source for boundary data which can be realized by a $W^{2,2}$ bending. It is easy to think of boundary data which cannot possibly be realized by a bending, e.g., any data which increase the distance between two points with respect to their distance in the reference configuration.

4.2 Equilibrium Equations

A derivation of the Euler-Lagrange (i.e., equilibrium) equations for infinitesimally narrow strips was given in [28]. In [30] the authors formally derive Euler-Lagrange equations for Möbius strips minimizing the bending energy functional. Their derivation is restricted to rectangular reference configurations and is based on a general formula derived in [2].

A more general (arbitrary reference domains) and self-contained derivation of the Euler-Lagrange equations can be found in [14]. However, the derivation of the equations in [30] and the one in [14] only work at a positive distance from the set Σ_0. A self-contained complete derivation of the Euler-Lagrange equations can be found in [16, Chap. 3].

As in [28] and here, also in [30] and in [22] a 'framed curve' formulation is used. However, [28, 30] use geodesic coordinates, rather than the line of curvature coordinates chosen in [16]. For the sake of brevity, we will not describe this derivation nor state the equations.

4.3 Qualitative Properties of Deformations Minimizing the Bending Energy

Proposition 4.1 asserts the existence of an energy minimizing shape which can be attained by a bent sheet of paper. Is the shape of such a minimizing bending smooth? Theorem 4.2 below (proven in [16]) answers this question; the result is essentially optimal.

In order to state it, we introduce the set of tangential segments Σ_τ: By definition, it consists of those $x \in S \setminus C_{\nabla u}$ such that $[x]$ intersects ∂S tangentially at one or both endpoints. We also introduce the constrained set Σ_c, which by definition is the closure of those points in $S \setminus C_{\nabla u}$ for which $\overline{[x]}$ intersects $\partial_c S$; here $\partial_c S$ is some fixed and given closed subset of the boundary as in Proposition 4.1. Finally, we introduce the set Σ_0, which consists of all

$$x \in S \setminus \overline{C}_{\nabla u} \setminus \Sigma_\tau \setminus \Sigma_c$$

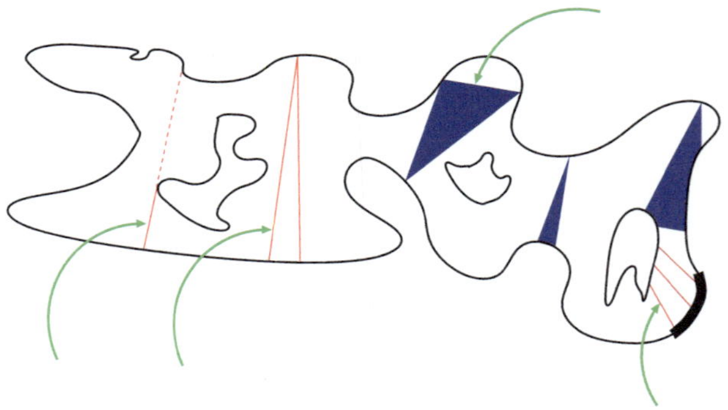

Fig. 8 The deformation gradient of an energy minimizing bending (Theorem 4.2): The *leftmost arrow* points at a segment in Σ_τ. The *arrow to its right* points at two segments in Σ_0 (but in truth only the degenerate case occurs, when both of these segments merge into one). The *lower arrow on the right* points at a segment in Σ_c. The *upper arrow* points at a segment in $S \cap \partial \widehat{C}_{\nabla u}$

such that a line of curvature $\Gamma : (-T, T) \to S$ of u with $\Gamma(0) = x$ satisfies

$$s_\Gamma^+(0)\kappa(0) = 1 \quad \text{or} \quad s_\Gamma^-(0)\kappa(0) = 1.$$

Locally, such a unique Γ always exists because x has positive distance from $C_{\nabla u}$, cf. [16, Sect. 2].

Theorem 4.2 *Let $S \subset \mathbb{R}^2$ be a bounded domain with C^∞ boundary, let $\partial_c S \subset \partial S$ be closed and let u be a $W^{2,2}$ bending of S that is a minimizer of Kirchhoff's bending energy functional among all deformations which on $\partial_c S$ agree with u up to the first derivatives. Then u is C^∞ away from the set*

$$\Sigma_\tau \cup \partial \widehat{C}_{\nabla u} \cup \Sigma_0 \cup \Sigma_c, \tag{14}$$

and u is C^3 (but not $C^{3,\varepsilon}$ for any $\varepsilon > 0$) at Σ_0.

The statement of Theorem 4.2 is illustrated in Fig. 8.

Several remarks regarding the optimality of Theorem 4.2 are in order:

(i) On the set Σ_c the bending u and the deformation gradient ∇u are fully determined by the boundary conditions. As we make no smoothness assumptions on the boundary data, u will not in general be smooth on Σ_c.
(ii) Smoothness of the boundary ∂S is a necessary condition. In fact, if ∂S is not smooth, then minimizers are, in general, not smooth either. This is related to the regularity of $s_\Gamma^\pm$.
(iii) For suitable domains and boundary conditions, the sets Σ_c, Σ_τ and $\partial \widehat{C}_{\nabla u}$ are indeed nonempty, and minimizers fail to be regular at these sets. We refer to [9] for details.
(iv) There is strong numerical evidence that Σ_0 can be nonempty as well, cf. Sect. 6.

The set (14) is closed. Moreover, each of the sets Σ_τ, Σ_0, Σ_c and $\partial \widehat{C}_{\nabla u}$ consists of straight line segments on which the deformation gradient is constant, and we have:

1. The set Σ_τ is closed and has empty interior. On convex domains, Σ_τ is clearly empty.
2. The set Σ_0 is contained in $\{H = 0\}$.
3. The set $\{H = 0\} \setminus C_{\nabla u}$ (in particular, Σ_0) consists of countably many line segments, which can only accumulate at ∂S, at Σ_τ, at $\partial \widehat{C}_{\nabla u}$ and at Σ_c.
4. In [9] it is shown that, under reasonable boundary conditions, minimizers are countably developable. So the set $S \cap \partial \widehat{C}_{\nabla u}$ consists of finitely many segments. This can even be made quantitative by estimating the number of segments in $S \cap \partial \widehat{C}_{\nabla u}$ by the number of maximal intervals in $\partial_c S$.

At the sets Σ_τ, Σ_c and $\partial\widehat{C}_{\nabla u}$ minimizers generically fail to be C^2, cf. [9]. The situation is different for Σ_0. Here u is C^3, and additional information about the behaviour at Σ_0 is available, cf. [16]:

1. Energy minimizing bendings u are locally nearly planar at Σ_0. In fact, Σ_0 can be characterized as the set of those $x_0 \in S \setminus (\partial C_{\nabla u} \cup \Sigma_\tau \cup \Sigma_c)$ for which
$$r^{4+\varepsilon} \ll \int_{B_r(x_0)} |A|^2 \ll r^4 \quad \text{as } r \downarrow 0$$
for all $\varepsilon > 0$.
2. The energy density concentrates at one endpoint z of each segment $[x] \subset \Sigma_0$: If $z_n \to z$ approaches z at a rate comparable to the distance of z_n from $[x]$, then $A(z_n)$ diverges. This explains the concentration observed in the figures in Sect. 6 (cf. also [30]), and perhaps also the observations in [28, Sect. 4].

Regularity on the Level of the Curvatures Theorem 4.2 is based on a regularity result on line of curvature patches, which asserts the following: Suppose $\Gamma : [0, T] \to S$ is admissible and such that all $[\Gamma(t)]$ with $t \in [0, T]$ intersect ∂S transversally at both endpoints. Let u be a $W^{2,2}$ bending of $[\Gamma(0, T)]$ such that Γ is a line of curvature for u. So there exists $\mu \in L^2(0, T)$ such that (7) holds (with κ as in (5)). Assume, moreover, that u is an energy minimizer among all bendings $\tilde{u}$ of $[\Gamma(0, T)]$ satisfying (11). Furthermore, define

$$I_0 = \big\{t \in [0, T] : s_\Gamma^+(t)\kappa(t) = 1 \text{ or } s_\Gamma^-(t)\kappa(t) = 1\big\},$$

which is a subset of $\{\mu = 0\}$; observe that $t \in I_0$ means that $\Gamma(t) \in \Sigma_0$. Then the curvatures κ and μ enjoy the following regularity properties, provided we exclude the degenerate (and trivial) case $\mu \equiv 0$:

- μ has only finitely many zeros in $(0, T)$ and changes its sign at each of them.
- $\mu \in C^\infty([-T, T] \setminus I_0) \cap C^1([-T, T])$. In particular, I_0 consists of finitely many points.
- $\kappa \in C^\infty([-T, T] \setminus I_0) \cap C^2([-T, T])$.
- The curvatures scale as follows near any point $t_0 \in I_0$:
$$\big|\kappa(t) - \kappa(t_0)\big| \sim \frac{(t-t_0)^2}{(\log|t-t_0|)^2},$$
$$\big|\mu(t)\big| \sim \left|\frac{t-t_0}{\log|t-t_0|}\right|.$$
In particular, $\kappa' = \kappa'' = \mu' = \mu = 0$ on I_0.
- Near zeros of μ which do not belong to I_0, the curvature μ grows linearly.

The fact that μ has only finitely many zeros in $(0, T)$ (in particular, that I_0 is finite) is a key property of minimizers. It is false in general for deformations u which are not stationary points of Kirchhoff's bending energy functional.

5 Developable Möbius Strips

In [27] a Möbius strip isometric to a rectangle in $\mathbb{R}^2$ was constructed explicitly. We denote this particular Möbius strip by $\mathcal{M}_0 \subset \mathbb{R}^3$. In fact, a family of such strips was constructed, depending on various parameters. We fix one of them.

By the construction in [27], the Möbius strip $\mathcal{M}_0$ contains exactly three planar subdomains. Two of them are triangles with all vertices on the boundary, but one of them, denoted by $\widetilde{U}$ in what follows, is a truncated triangle, cf. Fig. 9. Let $l \subset \widetilde{U}$ be the segment with endpoints on $\partial\mathcal{M}_0$ which is orthogonal (in the plane containing $\widetilde{U}$) to $\partial\mathcal{M}_0$, and (for definiteness) which divides $\widetilde{U}$ into two planar sets of equal area. Cutting $\mathcal{M}_0$ along l and rolling it onto the plane, we obtain a rectangle S. (The result is a rectangle and not an arbitrary trapezium, because $\widetilde{U}$ is planar and $l \subset \widetilde{U}$ is perpendicular to the boundary of $\partial\mathcal{M}_0$.) By suitably choosing the parameters in Sadowsky's construction (and possibly scaling), we may assume that this rectangle is given by

$$S = (-L, L) \times (-1, 1) \subset \mathbb{R}^2 \quad \text{for some } L > 5. \tag{15}$$

Reversing this procedure, we are in the setting of nonlinear elasticity: the domain S is the reference configuration, and there exists a deformation u_0 (obtained by reversing the previous procedure) which is a $W^{2,2}$—in Sadowsky's construction even a $W^{2,\infty}$ piecewise analytic—bending of S, and for which $u_0(S) = \mathcal{M}_0 \setminus l$. After rigid motion we may assume that u_0 is a normalized Möbius bending (cf. Sect. 5.1), so that

$$l = \{-L\} \times (-1, 1) \times \{0\}.$$

The level set structure of ∇u_0 is depicted in Fig. 9; this figure is shifted with respect to the figure in [27].

5.1 Möbius Bendings

For the rest of this section, S denotes a rectangle as in (15). A $W^{2,2}$ bending of S will be called a (normalized) Möbius bending of S if

$$\begin{aligned} u(\pm L, x_2) &= (\pm L, \mp x_2) \quad \text{for all } x_2 \in (-1, 1), \\ \nabla u(\pm L, x_2) &= (e_1 \mid \mp e_2) \quad \text{for all } x_2 \in (-1, 1); \end{aligned} \tag{16}$$

here $e_1 = (1, 0, 0)^T$, $e_2 = (0, 1, 0)^T \in \mathbb{R}^3$. In particular, the normal satisfies

$$n(-L, x_2) = -n(L, x_2) = e_3 \quad \text{for all } x_2 \in (-1, 1).$$

After a rigid motion (ensuring that (16) hold) and uniform scaling (ensuring that the strip has width 2), any developable Möbius can be described by a normalized Möbius bending.

Observe that if u is a Möbius bending of a reference domain S as in (15), then we can extend u to $\mathbb{R} \times (-1, 1)$ via the identity

$$u(x_1 + 2L, x_2) = u(x_1, -x_2). \tag{17}$$

Then u is $W^{2,2}$ on each rectangle $(-M, M) \times (-1, 1)$ with $M < \infty$; if M is an integer multiple of L then u defines a multiply covered Möbius strip. The following lemma is proven in the same way as the oscillation estimates in [19, 23]. In its statement, we use the notation $B_r^-(z) = \{x \in B_r(z) : x_2 < z_2\}$, and it is understood that we extend u via (17).

Lemma 5.1 *There exists a universal constant C such that, for any $z \in [-L, L] \times [0, 1]$, we have*

$$\operatorname{osc}_{B_r^-(z)} \nabla u \le C |\log r|^{-1/2} \|\nabla^2 u\|_{L^2(B_{1/2}^-(z))}$$

for all $r \in (0, \frac{1}{2})$ and all Möbius bendings of S.

Proof By the hypotheses clearly $B^-_{1/2}(z) \subset (-4L, 4L) \times (-1, 1)$. Moreover, for all $\rho \in (0, \frac{1}{2})$ the following is true: if $x \in B^-_\rho(z)$ then there exists a point in the lower half-circle,

$$y \in \partial B_\rho(z) \cap \partial B^-_\rho(z),$$

with $\nabla u(y) = \nabla u(x)$. (This is a consequence of Theorem 2.1 and the geometry of the reference domain.) Thus

$$\operatorname{osc}_{B^-_\rho(z)} \nabla u = \operatorname{osc}_{\partial B_\rho(z) \cap \partial B^-_\rho(z)} \nabla u.$$

Applying Sobolev embedding on each half-circle and then integrating in ρ, the claim follows as in [23]. □

An immediate consequence of Lemma 5.1 is the following:

Proposition 5.2 *For every $C_1 > 0$ there exists $\delta > 0$ such that if u is a Möbius bending with $\int_S |\nabla^2 u|^2 \leq C_1$, then*

$$\big|u(x) - u(y)\big| \geq \frac{1}{4}|x - y|$$

for any pair $x, y \in S$ with $|x - y| \leq \delta$.

Proof By Lemma 5.1 there exists $r > 0$ such that

$$\operatorname{osc}_{B^-_r(z)} \nabla u < \frac{1}{10}$$

for all $z \in [-L, L] \times [0, 1]$. Choosing δ small enough, after a possible reflection across the x_1-axis, there exists $z \in [-L, L] \times [0, 1]$ such that the segment with endpoints x and y is contained in $B^-_r(z)$. Now the claim follows by applying the fundamental theorem of calculus along the segment with endpoints x and y and using that $\nabla u(x) \in O(2, 3)$. □

Another immediate consequence of Lemma 5.1 is the following:

Corollary 5.3 *Möbius bendings of S belong to $C^1(\overline{S}, \mathbb{R}^3)$. Sequences of Möbius bendings with uniformly bounded energy $\mathcal{E}$ have a subsequence that converges in $C^1(\overline{S}, \mathbb{R}^3)$.*

5.2 Smooth Approximation of Sadowsky's Möbius Strip

Theorem 5.4 *Let S as in* (15). *For each $\delta > 0$ there exists a Möbius bending $u \in C^\infty(\overline{S}; \mathbb{R}^3)$ which is arbitrarily close in $W^{2,2}$ to u_0. In particular, u is an embedding, i.e., has no self-intersections.*

Remarks

1. Our construction is such that u agrees with u_0 on an arbitrarily large fraction of the planar regions of u_0.
2. Observe that, in order to construct smooth approximations on closed strips such as the Möbius strip, it is essential to employ an approximation method (such as in Sect. 3.3) which preserves the boundary conditions of the original deformation.

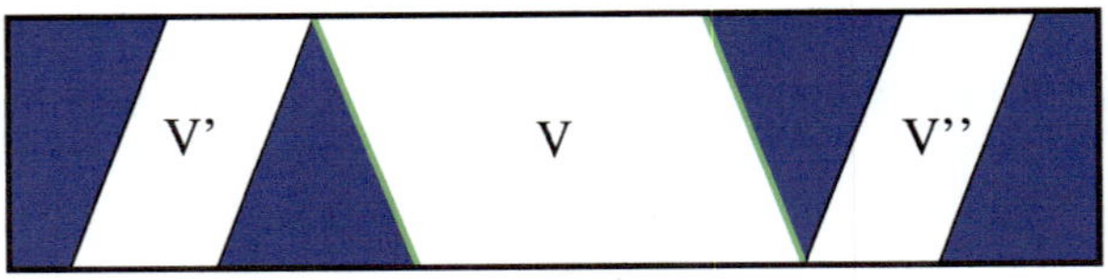

Fig. 9 Level set structure of Sadowsky's Möbius bending u_0. The *blue regions* constitute the set $C_{\nabla u_0}$ on which the deformation gradient is locally constant. The domains V, V', V'' are the three components of $S \setminus \overline{C}_{\nabla u_0}$. The union of the *two green segments* is $S \cap \partial V$. The smooth approximation of Sadowsky's Möbius bending constructed in Theorem 5.4 has a similar level set structure, except that the *blue regions* are slightly 'fattened'

Proof Fix $\delta > 0$. The construction from [15, 17] outlined in Chapter 3 allows to replace u_0 on each of the three subdomains of $S \setminus \overline{C_{\nabla u_0}}$ by a smooth deformation. To be more precise, let V be such a subdomain, cf. Fig. 9.

Then there exists a bending $\widetilde{u} \in C^\infty$ of V such that

- $\|u - \widetilde{u}\|_{W^{2,2}(V)} < \delta$.
- $\widetilde{u}$ is affine on a δ-neighbourhood of $S \cap \partial V$.
- $(\widetilde{u}, \nabla\widetilde{u}) = (u, \nabla u)$ on $S \cap \partial V$.

As u_0 is affine on the neighbouring regions, it follows that

$$u = \begin{cases} \widetilde{u} & \text{on } V, \\ u_0 & \text{on } S \setminus V \end{cases}$$

is C^∞ on a neighbourhood of $\overline{V}$ in S. Replacing u_0 by smooth pieces also on the other two connected components of $S \setminus \overline{C_{\nabla u_0}}$ (denoted by V' and V'' in Fig. 9), we obtain the desired smooth bending defined on all of S. The above arguments can be made more precise; we refer to [17] for the appropriate technical tools and leave the details to the interested reader.

It remains to verify that u is a Möbius deformation and that it is an embedding. As (for small $\delta > 0$) clearly u and its gradient agree with u_0 and its gradient on $\{\pm L\} \times (-1, 1)$, we see that u is indeed a Möbius bending of S. Moreover, by choosing δ small enough, u is arbitrarily close to u_0 in $W^{2,2}(S)$, hence in $C^0(\overline{S})$. Since by construction u_0 is an embedding and u is uniformly close to u_0, it follows that, for any $\varepsilon > 0$, the implication

$$|x - y| > \frac{\varepsilon}{2} \quad \Longrightarrow \quad u(x) \neq u(y) \tag{18}$$

holds for all x, $y \in S$ and all δ small enough (depending on ε). On the other hand, by Proposition 5.2 and since $\mathcal{E}(u_0) < \infty$, there exists $\varepsilon > 0$ such that for all Möbius bendings u of S which are $W^{2,2}$-close to u_0 (so that $\mathcal{E}(u) \le \mathcal{E}(u_0) + 1$), we have the implication

$$|x - y| \le \varepsilon \quad \text{and} \quad u(x) = u(y) \quad \Longrightarrow \quad x = y \tag{19}$$

for all x, $y \in S$. Combining (18) and (19), we conclude that indeed u is injective on S, provided we choose δ small enough. □

5.3 Möbius Strips Minimizing the Bending Energy

Theorem 5.5 *Let S as in* (15). *The functional $\mathcal{E}$ attains its minimum among all Möbius bendings $u : S \to \mathbb{R}^3$. Any such minimizing Möbius bending u satisfies $u \in C^3(S, \mathbb{R}^3)$ and*

$$C_{\nabla u} = \Sigma_\tau = \Sigma_c = \emptyset.$$

Moreover, the set $\{x \in S : \nabla^2 u(x) = 0\}$ *is the union of finitely many disjoint segments* $[x_1], \ldots, [x_N]$ *of constancy of* ∇u*, and we have* $u \in C^\infty(S \setminus \Sigma_0, \mathbb{R}^3)$*. In particular,*

$$u \in C^\infty\left(S \Big\backslash \bigcup_{k=1}^{N} [x_k], \mathbb{R}^3\right).$$

Remarks

(i) Theorem 5.5 is based on Theorem 4.2, which is a local result. Therefore, similar conclusions apply as well to simple strips without any twist.
(ii) Since $\Sigma_\tau = \Sigma_c = C_{\nabla u} = \emptyset$, the only singular set that can occur is Σ_0. Since $\Sigma_0 \subset \{\nabla^2 u = 0\}$, we see that Σ_0 consists of at most finitely many segments.
(iii) Theorem 4.2 and the remarks following it describe how u behaves near Σ_0.

Proof of Theorem 5.5 Set $\partial_l S = \{-L\} \times (-1, 1)$ and $\partial_r S = \{L\} \times (-1, 1)$. Applying Proposition 4.1 with $\partial_c S = \partial_l S \cup \partial_r S$ and u_0 as above, we see that there exists a minimizer u of $\mathcal{E}$ among all bendings satisfying $(u, \nabla u) = (u_0, \nabla u_0)$ on $\partial_c S$. (These boundary conditions mean that such bendings are normalized Möbius bendings.) In particular, u satisfies the hypotheses of Theorem 4.2.

Let us see what Theorem 4.2 asserts about u. Firstly, note that $\Sigma_\tau = \emptyset$ because S is convex. In addition, $\Sigma_c = \emptyset$ because two segments of constancy of the deformation gradient never intersect, and $\partial_r S$ and $\partial_l S$ in fact are such segments; this can be seen, e.g., by considering the extension (17) of u to $\mathbb{R} \times (-1, 1)$, or we could consider two overlapping charts of the Möbius strip.

The results in [9, Sect. 3] show that also $\widehat{C}_{\nabla u} = \emptyset$. In fact, we do not even need those results here: since S happens to be a rectangle and since no segment intersects $\partial_c S$ (i.e., all segments $[x]$ have one endpoint on the upper boundary and one on the lower boundary of the rectangle S), it is clear that there can be no connected component of $C_{\nabla u}$ with more that two boundary segments.

Therefore Theorem 4.2 shows that Σ_0 is the only possible singular set. And u is C^∞ away from Σ_0, which is a subset of $\{\nabla^2 u = 0\}$. The remarks following Theorem 4.2 show that the set $\{\nabla^2 u = 0\}$ consists of finitely many straight line segments. In fact, segments in $\{\nabla^2 u = 0\}$ could only accumulate at $\partial_r S$ or $\partial_l S$. But they cannot do so, again because these are in fact interior segments (for a shifted chart). □

For greater clarity we also provide a formulation of Theorem 5.5 in terms of the energy minimizing strip (i.e. the deformed configuration) itself.

Corollary 5.6 *There exists a Möbius strip* $\mathcal{M} \subset \mathbb{R}^3$ *minimizing the functional* $\int_{\mathcal{M}} H^2$ *among all Möbius strips which are isometric to* $\mathcal{M}_0$*. Any such minimizer* $\mathcal{M}$ *has the following properties*:

(i) $\mathcal{M}$ *is a* C^3 *surface. In particular, the mean curvature is continuously differentiable across the whole surface.*
(ii) $\mathcal{M}$ *contains no planar open set*: *the set of points where the mean curvature of* $\mathcal{M}$ *vanishes consists of a finite number* N *of disjoint straight line segments* $l_1, \ldots, l_N \subset \mathcal{M}$ *with endpoints on* $\partial\mathcal{M}$*.*
(iii) $\mathcal{M}$ *is* C^∞ *away from* $l_1 \cup \ldots \cup l_N$*.*

Remarks

(i) Corollary 5.6 makes no assertion about the uniqueness of $\mathcal{M}$.
(ii) In [28, Sect. 4] Sadowsky points out that if the experimental observations leading to his equations (32, 33) are correct, then his equilibrium equations for the infinitely narrow Möbius strip would suggest the 'strange fact' ('merkwürdige Tatsache') that the energy minimizing Möbius strip consists of a planar triangle to which two analytic pieces of the strip are connected in a C^1 but not C^2 fashion (discontinuous curvature).

For energy minimizing bendings, the (generic) failure of C^2 regularity at the boundary of planar regions is indeed confirmed in [9]. However, Corollary 5.6 asserts that the minimizing Möbius strip $\mathcal{M}$ contains no such planar region at all, and that the curvature of $\mathcal{M}$ is has no discontinuities (it even is C^1), and that it is C^∞ away from finitely many segments.
(iii) Observe that competitors of $\mathcal{M}$, i.e., non-energy minimizing Möbius strips, may well contain planar regions. So the set where the mean curvature vanishes may a priori consist of uncountably many segments. This cannot be excluded by any topological argument, and indeed $\mathcal{M}_0$ has this feature. The fact that the mean curvature only vanishes on finitely many segments is a nontrivial consequence of energy minimization.

Numerical Observations The numerical simulations in Sect. 6 suggest the following facts about the energy minimizing Möbius strip $\mathcal{M}$:

(i) $\mathcal{M}$ has a single (half-)twist and is embedded.
(ii) We have $N = 1$ in the statement of Corollary 5.6.
(iii) The segment l_1 belongs to Σ_0.

Then Corollary 5.6 asserts that $\mathcal{M}$ is C^∞ away from a single line segment l_1 and $\mathcal{M}$ is precisely C^3 at l_1. Moreover, Theorem 4.2 and the remarks following it describe rigorously how $\mathcal{M}$ behaves near l_1. In particular, $\mathcal{M}$ is 'almost' planar near l_1, and the stress concentrates at one endpoint of l_1. This agrees with the results of the numerical experiments.

6 Numerical Experiments

The numerical computation of minimizers of the bending energy within isometric deformations is challenging for several reasons. The fourth order nature of the problem prohibits the use of standard, conforming piecewise polynomial finite element functions and appropriately chosen discontinuous elements have to be employed. The nonlinear pointwise isometry constraint on the deformation gradient cannot be imposed almost everywhere since this would lead to a numerical locking effect that limits the flexibility of the numerical method. In [4] it was shown that the space of nonconforming discrete Kirchhoff triangles provides a useful discrete function space to approximate bending isometries. These elements are well established for linear unconstrained bending problems, see, e.g., [7], and replace the continuous second order derivatives by discrete operators with optimal convergence properties. Since the method provides approximations of the deformation gradient that are continuous at the nodes of an underlying triangulation, it is possible to impose the isometry constraint on this discrete subset. Due to the boundedness of discrete minimizers in $W^{2,2}(S,\mathbb{R}^3)$ this is sufficient to deduce that every accumulation point is an isometry as the mesh size tends to zero. An alternative approach that is based on a splitting and corresponding penalization of the second order derivatives is discussed in [5].

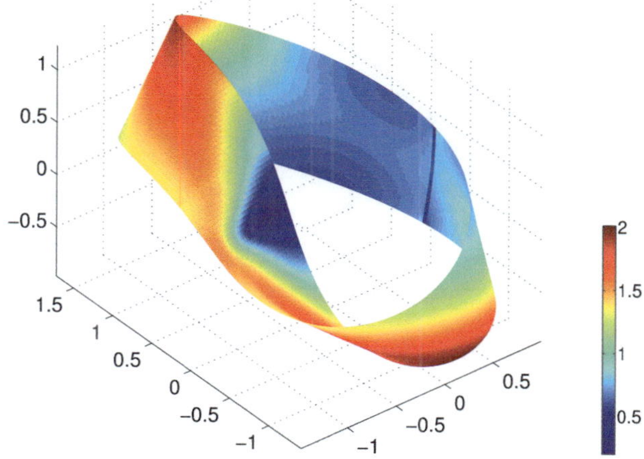

Fig. 10 Numerically computed finite element approximation of the Möbius strip with aspect ratio 8:1 colored by the squared discrete mean curvature. The underlying triangulation consists of 28.800 congruent right-angled triangles

Another computational challenge is the iterative solution of the discretized problem. The restricted validity of the Euler-Lagrange equations is reflected in a poor performance of the Newton scheme. Instead, gradient flows with respect to sufficiently strong metrics and a linearized treatment of the constraint can be shown to be energy decreasing and convergent to critical points. This procedure follows recent developments in the approximation of harmonic maps [1, 3]. Although we are not able to exclude the possibility of local minimizers, we never encountered related difficulties in our experiments. For details about the iterative scheme, its convergence, and its implementation we refer the reader to [4].

Figure 10 shows the result of a numerical experiment with a strip of aspect ratio 8:1 and overlapping, Möbius-periodic boundary conditions on the ends that lead to the formation of the shape of a Möbius strip. No special a priori knowledge about the expected optimal shape was used to define the initial configuration needed to start the globally convergent iterative scheme. For the displayed partitioning of the strip into 28.800 many congruent triangles, the iterates became stationary in $W^{2,2}(S, \mathbb{R}^3)$ up to a tolerance 1 % of the maximal mesh-size after about 1.000 outer and about 4.000 inner subdomain iterations of the discrete gradient flow combined with a domain decomposition technique into 2 subdomains. We remark that the irregularities occurring in the coloring at the ends of the strip are related to the visualization.

The discrete deformation shown in Fig. 10 resembles the expected shape of a Möbius strip and is colored according to a discrete mean curvature quantity. It is not an exact isometry, but the maximal deviation of the discrete first fundamental form from the unit matrix was comparable to the maximal mesh-size. The curvature lines shown in Fig. 11 were computed by numerically solving an ordinary differential equation defined by appropriately chosen eigenvectors of a discrete second fundamental form related to the nodes of the triangulation.

The quantitative properties of the Möbius strip change significantly when we modify the aspect ratio of the undeformed strip. Figures 12 and 13 show the numerical results for the aspect ratio 16:1.

The convergence proof of the numerical scheme used for the above simulations depends upon the existence of the smooth approximations ensured by Theorem 3.1. However, the scheme does not make any explicit use of the developable structure of bendings, i.e., of Theorem 2.1.

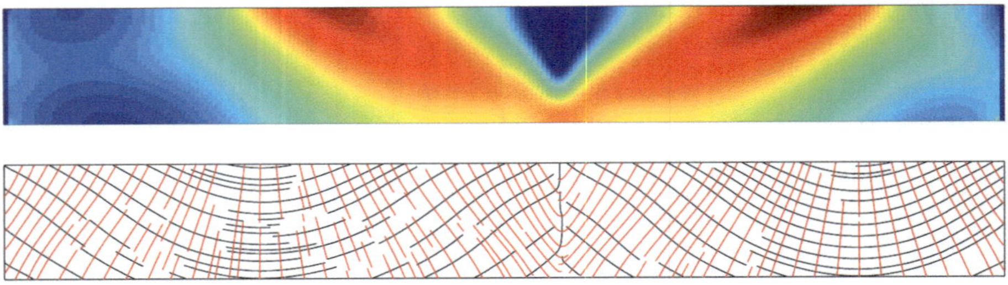

Fig. 11 Squared discrete mean curvature and *curvature lines* of the numerically computed Möbius strip with aspect ratio 8:1

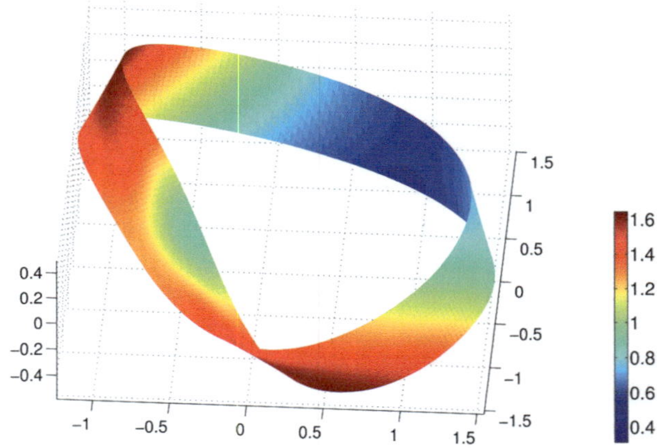

Fig. 12 Numerically computed finite element approximation of the Möbius strip with aspect ratio 16:1 colored by the squared discrete mean curvature

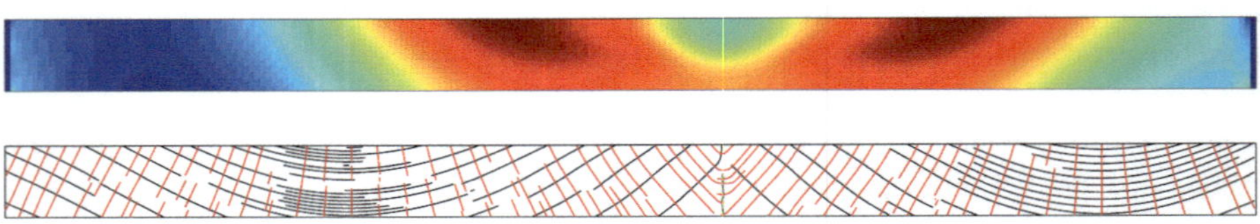

Fig. 13 Squared discrete mean curvature and *curvature lines* of the numerically computed Möbius strip with aspect ratio 16:1

In contrast, the numerical simulations e.g. in [30] are based on the fact that the surfaces in question are developable and can therefore be described in terms of a framed curve. Observe, however, that their choice of reference curve is a geodesic (i.e. a straight line), whereas the curves that are used in the analysis parts of the present paper are lines of curvature.

Both the simulations in [30] and those presented here clearly show that the mean curvature vanishes only on one single line segment $[x_1]$. (This is clearly visible from the coloring in Fig. 10.) In particular, this agrees with the conclusion of Corollary 5.6; indeed it even shows that $N = 1$ in its conclusion.

Moreover, this segment displays several features that are characteristic properties of the singular set Σ_0 identified in Sect. 4.3. In fact, at the segment $[x_1]$ one can see the 'conical' accumulation of segments which characterizes the set Σ_0; this is more clearly visible in [30] than in the figures above. Moreover, one observes stress concentration at one endpoint of $[x_1]$; this is represented by the dark red color in Fig. 10. In addition, the surface is almost affine near each $[x_1]$; this is clearly visible in the coloring of Fig. 10, which is dark blue near the segment in question.

Acknowledgements The authors are grateful to Alexis Papathanassopoulos for providing the results of the numerical simulations. The second author was supported by Deutsche Forschungsgemeinschaft grant no. HO-4697/1-1.

References

1. Alouges, F.: A new finite element scheme for Landau-Lifchitz equations. Discrete Contin. Dyn. Syst., Ser. S **1**(2), 187–196 (2008). MR 2379897 (2009a:65244)
2. Anderson, I.M.: The variational bicomplex. Technical Report, Utah State Univ. (1989)
3. Bartels, S.: Numerical analysis of a finite element scheme for the approximation of harmonic maps into surfaces. Math. Comput. **79**(271), 1263–1301 (2010). MR 2629993 (2011b:65217)
4. Bartels, S.: Approximation of large bending isometries with discrete Kirchhoff triangles. SIAM J. Numer. Anal. **51**(1), 516–525 (2013). MR 3033021
5. Bartels, S.: Finite element approximation of large bending isometries. Numer. Math. **124**(3), 415–440 (2013). MR 3066035
6. Borisov, Ju.F.: $C^{1,\alpha}$-isometric immersions of Riemannian spaces. Dokl. Akad. Nauk SSSR **163**, 11–13 (1965). MR 0192449 (33 #674)
7. Braess, D.: Finite Elements, 3rd edn. Cambridge University Press, Cambridge (2007). Theory, fast solvers, and applications in elasticity theory. Translated from the German by Larry L. Schumaker, MR2322235 (2008b:65142)
8. Conti, S., Lellis, C., Szekelyhidi, L. Jr.: h-principle and rigidity for $C^{1,\alpha}$ isometric embeddings. In: Holden, H., Karlsen, K.H. (eds.) Nonlinear Partial Differential Equations. Abel Symposia, vol. 7, pp. 83–116. Springer, Berlin, Heidelberg (2012) (in English)
9. Dall'Acqua, A., Hornung, P.: Structural results about the singular set of energy minimizing bendings. Preprint (2014)
10. Friesecke, G., James, R.D., Müller, S.: A theorem on geometric rigidity and the derivation of nonlinear plate theory from three-dimensional elasticity. Commun. Pure Appl. Math. **55**(11), 1461–1506 (2002). MR MR1916989 (2003j:74034)
11. Friesecke, G., James, R.D., Müller, S.: A hierarchy of plate models derived from nonlinear elasticity by gamma-convergence. Arch. Ration. Mech. Anal. **180**(2), 183–236 (2006). MR MR2210909 (2006k:74061)
12. Gromov, M.: Partial Differential Relations. Ergebnisse der Mathematik und ihrer Grenzgebiete (3) [Results in Mathematics and Related Areas (3)], vol. 9. Springer, Berlin (1986). MR 864505 (90a:58201)
13. Hartman, P., Nirenberg, L.: On spherical image maps whose Jacobians do not change sign. Am. J. Math. **81**, 901–920 (1959). MR MR0126812 (23 #A4106)
14. Hornung, P.: Analysis of thin elastic films. Leipzig Univ., Diss., Leipzig (2006)
15. Hornung, P.: Approximation of flat $W^{2,2}$ isometric immersions by smooth ones. Arch. Ration. Mech. Anal. **199**(3), 1015–1067 (2011). MR 2771672
16. Hornung, P.: Euler-Lagrange equation and regularity for flat minimizers of the Willmore functional. Commun. Pure Appl. Math. **64**(3), 367–441 (2011). MR 2779088 (2011m:49080)
17. Hornung, P.: Fine level set structure of flat isometric immersions. Arch. Ration. Mech. Anal. **199**(3), 943–1014 (2011). MR 2771671 (2012d:53194)
18. Jerrard, R.L.: Some rigidity results related to Monge-Ampère functions. Can. J. Math. **62**(2), 320–354 (2010). MR 2643046 (2011c:49082)
19. Kirchheim, B.: Geometry and rigidity of microstructures. Habilitation Thesis, University of Leipzig, Leipzig (2001)
20. Kuiper, N.H.: On C^1-isometric imbeddings. I, II. Nederl. Akad. Wetensch. Proc. Ser. A. **58**, 545–556 (1955), 683–689. Indag. Math. **17**. MR0075640 (17,782c)
21. Massey, W.S.: Surfaces of Gaussian curvature zero in Euclidean 3-space. Tôhoku Math. J. (2) **14**, 73–79 (1962). MR MR0139088 (25 #2527)
22. Moore, A., Healey, T.J.: The shape of a Möbius strip via elastic rod theory revisited (2014). arXiv: 1407.0571
23. Müller, S., Pakzad, M.R.: Regularity properties of isometric immersions. Math. Z. **251**(2), 313–331 (2005). MR MR2191030 (2006g:49006)
24. Nash, J.: C^1 isometric imbeddings. Ann. Math. (2) **60**, 383–396 (1954). MR 0065993 (16,515e)

25. Pakzad, M.R.: On the Sobolev space of isometric immersions. J. Differ. Geom. **66**(1), 47–69 (2004). MR MR2128713 (2006k:58012)
26. Pogorelov, A.V.: Extrinsic Geometry of Convex Surfaces. American Mathematical Society, Providence (1973). MR MR0346714 (49 #11439)
27. Sadowsky, M.: Ein elementarer Beweis für die Existenz eines abwickelbaren Möbius'schen Bandes und Zurückführung des geometrischen Problems auf ein Variationsproblem. Sitzungsber. Preuss. Akad. Wiss. (1930)
28. Sadowsky, M.: Theorie der elastisch biegsamen undehnbaren Bänder mit Anwendungen auf das Möbiussche band. Verhandl. des 3. Intern. Kongr. f. Techn. Mechanik (2), 444–451 (1930)
29. Schmidt, B.: Plate theory for stressed heterogeneous multilayers of finite bending energy. J. Math. Pures Appl. (9) **88**(1), 107–122 (2007). MR 2334775 (2008j:74043)
30. Starostin, E.L., van der Heijden, G.H.M.: The shape of a Möbius strip. Nat. Mater. **6**, 563–567 (2007)

DOI 10.1007/978-94-017-7300-3_10
Reprinted from *Journal of Elasticity* Journal, DOI 10.1007/s10659-014-9498-x

Roadmap to the Morphological Instabilities of a Stretched Twisted Ribbon

Julien Chopin · Vincent Démery · Benny Davidovitch

Received: 28 February 2014 / Published online: 19 September 2014

Abstract We address the mechanics of an elastic ribbon subjected to twist and tensile load. Motivated by the classical work of Green (Proc. R. Soc. Lond. Ser. A Math. Phys. Sci. 154(882):430, 1936; 161(905):197, 1937) and a recent experiment (Chopin and Kudrolli in Phys. Rev. Lett. 111(17):174302, 2013) that discovered a plethora of morphological instabilities, we introduce a comprehensive theoretical framework through which we construct a 4D phase diagram of this basic system, spanned by the exerted twist and tension, as well as the thickness and length of the ribbon. Different types of instabilities appear in various "corners" of this 4D parameter space, and are addressed through distinct types of asymptotic methods. Our theory employs three instruments, whose concerted implementation is necessary to provide an exhaustive study of the various parameter regimes: (i) a covariant form of the Föppl–von Kármán (cFvK) equations to the helicoidal state—necessary to account for the large deflection of the highly-symmetric helicoidal shape from planarity, and the buckling instability of the ribbon in the transverse direction; (ii) a far from threshold (FT) analysis—which describes a state in which a longitudinally-wrinkled zone expands throughout the ribbon and allows it to retain a helicoidal shape with negligible compression; (iii) finally, we introduce an asymptotic isometry equation that characterizes the energetic competition between various types of states through which a twisted ribbon becomes strainless in the singular limit of zero thickness and no tension.

J.C. and V.D. have contributed equally to this work.

J. Chopin
Civil Engineering Department, COPPE, Universidade Federal do Rio de Janeiro, 21941-972, Rio de Janeiro, RJ, Brazil
e-mail: jchopin@coc.ufrj.br

J. Chopin
Department of Physics, Clark University, Worcester, MA 01610, USA

V. Démery · B. Davidovitch (✉)
Physics Department, University of Massachusetts, Amherst, MA 01003, USA
e-mail: bdavidov@physics.umass.edu

V. Démery
e-mail: vdemery@physics.umass.edu

Keywords Buckling and wrinkling · Far from threshold · Isometry · Helicoid

Mathematics Subject Classification (2010) 74K20 · 53Z05 · 35Q74 · 74K35

Abbreviations

ssFvK equations	"small-slope" (standard) Föppl–von Kármán equations
cFvK equations	covariant Föppl–von Kármán equations
t, W, L	thickness, width and length of the ribbon (non-italicized quantities are dimensional)
t, $W=1$, L	thickness, width and length normalized by the width
ν	Poisson ratio
$\mathrm{E}, \mathrm{Y}, \mathrm{B}=\frac{\mathrm{Yt}^2}{12(1-\nu^2)}$	Young, stretching and bending modulus
$Y=1$, $B=\frac{t^2}{12(1-\nu^2)}$	stretching and bending modulus, normalized by the stretching modulus
$T=\mathrm{T}/\mathrm{Y}$	tensile strain (tensile load normalized by stretching modulus)
$\theta, \eta=\theta/L$	twist angle and normalized twist
$(\hat{\boldsymbol{x}}, \hat{\boldsymbol{y}}, \hat{\boldsymbol{z}})$	Cartesian basis
s, r	material coordinates (longitudinal and transverse)
$z(s,r)$	out of plane displacement (of the helicoid) in the small-slope approximation
$\boldsymbol{X}(s,r)$	surface vector
$\hat{\boldsymbol{n}}$	unit normal to the surface
$\sigma^{\alpha\beta}$	stress tensor
$\varepsilon_{\alpha\beta}$	strain tensor
$g_{\alpha\beta}$	metric tensor
$c_{\alpha\beta}$	curvature tensor
$\mathcal{A}^{\alpha\beta\gamma\delta}$	elastic tensor
∂_α, D_α	partial and covariant derivatives
H, K	mean and Gaussian curvatures
ζ	infinitesimal amplitude of the perturbation in linear stability analysis
$z_1(s,r)$	normal component of an infinitesimal perturbation to the helicoidal shape
$\eta_{\text{lon}}, \lambda_{\text{lon}}$	longitudinal instability threshold and wavelength
$\eta_{\text{tr}}, \lambda_{\text{tr}}$	transverse instability threshold and wavelength
$\alpha=\eta^2/T$	confinement parameter
α_{lon}	threshold confinement for the longitudinal instability
r_{wr}	(half the) width of the longitudinally wrinkled zone
$\Delta\alpha=\alpha-24$	distance to the threshold confinement
$f(r)$	amplitude of the longitudinal wrinkles
$U_{\text{hel}}, U_{\text{FT}}$	elastic energies (per length) of the helicoid and the far from threshold longitudinally wrinkled state
$U_{\text{dom}}, U_{\text{sub}}$	dominant and subdominant (with respect to t) parts of U_{FT}
$\boldsymbol{X}_{\text{cl}}(s)$	ribbon centerline
$\hat{\boldsymbol{t}}=d\boldsymbol{X}_{\text{cl}}(s)/ds$	tangent vector in the ribbon midplane
$\hat{\boldsymbol{r}}(s)$	normal to the tangent vector
$\hat{\boldsymbol{b}}(s)$	Frenet binormal to the curve $\boldsymbol{X}_{\text{cl}}(s)$
$\tau(s), \kappa(s)$	torsion and curvature of $\boldsymbol{X}_{\text{cl}}(s)$

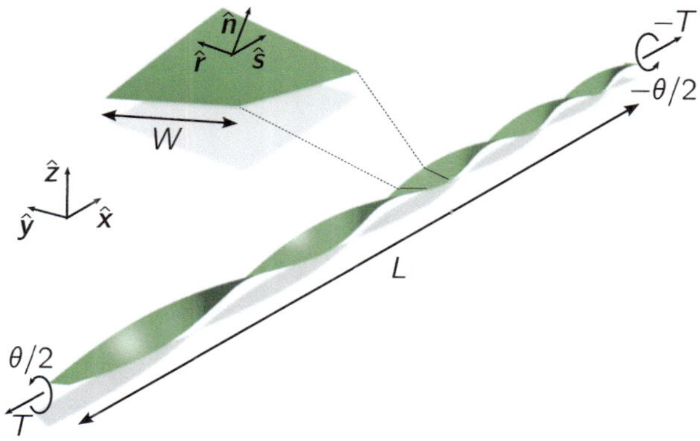

Fig. 1 A ribbon of length L and width W (and thickness t, not shown) is submitted to a tension T and a twist angle θ; the twist parameter is defined as $\eta = \theta/L = \theta$W/L. The longitudinal and transverse material coordinates are s and r, respectively. $\hat{\boldsymbol{n}}$ is the unit normal to the surface, $(\hat{x}, \hat{y}, \hat{z})$ is the standard basis, $(\hat{x}, \hat{y})$ being the plane of the untwisted ribbon

1 Introduction

1.1 Overview

A ribbon is a thin, long solid sheet, whose thickness and length, normalized by the width, satisfy:

$$\text{thickness:} \quad t \ll 1; \qquad \text{length:} \quad L \gg 1. \tag{1}$$

The large contrast between thickness, width, and length, distinguishes ribbons from other types of thin objects, such as rods ($t \sim 1$, $L \gg 1$) and plates ($t \ll 1$, $L \sim 1$), and underlies their complex response to simple mechanical loads. The unique nature of the mechanics of elastic ribbons is demonstrated by subjecting them to elementary loads—twisting and stretching—as shown in Fig. 1. This basic loading, which leads to surprisingly rich plethora of patterns, a few of which are shown in Fig. 2, is characterized by two small dimensionless parameters:

$$\text{twist:} \quad \eta \ll 1; \qquad \text{tension:} \quad T \ll 1, \tag{2}$$

where η is the average twist (per length), and T is the tension, normalized by the stretching modulus.[1]

Most theoretical approaches to this problem consider the behavior of a real ribbon through the asymptotic "ribbon limit", of an ideal ribbon with infinitesimal thickness and infinite length: $t \to 0$, $L \to \infty$. A first approach, introduced by Green [1, 2], assumes that the ribbon shape is close to a helicoid (Fig. 2a), such that the ribbon is strained, and may therefore become wrinkled or buckled at certain values of η and T (Fig. 2b,c,g,h) [4, 5]. A second approach to the ribbon limit, initiated by Sadowsky [6] and revived recently by Korte et al. [7], considers the ribbon as an "inextensible" strip, whose shape is close to a *creased helicoid*—an isometric (i.e., strainless) map of the unstretched, untwisted ribbon (Fig. 2d). A third approach, which may be valid for sufficiently small twist, assumes that the stretched-twisted ribbon is similar to the wrinkled shape of a planar, purely stretched rect-

[1] Our convention in this paper is to normalize lengths by the ribbon width W, and stresses by the stretching modulus Y, which is the product of the Young modulus and the ribbon thickness (non-italicized fonts are used for dimensional parameters and italicized fonts for dimensionless parameters). Thus, the actual thickness and length of the ribbon are, respectively, t $= t \cdot$ W and L $= L \cdot$ W, the actual force that pulls on the short edges is $T \cdot$ YW, and the actual tension due to this pulling force is T $= T \cdot$ Y.

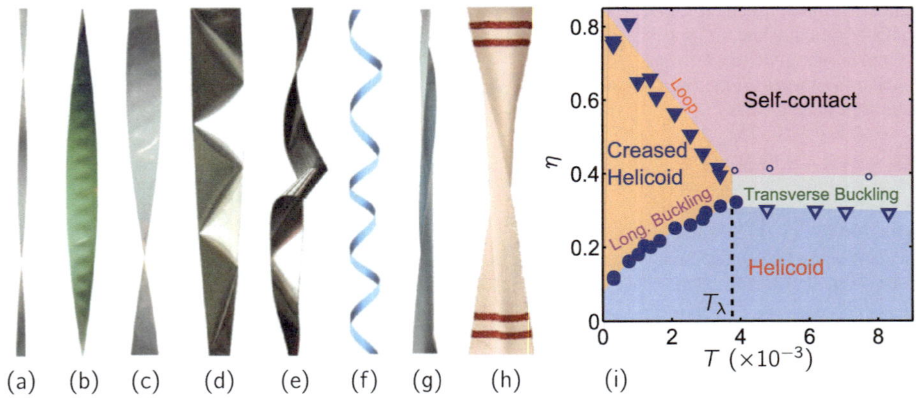

Fig. 2 *Left:* Typical morphologies of ribbons subjected to twist and stretching: (**a**) helicoid, (**b**, **c**) longitudinally wrinkled helicoid, (**d**) creased helicoid, (**e**) formation of loops and self-contact zones, (**f**) cylindrical wrapping, (**g**) transverse buckling and (**h**) twisted towel shows transverse buckling/wrinkling. *Right:* (**i**) Experimental phase diagram in the tension-twist plane, adapted from [3]. The descriptive words are from the original diagram [3]. Note that the twist used in the experiment is not very small; this apparent contradiction with our hypothesis $\eta \ll 1$ (Eq. (2)) is clarified in Appendix A

angular sheet, with a wrinkle's wavelength that vanishes as $t \to 0$ and increases with L [8]. Finally, considering the ribbon as a rod with highly anisotropic cross section, one may approach the problem by solving the Kirchoff's rod equations and carrying out stability analysis of the solution, obtaining unstable modes that resemble the looped shape (Fig. 2e) [9].

A recent experiment [3], which we briefly describe in Sect. 1.2, revealed some of the predicted patterns and indicated the validity of the corresponding theoretical approaches at certain regimes of the parameter plane (T, η) (Fig. 2). Motivated by this development, we introduce in this paper a unifying framework that clarifies the hidden assumptions underlying each theoretical approach, and identifies its validity range in the (T, η) plane for given values of t and L. Specifically, we show that a single theory, based on a covariant form of the Föppl–von Kármán (FvK) equations of elastic sheets, describes the parameter space (T, η, t, L^{-1}) of a stretched twisted ribbon where all parameters in Eqs. (1) and (2) are assumed small. Various "corners" of this 4D parameter space are described by distinct singular limits of the governing equations of this theory, which yield qualitatively different types of patterns. This realization is illustrated in Fig. 3, which depicts the projection of the 4D parameter space on the (T, η) plane, and indicates several regimes that are governed by different types of asymptotic expansions.

1.2 Experimental Observations

The authors of [3] used Mylar ribbons, subjected them to various levels of tensile load and twist, and recorded the observed patterns in the parameter plane (T, η), which we reproduce in Fig. 2. The experimental results indicate the existence of three major regimes that meet at a "λ-point" $(T_\lambda, \eta_\lambda)$. We describe below the morphology in each of the three regimes and the behavior of the curves that separate them:

- The helicoidal shape (Fig. 2a) is observed if the twist η is sufficiently small. For $T < T_\lambda$, the helicoid is observed for $\eta < \eta_{\text{lon}}$, where $\eta_{\text{lon}} \approx \sqrt{24T}$ is nearly independent on the ribbon thickness t. For $T > T_\lambda$, the helicoid is observed for $\eta < \eta_{\text{tr}}$, where η_{tr} exhibits a strong dependence on the thickness ($\eta_{\text{tr}} \sim \sqrt{t}$) and a weak (or none) dependence on the

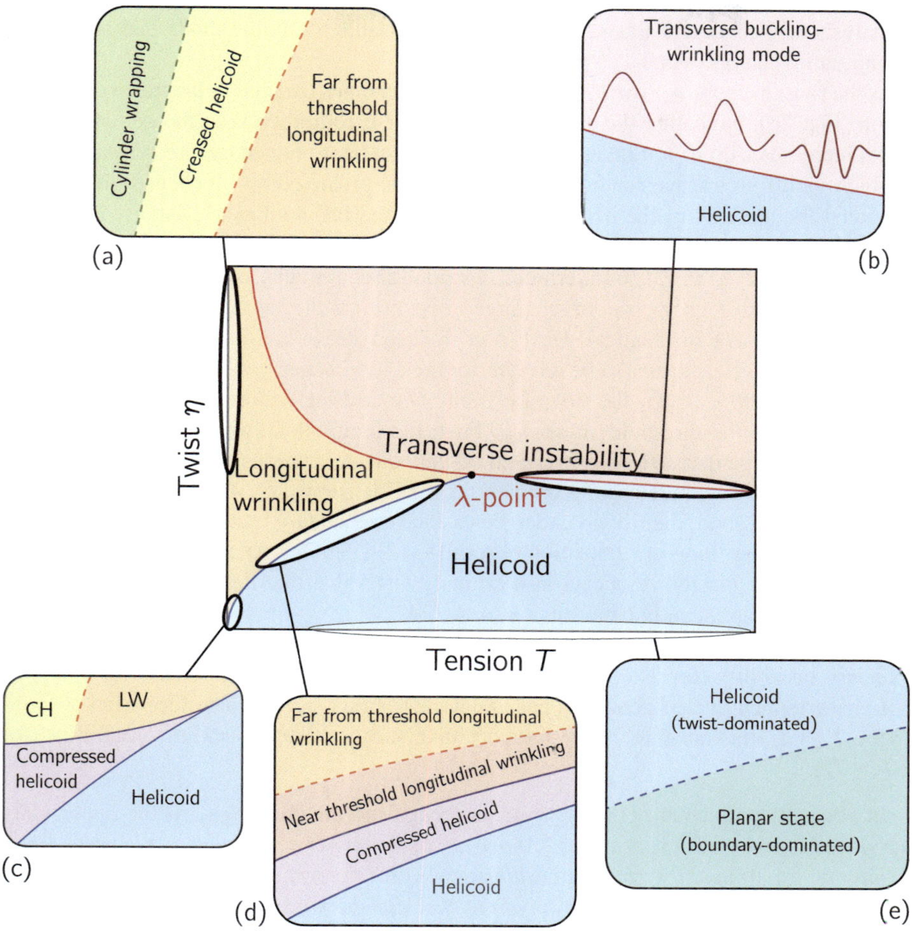

Fig. 3 The phase diagram in the tension-twist plane consists of three main regions: the helicoid, the longitudinally-wrinkled helicoid and a region delimited from below by a transverse instability. These regions meet at the λ-point. The complete phase diagram is more subtle and the following parts are magnified: (**a**) At vanishing tension, the ribbon shape becomes closer and closer to an (asymptotic) isometry; this is investigated in Sect. 3.5. (**b**) The transverse buckling instability is the focus of Sect. 4, where a transition from buckling to wrinkling is predicted. (**c**) At very low tension and twist, the longitudinal instability is described by Green's theory [2] (see Sect. 3.2). (**d**) The transition from the helicoid to the far from threshold longitudinally-wrinkled helicoid is detailed in Sect. 3. (**e**) At very low twist, the transverse compression due to the clamped edges overcomes the one due to the twist (see Sect. 5.4). *Solid lines* are for quantitative predictions, *dashed lines* indicate scaling laws or unknown thresholds

tension T. The qualitative change at the λ-point reflects two sharply different mechanisms by which the helicoidal shape becomes unstable.

- As the twist exceeds η_{lon} (for $T < T_\lambda$), the ribbon develops longitudinal wrinkles in a narrow zone around its centerline (Fig. 2b,c). Observations that are made close to the emergence of this wrinkle pattern revealed that both the wrinkle's wavelength and the width of the wrinkled zone scale as $\sim(t/\sqrt{T})^{1/2}$. This observation is in excellent agreement with Green's characterization of the helicoidal state, based on the familiar FvK equations of elastic sheets [2]. Green's solution shows that the longitudinal stress at the helicoidal state becomes compressive around the ribbon centerline if $\eta > \sqrt{24T}$, and the linear stability analysis of Coman and Bassom [5] yields the unstable wrinkling mode that relaxes the longitudinal compression.

- As the twist exceeds $\eta_{\rm tr}$ (for $T > T_\lambda$), the ribbon becomes buckled in the transverse direction (Fig. 2g), indicating the existence of transverse compression at the helicoidal state that increases with η. A transverse instability cannot be explained by Green's calculation, which yields no transverse stress [2], but has been predicted by Mockensturm [4], who studied the stability of the helicoidal state using the full nonlinear elasticity equations. Alas, Mockensturm's results were only numerical and did not reveal the scaling behavior $\eta_{\rm tr} \sim \sqrt{t}$ observed in [3]. Furthermore, the nonlinear elasticity equations in [4] account for the inevitable geometric effect (large deflection of the twisted ribbon from its flat state), as well as a mechanical effect (non-Hookean stress-strain relation), whereas only the geometric effect seems to be relevant for the experimental conditions of [3].
- Turning back to $T < T_\lambda$, the ribbon exhibits two striking features as the twist η is increased above the threshold value $\eta_{\rm lon}$. First, the longitudinally-wrinkled ribbon transforms to a shape that resembles the creased helicoid state predicted by [7] (Fig. 2d); this transformation becomes more prominent at small tension (i.e., decreasing T at a fixed value of η). Second, the ribbon undergoes a sharp, secondary transition, described in [3] as similar to the "looping" transition of rods [9–12] (Fig. 2e). At a given tension $T < T_\lambda$, this secondary instability occurs at a critical twist value that *decreases* with T, but is nevertheless significantly larger than $\eta_{\rm lon} \approx \sqrt{24T}$.
- Finally, the parameter regime in the (T, η) plane bounded from below by this secondary instability (for $T < T_\lambda$) and by the transverse buckling instability (for $T > T_\lambda$), is characterized by self-contact zones along the ribbon (Fig. 2e). The formation of loops (for $T < T_\lambda$) is found to be hysteretic unlike the transverse buckling instability (for $T > T_\lambda$).

In a recent commentary [13], Santangelo recognized the challenge and the opportunity introduced to us by this experiment: *"Above all, this paper is a challenge to theorists. Here, we have an experimental system that exhibits a wealth of morphological behavior as a function of a few parameters. Is there anything that can be said beyond the linear stability analysis of a uniform state? How does a smooth, wrinkled state become sharply creased? These are questions that have been asked before, but maybe now there is a possibility to answer them—at least in one system"*. The current paper is motivated by four specific puzzles:

(A) What is the minimal generalization of the standard FvK equations (i.e., beyond Green's calculation) that accounts for the transverse compression of the helicoidal state, and allows a quantitative description of the transverse instability of a ribbon with Hookean stress-strain relationship (i.e., linear material response)?

(B) How does the longitudinally-wrinkled pattern evolve upon exerting a twist η larger than the threshold $\eta_{\rm lon}$, where the state cannot be described any longer as a small perturbation to the compressed helicoidal shape?

(C) Why do the three curves, that mark the thresholds for the secondary, "looping" instability of the helicoidal state, and the two primary instabilities (longitudinal wrinkling and transverse buckling), meet at a single triple point $(T_\lambda, \eta_\lambda)$? If the three thresholds are associated with distinct physical mechanisms, as was conjectured in [3], it would have been natural for them to cross at two points (at least), rather than to meet at a single point.

(D) What is the physical mechanism underlying the transformation of the ribbon from the longitudinally-wrinkled pattern to the creased helicoid shape upon reducing the tension T? Is this a smooth crossover, or a sharp "phase transition" that occurs at some threshold curve in the (T, η) plane?

1.3 Main Results and Outline

Motivated by the above questions, we develop a unified theoretical framework that addresses the rich phenomenology exhibited by the stretched-twisted ribbon in the 4D parameter space spanned by the ribbon length L, its thickness t, the twist η, and the tension T, where we focus on the asymptotic regime defined by Eqs. (1), (2). Our theory leads to a phase diagram whose projection on the tension-twist plane is plotted schematically in Fig. 3, and reveals three major morphological phases: the helicoid, the longitudinally wrinkled state, and a region delimited by the transverse instability. This development is based on three fundamental elements:

(i) A covariant version of FvK equations of elastic sheets, dubbed here "cFvK", which is needed to describe the large deflection (from planarity) of the twisted state of the ribbon.
(ii) A far-from-threshold (FT) expansion of the cFvK equations that describes the state of the ribbon when the twist exceeds the threshold value η_{lon} for the longitudinal wrinkling instability.
(iii) A new, *asymptotic isometry equation* (Eq. (42)), that describes the elastic energies of admissible states of the ribbon in the vicinity of the vertical axis in the parameter plane (T, η). We use the notion of "*asymptotic isometry*" to indicate the unique nature by which the ribbon shape approaches the singular limit of vanishing thickness and tension ($t \to 0, T \to 0$ and fixed η and L).

We commence our study in Sect. 2 with the helicoidal state of the ribbon (Fig. 2a)—a highly symmetric state whose mechanics was addressed by Green through the standard FvK equations [2], which is valid for describing small deviations of an elastic sheet from its planar state. We employ a covariant form of the FvK theory for Hookean sheets (cFvK equations), which takes into full consideration the large deflection of the helicoidal shape from planarity. Our analysis of the cFvK equations provides an answer to question (**A**) above, curing a central shortcoming of Green's approach, which provides the longitudinal stress but predicts a vanishing transverse stress. The cFvK equations of the helicoidal state yield both components of the stress tensor, and show that the magnitude of the transverse stress is nonzero, albeit much smaller than the longitudinal one. Another crucial difference between the two stress components of the helicoidal state pertains to their sign: the transverse stress is compressive throughout the whole ribbon, everywhere in the parameter plane (T, η); in contrast, the longitudinal stress is compressive in a zone around the ribbon centerline only for $\eta > \eta_{\text{lon}} \approx \sqrt{24T}$. The compressive nature of the stress components gives rise to buckling and wrinkling instabilities that we address in Sects. 3 and 4.

In Sect. 3 we address the wrinkling instability that relaxes the longitudinal compression for $\eta > \sqrt{24T}$. Noticing that the longitudinally-compressed zone of the helicoidal state broadens upon increasing the ratio $\alpha = \eta^2/T$, we recognize a close analogy between the longitudinally-wrinkled state of the ribbon and wrinkling phenomena in radially-stretched sheets [14–16], where the size of the wrinkled zone depends on a *confinement* parameter, defined by a ratio between the loads exerted on the sheet. Exploiting this analogy further, we find that the longitudinally-wrinkled ribbon at $\eta > \eta_{\text{lon}}$ is described by a *far-from-threshold (FT)* expansion of the cFvK equations, where the longitudinal stress (at any given $\alpha > 24$) becomes compression-free in the singular limit of an infinitely thin ribbon, $t \to 0$. The FT theory predicts that the broadening of the wrinkled zone with the confinement α is dramatically larger than the prediction of a near-threshold (NT) approach, which is based on a perturbative (amplitude) expansion around the compressive helicoidal state. Our FT theory of the longitudinally wrinkled state provides an answer to question (**B**) in the above list.

Analyzing the FT expansion in the two limits $\alpha \to 24$ (i.e., $\eta \to \sqrt{24T}$), and $\alpha \to \infty$ (i.e., fixed η and $T \to 0$), elucidates further the nature of the longitudinally wrinkled state. In the limit $\alpha \to 24$, plotted schematically in Fig. 3d, we find that the FT regime prevails in the domain $\eta > \sqrt{24T}$ in the (T, η) plane, whereas the NT parameter regime, at which the state is described as a perturbation to the unwrinkled helicoidal state, shrinks to a narrow sliver close to the threshold curve as the thickness vanishes, $t \to 0$. Analyzing the other limit, $\alpha \to \infty$, we show that the longitudinally-wrinkled state becomes an asymptotic isometry, where the strain vanishes throughout the twisted ribbon. In Sect. 5 we expand more on the meaning and implications of asymptotic isometries for a stretched-twisted ribbon. The FT analysis of the two limits, $\alpha \to 24$ and $\alpha \to \infty$, reveals the intricate mechanics of a ribbon subjected to twist η, whereby the longitudinally wrinkled state entails a continuous trajectory in the (T, η) plane, from a strainless deformation (at $T \to 0$) to a fully strained helicoidal shape (at $T \geq \eta^2/24$).

In Sect. 4 we turn to the transverse instability, capitalizing on our results from Sects. 2 and 3. First, we note that the transverse stress is compressive everywhere in the (T, η) plane; second, we note that it is obscured by the longitudinal stress. These two features imply that the threshold for the transverse instability occurs at a curve $\eta_{\text{tr}}(T)$ in the (T, η) plane that divides it into two parts: In the first part, defined by the inequality $\eta_{\text{tr}}(T) < \sqrt{24T}$, the longitudinal stress is purely tensile, and the transverse instability appears as a primary instability of the helicoidal state; in the second part, defined by $\eta_{\text{tr}}(T) > \sqrt{24T}$, the transverse instability is preceded by the longitudinal instability, and thus materializes as a secondary instability of the helicoidal state. We conclude that the "looping" instability observed in [3] does not stem from a new physical mechanism, but simply reflects the change in nature of the transverse instability when the threshold line $\eta_{\text{tr}}(T)$ crosses the curve $\eta_{\text{lon}} = \sqrt{24T}$ that separates the longitudinally-compressed and longitudinally-tensed domains of the (T, η) plane. Thus, the emergence of a single "triple" point $(T_\lambda, \eta_\lambda)$ is not mysterious, but comes naturally as the intersection of these two curves in the (T, η) plane. This result answers question (**C**) in our list.

The cFvK equations, together with the FT analysis of the longitudinally-wrinkled state in Sect. 3, allow us to compute the deformation modes that relax the transverse compression. Two results from this stability analysis are noteworthy. First, assuming an infinitely long ribbon, we find that the threshold curve satisfies $\eta_{\text{tr}}(T) \sim t/\sqrt{T}$ in both the "low"-tension regime ($T < T_\lambda$) and "large"-tension regime ($T > T_\lambda$), albeit with different numerical prefactors. This theoretical prediction is in strong accord with the experimental data for the transverse buckling instability and the "looping" instability in [3]. Second, we find that the length of the ribbon has a dramatic effect on the dependence of the λ-point on the ribbon thickness t, and—more importantly—on the spatial structure of the transverse instability. Specifically, we predict that if $L^{-2} \ll t \ll 1$, the transverse instability is buckling, and if $t \ll L^{-2} \ll 1$, it may give rise to a wrinkling pattern, similarly to a stretched, untwisted ribbon [8], with a characteristic wavelength $\lambda_{\text{tr}} < 1$ that becomes smaller as T increases. This "buckling to wrinkling" transition is depicted in Fig. 3b.

In Sect. 5 we turn to the edges of the (T, η) plane, namely, the vicinity of the vertical and horizontal axes: $(T = 0, \eta)$ and $(T, \eta = 0)$, respectively. In order to address the first limit, we briefly review the work of Korte et al. [7] that predicted and analyzed the creased helicoid state. We discuss the asymptotic isometry exhibited by the creased helicoid state in the singular limit $t \to 0$, $T \to 0$, and contrast it with the asymptotic isometry of the longitudinally wrinkled state, which was noted first in Sect. 3. We elucidate a general framework for analyzing morphological transitions between various types of asymptotic isometries in the neighborhood of the singular hyper-plane $t = 0$, $T = 0$ in the 4D parameter space (T, η, t, L). As a consequence of this discussion, we propose the scenario illustrated in Fig. 3a, where the longitudinally wrinkled state undergoes a sharp transition to the creased helicoid state in the vicinity of the $(T = 0, \eta)$ line. Thus, while our discussion here is less rigorous than in the previous sections (due to the complexity of the creased helicoid state [7]), we nevertheless provide a heuristic answer to question (**D**) in our list.

Since the characterization of the creased helicoid state in [7] is based on the Sadowsky's formalism of inextensible strips rather than on the FvK theory of elastic sheets, we use this opportunity to elaborate on the basic difference between the "rod-like" and "plate-like" approaches to the mechanics of ribbons. We also recall another rod-like approach, based on implementation of the classical Kirchoff equations for a rod with anisotropic cross section [9–12], and explain why it is not suitable to study the ribbon limit (Eq. (1)) that corresponds to a rod with highly anisotropic cross section. Finally, we turn to the vicinity of the pure-stretching line, $(\eta = 0, T)$, and address the parameter regime where the twist η is so small that the ribbon does not accommodate a helicoidal shape. We provide a heuristic, energy-based argument, which indicates that the helicoidal state is established if the twist η is larger than a minimal value that is proportional to the Poisson ratio, and scales as $1/\sqrt{L}$.

Each section (Sects. 2–5) starts with an overview that provides a detailed description of the main results in that section. Given the considerable length of this manuscript, a first reading may be focused on these overview subsections only (Sects. 2.1, 3.1, 4.1, 5.1), followed by Sect. 6, where we describe experimental challenges and propose a list of theoretical questions inspired by our work.

2 Helicoidal State

2.1 Overview

The helicoidal state has been studied by Green [2], who computed its stress field using the standard version of the FvK equations (8), (9). This familiar form, to which we refer here as the ss-FvK equations ("ss" stands for "small slope") is valid for small deflections of elastic sheets from their planar state [17]. The Green's stress, Eqs. (21), (22), has a longitudinal component that contains terms proportional to T and to η^2, and no transverse component. However, the experiments of [3], as well as numerical simulations [18, 19], have exhibited a buckling instability of the helicoidal state in the transverse direction, indicating the presence of transverse compression. One may suspect that the absence of transverse component in Green's stress indicates that the magnitude of this component is small, being proportional to a high power of the twist η, which cannot be captured by the ss-FvK equations.

Here we resort to a covariant form of the FvK equations, which we call "cFvK" [20–22], that does not assume a planar reference state, and is thus capable of describing large deviations from a planar state. Notably, the large deflection of the helicoidal state from planarity does not involve large strains. Hence, as long as $T, \eta \ll 1$, we consider a ribbon with Hookean response, namely—linear stress-strain relationship. This approach is simpler than Mockensturm's [4] (which assumes a non-Hookean, material-dependent response), and enables the analytical progress in this section and the following ones. Solving the cFvK equations for the helicoidal state, we get the following expressions for the stress field in the longitudinal ($\hat{s}$) and transverse ($\hat{r}$) directions:

$$\sigma^{ss}_{\text{hel}}(r) = T + \frac{\eta^2}{2}\left(r^2 - \frac{1}{12}\right), \tag{3}$$

$$\sigma^{rr}_{\text{hel}}(r) = \frac{\eta^2}{2}\left(r^2 - \frac{1}{4}\right)\left[T + \frac{\eta^2}{4}\left(r^2 + \frac{1}{12}\right)\right], \tag{4}$$

where $r \in [-1/2, 1/2]$ is the dimensionless transverse coordinate. The longitudinal component is exactly the one found by Green [2], whereas the transverse component is nonzero, albeit of small magnitude: $\sigma^{rr}_{\text{hel}} \sim \eta^2 \sigma^{ss}_{\text{hel}}$, which explains why it is missed by the ss-FvK equations. The transverse stress arises from a subtle coupling between the longitudinal stress and the geometry of the ribbon.

As Eqs. (3), (4) show, the longitudinal stress $\sigma^{ss}_{\text{hel}}(r)$ is compressive close to the centerline $r = 0$ if $\eta^2 > 24T$, whereas the transverse stress $\sigma^{rr}_{\text{hel}}(r)$ is compressive everywhere in the ribbon for any (T, η). The compressive nature of $\sigma^{ss}_{\text{hel}}(r)$ and $\sigma^{rr}_{\text{hel}}(r)$ leads to buckling and wrinkling instabilities that we address in the next sections.

In Sect. 2.2 we review the (standard) ss-FvK equations, and their helicoidal solution found by Green [2]. In Sect. 2.3 we proceed to derive the cFvK equations, following [22], and use this covariant formalism to determine the stress in the helicoidal state.

2.2 Small Slope Approximation and the Green's Solution

2.2.1 Small-Slope FvK Equations

We review briefly the standard ss-FvK equations, using some basic concepts of differential geometry that will allow us to introduce their covariant version in the next subsection. Assuming a small deviation from a plane, a sheet is defined by its out-of-plane displacement $z(s, r)$ and its in-plane displacements $u_s(s, r)$ and $u_r(s, r)$; where s and r are the material coordinates. In this configuration, the strain is given by

$$\varepsilon_{\alpha\beta} = \frac{1}{2}\left[\partial_\alpha u_\beta + \partial_\beta u_\alpha + (\partial_\alpha z)(\partial_\beta z)\right]. \tag{5}$$

The Greek indices α and β take the values s or r. We define the curvature tensor and the mean curvature as:

$$c_{\alpha\beta} = \partial_\alpha \partial_\beta z, \tag{6}$$

$$H = \frac{1}{2} c^\alpha_\alpha = \frac{1}{2}\Delta z, \tag{7}$$

where we use the Einstein summation convention, such that c^α_α is the trace of the curvature tensor. The use of upper or lower indices corresponds to the nature of the tensor (contravariant or covariant, respectively), which will become relevant in the next subsection. The ss-FvK equations express the force balance in the normal direction ($\hat{z}$) and the in-plane directions ($\hat{s}, \hat{r}$), and involve the curvature tensor and the stress tensor $\sigma^{\alpha\beta}(s, r)$:

$$c_{\alpha\beta}\sigma^{\alpha\beta} = 2B\Delta H, \tag{8}$$

$$\partial_\alpha \sigma^{\alpha\beta} = 0, \tag{9}$$

where $B = t^2/[12(1-\nu^2)]$ is the bending modulus of the sheet.[2] The stress-strain relationship is given by Hooke's law (linear material response):

$$\sigma^{\alpha\beta} = \frac{1}{1+\nu}\varepsilon^{\alpha\beta} + \frac{\nu}{1-\nu^2}\varepsilon^{\gamma}_{\gamma}\delta^{\alpha\beta}, \tag{10}$$

where we used the Kronecker symbol $\delta^{\alpha\beta}$.

2.2.2 Green's Solution for the Helicoid

We now apply the ss-FvK equations (8), (9) to find the stress in the helicoidal state. Since this formalism assumes a small deviation of the ribbon from the plane, we approximate the helicoidal shape through:

$$z(s,r) = \eta s r, \tag{11}$$

obtained by Taylor expansion of the z-coordinate of the full helicoidal shape, given below in Eq. (29), for $|s| \ll \eta^{-1}$. Its corresponding curvature tensor is:

$$c_{\alpha\beta} = \eta \begin{pmatrix} 0 & 1 \\ 1 & 0 \end{pmatrix}, \tag{12}$$

leading to the mean curvature $H = 0$. The force balance equations (8)–(9) now read:

$$\sigma^{sr} = 0, \tag{13}$$

$$\partial_s \sigma^{ss} = 0, \tag{14}$$

$$\partial_r \sigma^{rr} = 0. \tag{15}$$

The ss-FvK equations are supplemented by two boundary conditions: The longitudinal stress must match the tensile load exerted on the short edges, whereas the long edges are free, namely:

$$\int_{-1/2}^{1/2} \sigma^{ss}(r)dr = T, \tag{16}$$

$$\sigma^{rr}(r = \pm 1/2) = 0. \tag{17}$$

Since Eq. (15) implies that the transverse stress is uniform across the ribbon, the boundary condition (17) implies that it is identically zero:

$$\sigma^{rr}(r) = 0. \tag{18}$$

With Hooke's law (10), this shows that $\sigma^{ss} = \varepsilon^{ss}$. Using the small slope expressions for the strain-displacement relationship (Eq. (5)) and for helicoidal shape (Eq. (11)), we obtain the longitudinal stress

$$\sigma^{ss}(r) = \frac{\eta^2 r^2}{2} - \chi, \tag{19}$$

[2]Recall that we normalize stresses by the stretching modulus Y and lengths by the ribbon width W. The dimensional bending modulus is thus: $Y(Wt)^2/(12(1-\nu^2))$.

where $\chi = -\partial_s u_s$ is the longitudinal contraction of the ribbon. Since χ does not depend on s (due to Eq. (14)) or on r (due to the translational symmetry of the helicoidal shape along $\hat{\boldsymbol{s}}$), its value is determined by the condition (16):

$$\chi = \frac{\eta^2}{24} - T. \tag{20}$$

We thus obtain the Green's stress [2]:

$$\sigma^{ss}(r) = T + \frac{\eta^2}{2}\left(r^2 - \frac{1}{12}\right), \tag{21}$$

$$\sigma^{rr}(r) = 0; \qquad \sigma^{rs} = 0. \tag{22}$$

2.3 Covariant FvK and the Helicoidal Solution

2.3.1 Covariant FvK Equations

In order to address sheet's configurations that are far from planarity, we must avoid any reference to a planar state. The shape of the sheet is now described by a surface $\boldsymbol{X}(s, r)$, and the covariant form of the force balance equations, which we call here the cFvK equations, requires us to revisit the definitions of the quantities invoked in our description of the ss-FvK equations: the strain, the curvature, and the derivative. We do this by following the general approach of [22].

First, we define the surface *metric* as a covariant tensor:

$$g_{\alpha\beta} = \partial_\alpha \boldsymbol{X} \cdot \partial_\beta \boldsymbol{X}, \tag{23}$$

where the inverse metric is a contravariant tensor, denoted with upper indices, that satisfies $g^{\alpha\beta} g_{\beta\gamma} = \delta^\alpha_\gamma$ (δ^α_γ is the Kronecker symbol). The strain is defined as the difference between the metric and the rest metric $\bar{g}_{\alpha\beta}$:

$$\varepsilon_{\alpha\beta} = \frac{1}{2}(g_{\alpha\beta} - \bar{g}_{\alpha\beta}). \tag{24}$$

The curvature tensor (6) is now defined by

$$c_{\alpha\beta} = \hat{\boldsymbol{n}} \cdot \partial_\alpha \partial_\beta \boldsymbol{X}, \tag{25}$$

where $\hat{\boldsymbol{n}}$ is the unit normal vector to the surface (the ss-FvK equations are based on the approximation $\hat{\boldsymbol{n}} \approx \hat{\boldsymbol{z}}$).

In this formulation, the covariant/contravariant nature of tensors does matter, for instance: $c_{\alpha\beta} \neq c^{\alpha\beta}$. To lower or raise the indices, one must use the metric or its inverse, respectively: $c^\alpha_\beta = g_{\beta\gamma} c^{\alpha\gamma} = g^{\alpha\gamma} c_{\gamma\beta}$.

The mean curvature now invokes the inverse metric, $H = c^\alpha_\alpha/2 = g^{\alpha\beta} c_{\alpha\beta}/2$, and the Gaussian curvature of the surface is: $K = \frac{1}{2}(c^\alpha_\alpha c^\beta_\beta - c^\alpha_\beta c^\beta_\alpha)$. Hooke's law (10) is only slightly changed:[3]

$$\sigma^{\alpha\beta} = \frac{1}{1+\nu}\varepsilon^{\alpha\beta} + \frac{\nu}{1-\nu^2}\varepsilon^\gamma_\gamma g^{\alpha\beta}, \tag{26}$$

[3] Other terms, proportional to t^2, may appear on the right hand side of Eq. (26) [22]; however, they are negligible here.

and the force balance equations (8)–(9) now read

$$c_{\alpha\beta}\sigma^{\alpha\beta} = 2B\big[D_\alpha D^\alpha H + 2H\big(H^2 - K\big)\big], \tag{27}$$

$$D_\alpha \sigma^{\alpha\beta} = 0. \tag{28}$$

There are two major differences between the ss-FvK equations (8)–(9) and the cFvK equations (27)–(28). First, there is a new term in the normal force balance (27); this term may be relevant when the equilibrium shape is characterized by a uniform, nonvanishing mean curvature (such that $|H^3|$ or $|HK|$ are comparable to or larger than $|D_\alpha D^\alpha H|$), but is negligible for a surface that can be described by small deviations from a plane or a helicoid, for which $H \approx 0$. Second—and central to our analysis—the usual derivative ∂_α is replaced by the *covariant derivative* D_α that takes into account the variation of the metric along the surface. The covariant derivative D_α is defined through the Christoffel symbols of the surface, and is given in Appendix B.

2.3.2 *Application to the Helicoid*

Here, we show that the helicoid is a solution of the cFvK equations and determine its stress and strain. The helicoidal shape is described by

$$\begin{aligned} \boldsymbol{X}(s,r) &= (1-\chi)s\hat{\boldsymbol{x}} + \big[r + u_r(r)\big]\cos(\eta s)\hat{\boldsymbol{y}} + \big[r + u_r(r)\big]\sin(\eta s)\hat{\boldsymbol{z}} \\ &= \begin{pmatrix} (1-\chi)s \\ [r+u_r(r)]\cos(\eta s) \\ [r+u_r(r)]\sin(\eta s) \end{pmatrix}, \end{aligned} \tag{29}$$

where $(\hat{\boldsymbol{x}}, \hat{\boldsymbol{y}}, \hat{\boldsymbol{z}})$ is the standard basis of the three-dimensional space. The longitudinal contraction χ and transverse displacement $u_r(r)$ are small (i.e., both vanish when $T = \eta = 0$), and must be determined by our solution. Expanding Eq. (29) to leading order in χ and $u_r(r)$ we obtain the metric:

$$g_{\alpha\beta} = \begin{pmatrix} 1 + \eta^2 r^2 - 2\chi + 2\eta^2 r u_r(r) & 0 \\ 0 & 1 + 2u_r'(r) \end{pmatrix}. \tag{30}$$

The curvature tensor is still given by (12), to leading order in χ and $u_r(r)$, and the mean curvature in this approximation is $H = 0$.

It must be understood that in deriving the metric tensor, Eq. (30), we assumed that both the twist and the exerted tension are small ($\eta \ll 1, T \ll 1$), such that χ and $u_r(r)$ (which appear explicitly in $g_{\alpha\beta}$) can be expressed as expansions in η and T that vanish for $\eta, T \to 0$. This natural assumption, which simplifies considerably the forthcoming analysis, implies that a consistent calculation of the stress components σ^{ss}, σ^{sr}, and σ^{rr}, must treat them as expansions in η and T (in Appendix A we discuss this issue further). With this in mind, we note that the force balance equations (27)–(28) become, to leading order in η:

$$\sigma^{sr} = 0, \tag{31}$$

$$\partial_s \sigma^{ss} = 0, \tag{32}$$

$$\partial_r \sigma^{rr} - \eta^2 r \sigma^{ss} = 0. \tag{33}$$

The second term in the left hand side of Eq. (33), which has no analog in the ss-FvK equations (14)–(15), encapsulates the coupling of the transverse and longitudinal stress components imposed by the non-planar helicoidal structure. Its derivation, which reflects the profound role of the covariant derivative in our study, is detailed in Appendix B. Now, comparing the two terms in Eq. (33) shows that for $\eta \ll 1$:

$$\sigma^{rr} \sim \eta^2 \sigma^{ss} \ll \sigma^{ss}. \tag{34}$$

Recalling that our computation of the stress components assumes an expansion in η and T, the inequality (34) implies that the expansion of σ^{rr} starts with a higher order term than the expansion of σ^{ss}. An immediate consequence of this observation is obtained by expressing σ^{ss} and σ^{rr} through Hooke's law. From the metric (30), we deduce the strain (24):

$$\varepsilon_{\alpha\beta} = \begin{pmatrix} \frac{\eta^2 r^2}{2} - \chi + \eta^2 r u_r(r) & 0 \\ 0 & u_r'(r) \end{pmatrix}, \tag{35}$$

where we substituted $\bar{g}_{\alpha\beta} = \delta_{\alpha\beta}$. Using Hooke's law to compute the stress components to leading order in η (where we anticipate that both χ and $u_r(r)$ vanish as $\eta \to 0$), we obtain:

$$\sigma^{ss} = \frac{1}{1-\nu^2}\left(\frac{\eta^2 r^2}{2} - \chi\right) + \frac{\nu}{1-\nu^2} u_r'(r), \tag{36}$$

$$\sigma^{rr} = \frac{1}{1-\nu^2} u_r'(r) + \frac{\nu}{1-\nu^2}\left(\frac{\eta^2 r^2}{2} - \chi\right). \tag{37}$$

Since the force balance Eq. (33) implies that an expansion in η and T is valid only if σ^{rr} starts with a higher order than σ^{ss}, Eqs. (36), (37) yield the solvability condition:

$$u_r'(r) = -\nu\left(\frac{\eta^2 r^2}{2} - \chi\right), \tag{38}$$

which guarantees that $\sigma^{ss} \sim O(T, \eta^2)$, whereas σ^{rr} has no terms of that order (such that $\sigma^{rr} \sim O(T\eta^2, \eta^4)$), consistently with Eq. (33). Inserting this result into Eq. (36) gives the same longitudinal stress (19) as the small-slope approximation; the longitudinal contraction (20) does not change either.

Now that the longitudinal stress is known, the transverse component is obtained by integrating Eq. (33) with the boundary condition (17), so that finally:

$$\sigma^{ss}(r) = T + \frac{\eta^2}{2}\left(r^2 - \frac{1}{12}\right), \tag{39}$$

$$\sigma^{rr}(r) = \frac{\eta^2}{2}\left(r^2 - \frac{1}{4}\right)\left[T + \frac{\eta^2}{4}\left(r^2 + \frac{1}{12}\right)\right]. \tag{40}$$

Comparing these equations to the Green's stress (21)–(22), which was obtained through the ss-FvK equations, we note two facts: First, the longitudinal component is unchanged. Second, we find a compressive transverse component that originates from the coupling of the transverse and longitudinal stress components by the helicoidal geometry of the ribbon. Since the transverse component is much smaller than the longitudinal one, the Green's stress is useful for studying certain phenomena, most importantly—the longitudinal instability of

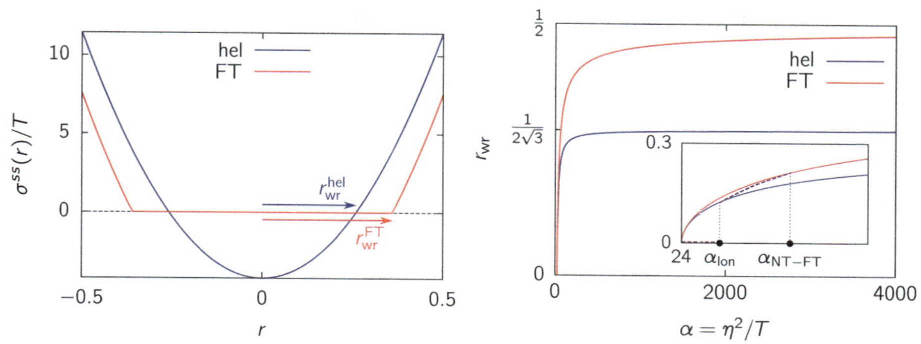

Fig. 4 *Left:* longitudinal stress of the helicoidal state (that approximates the stress in the NT regime), and of the far from threshold (FT) longitudinally wrinkled state, where r_{wr} is the extent of the respective wrinkled zone. The confinement is $\alpha = 125$. *Right:* extent of the wrinkled zone r_{wr} in the NT regime, where it is approximated through the helicoidal state (where r_{wr} is defined as the width of the zone under longitudinal compression), and in the FT regime. *Inset:* the ribbon supports compression without wrinkling for $24 < \alpha < \alpha_{lon}$, and then the extent of the wrinkled zone interpolates between the NT and FT predictions for $\alpha_{lon} < \alpha < \alpha_{NT\text{-}FT}$. Above $\alpha_{NT\text{-}FT}$, the state is described by the FT approach

the helicoidal state [5]. However, the instability of the ribbon that stems from the compressive transverse stress is totally overlooked in Green's approach. Furthermore, the covariant formalism provides a considerable conceptual improvement to our understanding since it allows to think of the helicoid (or any other shape) without assuming a planar reference shape. Finally, let us re-emphasize that, although the transverse stress $\sigma^{rr}(r)$ is proportional to products of the small exerted strains $(T\eta^2, \eta^4)$, it originates from Hookean response of the material; its small magnitude simply reflects the small transverse strain in the helicoidal shape.

3 Longitudinal Wrinkling

3.1 Overview

If the twist is sufficiently large with respect to the exerted tension, the stress in the helicoidal state becomes compressive in the longitudinal direction in a zone around the ribbon centerline. This can be easily seen from the expression (3): if $\eta > \eta_{lon}(T) = \sqrt{24T}$, then $\sigma_{hel}^{ss}(r) < 0$ for $|r| < r_{wr}$, where the width r_{wr} increases with the ratio η^2/T (see Fig. 4). This effect reflects the helicoidal geometry, where the long edges are extended with respect to the centerline, such that the longitudinally compressive zone expands outward upon reducing the exerted tension. The ratio $\alpha = \eta^2/T$, whose critical value $\alpha = 24$ signifies the emergence of longitudinal compression, plays a central role in this section and we call it the *confinement* parameter:

$$\text{Confinement:} \quad \alpha \equiv \frac{\eta^2}{T}. \tag{41}$$

Near Threshold (NT) and Far from Threshold (FT) Regimes The longitudinal compression may induce a wrinkling instability, where periodic undulations of the helicoidal shape relax the compression in the zone $|r| < r_{wr}$. A natural way to study this instability is through linear

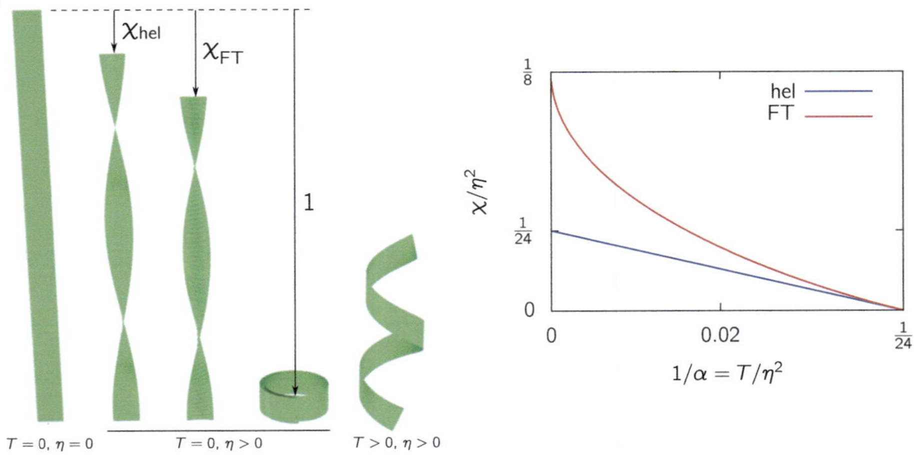

Fig. 5 *Left:* Longitudinal contraction (defined with respect to the untwisted ribbon without any tension) of the helicoidal (unwrinkled) state, the FT-longitudinally-wrinkled state, and the cylindrical wrapping state as $T \to 0$. *Right:* Longitudinal contractions of the helicoidal state and the FT-longitudinally-wrinkled state as a function of $1/\alpha$

stability analysis, which assumes that the longitudinally-wrinkled state of the ribbon can be described as a small perturbation to the compressed helicoidal state [5]. While this perturbative approach is useful to address the wavelength λ_{lon} of the wrinkle pattern at threshold [3], we argue that it describes the ribbon state only at a narrow, *near threshold (NT)* regime in the (η, T) plane, above which we must invoke a qualitatively different, *far from threshold (FT)* approach (see Fig. 3d). The fundamental difference between the NT and FT theories is elucidated in Fig. 4, which plots the approximated profiles of the longitudinal stress, $\sigma_{\text{hel}}^{ss}(r)$ and $\sigma_{\text{FT}}^{ss}(r)$, respectively, for a given confinement $\alpha > 24$. The NT theory assumes that the wrinkles relax slightly the compression in $\sigma_{\text{hel}}^{ss}(r)$, whereas the FT theory assumes that at a given $\alpha > 24$ the stress in the longitudinally-wrinkled ribbon approaches a compression-free profile as $t \to 0$.[4] For a very thin ribbon, which can support only negligible level of compression, the transition between the NT and FT regimes converges to the threshold curve $\eta_{\text{lon}}(T)$ (see Fig. 3d).

The sharp contrast between the NT and FT theories is further elucidated in Figs. 4, 5, and 6, where the respective predictions for the spatial width of the longitudinally-wrinkled zone, the longitudinal contraction, and the energy stored in the ribbon are compared. Figure 4 shows that the wrinkled zone predicted by the FT theory expands beyond the compressed zone of the helicoidal state. Furthermore, as the confinement α increases, the FT theory predicts that the wrinkled zone invades the whole ribbon (except narrow strips that accommodate the exerted tension), whereas the compressed zone of the helicoidal state covers only a finite fraction $(1/\sqrt{3})$ of the ribbon width. Figure 5 shows that the longidudinal contraction predicted by the FT theory is larger than the contraction of the unwrinkled helicoidal state, and the ratio between the respective contractions $\chi_{\text{FT}}/\chi_{\text{hel}} \to 3$ as $\alpha \to \infty$. Figure 6

[4]More precisely, the NT method is an amplitude expansion of FvK equations around the compressed helicoidal state, whereas the FT theory is an asymptotic expansion of the FvK equations around the singular limit $t \to 0$, carried at a fixed confinement α. In this limit, the longitudinally wrinkled state of the ribbon approaches the compression-free stress $\sigma_{\text{FT}}^{ss}(r)$.

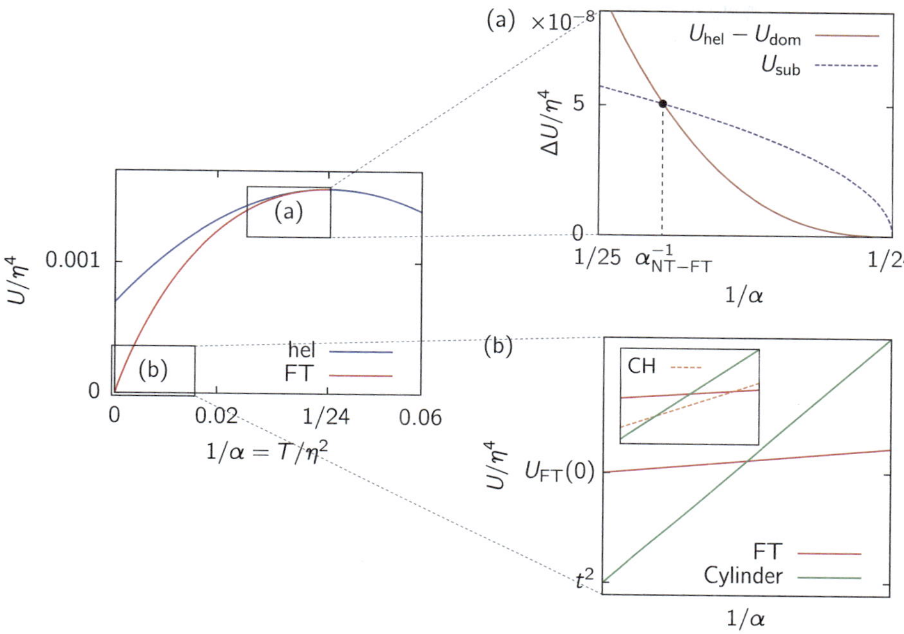

Fig. 6 *Left:* Dominant energy stored in the stress field of the ribbon as a function of the inverse confinement $1/\alpha = T/\eta^2$ at the helicoidal state and at the far from threshold (FT) longitudinally wrinkled state. *Right:* (**a**) Energy difference $U_{\text{hel}} - U_{\text{dom}}$ and the subdominant energy U_{sub} due to the wrinkles close to the threshold. (**b**) Energies of the FT longitudinally wrinkled helicoid and the cylinder wrapping at vanishing tension ($1/\alpha \to 0$); *Inset:* energy of the creased helicoid (CH) is added (see Sect. 5.3)

plots the energies stored in the compressive helicoidal state (U_{hel}) and in the compression-free state (U_{dom}), demonstrating the significant gain of elastic energy enabled by the collapse of compression. Focusing on the vicinity of $\alpha = 24$, we illustrate in Fig. 6a how the vanishing size of the NT parameter regime for $t \ll 1$ results from the small (amplitude-dependent) reduction of the energy U_{hel} versus the sub-dominant (t-dependent) addition to the energy U_{dom}. The subdominant energy stems from the small bending resistance of the ribbon in the limit $t \to 0$.

Asymptotically Isometric States Focusing on the limit $\alpha^{-1} \to 0$ in Fig. 6, which describes the ribbon under twist η and infinitely small T, one observes that the dominant energy U_{dom} becomes proportional to T and vanishes as $T \to 0$. This result reflects the remarkable geometrical nature of the FT-longitudinally-wrinkled state, which becomes infinitely close to an isometric (i.e., strainless) map of a ribbon under finite twist η, in the singular limit $t, T \to 0$. At the singular hyper-plane ($t = 0, T = 0$), which corresponds to an ideal ribbon with no bending resistance and no exerted tension, the FT-longitudinally-wrinkled state is energetically equivalent to simpler, twist-accommodating isometries of the ribbon: the cylindrical shape (Fig. 5) and the creased helicoid shape (Fig. 2d, [7]). We argue that this degeneracy is removed in an infinitesimal neighborhood of the singular hyper-plane (i.e., $t > 0, T > 0$), where the energy of each *asymptotically isometric* state is described by a linear function of T with a t-independent slope and a t-dependent intercept. Specifically:

$$U_j(t, T) = A_j T + B_j t^{2\beta_j}, \tag{42}$$

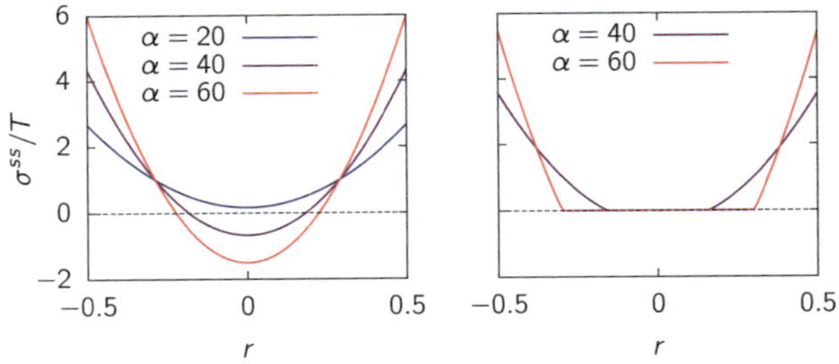

Fig. 7 Longitudinal stress along the width of the ribbon in the helicoidal state (*left*) and in the far from threshold longitudinally wrinkled state (*right*) for different values of the confinement parameter α

where j labels the asymptotic isometry type (cylindrical, creased helicoid, longitudinal wrinkles), and $0 < \beta_j < 1$. For a fixed twist $\eta \ll 1$, we argue that the intercept $(Bt^{2\beta})$ is smallest for the cylindrical state, whereas the slope (A) is smallest for the FT-longitudinally-wrinkled state. This scenario, which is depicted in Fig. 3a, underlies the instability of the longitudinally wrinkled state in the vicinity of the axis $T = 0$ in the (T, η) plane.

The concept of asymptotic isometries has been inspired by a recent study of an elastic sheet attached to a curved substrate [23]. We conjecture that the form of Eq. (42) is rather generic, and underlies morphological transitions also in other problems, where thin elastic sheets under geometric confinement (e.g., twist or imposed curvature) are subjected to small tensile loads.

We start in Sect. 3.2 with a brief review of the linear stability analysis. In Sect. 3.3 we introduce the FT theory, and discuss in detail the compression-free stress σ_{FT} and its energy U_{dom}. In Sect. 3.4 we address the transition from the NT to the FT regime. In Sect. 3.5 we introduce the asymptotic isometries, where we explain the origin of Eq. (42) and compare the energetic costs of the cylindrical and the longitudinally-wrinkled states.

3.2 Linear Stability Analysis

In this subsection we develop a linear stability analysis of the longitudinal wrinkling, following [2, 3, 5] and focusing on scaling-type arguments rather than on exact solutions. We use the small slope approximation of the FvK equations (see Sect. 2.2) and its Green's solution (11), (21)–(22). This approximation is justified here since the transverse stress σ_{hel}^{rr} is smaller by a factor η^2 than the longitudinal stress σ_{hel}^{ss} which is responsible for the instability.

Dividing σ_{hel}^{ss} by the tension T we obtain a function of the transverse coordinate r that depends only on the confinement parameter α (41) and is plotted in Fig. 7 for three representative values of α. For $\alpha > 24$ a zone $|r| < r_{\text{wr}}(\alpha)$ around the ribbon centerline is under compression, and we thus expect that for a thin ribbon the threshold value for the longitudinal instability follows $\alpha_{\text{lon}}(t) \to 24$ when $t \to 0$. A simple analysis of the function $\sigma_{\text{hel}}^{ss}(r)$ leads to the following scalings for the magnitude of the compression $\sigma_{\text{hel}}^{ss}(r = 0)$ and the width r_{wr} of the compressed zone near the threshold:

$$\begin{aligned} \sigma_{\text{hel}}^{ss}(r = 0) &\sim T(\alpha - 24), \\ r_{\text{wr}} &\sim \sqrt{\alpha - 24}. \end{aligned} \tag{43}$$

Consider now a small perturbation of the planar approximation (11) of the helicoidal state such that $z(s, r) \simeq \eta s r + \zeta z_1(s, r)$, where ζ is a small parameter. Substituting this expression in the normal force balance (8), we obtain a linear equation for $z_1(s, r)$

$$\sigma_{\text{hel}}^{ss} \partial_s^2 z_1 = B \Delta^2 z_1. \tag{44}$$

Equation (44) should be understood as the leading order equation in an *amplitude expansion* of the ss-FvK equations (8)–(9) around the helicoidal state, where the small parameter is the amplitude ζ of the wrinkle pattern. The absence of s-dependent terms in Eq. (44) stems from the *translational symmetry* in the longitudinal direction of the helicoidal state that is broken by the wrinkling instability. The natural modes are thus: $z_1(s,r)=\cos(2\pi s/\lambda_{\text{lon}})f(r)$, where λ_{lon} is the wrinkles wavelength and $f(r)$ is a function that vanishes outside the compressive zone of σ_{ss}^{hel}. An exact calculation of λ_{lon}, $f(r)$ and the threshold $\alpha_{\text{lon}}(t)$ can be found in [5], but the scaling behavior with t can be obtained (as was done in [3]) by noticing that the most unstable mode is characterized by a "dominant balance" of all forces in Eq. (44): The restoring forces, which are associated here with the bending resistance to deflection in the two directions, $B\partial_s^4 z_1$ and $B\partial_r^4 z_1$, as well as the destabilizing force $\sigma_{\text{hel}}^{ss}\partial_s^2 z_1$. Equating these forces yields the two scaling relations: $\lambda_{\text{lon}}\sim r_{\text{wr}}$, and $B/\lambda_{\text{lon}}^2\sim\sigma_{\text{hel}}^{ss}(r=0)$. With the aid of Eq. (43) we obtain the NT scaling laws:

$$\begin{aligned}\Delta\alpha_{\text{lon}} &= \alpha_{\text{lon}}-24\sim\frac{t}{\sqrt{T}},\\ \lambda_{\text{lon}} &\sim r_{\text{wr}}\sim\frac{\sqrt{t}}{T^{1/4}}.\end{aligned}\tag{45}$$

These scaling laws which are based upon Eq. (43) are only valid for $\Delta\alpha_{\text{lon}}\ll 1$ or, equivalently for $t^2\ll T$. In this regime, the ribbon is so thin that the thresholds for developing a compressive zone and for wrinkling become infinitely close to each other as $t\to 0$. In contrast, in the regime where $t^2\gg T$, the ribbon is too thick compared to the exerted tension and the threshold for wrinkling is much larger (in terms of α) than the threshold for developing compression. In this regime of very small tension, the linear stability analysis of the helicoid is different from the one presented above and has been performed by Green [2]. It resulted in a plateau in the threshold $\eta_{\text{lon}}(T)$, that we refer to as the "Green's plateau":

$$\eta_{\text{lon}}(T)\to 10t\quad\text{for } T\ll T_{\text{sm}},\tag{46}$$

where

$$T_{\text{sm}}\sim t^2.\tag{47}$$

This plateau is pictured in Fig. 3c. It can be obtained by a simple scaling argument, which balances, as before, the longitudinal compression and bending in the longitudinal and transverse directions: $\eta^2/\lambda^2\sim t^2\sim t^2/\lambda_{\text{lon}}^4$, giving $\lambda_{\text{lon}}\sim 1$ and $\eta_{\text{lon}}\sim t$.

3.3 Far-from-Threshold Analysis

As the confinement gets farther from its threshold value, the wrinkle pattern starts to affect considerably the longitudinal stress and can eventually relax completely the compression. The emergence of a compression-free stress field underlying wrinkle patterns has been recognized long ago in the solid mechanics and applied mathematics literature [24–26]. More recently, it has been shown that such a compression-free stress field reflects the leading order of an expansion of the FvK equations under given tensile load conditions [14, 27, 28]. In contrast to the NT analysis, which is based on amplitude expansion of FvK equations around a compressed (helicoidal) state, and whose validity is therefore limited to values of (T,η) at the vicinity of the threshold curve, the FT analysis is an expansion of the FvK equations around the compression-free stress, which is approached in the singular limit $t=0$. For a sufficiently small thickness t, the FT expansion is thus valid for *any* point (T,η) with confinement $\alpha=\eta^2/T>24$ (see footnote 4). The leading order in the FT expansion captures the compression-free stress field $\sigma_{\rm FT}^{\alpha\beta}$, which is independent on t, in the asymptotic limit $t\to 0$. The wrinkled part of the sheet (here $r<|r_{\rm wr}|$) is identified as the zone where a principal component of the stress (here $\sigma_{\rm FT}^{ss}(r)$) vanishes.

Underlying the FT expansion there is a hierarchical energetic structure:

$$U_{\rm FT}(\alpha,t)=U_{\rm dom}(\alpha)+t^{2\beta}F(\alpha), \tag{48}$$

where $0<\beta<1$. The *dominant* term $U_{\rm dom}(\alpha)$ is the elastic energy stored in the compression-free stress field, which depends on the loading conditions (through α) but not on t, and the *sub-dominant* term $t^{2\beta}F(\alpha)$ stems from the small bending resistance of the sheet, which vanishes as $t\to 0$. A nontrivial feature of the FT expansion, which is implicit in Eq. (48), is the singular, degenerate nature of the limit $t\to 0$. There may be multiple wrinkled states, all of which give rise to the same $U_{\rm dom}(\alpha)$ and $\sigma_{\rm FT}^{\alpha\beta}(\alpha)$ and therefore share the same width $r_{\rm wr}(\alpha)$ of the wrinkled zone. The sub-dominant term $t^{2\beta}F(\alpha)$ lifts this degeneracy by selecting the energetically-favorable state, and therefore determines the fine-scale features of the wrinkle pattern, namely: the wavelength $\lambda_{\rm lon}$ [27], the possible emergence of wrinkle cascades [28–31], and so on. In this paper, we focus on the dominant energy $U_{\rm dom}$, and will make only a brief, heuristic comment on the sub-dominant energy and the fine-scale features of the wrinkle pattern.

In the first part of this subsection we find the compression-free stress, and in the second part we study the energy $U_{\rm dom}$ associated with it.

3.3.1 The Compression-Free Stress Field

One may think of the compression-free stress field by imagining a hypothetic ribbon with finite stretching modulus but zero bending resistance. When such a ribbon is twisted (with $\alpha>24$), the helicoidal shape can be retained up to wrinkly undulations of infinitesimal amplitude and wavelength, that fully relax any compression. This hypothetic ribbon is exactly the singular point, $t=0$, around which we carry out the FT expansion. Considering the FvK equations (27), (28), this means that the compression-free stress could be found by assuming the helicoidal shape (29) and searching for a stress whose longitudinal component is non-negative. (Since the magnitude of the compressive transverse component $\sigma_{\rm FT}^{rr}$ is smaller by a factor of η^2 than the longitudinal stress, it has a negligible effect on the longitudinal instability; its effect on the transverse instability will be the subject of the next section.) It must be understood though, that the longitudinal wrinkles, no matter how small their amplitude is, contain a finite fraction of the ribbon's length, which is required to eliminate compression. This effect must be taken into consideration when analyzing the stress-strain relations, Eq. (26), and leads to a "slaving" condition on the amplitude and wavelength of the wrinkles [14].

The above paragraph translates into a straightforward computation of the compression-free stress. We assume a continuous $\sigma_{\rm FT}^{ss}(r)$, which is zero for $|r|<r_{\rm wr}$ and positive for $|r|>r_{\rm wr}$ (see [14–16, 32] for analogous derivations of FT wrinkle patterns in radial stretching set-ups). In the tensile zone there are no wrinkles that modify the helicoidal shape, and inspection of the strain (35) shows that the longitudinal stress must be of the form $\eta^2r^2/2+\text{cst}$. This leads to:

$$\sigma_{\rm FT}^{ss}(r)=\begin{cases}0 & \text{for } |r|<r_{\rm wr},\\ \frac{\eta^2}{2}(r^2-r_{\rm wr}^2) & \text{for } |r|>r_{\rm wr}.\end{cases} \tag{49}$$

Recalling that the integral of $\sigma_{\mathrm{FT}}^{ss}(r)$ over r must equal the exerted force, we obtain an implicit equation for the width $r_{\mathrm{wr}}(\alpha)$:

$$(1-2r_{\mathrm{wr}})^2(1+4r_{\mathrm{wr}}) = \frac{24}{\alpha}. \tag{50}$$

Figure 7 shows the longitudinal stress profile (49) for different values of the confinement α. The wrinkle's width $r_{\mathrm{wr}}(\alpha)$, derived from Eq. (50), is shown in Fig. 4 and compared to the width of the compressive zone in the helicoidal state for the corresponding values of α.

We obtained the stress field (49), (50) by requiring that, in the tensile zone $|r| > r_{\mathrm{wr}}$, the helicoidal shape with the stress $\sigma_{\mathrm{FT}}^{ss}(r)$ form a solution of the cFvK equations (26), (27), (28), subjected to the constraint that $\sigma_{\mathrm{FT}}^{ss}(r) = 0$ at $|r| < r_{\mathrm{wr}}$. In order to understand how the FvK equations are satisfied also in the wrinkled zone $|r| < r_{\mathrm{wr}}$ it is useful to assume the simplest type of wrinkles where the helicoidal shape is decorated with periodic undulations of wavelength $2\pi/k$ and amplitude $f(r)$:

$$\boldsymbol{X}^{(\mathrm{wr})}(s,r) = \begin{pmatrix} (1-\chi_{\mathrm{FT}})s \\ r\cos(\eta s) - f(r)\cos(ks)\sin(\eta s) \\ r\sin(\eta s) + f(r)\cos(ks)\cos(\eta s) \end{pmatrix}, \tag{51}$$

where the longitudinal contraction is given by

$$\chi_{\mathrm{FT}} = \frac{1}{2}\eta^2 r_{\mathrm{wr}}^2, \tag{52}$$

which follows from Eq. (19) and the continuity of $\sigma_{\mathrm{FT}}^{rr}(r)$ at $r = r_{\mathrm{wr}}$. In the limit of small wrinkles amplitude and wavelength, the translationally invariant (i.e., s-independent) longitudinal strain is

$$\varepsilon_{ss}(r) = \frac{\eta^2}{2}\left(r^2 - r_{\mathrm{wr}}^2\right) + \frac{1}{4}k^2 f(r)^2. \tag{53}$$

Using Hookean stress-strain relation (26) together with the requirement $\sigma_{\mathrm{FT}}^{ss}(r) = 0$ for $|r| < r_{\mathrm{wr}}$ yields

$$k^2 f(r)^2 = 2\eta^2\left(r_{\mathrm{wr}}^2 - r^2\right). \tag{54}$$

Equation (54) is a "slaving" condition (in the terminology of [14]) imposed on the wrinkle pattern by the necessity to collapse compression, which reflects the singular nature of the FT expansion. Although $k \to \infty$ and $f(r) \to 0$ as $t \to 0$, and k and f cannot be extracted from our leading order analysis, their product remains constant and is determined solely by the confinement α. In Appendix C we show that the oscillatory (s-dependent) part of the strain $\varepsilon_{ss}(r)$, as well as other components of the strain tensor, can also be eliminated in the limit $t \to 0$ by modifying the deformed shape, Eq. (51), with a wrinkle-induced longitudinal displacement $u_s(s,r)$.

Finally, we use the in-plane force balance (33) to deduce the transverse component of the stress:

$$\sigma_{\mathrm{FT}}^{rr}(r) = \begin{cases} -\frac{\eta^4}{8}\left(\frac{1}{4} - r_{\mathrm{wr}}^2\right)^2 & \text{for } |r| < r_{\mathrm{wr}}, \\ -\frac{\eta^4}{8}\left(\frac{1}{4} - r^2\right)\left(\frac{1}{4} + r^2 - 2r_{\mathrm{wr}}^2\right) & \text{for } |r| > r_{\mathrm{wr}}. \end{cases} \tag{55}$$

In Sect. 4 we will employ both longitudinal and transverse components of the stress to study the transverse instability of the longitudinally wrinkled helicoid.

3.3.2 The FT Energy

The Dominant Energy The dominant energy U_{dom} of the FT longitudinally wrinkled state is simply the energy associated with the compression-free stress and is given by

$$U_{\mathrm{dom}} = \frac{1}{2}\int_{-1/2}^{1/2} \sigma_{\mathrm{FT}}^{ss}(r)^2 dr + T\chi_{\mathrm{FT}}, \tag{56}$$

where the first term results from the strain in the ribbon and the second one is the work done by the exerted tension upon pulling apart the short edges.[5] The right hand side of Eq. (56) is easily evaluated using Eqs. (49), (50), (52), yielding

$$\frac{U_{\mathrm{dom}}}{T^2} = \frac{\alpha^2}{1920}(1-2r_{\mathrm{wr}})^3\left(3+18r_{\mathrm{wr}}+32r_{\mathrm{wr}}^2\right) + \frac{\alpha r_{\mathrm{wr}}^2}{2}, \tag{57}$$

where the extent of the wrinkled zone is given by Eq. (50). The energy U_{hel} of the compressed helicoidal state is evaluated by an equation analogous to (56), where $\sigma_{\mathrm{FT}}^{ss}$ and χ_{FT} are replaced, respectively, by Eqs. (3), (20), yielding:

$$\frac{U_{\mathrm{hel}}}{T^2} = \frac{\alpha^2}{1440} + \frac{\alpha}{24} - \frac{1}{2}. \tag{58}$$

The two energies U_{dom} and U_{hel} are plotted in Fig. 6, demonstrating the dramatic effect associated with the formation of wrinkles and the consequent collapse of compression on the elastic energy of a stretched-twisted ribbon. A notable feature, clearly visible in Fig. 6, is the vanishing of U_{dom} as $T \to 0$ for a fixed twist η. This is elucidated by an inspection of the terms in Eq. (56): assuming a fixed twist η (such that $T \sim \alpha^{-1}$), the stress integral vanishes as $\sim T^2$, whereas the longitudinal contraction $\chi_{\mathrm{FT}} \sim \eta^2$ is independent on T and hence the work term scales as $\sim T$. This low-T scaling of U_{dom}, together with the behavior of the sub-dominant energy that we describe below, underlies the asymptotic isometry equation (42). In Sect. 3.5, we will argue that the linear dependence of the energy on the tension T is a general feature, shared also by other types of asymptotic isometries.

The Sub-Dominant Energy As we noted already, computation of the sub-dominant energy requires one to consider all the wrinkled states whose energy approaches the dominant energy $U_{\mathrm{dom}}(\alpha)$ (57) in the limit $t \to 0$. A complete analysis of the sub-dominant energy is beyond the scope of this paper. However, we can obtain a good idea on the scaling behavior by considering a fixed confinement $\alpha > 24$ and assuming that the energetically favorable pattern consists of simply-periodic wrinkles (Eqs. (51), (52)) with $1 \ll k \ll t^{-1}$. We will use the bending energy of such a pattern to estimate the subdominant energy at the two limits of the confinement parameter: (a) α is slightly larger than 24, which we denote as $\Delta\alpha = \alpha - 24 \ll 1$, (b) large confinement, $\alpha \gg 1$.

(a) Here the wrinkles are confined to a narrow zone of width $r_{\mathrm{wr}} \sim \sqrt{\Delta\alpha}$ (which follows from the Taylor expansion of Eq. (50) around $\alpha = 24$). Hence, the curvature of the wrinkles in both transverse and longitudinal directions is significant, and a similar argument

[5] For simplicity, we assume that the Poisson ratio $\nu = 0$. This does not affect any of the basic results. Also, note that we neglected the contribution of the transverse stress ($\sim\sigma_{\mathrm{FT}}^{rr}(r)^2$) since it comes with a factor $O(\eta^4)$ with respect to the terms in Eq. (56).

to Sect. 3.2, which relies on balancing the normal forces proportional to the wrinkle amplitude $f(r)$, implies: $k \sim 1/r_{\mathrm{wr}}$. The excess bending energy (per unit of length in the longitudinal direction) is: $U_{\mathrm{B}} \sim (B/2)\int_{-r_{\mathrm{wr}}}^{r_{\mathrm{wr}}}[k^2 f(r)]^2 dr$. Using the slaving condition (54) we obtain: $U_{\mathrm{B}} \sim \eta^2 t^2 (\Delta\alpha)^{1/2}$.

(b) As $\alpha \gg 1$ (corresponding to the limit $T \to 0$ for fixed twist η), the exerted tension is felt only at infinitesimal strips near the long edges, and we may therefore assume that $k \sim t^{\beta-1}$, where $0 < \beta < 1$ is independent on T. A similar calculation to the above paragraph, where now $r_{\mathrm{wr}} \approx 1/2$, yields: $U_{\mathrm{B}} \sim t^{2\beta}\alpha^2$.

We thus obtain the scaling estimates for the sub-dominant energy:

$$U_{\mathrm{sub}} \sim \begin{cases} \eta^2 t^2 \Delta\alpha^{1/2} & \text{for } \Delta\alpha \ll 1, \\ t^{2\beta}\alpha^2 & \text{for } \alpha \gg 1. \end{cases} \tag{59}$$

3.4 Transition from the Near-Threshold to the Far-from-Threshold Regime

As the confinement α is increased above the threshold value α_{lon} given in Eq. (45), we expect a transition of the width $r_{\mathrm{wr}}(\alpha)$ of the wrinkled zone from the extent of the compressive zone of the helicoidal state (43) to the FT result (50). This transition is depicted in the inset to Fig. 4 (*right*).

The energetic mechanism underlying the NT-FT transition is described schematically in Fig. 6a: In the NT regime, the energy of the wrinkled state is reduced from $U_{\mathrm{hel}}(\alpha)$ (Eq. (58)) by a small amount, proportional to the wrinkle's amplitude. In the FT regime, the energy U_{FT} is expressed by Eq. (48) where the t-independent part U_{dom} is given by Eq. (57) and the t-dependent part U_{sub} is given by the first line of Eq. (59). Expanding the various energies for $\Delta\alpha \ll 1$, we find that the energy gain due to the collapsed compression scales as: $U_{\mathrm{hel}} - U_{\mathrm{dom}} \sim T^2\Delta\alpha^{5/2}$ (solid brown curve in Fig. 6a), whereas the energetic cost due to the finite-amplitude wrinkles scales as $\sim t^2\eta^2\Delta\alpha^{1/2}$ (dashed purple curve). Plotting these curves as a function of $\Delta\alpha$ we find that the FT behavior becomes energetically favorable for $\Delta\alpha$ above a characteristic confinement

$$\Delta\alpha_{\text{NT-FT}} \sim \frac{t}{\sqrt{T}}, \tag{60}$$

where we used the fact that $\eta \approx \sqrt{24T}$ for $\Delta\alpha \ll 1$. We note that $\Delta\alpha_{\text{NT-FT}}$ exhibits a scaling behavior that is similar to the wrinkling threshold $\Delta\alpha_{\mathrm{lon}}$, Eq. (45). This scenario, which is similar to tensional wrinkling phenomena [14, 27], is depicted in Fig. 4. The dashed curve describes the expected behavior of the width of the wrinkled zone as α increases above 24. For $\Delta\alpha < \Delta\alpha_{\mathrm{lon}}$ the ribbon remains in the helicoidal (unwrinkled) state; at onset, the width matches the compressed zone of the helicoidal state; as the confinement is increased further, the width overshoots the compressed zone of $\sigma^{ss}_{\mathrm{hel}}(r)$, signifying the transformation, over a confinement interval that it comparable to $\Delta\alpha_{\mathrm{lon}}$, to the compression-free stress $\sigma^{ss}_{\mathrm{FT}}(r)$.

3.5 Asymptotic Isometries at $T \to 0$

We now turn to study the vicinity of the singular hyper-plane ($T = 0, t = 0$) in the 4D parameter space, assuming fixed, small values of η and L^{-1}. Obviously, for a fixed twist η, the helicoidal shape contains a finite amount of strain that does not go away even if the exerted tensile load $T \to 0$. This is seen in the behavior of U_{hel}, which approaches in this limit (i.e., $\alpha^{-1} \to 0$ in Fig. 6) $\frac{2}{7}$ of its value at the onset of the longitudinal instability ($\alpha = 24$). This result is consistent with our intuitive picture of the helicoid, as well as from Green's stress,

Eqs. (19), (20), which shows that longitudes (i.e., material lines $\boldsymbol{X}(s, r = \text{cst})$) are strained in the limit $T \to 0$ by $\eta^2(\frac{1}{2}r^2 - \frac{1}{24})$. This strain stems from the helicoidal structure rather than from a tensile load, and we thus call it "geometric strain".

At first, one may expect that such a T-independent geometric strain is inherent to the helicoidal structure and cannot be removed by wrinkly decorations of the helicoid. However, the energy U_{dom} of the FT-longitudinally-wrinkled state, Eq. (57), invalidates this intuitive expectation. As Fig. 6 shows, $U_{\text{dom}}/U_{\text{hel}}$ vanishes as $T \to 0$, indicating that the wrinkled state becomes an *asymptotic isometry* of the ribbon, which can accommodate an imposed twist η with no strain. Importantly, the subdominant energy (59) shows that, although the asymptotic isometry requires a diverging curvature of wrinkles, its bending cost eventually vanishes as $t \to 0$. Hence, the longitudinal wrinkling leads to a physically admissible, nearly strainless state for the stretched-twisted ribbon, at an infinitely small neighborhood of the hyper-plane $(T = 0, t = 0)$.

Equation (56) shows that the actual energetic cost of U_{dom} as $T \to 0$ is proportional to T, and stems from the work done on the ribbon by the (small) tensile load, where the prefactor is the longitudinal contraction χ_{FT} that approaches the value $\eta^2/8$ in this limit. Notably, the contraction χ_{FT} is larger than the analogous contraction χ_{hel} of the unwrinkled helicoidal state (see Fig. 5). This observation shows that the formation of wrinkles necessitates a slight increase in the contraction of the helicoidal shape, which implies a corresponding increase of the work done by the tensile load, but gives much more in return: an elimination of the geometric strain from the helicoidal shape.

The asymptotic behavior of U_{FT} in the limit $(T \to 0, t \to 0)$ leads us to propose the general form of the *asymptotic isometry equation* (42), which applies to all physically admissible states of the stretched-twisted ribbon in this limit. Since such states become strainless in this limit, we expect that the strain at a small finite T is proportional to T, such that the integral in the energetic term analogous to Eq. (56) is proportional to T^2, and is negligible in comparison to the work term that is linear in T. The prefactor (A_j) is nothing but the corresponding longitudinal contraction in the limit $T \to 0$. The second term in Eq. (42) reflects the bending cost, and the physical admissibility of the state implies the scaling $t^{2\beta_j}$ with $\beta_j > 0$ and a prefactor B_j that approaches a finite value as $T \to 0$.[6]

We demonstrate this idea by considering the simple deformation of a long, twisted ribbon: a cylindrical wrapping (Fig. 5), where the centerline, along with all other longitudes, are mapped into parallel helices. Considering first the case $T = 0$, we see that the bending energy of this state is minimized by the smallest possible curvature that allows conversion of the imposed twist into a writhe. This minimal curvature is η^2, and is obtained when the twisted, unstretched ribbon, "collapses" onto a plane perpendicular to its long axis, such that the longitudinal contraction is the maximal possible: $\chi_{\text{cyl}} = 1$ (see Fig. 5). For small T and t, we obtain the energy:

$$U_{\text{cyl}} \simeq T + \eta^4 t^2. \tag{61}$$

Comparing U_{cyl} to the energy U_{FT} of the longitudinally-wrinkled helicoidal shape, we note the basic difference between these states, which is depicted in Fig. 6b. The formation of longitudinal wrinkles is associated with a larger cost of bending energy (i.e., $\beta < 1$ in Eq. (59)), and is thus less favorable at very small T. However, the small longitudinal contraction of the longitudinally-wrinkled state allows an energetically efficient mechanism to accommodate

[6]The upper bound $\beta_j \leq 1$ stems from the bending modulus, and assuming that the minimal curvature of any nontrivial state is $O(1)$.

the exerted tensile load, and makes it favorable if $T > \eta^4 t^{2\beta}$. Notably, the transition between the two states occurs at $T \sim \eta^4 t^{2\beta}$, approaching the vertical axis in the (T, η) plane when $t \to 0$. This scenario, on which we will elaborate more in Sect. 5.3, underlies the secondary instabilities of the helicoidal state depicted in Fig. 3a.

The relevance of isometric maps (of 2D sheets embedded in 3D space) to the behavior of thin sheets with small but finite thickness, has been recognized and exploited in numerous studies [33–41]. Most studies, however, consider confining conditions that do not involve an exerted tension (i.e., $T = 0$), such that the only limit being considered is $t \to 0$. The *asymptotic isometry equation* (42) reveals the relevance of this concept even when a small tensile load is exerted on the sheet, and provides a quantitative tool to study the energetic competition between various types of asymptotic isometries at the presence of small tension.

4 Transverse Buckling and Wrinkling

4.1 Overview

The longitudinal wrinkling instability addressed in Sect. 3 occurs when $\sigma^{ss}(r)$ has a compressive zone. In this section we address a different instability, whereby the ribbon buckles or wrinkles due to the compression of the transverse stress component $\sigma^{rr}(r)$. The transverse instability emerges when the exerted twist exceeds a threshold $\eta_{\text{tr}}(T)$, whereby the ribbon develops periodic undulations in the transverse direction (with wavelength $\lambda_{\text{tr}} \ll W$) or a single buckle ($\lambda_{\text{tr}} \sim W$).

Our analysis highlights two principal differences between the longitudinal and transverse instabilities, which are intimately related to the experimental observation in [3]. First, in contrast to the longitudinal threshold, which occurs near a curve, $\eta_{\text{lon}}(T) \approx \sqrt{24T}$, that is independent on the thickness and length of the ribbon, the threshold $\eta_{\text{tr}}(T)$ and the nature of the transverse instability exhibit a strong, nontrivial dependence on t and L. Second, in contrast to the longitudinal instability, which emerges as a primary instability of the helicoidal state, the transverse instability underlies two qualitatively distinct phenomena: a primary instability of the helicoid in a "large" tension regime ($T > T_\lambda$), where the longitudinal stress is purely tensile, and a secondary instability of the helicoid preceded by the longitudinal instability at a low tension regime ($T < T_\lambda$). We placed the word "large" in quotation marks since $T_\lambda(t, L) \ll 1$ (see Fig. 8), hence it is fully justified to assume a Hookean response for $T_\lambda(t, L) \ll T \ll 1$.

This scenario implies that the tension-twist parameter space (T, η) consists of three major phases: A helicoidal state, a FT-longitudinally-wrinkled state, and a state delimited from below by the transverse instability. This division is shown in Fig. 8 and strongly resembles the experimental phase diagram reported in [3]. In [3], the instability of the longitudinally-wrinkled state upon increasing twist was attributed to a "looping" mechanism and was described as a new, third type of instability, separate from the longitudinal and transverse instabilities. In our picture, this instability emerges simply as the transverse instability in the low tension regime, where it is superimposed on the FT longitudinally wrinkled state. This insight provides a natural explanation to the appearance of a single "triple" λ-point $(T_\lambda, \eta_\lambda = \sqrt{24T_\lambda})$ in the tension-twist plane, where the threshold curve $\eta_{\text{lon}}(T)$ divides $\eta_{\text{tr}}(T)$ into a low-tension branch and a large-tension branch.

Beyond this central result, we predict that the threshold curve $\eta_{\text{tr}}(T)$, the λ-point $(T_\lambda, \eta_\lambda)$, and the wavelength λ_{tr}, exhibit a remarkable dependence on the mutual ratios of the thickness, width, and length of the ribbon. This complex phenomenology is depicted in Fig. 9 and is summarized in the following paragraph:

Fig. 8 The parameter plane (T, η) exhibits the helicoid, the far from threshold longitudinal wrinkling and the transverse instability (that is buckling here) when $Lt \gg 1$, plotted here for $t = 0.005$. The coordinates of the triple λ-point are denoted $(T_\lambda, \eta_\lambda)$

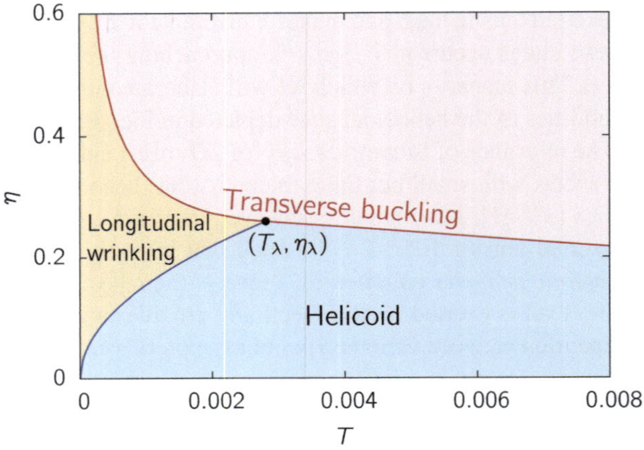

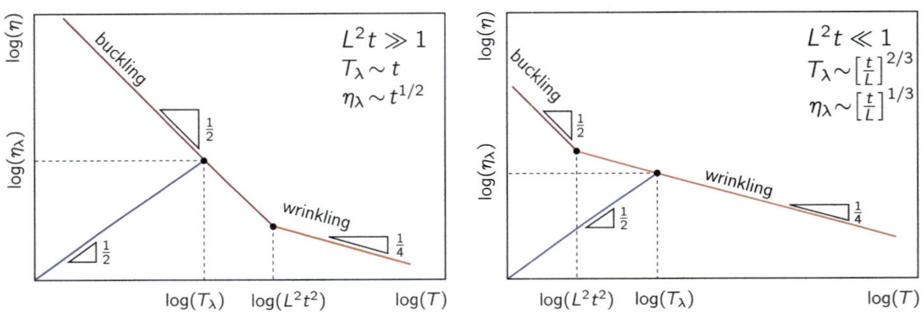

Fig. 9 Schematic phase diagram representing the two regimes $L^2t \gg 1$ (*Left*) and $L^2t \ll 1$ (*Right*) with the corresponding scaling laws for the coordinates T_λ and η_λ of the λ-point

- The threshold twist $\eta_{\text{tr}}(T)$ vanishes as the ribbon thickness vanishes, $t \to 0$.
- The threshold twist $\eta_{\text{tr}}(T)$ *diverges* as $T \to 0$.
- The tension $T_\lambda(t, L)$, which separates the regimes of low and "large" tension, vanishes in the ribbon limit at a rate that depends in a nontrivial manner on the mutual ratios of the length, width, and thickness of the ribbon: If $t \ll L^{-2}$ we find that $T_\lambda \sim (t/L)^{2/3}$, whereas if $L^{-2} \ll t$ we find that $T_\lambda \sim t$.
- The mutual ratios between the length, width, and thickness in the ribbon limit (Eq. (1)) affect also the *type* of the transverse instability. Specializing for the "large" tension regime, we find that the transverse instability may appear as a single buckle or as a periodic array of wrinkles with wavelength λ_{tr} that decreases as $T^{-1/4}$ upon increasing the tension: (a) If $L^{-1} \ll t$, the transverse instability appears as a single buckle of the helicoidal state. (b) If $L^{-2} \ll t \ll L^{-1}$ the transverse instability appears as a single buckle for $T \ll (Lt)^2$ and as a wrinkle pattern for $(Lt)^2 \ll T \ll 1$. (c) If $t \ll L^{-2}$, the transverse instability appears as a wrinkle pattern throughout the whole "large" tension regime.

We start by a scaling analysis of the parameter regime that explains the above scenario. Then we turn to a quantitative linear stability analysis that yields the transverse buckling threshold, as well as the shape of the buckled state for an infinitely long ribbon (or, more precisely, $L^{-2} \ll t$). Finally, we address at some detail the transverse instability of a ribbon with a finite length ($t \ll L^{-2} \ll 1$).

4.2 Scaling Analysis

Similarly to the longitudinal wrinkling, the basic mechanism of the transverse instability is simply the relaxation of compression (which is now σ^{rr}), by appropriate deformation of the helicoidal shape. Taking a similar approach to Sect. 3, we can find the scaling relations for the threshold η_{tr} and the wavelength λ_{tr}, by identifying the dominant destabilizing and stabilizing normal forces associated with such shape deformation.

The transverse compression gives rise to a destabilizing force $\sim\sigma^{rr}/\lambda_{\text{tr}}^2$. The normal restoring forces are similar to the respective forces that underlie the wrinkling of a stretched (untwisted) ribbon [8]: bending resistance to deformation in the transverse direction ($\sim B/\lambda_{\text{tr}}^4$), and tension-induced stiffness due to the spatial variation of the deformation in the longitudinal direction ($\sim T/L^2$). All other normal restoring forces, in particular the bending resistance to deformation in the longitudinal direction (that scales as $\sim B/L^4$) are negligible with respect to these two forces. The balance between these two dominant restoring forces and the destabilizing normal force due to the compression σ^{rr} may lead to buckling, namely $\lambda_{\text{tr}} \sim W = 1$, if the ribbon is extremely long ($t \ll L^{-1}$), in which case the tension-induced stiffness is negligible, or to wrinkling ($\lambda_{\text{tr}} \ll W$), where the bending and tension-induced forces are comparable.

In the following paragraphs we address first the case of an extremely long ribbon, where the only dominant restoring force is associated with bending, and then show how a finite value of L affects a transition from buckling to wrinkling. In each case we will discuss separately the regimes of low and "large" tension, and derive the scaling relation for the λ-point $(T_\lambda, \eta_\lambda)$ that separates these regimes.

Extremely Long Ribbon In this case, the tension-induced stiffness is negligible, and the only significant restoring force to shape deformations is the bending resistance. The transverse instability is then similar to the Euler buckling of a beam of width W and thickness t, and the instability mode is consequently buckling, i.e., $\lambda_{\text{tr}} \sim W$. The instability threshold $\eta_{\text{tr}}(T)$ is obtained when the destabilizing force becomes comparable to the stabilizing bending force, namely:

$$\frac{\sigma^{rr}}{W^2} \sim \frac{B}{W^4}. \tag{62}$$

The transverse compression $\sigma^{rr}(r)$ is given by Eq. (4) in the helicoidal state (for $\eta^2 < 24T$), and by Eq. (55) in the FT longitudinally wrinkled state (for $\eta^2 > 24T$). Considering the asymptotic regimes $T \gg \eta^2$ in Eq. (4) and $T \ll \eta^2$ in Eq. (55), we see that in both "large" and low tension regimes, the transverse stress scales similarly with η and T:

$$\sigma^{rr} \sim \eta^2 T. \tag{63}$$

Substituting Eq. (63) in Eq. (62), we obtain the scaling of the instability threshold $\eta_{\text{tr}}(T)$ for an extremely long ribbon:

$$\eta_{\text{tr}}(T) \sim \frac{t}{\sqrt{T}}. \tag{64}$$

This scaling relation (with different numerical prefactors in the limits $T \ll T_\lambda$ and $T_\lambda \ll T \ll 1$), is confirmed by our detailed calculations in Sect. 4.3, that are shown in Fig. 8. The relation (64) demonstrates the singular nature of the transverse instability: the threshold decreases with the ribbon thickness ($t \to 0$) and diverges as the exerted tension vanishes ($T \to 0$).

The tension T_λ is the horizontal coordinate of the "triple" λ-point in the tension-twist parameter plane (Fig. 8) at which the transverse buckling changes its character from a primary instability at "large" tension to a secondary instability of the helicoidal state at low tension, which is preceded by the longitudinal wrinkling instability. We find T_λ from Eq. (64) and the relation $\eta_{\text{lon}}(T) \sim \sqrt{T}$:

$$T_\lambda \sim t. \tag{65}$$

Ribbon of Finite Length Let us assume now that both tension-induced stiffness and bending resistance are significant restoring forces, which balance the destabilizing force due to transverse compression. The instability onset condition (62) is then replaced by

$$\frac{\sigma^{rr}}{\lambda_{\text{tr}}^2} \sim \frac{B}{\lambda_{\text{tr}}^4} \sim \frac{\sigma^{ss}}{L^2}. \tag{66}$$

Using the scaling law (63) for σ^{rr} and estimating $\sigma^{ss} \sim T$, we obtain the following scaling relations for the threshold and wavelength:

$$\eta_{\text{tr}} \sim \sqrt{\frac{t}{L}} T^{-1/4}, \tag{67}$$

$$\lambda_{\text{tr}} \sim \sqrt{Lt} T^{-1/4}. \tag{68}$$

In a similar way to the above paragraph, we find the coordinate T_λ of the λ-point by equating Eq. (67) with the relation $\eta_{\text{lon}}(T) \sim \sqrt{T}$, yielding:

$$T_\lambda \sim \left(\frac{t}{L}\right)^{2/3}. \tag{69}$$

From Buckling to Wrinkling Realizing the important effect of the ribbon length on the nature of the transverse instability, a natural question is: How long must a ribbon be such that the tension-induced stiffness becomes negligible and the scaling relations (64), (65) are valid?

A key to address this question is the obvious inequality $\lambda_{\text{tr}} \le W$. Substituting the scaling relation (69) for T_λ in Eq. (67), and requiring $\lambda_{\text{tr}} \ll W$, we find that T_λ is characterized by the scaling relation (69) if $t \ll L^{-2}$, and by the relation (65) if $L^{-2} \ll t$. This nontrivial dependence of T_λ on the thickness and length of the ribbon is depicted in Fig. 9. The behavior of T_λ indicates the complex nature of the ribbon limit, but a closer inspection of Eq. (67), subjected to the condition $\lambda_{\text{tr}} \le W$, reveals an even higher level of complexity. Focusing on the large tension regime $T_\lambda < T < 1$, and recalling that $\lambda_{\text{tr}} \le W$, we find that the ribbon limit is divided into three sub-regimes that exhibit qualitatively distinct types of transverse instabilities. This behavior is depicted in Fig. 9, and summarized below:

(a) If $t \ll L^{-2} \ll 1$, then T_λ satisfies the scaling relation (69) and the transverse instability appears as wrinkling, where the threshold η_{tr} and the wavelength λ_{tr} satisfy the scaling relations (67) and (68).
(b) If $L^{-2} \ll t \ll L^{-1} \ll 1$, then T_λ satisfies the scaling relation (65), but the large tension regime splits into two parts. For sufficiently small T, the transverse instability appears as a buckling mode ($\lambda_{\text{tr}} \sim W$), and the threshold η_{tr} satisfies the scaling (64); for larger values of T (which are nevertheless $\ll 1$), the instability appears as a wrinkling mode, described by the scaling relations (67).

(c) Finally, if $L^{-1} \ll t \ll 1$, then T_λ satisfies the scaling relation (65), and the transverse instability appears as a buckling mode, with the scaling (64), throughout the whole regime of large tension.

4.3 Linear Stability Analysis

In this subsection we present in detail the linear stability analysis for the case of an extremely long ribbon, assuming that the ribbon shape close to the transverse instability is well approximated by the form:

$$\boldsymbol{X}(s,r) = \begin{pmatrix} (1-\chi)s + \zeta\eta u_{s1}(r) \\ [r + u_r(r)]\cos(\eta s) - \zeta z_1(r)\sin(\eta s) \\ [r + u_r(r)]\sin(\eta s) + \zeta z_1(r)\cos(\eta s) \end{pmatrix}, \tag{70}$$

where the amplitude ζ is assumed to be infinitesimal, and the functions $u_{s1}(r)$ and $z_1(r)$ correspond to the two degrees of freedom that characterize the perturbed shape and strain. More precisely, it is reasonable to assume that the boundary conditions at the short edges ($s = \pm L/2$) have a prominent effect only at a zone of size $W = 1$ near those edges, and barely disturb the translational symmetry of the reference state in the longitudinal direction. (We will comment on this assumption later in Sect. 6.) Therefore, the eigenmodes of the system are approximated by:

$$\left[u_{sj}(r)\cos\left(\frac{\pi j s + \gamma_j}{L}\right), z_j(r)\cos\left(\frac{\pi j s}{L}\right)\right], \tag{71}$$

where $1 \leq j \ll L$. Since we address here an instability that relaxes the transverse compression, the variation in the longitudinal direction should be minimal to avoid any energetic costs, and hence we assume that the first eigenmode to become unstable is $j = 1$ (see also [8]). In the next subsection we present an approximate analysis of the $j = 1$ mode, but here we simplify further by neglecting the longitudinal variation altogether (i.e., replacing $\cos(\pi j s/L) \to 1$). Since all derivatives with respect to the variable s come with negative powers of L, we anticipate that this approximation is valid for sufficiently large L (such that $L^{-2} \ll t$, as found in Sect. 4.2).

Our linear stability analysis follows a classical approach, whose first use in elasticity theory has been attributed to Michell [42, 43].[7] First, we assume a small perturbation of the form (70) to the reference state, and expand the generalized FvK equations (27)–(28) to linear order in the amplitude ζ, obtaining a set of linear homogeneous equations. In general, these equations have no solution but the trivial one, $u_{s1}(r) = z_1(r) = 0$. For a given tension T, the transverse instability occurs at the lowest value $\eta_{tr}(T)$ of the twist for which the buckling equations admit a nontrivial solution. This solution is identified as the unstable transverse mode.

We must compute the perturbed curvature and stress tensors that enter the generalized FvK equations (27)–(28) upon substituting the shape (70), and retaining only the terms that are linear in ζ. We will limit our analysis to the simple case $\nu = 0$. Since the stress field (3)–(4) has been shown to be independent of the Poisson ratio, we expect that the

[7] The introduction of [44] contains a useful summary of the various approaches for linear stability analysis of elastic systems.

same is true for the transverse instability.[8] The perturbed curvature tensor is given by $c_{\alpha\beta} = c^{(0)}_{\alpha\beta} + \zeta c^{(1)}_{\alpha\beta} + O(\zeta^2)$, where

$$c^{(0)}_{\alpha\beta} = \begin{pmatrix} 0 & \eta \\ \eta & 0 \end{pmatrix}, \tag{72}$$

$$c^{(1)}_{\alpha\beta} = \begin{pmatrix} -\eta^2[z_1(r) - rz_1'(r)] & 0 \\ 0 & z_1''(r) \end{pmatrix}, \tag{73}$$

and the perturbed stress tensor is given by $\sigma^{\alpha\beta} = \sigma^{\alpha\beta}_{(0)} + \zeta\sigma^{\alpha\beta}_{(1)} + \mathcal{O}(\zeta^2)$, where

$$\sigma^{\alpha\beta}_{(0)} = \begin{pmatrix} T + \frac{\eta^2}{2}(r^2 - \frac{1}{12}) & 0 \\ 0 & \frac{\eta^2}{2}(r^2 - \frac{1}{4})[T + \frac{\eta^2}{4}(r^2 + \frac{1}{12})] \end{pmatrix}, \tag{74}$$

$$\sigma^{\alpha\beta}_{(1)} = \frac{\eta}{2}\big[u_{s1}'(r) - z_1(r) + rz_1'(r)\big]\begin{pmatrix} 0 & 1 \\ 1 & 0 \end{pmatrix}. \tag{75}$$

Notice that, to $O(\zeta)$, the diagonal stress components are not perturbed. Furthermore, the force balance in the transverse direction (28), which we evaluate as usual to $O(\eta^4)$, yields the equations:

$$0 = \frac{\eta}{2}\big[u_{s1}''(r) + rz_1''(r)\big], \tag{76}$$

$$0 = \frac{\eta^3}{2}\big(u_{s1}'(r) - z_1(r) + rz_1'(r) + 2r\big[u_{s1}''(r) + rz_1''(r)\big]\big). \tag{77}$$

These equations are solved by

$$u_{s1}'(r) = z_1(r) - rz_1'(r), \tag{78}$$

implying that the stress tensor does not deviate from its value at the reference state.

Turning now to the normal force balance, we substitute Eqs. (73), (75), (78) in Eq. (27), retain the linear order in the amplitude ζ, and obtain a 4th order differential equation for $z_1(r)$:

$$\frac{t^2}{12}z_1^{(4)}(r) = -\eta^2\sigma^{ss}_{(0)}\big[z_1(r) - rz_1'(r)\big] + \sigma^{rr}_{(0)}z_1''(r). \tag{79}$$

We recognize this equation as similar to the Euler buckling equation, whereby a destabilizing force $\sigma^{rr}_{(0)}z_1''(r)$ that originates from the relaxation of compression by deflection is balanced by the stabilizing bending force $\frac{t^2}{12}z_1^{(4)}(r)$ that opposes any deflection. However, Eq. (79) carries some differences from the simple Euler buckling instability. First, the compression $\sigma^{rr}_{(0)}(r)$ is not uniform across the ribbon width. Second, we note the existence of another normal force, that is proportional to the longitudinal stress $\sigma^{ss}_{(0)}$, and originates from the fact that the reference state is non-planar. Since we found that $\sigma^{rr}_{(0)} \sim -\eta^2\sigma^{ss}_{(0)}$ (see Eq. (33)) we can view the right hand side of Eq. (79) as a renormalized version of the normal force

[8] We just note that a non zero Poisson ratio would require another degree of freedom in the perturbative analysis, namely a transverse in-plane displacement $\zeta u_{r1}(r)$.

$\sigma_{(0)}^{rr} z_1''(r)$ that couples the compression to the curvature. We are poised to solve this equation subjected to the homogenous boundary conditions:

$$z_1''(\pm 1/2) = 0, \tag{80}$$

$$z_1^{(3)}(\pm 1/2) = 0. \tag{81}$$

It is noteworthy that the buckling equation (79) is general and does not depend on the particular form (74) of the stress of the reference state. As a consequence, the linear stability analysis can be performed over the helicoidal state, where the stress is given by Eqs. (3)–(4), as well as over the FT-longitudinally-wrinkled state, where it is given by Eqs. (49), (55).[9] The stress field that we have to use depends on our position in the tension-twist plane (T, η) with respect to the longitudinal instability line $\eta^2 = 24T$.

Before turning to numerical analysis, it is useful to consider the limit $T \ll \eta^2$ ($\alpha \to \infty$ in the terminology of Sect. 3), where an analytic solution of the buckling equations (79)–(81) is available. In this limit, the exerted tension is supported by two narrow strips near the long edges of the ribbon, and the stress field (49), (55) becomes:

$$\sigma^{ss}(r) = \frac{T}{2}\left[\delta\left(r + \frac{1}{2}\right) + \delta\left(r - \frac{1}{2}\right)\right], \tag{82}$$

$$\sigma^{rr}(r) = -\frac{\eta^2 T}{4}, \tag{83}$$

where $\delta(r)$ is the Dirac-delta function. Since the term $-\eta^2 \sigma_{(0)}^{ss}[z_1(r) - r z_1'(r)]$ is non-zero only at an infinitesimal strip near $r = \pm 1/2$, we can eliminate it from (79) by modifying the boundary condition (81) that becomes:

$$\frac{t^2}{12} z_1^{(3)}(-1/2) = -\frac{\eta^2 T}{2}\left[z_1(-1/2) + \frac{1}{2} z_1'(-1/2)\right], \tag{84}$$

with an analogous condition at the other edge, $r = 1/2$. The buckling equation (79) now simply reads

$$z_1^{(4)}(r) = -\frac{3\eta^2 T}{t^2} z_1''(r), \tag{85}$$

which is the familiar Euler buckling equation under uniform compression. It admits a non-zero solution $z_1(r) = \cos(\pi r)$ when η reaches its threshold value

$$\text{small } T: \quad \eta_{\text{tr}}(T) = \frac{\pi}{\sqrt{3}} \frac{t}{\sqrt{T}}. \tag{86}$$

In the opposite limit, $T \gg \eta^2$, an analytic solution of Eq. (79) is not available and a numerical solution of Eqs. (79)–(81) yields the threshold in this limit:

$$\text{large } T: \quad \eta_{\text{tr}}(T) = 4.4 \frac{t}{\sqrt{T}}. \tag{87}$$

[9] Recall that for a given (η, T) in the regime $\eta^2/T > 24$, the FT approach in Sect. 3.3 provides a longitudinally-wrinkled state whose shape is close to the helicoid, up to deviations whose amplitude vanishes as $t \to 0$, and whose stress is given by (49), (55), up to corrections that also vanish as $t \to 0$. Therefore, the transverse linear stability analysis in this regime provides expressions for the threshold $\eta_{\text{tr}}(T)$ and the unstable mode $[u_{s1}(r), z_1[(r)]]$ that become accurate as $t \to 0$.

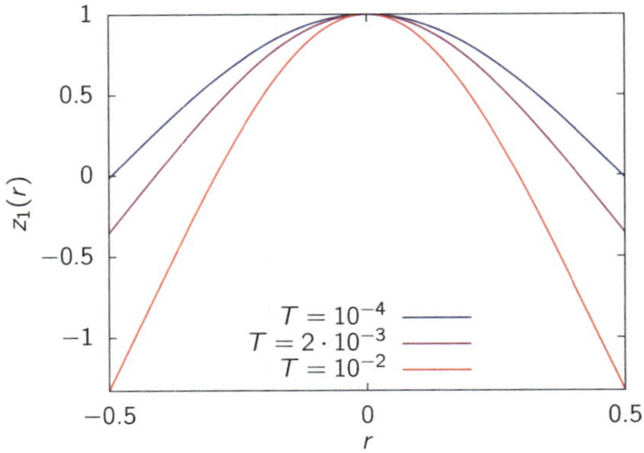

Fig. 10 Shape of the transverse unstable mode in the limit of an infinitely long ribbon ($L \to \infty$) of thickness $t = 0.005$, as a function of the exerted tension. For the range of tension applied here, the limit $L \to \infty$ is relevant for lengths $L > 20$ (see Eq. (68))

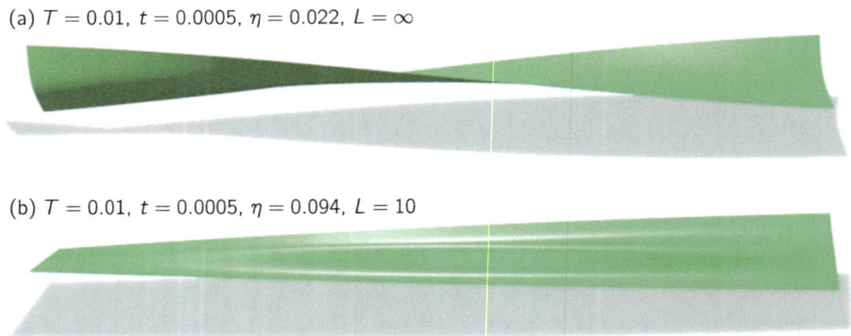

Fig. 11 Shape of the ribbon undergoing a transverse instability: (**a**) single mode buckling of a very long ribbon, from a numerical solution of Eq. (79) with an arbitrary amplitude (**b**) wrinkling of a ribbon with $Lt \ll 1$, from Eq. (88)

Interestingly, the two asymptotic expressions (86), (87) exhibit the scaling law (64), not only with the ribbon thickness t, but also with the tension T.

Our numerical results and the subsequent division of the tension-twist plane into three major phases (the helicoid, the longitudinal wrinkling and the region above the transverse instability) are shown in Fig. 8 for the thickness $t = 0.005$. This phase diagram exhibits a striking similarity, at a quantitative level, with the phase diagram found experimentally in [3] for the same thickness.[10]

The numerical analysis of the buckling equation gives also the shape of the buckling mode, which we show in Fig. 10 for a few representative values of T. Choosing some (arbitrary) small amplitude, we draw the shape of the buckled ribbon in Fig. 11.

4.4 Effect of a Finite Length

Our analysis in this paper assumes that the ribbon is long, such that the effect of boundary conditions at $s = \pm L/2$ is limited to the vicinity of the short edges, and the linear eigen-

[10] The experimental value of the length in [3] is $L = 20$. The maximal tension in the experiment is $T_{\text{max}} = 0.01$; from Eq. (68), we deduce that the minimal transverse wavelength is $\lambda_{\text{min}} \sim \sqrt{Lt}T_{\text{max}}^{-1/4} = 1$. This explains why buckling is observed and why the infinite length approximation is in good agreement with the experiment.

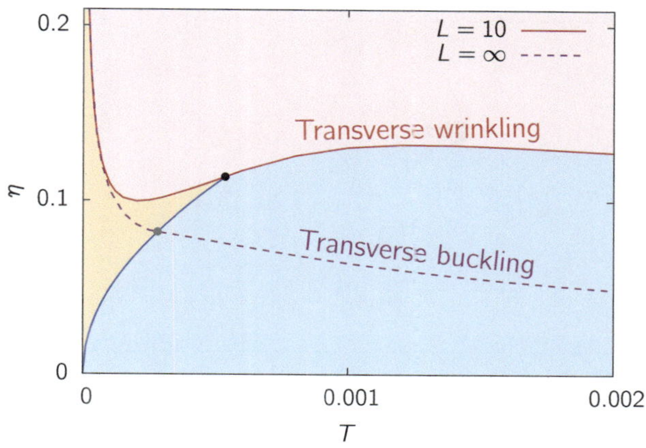

Fig. 12 Effect of a finite ribbon length on the phase diagram: for a thickness $t = 5 \times 10^{-4}$, the transverse buckling threshold is plotted for $L = \infty$ (*dashed line*) and $L = 10$ (*solid line*). For this thickness and this tension range, the infinite length approximation is relevant for lengths such that $L > t^{-1} T_{\max}^{1/2} \simeq 100$ (from the requirement that $\lambda > 1$ in Eq. (68))

modes can be expressed through a Fourier series, Eq. (71), where the most unstable one is $j = 1$. In Sect. 4.3 we went beyond this assumption and neglected the spatial variation of this unstable mode in the longitudinal direction, expecting it to make a negligible contribution to the force balance if L is sufficiently large. In this subsection we relax this last assumption, by taking into consideration the longitudinal variation of the perturbation. This means that the ribbon is not sufficiently long to justify a complete neglecting of the spatial variation, but it is long enough such that the mode structure is given by Eq. $(71, j = 1)$.[11]

Despite its simple form, a complete analysis of the mode $j = 1$ in Eq. (71) is rather cumbersome. In order to simplify our calculation we will retain only the term in the normal force balance that couples the longitudinal stress and the longitudinal curvature. Such a term was found to be crucial for the wrinkling of a stretched, untwisted sheet [8]. We thus obtain the buckling equation

$$\frac{t^2}{12} z_1^{(4)}(r) = -\eta^2 \sigma^{ss}_{(0)}\left[z_1(r) - r z_1'(r)\right] - \kappa^2 \sigma^{ss}_{(0)} z_1(r) + \sigma^{rr}_{(C)} z_1''(r), \tag{88}$$

where $\kappa = \pi / L$, subjected to the same boundary conditions that we described in the previous subsection.

The buckling equation can be solved numerically, and the effect of the finite ribbon length on the phase diagram is shown on Fig. 12. The main effect of the finite length is to increase the transverse buckling threshold; this is expected, since the new term is a stabilizing term. We also note that this effect is not important at small tension. Another effect is that the threshold $\eta_{\text{tr}}(T)$ may become a non-monotonic function, due to the fact that the tension enhances both the compressive (destabilizing) term and the stretching (stabilizing) term, which are represented, respectively, by the third and second terms of Eq. (88).

Finally, as was noted already in our scaling analysis, the unstable transverse mode transforms from buckling to wrinkling as the tension increases. This transformation is shown in Fig. 13.

[11] If the ribbon becomes even shorter, it is possible that the boundary effect is sufficiently strong and the assumed mode structure Eq. (71) is not valid. In Sect. 6 we discuss the possibility that this might happen even in the ribbon limit (i.e., $L \gg 1$) provided t is small enough.

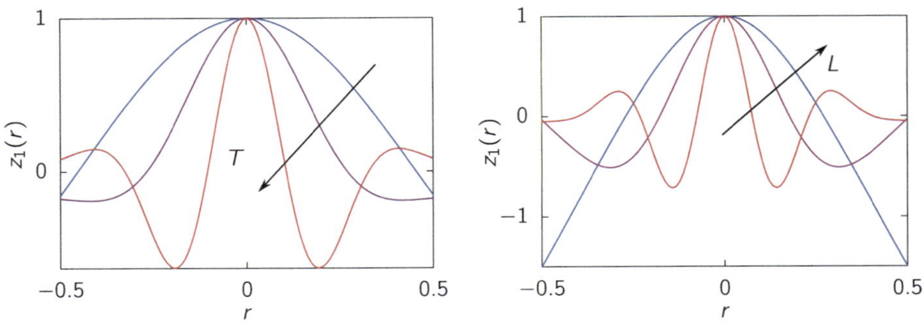

Fig. 13 Transition from transverse buckling to transverse wrinkling as a function of the tension and length. Shape of the transverse unstable modes: *Left*: $t = 5 \times 10^{-4}$, $L = 10$ and tensions $T = 10^{-4}$, 10^{-3} and 10^{-2} (arrow indicates increasing tension values). *Right*: $t = 0.0005$, $T = 0.01$ and ribbon lengths $L = 5$, 35 and 80 (arrow indicates increasing length values)

5 Twisting with Little Tension, Stretching with Little Twist

5.1 Overview

In Sects. 2–4 we assumed that the shape of the stretched-twisted ribbon is close to a helicoid, and employed asymptotic methods to characterize the deviations from this shape. This approach allowed us to compute the curves $\eta_{\text{lon}}(T)$ and $\eta_{\text{tr}}(T)$ that underlie the division of the (T, η) plane into three major regimes (Fig. 3), and to characterize the helicoidal state (blue), the longitudinally wrinkled state (orange), and the margins of the third regime (pink), close to the transverse instability threshold. The proximity to a helicoidal shape is violated at the bulk of the pink regime, which we do not address in this paper, where the self-contact zones emerge and the helicoidal shape is greatly mutilated [3]. Other two parameter regimes where the ribbon shape may become very different from a helicoid are the edge of the blue regime (i.e., close to the horizontal line $\eta = 0$), where the ribbon is stretched with little twisting, and the edge of the orange regime (i.e., close to the vertical line $T = 0$), where the ribbon is twisted with little tensile load. In this section we discuss the expected transformations to non-helicoidal morphologies in these parameter regimes: a nearly planar shape as $\eta \to 0$, where the twist is absorbed in the vicinity of the short edges; and the formation of a creased helicoidal shape and cylindrical wrapping as $T \to 0$.

In contrast to previous sections (Sects. 2–4), where we carried out a rigorous study based on the helicoidal solution to the cFvK equations, its linear stability analysis, and an FT analysis, our discussion in this section is more heuristic, and is based on energetic estimates and scaling arguments. We start in Sect. 5.2 with a general discussion of the difference between plate-like and rod-like approaches to the mechanics of ribbons. The first approach, which we employed in previous sections, is based on the cFvK equations; the second approach consists of Kirchoff's rod equations or Sadowsky equation, the last one provides the mathematical basis for a theoretical description of the creased helicoidal state [7]. We take this opportunity to explain why the Kirchoff's rod equations cannot be used to study the ribbon limit (Eq. (1)). In Sect. 5.3 we briefly describe the work of [7] on the creased helicoidal state, and explain how it gives rise to another type of asymptotic isometry, different from the longitudinally-wrinkled helicoidal shape and the cylindrical wrapping state. In Sect. 5.4 we turn to the vicinity of the horizontal line $\eta = 0$, and introduce an energetic comparison that allows us to estimate the minimal twist necessary to developing a helicoidal shape for a long ribbon ($L \gg 1$) subjected to tension T and clamping of its short edges.

5.2 Theory of Elastic Ribbons: Plate-Like or Rod-Like?

In the plate-like approach, one employs the cFvK equations for elastic plates with Hookean material response to find the shape of the ribbon midplane $\boldsymbol{X}(s,r)$. Except the restriction to small strains, no further assumptions are made on the deformation of the cross section or on the stress profile in the transverse direction $\hat{\boldsymbol{r}}$. This "transversal freedom" was reflected in our analysis of the cFvK equations in Sects. 3–4 through the r-dependence of the stresses and the consequent shape deformations, which underlie both longitudinal and transverse instabilities of the helicoidal state. The transversal freedom encapsulates the conceptual difference between the plate-like approach and the rod-like approach, wherein the ribbon shape is derived from a curve $\boldsymbol{X}_{\text{cl}}(s)$ that characterizes the shape of the centerline. In the Kirchoff's method, which addresses the ribbon as a rod with highly anisotropic cross section, the ribbon is allowed to have a tensile strain and the cross section (i.e., the ribbon shape in the plane perpendicular to the centerline) is assumed to retain its shape [29]. The relation between the midplane shape and the centerline is simply:

$$\text{Kirchoff rod:} \quad \boldsymbol{X}(s,r) = \boldsymbol{X}_{\text{cl}}(s) + r\hat{\boldsymbol{r}}(s), \tag{89}$$

where $\hat{\boldsymbol{r}}(s)$ is the normal to the tangent vector $\hat{\boldsymbol{t}} = d\boldsymbol{X}_{\text{cl}}(s)/ds$ in the ribbon midplane. In the Sadowsky's method, the ribbon is assumed to be *strainless*, and the shape of the midplane is related to the centerline by the following relation:

$$\text{Sadowsky strip:} \quad \boldsymbol{X}(s,r) = \boldsymbol{X}_{\text{cl}}(s) + r\left[\hat{\boldsymbol{b}}(s) + \frac{\tau(s)}{\kappa(s)}\hat{\boldsymbol{t}}(s)\right], \tag{90}$$

where $\hat{\boldsymbol{b}}(s)$ is the Frenet binormal to the curve $\boldsymbol{X}_{\text{cl}}(s)$ and $\tau(s)$. $\kappa(s)$, are its torsion and curvature [7]. Assuming a ribbon at mechanical equilibrium, the two methods yield strictly different sets of force balance equations that yield the centerline $\boldsymbol{X}_{\text{cl}}(s)$. In the rest of this subsection, we briefly recall recent studies of Kirchoff equations of stretched-twisted rods with anisotropic cross section, and explain why these analyses do not pertain to the ribbon limit, Eq. (1). In the next subsection we review a recent work that employed the Sadowsky strip to describe the creased helicoidal state of a stretched twisted ribbon, and discuss the regime in the (T, η) plane describable by this method.

Anisotropic Kirchoff's Rod The instability of a rod with circular cross section that is subjected to tension and twist, and the consequent formation of loops, has been studied already by Love [45], using the Kirchoff's rod equations. The theoretical works of Champneys, Thompson and van der Heijden [10–12] and of Goriely et al. [9], employed the Kirchoff's rod equations to study the response to tension and twist of a rod with asymmetric (i.e., non-circular) cross section. An important finding of these studies was the existence of instabilities (termed "thick" and "tapelike" [9]), through which the straight centerline that defines a helicoidal state of the ribbon becomes unstable (see Fig. 3 of [9]). The visual similarity of the "thick" mode to the secondary instability of a stretched-twisted ribbon at low tension (which we described in Sect. 4 as a transverse instability superimposed on the longitudinally wrinkled ribbon), motivated the original portraying of that instability as "looping" [3].

However, a close inspection of the phase diagram of [3] (Fig. 3), shows no signs of the instability predicted by [10–12]. Translating the results of [9] to our notations (see Appendix D) and considering the ribbon limit ($a \ll 1$ in the terminology of [9]), we find that the theoretical prediction suggests an instability of the helicoidal state around the curve $\eta \approx \sqrt{cT}$ with $1/2 < c < 2.8$, at which range of parameters the experiments of [3] show a stable helicoidal state. This observation indicates that using the Kirchhoff's rod equation may not be suitable at the ribbon limit (Eq. (1)), where the cross section is highly anisotropic. Indeed, the Kirchhoff's rod equations assume fixed values of the torsion and bending moduli (i.e., independent on the exerted loads), which characterize the response of unstretched, untwisted ribbon to infinitesimal loads. As was noted by Green, who considered the twisted ribbon as a 3D solid body, this assumption becomes invalid if the exerted twist $\eta \gg t$ (see Eq. (21) of [1]). As a consequence, the Kirchhoff's rod equation cannot be used to describe the helicoidal state of a twisted ribbon without appropriate renormalization of the rod's moduli (that reflect the exerted twist and possibly also the tensile load).

5.3 The Creased Helicoidal State: A Second Look at Asymptotic Isometries

The shape of a perfectly inextensible ribbon (Eq. (90)), has been addressed by Korte et al. [7], who built upon earlier studies [6, 26, 46]. These authors found that under given twist η and a range of tensile loads, the ribbon admits a strainless state, whose morphology is similar to the creased helicoid state found in the experiments of [3]. In fact, the theory of [7] yields a family of such states, parameterized by the angle between triangular facets. Importantly, the construction of [7] consists of "true" creases, with infinitely large curvature, whose bending energy would have been infinite if the ribbon had any thickness. The underlying assumption in [7] is that at a small, finite thickness, these creases are slightly smoothed (i.e., the curvature diverges as $t \to 0$ at a narrow zone whose size vanishes at the same limit), such that the overall bending energy of the crease vanishes as $t \to 0$. Such a "stress focusing" mechanism has been instrumental in studies of crumpled sheets [34].

In order to identify the regime in the (T, η) plane in which the ribbon morphology is describable by this approach we must clarify the meaning of "tensile load" on a purely inextensible (i.e., strainless) ribbon. For this purpose, we will use in this paragraph dimensional parameters (denoted by non-italicized fonts), introducing explicitly the ribbon width W and stretching modulus Y, which are taken to define, respectively, the units of length and stress throughout this paper. The dimensional bending modulus is $\mathrm{B} \sim \mathrm{YW}^2 t^2$. (Recall that we defined t, Eq. (1) as the ratio between the ribbon thickness and its width.) A thin elastic ribbon of width W has two characteristic scales for stress exerted on the midplane (i.e., force/length): The first one is just the stretching modulus Y, which is the product of Young's modulus of the material (E) and the ribbon thickness (tW); the second scale for stress is related to the bending modulus $\mathrm{B}/\mathrm{W}^2 \sim \mathrm{Y}t^2$. Since both scales are proportional to Young's modulus E, it is impossible to assume a "perfectly inextensible" ribbon (i.e., $\mathrm{Y} = \infty$) that is nevertheless bendable (i.e., $\mathrm{B} < \infty$). Hence, attributing an "inextensibility" feature to an elastic ribbon must be understood as assuming the asymptotic limit $t \to 0$, such that the exerted tensile load vanishes in comparison to Y but not in comparison to $\mathrm{B}/\mathrm{W}^2 = \mathrm{Y}t^2$. Returning to the parameter plane (T, η), we may expect that for a given twist η, the family of creased helicoid states predicted by [7] exists at a parameter regime $C_1 t^2 < T < C_2 t^2$ (where $C_{1,2}$ depend on η). In this regime, the exerted tension is sufficiently small with respect to the stress scale set by the stretching modulus Y, such that the state of the ribbon is nearly strainless and is thus close to an *isometry* of the untwisted ribbon; at the same time, the tension is sufficiently large in comparison to the other stress scale set by the bending modulus and the ribbon width, $\mathrm{Y}t^2$, such that the necessary conditions for constructing a creased helicoid state by the method of [7] are satisfied.

The Creased Helicoid State as an Asymptotic Isometry The above paragraph indicates that the analysis of [7] addresses the stretched, twisted ribbon, in the vicinity of the hyper-plane

Fig. 14 Picture of the ribbon shape at very small twist, (**a**) Helicoidal state, given for comparison with (**b**) Boundary dominated state where the twist is confined in the vicinity of the short edges, the central part being flat

$(T = 0, t = 0)$ in the 4D parameter space spanned by the dimensionless parameters t, L, T and η (Eq. (1)). As we argued in Sect. 3.5, the ribbon mechanics in this regime reflects a competition between distinct types of asymptotic isometries—namely, between states whose elastic energy at a small neighborhood of that singular hyper-plane is described by Eq. (42). The creased helicoid of [7] is yet another example of an asymptotic isometry, and its energy in that limit can also be expressed, as a linear function of T: the intercept, T-independent term of the energy (Eq. (42)), is governed by the bending energy of the creases at small finite t; the term that is linear in T originates from the work done by the tensile load, where the prefactor is the t-independent longitudinal contraction χ_{CH} of the ribbon in the creased helicoid state.

In order to carry out a quantitative comparison between the energies of the cylindrical wrapping, longitudinal wrinkling, and creased helicoid states, we must know the longitudinal contractions (A_j), as well as the exponents (β_j) that characterize the bending energy of the states at the vicinity of the hyper-plane $(T = 0, t = 0)$. While we do not have yet the complete set of those values for all three types of asymptotic isometries, the schematic plot in the inset of Fig. 6b seems as a plausible scenario to us: At a given η, the bending energy of the cylindrical wrapping (i.e., the intercept of the linear function in Fig. 6b) is minimal, and therefore this state should be observed if the exerted tension T is very small; upon increasing T, the experimental observations of [3] indicate that a creased helicoid state is formed and then gives way to a longitudinally wrinkled state, suggesting that creased helicoids are characterized by lower bending energy and larger longitudinal contraction in comparison to the longitudinal wrinkles.

5.4 Helicoid Versus Planar State: From Boundary-Dominated to Twist-Dominated

So far, we assumed that the highly-symmetric state of a stretched-twisted ribbon, characterized by translational symmetry along the longitudinal direction $\hat{s}$, is the helicoid. We carried out stability analysis of the helicoidal state and studied the transitions to states that break its translational symmetry. However, if the ribbon is sufficiently long ($L \gg 1$) and the exerted twist η is sufficiently small with respect to T, one may envision that the unbuckled state of the ribbon is not a helicoid but rather a stretched, planar ribbon, where the exerted twist remains confined to the vicinity of the short edges (see Fig. 14). Such a localized-twist state can be described as a perturbation to the well-known problem of purely stretching an elongated ribbon, where $0 < T \ll 1$ and $\eta = 0$ [8]. For that problem, it was found that the clamping of the short edges together with the Poisson ratio effect gives rise to contraction of the ribbon in the transverse direction not only in the vicinity of the clamped edges, but rather throughout most of the length of the ribbon (see [8]). Recalling our discussion in previous sections, one may conclude that there exist two distinct causes for transverse buckling and wrinkling in a ribbon: The first one, which we described above (Sect. 4), is a "bulk" mechanism, where σ^{rr} becomes compressive due to the simultaneous effect of uniaxial stretching along $\hat{s}$ and uniform twist η, experienced by any piece of the ribbon. The second mechanism is essentially a boundary-generated effect, whereby the tendency of the ribbon to transverse contraction ($u_r(1/2) = -\nu T/2$ in Eq. (29)) *away* from the clamped edges gives rise to a small transverse compression $\sigma^{rr} < 0$, which is also relieved by wrinkles. A natural question is whether, for a given set of parameters the emergence of a compressive transverse stress σ^{rr} and the consequent buckling/wrinkling instability, are governed by the bulk effect (twist) or rather by the boundary effect (clamping).

In order to address this question, we have to compare the compressive stresses ($\sigma^{rr} < 0$) associated with twist and with clamping of the short edges. However, while the first one was derived above (Eq. (4)), we are not aware of a similar analytic expression for the transverse stress due to clamped edges.[12] Hence, we will proceed by estimating the relevant energies. We will do this by denoting $U_0 = T^2$ the energy per length of a stretched ribbon that is not clamped and not twisted, and estimating the excess energies associated with twist and clamping, which we denote, respectively, by ΔU_{twist} and ΔU_{clamp}. Our purpose is to find the curve $\eta^*(T)$ in the (T, η) plane, below which the clamped-edge effect is significant.

The Excess Energy ΔU_{twist} Expecting the transition from twist-dominated to clamping-dominated instability to occur at a small value of η, we neglect terms of order $O(\eta^4)$ in comparison to terms of order $O(\eta^2 T)$, and thus estimate ΔU_{twist} by considering a stretched, twisted, unclamped ribbon:

$$\Delta U_{\text{twist}} \approx \int_{-1/2}^{1/2} \varepsilon_{ss}(r)^2 dr - U_0 \sim T\eta^2, \tag{91}$$

where we considered only the leading order in η^2, and therefore neglected the energy due to the strain ε_{rr}^2.

The Excess Energy ΔU_{clamp} Consider now a stretched, clamped, untwisted ribbon. In Appendix E we show that the deviation of the longitudinal strain ε_{ss} from the "base" value T is proportional to the Poisson ratio ν and is restricted to distances ~ 1 from the clamped edges, at which zone the transverse and shear strain components also have nonvanishing values that are proportional to ν. This allows us to estimate:

$$\Delta U_{\text{clamp}} \sim \frac{\nu F(\nu) T^2}{L}, \tag{92}$$

where $F(\nu)$ is some smooth function of ν that satisfies $F(\nu) \to \text{cst}$ for $\nu \to 0$.

Comparing now our estimates for the excess energies ΔU_{twist} and ΔU_{clamp}, we find that the transition from the clamping-dominated zone to the twist-dominated zone is expected to occur around:

$$\eta^* \sim \sqrt{\frac{\nu T}{L}}, \tag{93}$$

[12] The well-known work of Cerda and Mahadevan [8] addressed this problem in the far-from-threshold regime, where wrinkles are fully developed and the transverse compression cannot be approximated by its value at threshold. The planar (unbuckled) state that underlies the wrinkling instability due to clamped boundaries was studied numerically in some recent works [47, 48]. However, these works did not address the ribbon limit $L \gg 1$ that we study here. A couple of papers [49] attempted to extend the far-from-threshold approach of [8] to the near-threshold regime (by invoking effective "inextensibility" constraints), but the justification of this approach has yet to be established.

confirming our expectation that $\eta^*(T)$ approaches the T axis for small Poisson ratio and large L. For $\eta < \eta^*$, we expect that a transverse buckling instability is triggered by the clamped boundaries, whereas for $\eta > \eta^*$ we expect the instability mechanism described in the previous sections.

We have to compare the above expression with the threshold for the transverse instability found in Sect. 4, $\eta_{\text{tr}} \sim \sqrt{t}T^{-1/4}$. We find that both values are comparable when

$$T = T_{\text{clamp}} \sim \left(\frac{Lt}{\nu}\right)^{2/3}. \tag{94}$$

If $T > T_{\text{clamp}}$, the tension is sufficiently large and the effect of clamping on the transverse instability cannot be neglected. Above this critical tension, the transverse instability is governed by the clamped boundaries rather than the helicoid geometry.[13]

6 Discussion

Our theoretical study identified distinct types of morphologies in different regimes of the 4D parameter space spanned by T, η, t and L^{-1}. These dimensionless parameters are assumed small (Eqs. (1), (2)), and one may be tempted to describe the ribbon as "thin" ($t \ll 1$), "long" ($L^{-1} \ll 1$), subjected to "small" tensile load ($T \ll 1$), and "slightly" twisted ($\eta \ll 1$). However, our analysis highlights the deceptive nature of such a colloquial description, since the ribbon exhibits markedly different behaviors in different "corners" of the 4D parameter space. In other words, the relevant parameters that govern the ribbon morphology are various *ratios* between the four control parameters T, η, t and L rather than the "bare" value of each control parameter.

In Table 1 we summarize the central phenomena predicted in our paper, as well as a few open questions raised by our analysis at some of those corners of the parameter space. The challenge for an experimenter, who may be motivated by the predictions in this paper, is to construct a set-up that allows access to those distinct regimes and precise measurements of observables which characterize the transitions between them. In this section we will focus on this experimental perspective, propose specific measurements, and describe a few open questions that await further theoretical and experimental study.

6.1 Experimental Considerations

6.1.1 Objectives

Let us assume a ribbon with a fixed thickness and width, such that the parameter t is fixed at a very small value (say, $t \approx 10^{-5}$), and address the desired range of the other control parameters.

- *Controlling* T: As Fig. 15a depicts, the tension T may vary between T_{min}, which is determined by the quality of the set-up, and $T_{\text{Hook}} \ll 1$, above which the material response can no longer be approximated through Hookean elasticity. Ideally, one would

[13] Note that if $L \sim 1$, as was the case in Fig. 3 of [3], this equation means that the transverse wrinkling reflects the clamping-induced instability mechanism of a stretched sheet [8] rather than the helicoidal mechanism described in Sect. 4.

Table 1 Central predictions and open questions raised in our paper. The bold letter **P** stands for "prediction", whereas **OQ** stands for "open question"

Parameter regime	Phenomenon	Proposed measurement/Specific predictions
$T \lesssim O(t^2)$, $\eta \sim t$	(**OQ**) nature of longitudinal instability (Fig. 3c)	(Sect. 6.2.2)
$T \lesssim O(t^2)$, fixed η (independent on t, T)	(**P**) transitions between distinct asymptotic isometries (Figs. 3a, 6)	(Sect. 6.2.3)
$O(t^2) < T \ll \max\{t, (\frac{t}{L})^{2/3}\}$, $\eta > \sqrt{24T}$	(**P**) FT longitudinal wrinkling (Figs. 3d, 4, 5)	(Sect. 6.2.1) $r_{wr}(\alpha)$ (Eq. (50)) $\chi_{FT}(\alpha)$ (Eq. (52)) $\Delta\alpha_{\text{NT-FT}} \sim t/\sqrt{T}$ (Eq. (60))
$T \gtrsim \max\{t, (\frac{t}{L})^{2/3}\}$	(**P**) strong dependence of transverse instability on L, T (Figs. 3b, 9)	(Sect. 6.2.4) $\eta_{\text{tr}}(T), \lambda_{\text{tr}}, T_\lambda$ (Eqs. (64)–(65), and (67)–(69))
$(\frac{Lt}{\nu})^{2/3} \lesssim T \ll 1$	(**OQ**) from boundary-dominated to helicoidal shape (Figs. 3e,14)	(Sect. 5.4) η^* (Eq. (93))
$\eta \gtrsim O(t/\sqrt{T})$	(**OQ**) nature of looping instability	(Sects. 6.2.5)

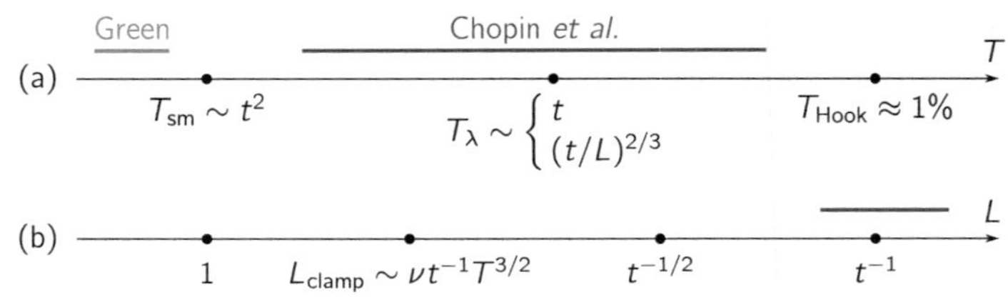

Fig. 15 Ranges of tension and length probed by the experiments [2, 3]. (**a**) Tension: the Green's plateau for $\eta_{\text{lon}}(T)$ pictured in Fig. 3c is expected for $T \ll T_{\text{sm}}$, longitudinal wrinkling (near threshold and far from threshold) is expected for $T_{\text{sm}} < T < T_\lambda$, and transverse buckling/wrinkling becomes a primary instability of the helicoid for $T > T_\lambda$. The upper limit T_{Hook} is the limit of linear Hookean response of the material. (**b**) Length: (in depicting this figure we assume $L_{\text{clamp}} = \nu t^{-1} T^{3/2} \ll t^{-1/2}$, see Eq. (94)). For lengths $L > t^{-1}$ the primary transverse instability is buckling; For $L < L_{\text{clamp}}$ the transverse instability is governed by the clamping of the short edges, similarly to [8]; For $L_{\text{clamp}} \ll L \ll t^{-1/2}$, the transverse instability is wrinkling; finally for lengths in the range $t^{-1/2} \ll L \ll t^{-1}$ we predict a crossover from buckling to wrinkling as the exerted tension T varies from T_λ to T_{Hook} (see Fig. 13(*Left*))

like $T_{\text{min}} \ll T_{\text{sm}}(t)$ and $T_{\text{Hook}} \gg T_\lambda$, where $T_{\text{sm}} \sim t^2$ (Eq. (46)) and $T_\lambda \sim \max\{t, (t/L)^{2/3}\}$ (Eqs. (65), (69)). As we describe in the next subsection, varying the tensile load in the range $(T_{\text{sm}}, T_\lambda)$ will allow probing most of the phenomena associated with the nature of the longitudinally-wrinkled pattern (Sects. 3.2–3.4, Fig. 3d), and the mechanism by which it becomes unstable at sufficiently low T and given η (Sect. 5.3, Fig. 3a); varying

Table 2 Typical experimental parameters, used by Green [1] and Chopin and Kudrolli [3], and the corresponding ratios that are relevant for our analysis

Exp.	E (GPa)	t	L	T_{min}/T_{sm}	T_{Hook}/T_λ	$L_{max}t$
Spring steel [1]	210	2×10^{-3}	20	2	2	5×10^{-2}
Mylar [3]	3.4	5×10^{-3}	15	21	2.4	2×10^{-2}

the tensile load in the range (T_λ, T_{max}) is necessary to understand the effect of T on the transverse instability (Sect. 4).

- *Controlling* L: As Fig. 15b depicts, a desired set-up should allow variation of the ribbon length L from L_{min} to L_{max}, where $1 < L_{min} \ll t^{-1/2}$ and $L_{max} \gg t^{-1}$. Varying L in such a range will provide an access to most of the predictions associated with the transverse instability: the wrinkling-buckling transition (Sect. 4.4, Fig. 3b), the scaling law of the triple point (Eqs. (65), (69)), and the possible existence of a localized transverse buckling mode (even for a very long ribbon), which we discuss in the next subsection.
- *Controlling* η: A good experimental set-up may allow a nearly-continuous variation of the imposed twist angle $\theta = \eta L$. For instance, if θ is varied by increments of 1° then the minimal twist that could be imposed is $\eta_{min} \approx 2\pi/(360\ L)$. A reliable control on η_{min} is required for two purposes: In the very low tension regime ($T < T_{sm}(t)$), it may allow to address Green's "plateau" of the longitudinal wrinkling instability $\eta_{lon}(T) \to 10t$ (Sect. 3.2, Fig. 3c); For larger values of exerted tension, it may be necessary to probe the predicted transition from a planar state with twist confined to the short edges to a helicoidal shape (Sect. 5.4, Fig. 3e).

6.1.2 Challenges

We are aware of two documented experiments that addressed the behavior of a stretched-twisted elastic ribbon: Green's experiment from 1937 [2], where ribbons were made of steel; and [3], which used Mylar. In Table 2, we compare the control parameters and their relevant mutual ratios in both experiments. Green, who used a material with very large Young's modulus, could address the "ultra-low" tension regime, $T \sim T_{sm}(t)$ (Fig. 3c), but a simple steel may exhibit a non-Hookean (or even inelastic) response at rather small T, which limits its usage for addressing the regime around and above the triple point (i.e., $T > T_\lambda$). In contrast, the experiment of [3] used a material with much lower Young's modulus, which allows investigation of the ribbon patterns in Fig. 2g, but the minimal exerted tension T_{min} (associated with the experimental set-up) was not sufficiently small to probe Green's threshold plateau $\eta_{lon}(T) \to 10t$ for $T \lesssim T_{sm}(t)$.

This comparison reveals the basic difficulty in building a single set-up that exhibits clearly the whole plethora of shapes shown in Fig. 3. In addition to the effect of T_{min} and T_{Hook}, there is an obvious restriction on L_{max} (at most few meters in a typical laboratory). Below we propose a couple of other materials, whose study—through experiment and numerical simulations—may enable a broader range of the ratios T_{min}/T_{sm}, T_{Hook}/T_λ and $L_{max}t$.

Graphene This novel 2D material is characterized by $t \sim 0.3$ nm (which we assume to approximate the "mechanical thickness", i.e., the ratio $\sqrt{B/Y}$), Young's modulus $E = 10^3$ GPa and a yield stress of ~100 GPa [50], whose ratio (≈0.1) we use as an approximation of T_{Hook}. Graphene sheets can be produced with lateral scales of up to 1 mm [51]. Assuming a graphene ribbon with length 1 mm and width 0.03 mm, we obtain $t \approx 10^{-5}$ and $L \approx 30$. A narrower ribbon may allow exploring a larger range of L, at the expense of smaller t. Tensile load may be exerted on graphene by optical tweezers, which allow a force (on the short edge) in the picoNewton range, such that: $T_{\min}/T_{\mathrm{sm}} \sim 1$, and $T_{\mathrm{Hook}}/T_{\lambda} \sim 10^3$.

Ultrathin PS Films Polymer films with thickness of 30–300 nm can be fabricated by spin coating and have been used extensively in studies of wrinkling and other elastic phenomena [15, 32]. Such sheets are characterized by E = 3.4 GPa, and their lateral scales may be few cm's. It may thus be possible, for instance, to create ultrathin PS ribbons with $L \approx 10^2$ and $t \approx 10^{-4}$. Capillary forces have been used to exert tension on floating ultrathin PS sheets, where the surface tension varies from a maximum of ≈70 mN/m to 3 times lesser than this value (by using surfactants). This corresponds to $T_{\min}/T_{\mathrm{sm}} \sim 10^4$, $T_{\mathrm{Hook}}/T_{\lambda} \sim 20$ (for $t \approx 300$ nm).

Thus, making ribbons from graphene or ultrathin PS films may allow a broad range of the three most relevant ratios (right columns of Table 2) that are necessary to explore the various asymptotic regimes of the system. We recall though, that both graphene and ultrathin polymer sheets are rarely used in a free-standing form, which is the one needed for the stretch-twist experiment that we address here.

6.2 Proposed Measurements and Open Questions

Let us now discuss the specific parameter regimes and the corresponding morphological instabilities addressed by our theory. In each of the following paragraphs we will propose measurements and mention open theoretical questions.

6.2.1 Longitudinal Wrinkling at $T_{\mathrm{sm}} < T < T_{\lambda}$

The parameter regime $T_{\mathrm{sm}} < T < T_{\lambda}$ (T_{sm} is defined in Eq. (47)) was addressed in the experiment of Chopin and Kudrolli [3], who found that the threshold $\eta_{\mathrm{lon}}(T)$ becomes close to the curve $\sqrt{24T}$ at which the Green's stress predicts longitudinal compression. In this regime, our discussion in Sect. 3 predicts the emergence of the FT regime rather close to the threshold curve.

Experiment Chopin and Kudrolli [3] used ribbons with $t = 5 \cdot 10^{-3}$ and confirmed that the threshold $\eta_{\mathrm{lon}}(T)$ is close to the curve $\sqrt{24T}$, at which the Green's longitudinal stress becomes compressive. Additionally, the dependence of $\Delta\alpha_{\mathrm{lon}}$ and λ_{lon} on t and T (Eq. (45)) was in excellent agreement with the prediction of the NT approach [5]. Such a value of t, however, may be too large to probe the transition from the NT to the FT regime predicted in Sect. 3, since the initial width of the wrinkled zone ($r_{\mathrm{wr}} \sim \lambda_{\mathrm{lon}}$) is already a relatively large fraction of the ribbon width. Using thinner ribbons (e.g., $t \sim 10^{-5}$) may lead to substantially smaller value of the initial width of the wrinkled zone, such that the different predictions of the NT and FT methods for $r_{\mathrm{wr}}(\alpha)$ (Fig. 4) may be more pronounced.

A useful probe for testing the predicted NT-FT transition may involve the longitudinal contraction $\chi(\alpha)$ (Eqs. (20), (52), Fig. 5). The function $\chi(\alpha)$ may be easier to measure in comparison to the width $r_{\mathrm{wr}}(\alpha)$ of the wrinkled zone, which requires the usage of optical tools. Notably, the NT and FT predictions for the longitudinal contraction are significantly different, wherein the last one becomes 3 times smaller than the first for sufficiently large α.

Theory Our FT analysis in Sect. 3.3 has been focused on the *dominant* part of the elastic energy of the longitudinally wrinkled state, stored in the asymptotic, compression-free stress field. The dominant energy is associated with macroscale features, which do not depend explicitly on t, and underlies the predictions for r_{wr} and χ_{FT}. However, as we emphasized in Sect. 3.3, a complete characterization of the wrinkle pattern requires evaluation of the *subdominant* energy, which includes the bending cost due to wrinkles, as well as the comparable cost due to the formation of a wrinkled structure on the stretched helicoidal shape. The calculation of the subdominant energy may involve some subtleties, such as internal boundary layers [27] and the possible formation of wrinkle cascades rather than a simply periodic structure [28]. Importantly, the actual subdominant mechanics that govern the wrinkle wavelength may depend on the confinement parameter α and therefore the exponent β that characterizes the subdominant energy (Eq. (59)) may take different values in the limit $\alpha \to 24$ (where the helicoidal shape is highly strained) and the limit $\alpha \to \infty$, at which the longitudinally-wrinkled helicoid becomes an asymptotic isometry of the ribbon. Therefore, evaluation of the subdominant energy is essential not only for finding the fine structure of the wrinkle pattern, but also to understand how it becomes unstable with respect to the creased helicoid shape as α becomes large (i.e., the limit $T \to 0$ for fixed r_i).

6.2.2 *Longitudinal Wrinkling at* $T < T_{sm}$

Green's theory [2] consists of a linear stability analysis of the helicoidal state (Eqs. (11), (21)–(22)) in the limit $T \to 0$, where the longitudinally buckled/wrinkled zone is not confined to the vicinity of the centerline, but rather expands throughout the whole width of the ribbon. This (NT) analysis yields the threshold plateau $\eta_{lon}(T) \to 10t$ as $T \to 0$, which was obtained (with deviation of 10%) in Green's experiment [2].

Experiment The experimental data in Green's paper [2] does show a good agreement with the theoretical prediction based on his linear stability analysis. However, the available data (Figs. 4 and 5 of [2]) may not be sufficient to determine whether the actual instability observed by Green was a longitudinal wrinkling or a creased helicoid state. This confusion is illustrated in Fig. 3c, which reflects our expectation that a creased helicoid pattern (or even a cylindrical wrapping state) may be observed sufficiently close to the vertical line $T = 0$. The possible emergence of a creased helicoid state directly from the helicoidal state (i.e., without an intervening wrinkle pattern) may indicate that the longitudinal instability changes its supercritical (continuous) character, becoming a subcritical bifurcation at sufficiently small T. A careful experiment may provide a conclusive answer to this question.

Theory The confusing nature of the longitudinal instability at the regime $T < T_{sm}$, depicted in Fig. 3c, is reflected also in the fuzziness of the NT-FT transition in this regime. The FT approach (Sect. 3.3) assumes that the wrinkle pattern near threshold is confined to a strip of width $r_{wr} < 1/2$ around the centerline, and describes how r_{wr} varies upon increasing the confinement α. However, for $T < T_{sm}$, Green's analysis [2] shows that the ribbon is deformed across its whole width as soon as the longitudinal instability sets in. A natural question is whether the FT regime of the longitudinally wrinkled helicoid terminates at a small $T \sim T_{sm}$. A related question is whether a disappearance of the NT-FT transition below a certain tensile load indicates a qualitative change of the longitudinal instability from a supercritical to a subcritical type.

6.2.3 *From Longitudinal Wrinkling to the Creased Helicoid State*

For a given twist η and sufficiently small tensile load T, we expect the formation of a creased helicoid state (Fig. 3), predicted in [7] and observed in [3]. Our energetic argument, based on the asymptotic isometry equation, Eq. (42), suggests the existence of a transition between the creased helicoid state and the FT-longitudinally-wrinkled state, which becomes sharply localized along a curve in the parameter plane (T, η) in the limit $t \to 0$, as is illustrated in Fig. 3a.

Experiment Chopin and Kudrolli [3] did observe such a transition, but noted that "*...the longitudinally buckled ribbon evolves continuously into a self-creased helicoid...*". The appearance of a smooth crossover, rather than a sharp transition between these states, could be attributed to the thickness parameter used in their experiment ($t \approx 5 \times 10^{-3}$), which may be sufficiently small to notice the "wake" of a morphological phase transition that becomes asymptotically sharp as $t \to 0$, but not the acute, critical nature of the transition. Future experiments may have to use significantly smaller values of t in order to study this transition.

A useful indirect probe of the transition from a longitudinal wrinkling to a creased helicoid state may again be the longitudinal contraction of the ribbon $\chi(T, \eta)$. The FT approach (Sect. 3.3) predicts the T-dependence through the function $\chi_{\rm FT}(\alpha)$ (Fig. 5) and we expect that the method of [7] yields another function $\chi_{\rm CH}(T, \eta)$. A signature of a sharp morphological transition between the two states may be a discontinuity of the measured derivative $(\partial\chi/\partial T)_\eta$ at a curve $T_c(\eta)$ in the (T, η) plane.

Theory As we pointed out in Sect. 5.3, our schematic plot of the energies of the longitudinal wrinkling and the creased helicoid states (Fig. 6b) reflects their asymptotic isometry in the vicinity of the hyper-plane $(T = 0, t = 0)$, but has a heuristic content, since the coefficients of those linear functions are yet unknown. The FT analysis of the longitudinally wrinkled state still lacks an exact evaluation of the subdominant energy, which underlies the intercept value in the corresponding linear function (red line in Fig. 6b); The corresponding plot for the creased helicoid state (dashed brown in inset) lacks the values of both the slope and the intercept. The slope of that line is simply the longitudinal contraction of the state, and can be computed from the construction of Korte et al. [7], which describes the ribbon through the Sadowsky equation of a strainless strip. The intercept of that linear function, however, cannot be computed through this framework, since the subdominant energy of the creased helicoid state stems from bending energy associated with broadening the creases into narrow zones in which the ribbon cannot be considered as strainless. Thus, a reliable evaluation of the subdominant energy of the creased helicoid state may require one to use the FvK framework, where the ribbon—at least in the vicinity of the creases—is allowed to have strain. It is possible that familiar concepts, such as the "minimal ridge" [34], can be invoked in order to approximate the subdominant energy of the creased helicoid state.

As we briefly described in Sect. 5.3, the creased helicoid shape characterizes in fact a family of states, parameterized by the angle between triangular facets. Therefore, a more realistic picture of the inset to Fig. 6b may consist of a series of lines that represent this family, rather than the single brown dashed line. The slope and intercept of each of those lines would result, respectively, from the longitudinal contraction and the subdominant energy of each creased helicoid state. Therefore, upon increasing the tensile load, a ribbon subjected to twist η may undergo a series of transitions between creased helicoid states before the transition to the wrinkled helicoid.

6.2.4 The Transverse Instability: Threshold

Our theoretical analysis of the transverse instability (Sect. 4) predicts that the threshold curve $\eta_{\rm tr}(T)$ in the (T, η) plane depends on the ribbon thickness t and the tension T (Eq. (64), Figs. 8, 9). Furthermore, if the ribbon is not "infinitely" long, such that $L \ll t^{-1}$, the threshold $\eta_{tr}(T)$ depends also on L (Fig. 12) and the nature of the transverse instability changes from buckling to wrinkling (i.e., wavelength $\lambda_{tr} < W$).

Experiment For a ribbon (where $L \gg 1$), Chopin and Kudrolli [3] observed only a buckling instability ($\lambda_{tr} > W$), and reported that $\eta_{tr}(T)$ (for $T > T_\lambda$) has a plateau value, which scales as $\sim\sqrt{t}$ but does not depend on the tension T, nor on the ribbon length L. The appearance of a buckling mode is consistent with the relatively long ribbon used in comparison to t^{-1} [3] (see Fig. 15), but the independence of η_{tr} on T disagrees with our result. It is quite possible though, that the dependence on tension, $\eta_{tr} \sim T^{-1/2}$, has simply been overlooked in those measurements. Future experiments that will examine our predictions will have to use an appropriate range of ribbon lengths, depicted in Fig. 15.

Theory Our predictions concerning the effect of the ribbon length on the transverse instability (Sects. 4.2, 4.4) assume that L enters through a single term in the normal force balance (σ^{ss}/L^2), which couples the longitudinal stress to the unavoidable, wrinkle-induced curvature in that direction. A complete analysis should include other terms, accounting for example for the strain induced by the longitudinal variation of the longitudinal in-plane displacement. These terms may affect the exact numerical values of our predictions (e.g., the location of the threshold $\eta_{tr}(T)$ in Fig. 12), but are unlikely to affect any of the scaling laws.

A nontrivial assumption in our linear stability analysis of the transverse instability is encapsulated in the form of the unstable mode (Eq. (70) in Sect. 4.3). If the ribbon was truly infinitely long, then the translational symmetry of the stress in the longitudinal direction is naturally broken through such Fourier modes, and the one which requires the least curvature (i.e., $j = 1$ in Eq. (71)) is the first to become unstable under given twist and tensile loads. The translational symmetry of the helicoidal shape is not perfect, however, due to the boundary conditions at the short edges.[14] It is natural to expect that the unstable mode "feels" these boundary conditions at a small region near the short edges whose size is comparable to the ribbon width. However, recent studies have shown that the boundary shape may have a long-range effect on the deformation of a thin sheet, with penetration length that diverges as $t \to 0$ [52, 53]. If the boundary conditions at $s = \pm L/2$ have such a long-range effect, the longitudinal structure of the unstable mode may exhibit strong deviation from the sinusoidal shape ($z_1 \sim \cos(\pi s/L)$) even if the ribbon is very long, as long as $L < t^{-x}$ with some $x > 0$. In order to address this question, one may have to carry out the transverse stability analysis, taking into full consideration the boundary conditions at $s = \pm L/2$, and finding the unstable mode through numerical analysis of the linear partial differential equation (i.e., for $z_1(s, r)$ and $u_{s1}(s, r)$).

6.2.5 The Transverse Instability: Beyond Threshold

Our analysis in Sect. 4 identified the threshold curve $\eta_{tr}(T)$ and characterized the nature of the unstable modes through linear stability analysis, but this approach cannot clarify the spatial structure of the ribbon above that threshold curve. The experiment of [3] found that the ribbon forms loops and self-contact zones very close to the threshold curve $\eta_{tr}(T)$, and furthermore—a strong hysteretic behavior has been observed, especially in the low tension regime ($T < T_\lambda$) in which the transverse instability emerges as a secondary instability of the

[14] The exact boundary conditions at $s = \pm L/2$ may depend on the specific set-up used to apply simultaneously tension and twist. These may include, for instance, complete clamping (i.e., $u = z = 0$) or partial clamping (i.e., only $z = 0$).

helicoidal state. The emergence of a transverse instability of twisted ribbons as a precursor to the formation of loops and coils has been recognized in recent numerical studies [18, 19].

While it may be possible to address the formation of loops and hysteresis phenomena through the cFvK equations, an effective theory that describes the ribbon through its centerline $\boldsymbol{X}_{\mathrm{cl}}(s)$ may provide deeper insight into this mechanics. Such an approach may be similar in spirit to the Sadowsky strip or Kirchoff rod equations (Sect. 5.2), but the effective equations that govern the mechanics of a stretched-twisted ribbon above the threshold curve $\eta_{\mathrm{tr}}(T)$ are likely to be markedly different from each of these approaches.

7 Summary

The central result of our paper is illustrated in Fig. 3: A phase diagram that describes the distinct morphologies of a ribbon in the parameter plane spanned by the exerted tension T and twist η. The separation of the (T, η) plane into three major parts (blue, orange, pink) that meet at a single triple point $(T_\lambda, \eta_\lambda)$ has been recognized by Chopin and Kudrolli [3], who attributed this peculiar property to the presence of three operative instability mechanisms. Our study reveals that this phenomenology is rooted at two basic instabilities only, whereby the ribbon responds by wrinkling/buckling to the compressive stresses in the longitudinal and transverse directions. The three major morphological phases correspond to a highly symmetric helicoidal state (blue), and two states that break this symmetry through instabilities that deform the shape in the longitudinal direction (orange), in the transverse direction (pink, at $T > T_\lambda$), or in both principal directions (pink, at $T < T_\lambda$). This insight is borne out by bringing together two theoretical elements: The cFvK equations that capture the transverse stress due to the non-planar, helicoidal shape of the twisted ribbon; and a far-from-threshold (FT) analysis of these equations, that describes the collapse of longitudinal compression enabled by the formation of wrinkles.

The far-from-threshold analysis of the cFvK equations revealed a profound feature of the wrinkling instability: assuming a fixed twist η, and reducing the exerted tension (along a horizontal line in Fig. 3), the formation of longitudinal wrinkles that decorate the helicoidal shape enables a continuous, gradual relaxation of the elastic stress from the strained helicoidal shape at $T > \eta^2/24$, to an asymptotically strainless state at $T \to 0$. This remarkable feature led us to propose a general form of the asymptotic isometry equation (42), which characterizes the wrinkled state of the ribbon (Fig. 2b, c), as well as other admissible states at the limit $T \to 0$, such as the cylindrical wrapping (Fig. 2e) and the creased helicoid state (Fig. 2d). The asymptotic isometry equation provides a simple framework, in which the transitions between those morphologies in the vicinity of the vertical line ($T = 0$ in Fig. 3) correspond to the intersection points between linear functions of T (Fig. 6b), whose intercepts and slopes are determined solely by the geometry of each state.

Beyond its role for the mechanics and morphological instabilities of ribbons, the asymptotic isometry equation may provide a valuable tool for studying the energetically favorable configurations of elastic sheets. Notably, Eq. (42) takes into consideration the deviations of the sheet's midplane from a perfectly strainless shape, not only due to the small thickness of the sheet but also due to a small tensile load. In the context of conventional elastic sheets, whose stress-free state is planar, such as the twisted ribbon or an adhesive sheet attached to a curved substrate [23], the tension T in Eq. (42) can easily be recognized as the tensile load exerted on the boundary of the sheet. The asymptotic isometry equation may be useful, however, also in studies of "non-Euclidean" sheets, whose stress-free state is determined by a "target metric", programmed by differential swelling techniques or other means [54, 55].

For such non-Euclidean sheets, the tension T in Eq. (42) may originate from a less direct source, such as imperfections in the prescribed metric. A puzzling experimental result in this emerging field has been the surprising wrinkled shape adopted by a sheet whose target metric was prescribed to be compatible with a hyperbolic shape (constant negative G) [37]. It has been noted in [38] that such a wrinkling pattern is consistent with an asymptotic isometry, whose bending energy, however, is higher than the simple hyperbolic shape. These two isometries, may be analogous, respectively, to the cylindrical wrapping state and the longitudinally wrinkled state of the twisted ribbon, whose energetic degeneracy is lifted not only by the thickness t but also by a tensile load T (Fig. 6b). The presence of a tension-like term in the corresponding asymptotic isometry equation that describe the energy of such a non-Euclidean sheet, may clarify the experimental conditions under which the hyperbolic shape may be observed.

Acknowledgements The authors would like to thank C. Santangelo for many enlightening discussions and particularly for educating us on the covariant FvK equations; and to A. Romaguera, F. Brau, B. Audoly, and two anonymous referees, for their critical reading and useful comments on the manuscript. B.D. would like to thank E. Hohlfeld for many inspiring discussions on asymptotic isometries and their use for elastic sheets subjected to geometric constraints and tensile loads. The authors acknowledge financial support by CNPq-Ciência sem fronteiras program, Brazil (J.C.), the KECK foundation Award 37086 (V.D.), and NSF CAREER Award DMR-11-51780 (B.D.).

Appendix A: Hookean Elasticity and Leading Order Stresses and Strains

Our theory addresses the "corner" in the 4d parameter space, defined by Eqs. (1), (2), and therefore most of the analysis in this paper employs expansions in these parameters. Why do we assume these parameters are small?

First, $t \ll 1$ and $L^{-1} \ll 1$ stem from the definition of a ribbon. Second, we focus our discussion on the universal, material-independent behavior of elastic ribbons and therefore we consider a Hookean response, whereby the stress-strain relationship is linear. Since Hookean response is valid only for small strains, and since the exerted tension T necessarily induces strain, we must require $T \ll 1$. Finally, the assumption $\eta \ll 1$ is more subtle. For the unwrinkled helicoidal state, we showed in Sect. 2.3 that the components of the strain and stress tensors are proportional to positive powers of η, and therefore Hookean response is valid only for sufficiently small values of η. In contrast, for the longitudinally-wrinkled helicoidal state (i.e., $\eta > \sqrt{24T}$), we showed in Sect. 3.3 that the ribbon may become nearly strainless (i.e., asymptotically isometric to the undeformed ribbon) even under finite η, therefore the Hookean response for the wrinkled state is not limited to small values of η. However, even for the wrinkled state the assumption $\eta \ll 1$ is very useful, since it enables an easy way to compute the various components of the stress tensor and allows us to characterize the wrinkled state as a sinusoidal undulation (Eq. (54) and Appendix C). Importantly, our finding that the threshold values $\eta_{\text{lon}}, \eta_{\text{tr}}$ vanish in the asymptotic limit $t \to 0$ (see Sect. 4.1 and Fig. 8), proves in a self-consistent manner that the basic morphological instabilities of a stretched-twisted ribbon are well described by assuming $\eta \ll 1$.

Our theory is thus valid at the leading order in t, L^{-1}, η and T, and any higher order terms are ruled out from the derivations. *Leading terms* should be understood with respect to these expansion parameters. Namely, denoting by A a scalar, or a component of a vector or tensor (e.g., longitudinal contraction, transverse or longitudinal stress or strain), then A is expanded as

$$A = \sum_{a_1,a_2,a_3,a_4 \geq 0} T^{a_1} \eta^{a_2} t^{a_3} L^{-a_4} A_{(a_1,a_2,a_3,a_4)}, \tag{95}$$

and the *order* of A is given by the four positive integers (a_i). Saying that the order (a_i') is higher than the order (a_i) means that

$$(a_i) < (a_i') \Longleftrightarrow \begin{cases} \forall i, a_i \leq a_i' \\ \text{and} \\ \exists i \text{ such that } a_i < a_i'. \end{cases} \tag{96}$$

The leading terms of A are the minimal orders for the relation (95) with non-zero coefficient. For example, the leading terms of the longitudinal stress in the helicoidal state are given in Eq. (3), $\sigma_{\text{hel}}^{ss} = T + \frac{\eta^2}{2}(r^2 - \frac{1}{12})$, they correspond to the orders $(1,0,0,0)$ and $(0,2,0,0)$. These two orders are minimal and cannot be compared. The transverse stress given in Eq. (4) has vanishing coefficients for these orders, and the minimal orders with non zero coefficients are $(1,2,0,0)$ and $(0,4,0,0)$ (i.e., $T\eta^2$ and η^4). The s-independent transverse buckling equation (79) contains terms of order $(1,2,0,0)$, $(0,4,0,0)$ and $(0,0,2,0)$ (respectively, $T\eta^2$, η^4 and t^2).

In a given equation, several orders may appear; in this case only the minimal ones should be considered. This happens in the computation of the stress in the helicoidal state. Equation (33) shows that σ^{rr} is of higher order than σ^{ss}; besides, one of the two terms in the r.h.s. of Eq. (37) has obviously the same order $(0,2,0,0)$ as σ^{ss} in Eq. (36). Thus, consistency of Eqs. (33), (36), (37) implies that the leading terms in the r.h.s. mutually cancel each other (which we call a "solvability condition"), leading to Eq. (38) for $u_r'(r)$. This last equation allows to compute the longitudinal stress, Eq. (39), from which we deduce the transverse stress, Eq. (40), using again Eq. (33).

Appendix B: Covariant Derivative: Definition and Application to the Helicoid

For an arbitrary surface with metric $g_{\alpha\beta}$, the covariant derivative is defined with the Christoffel symbols, that are given by

$$\Gamma^{\alpha}_{\beta\gamma} = \frac{1}{2} g^{\alpha\delta}(\partial_\beta g_{\gamma\delta} + \partial_\gamma g_{\beta\delta} - \partial_\delta g_{\beta\gamma}). \tag{97}$$

The covariant derivative of a vector u^α is then defined as

$$D_\alpha u^\beta = \partial_\alpha u^\beta + \Gamma^{\beta}_{\alpha\gamma} u^\gamma. \tag{98}$$

The strain tensor has two indices, so that its covariant derivative is

$$D_\alpha \sigma^{\beta\gamma} = \partial_\alpha \sigma^{\beta\gamma} + \Gamma^{\beta}_{\alpha\delta}\sigma^{\delta\gamma} + \Gamma^{\gamma}_{\alpha\delta}\sigma^{\beta\delta}. \tag{99}$$

For the helicoid with metric (30)

$$g_{\alpha\beta} = \begin{pmatrix} 1 + \eta^2 r^2 - 2\chi + 2\eta^2 r u_r(r) & 0 \\ 0 & 1 + 2u_r'(r) \end{pmatrix}, \tag{100}$$

the non-zero Christoffel symbols are (to the leading order):

$$\Gamma^{r}_{ss} = -\eta^2 r, \tag{101}$$

$$\Gamma^{s}_{sr} = \eta^2 r, \tag{102}$$

$$\Gamma^{r}_{rr} = u_r''(r). \tag{103}$$

Appendix C: Shape of the Longitudinally Wrinkled Helicoid Far from Threshold

Far from threshold, longitudinal wrinkles relax the longitudinal compression. In the main text, we propose the following form for the wrinkles:

$$\boldsymbol{X}^{(\mathrm{wr})}(s,r)=\begin{pmatrix}(1-\chi_{\mathrm{FT}})s\\ r\cos(\eta s)-f(r)\cos(ks)\sin(\eta s)\\ r\sin(\eta s)+f(r)\cos(ks)\cos(\eta s)\end{pmatrix}, \tag{104}$$

where the longitudinal contraction is given by $\chi_{\mathrm{FT}}=\frac{1}{2}\eta^2 r_{\mathrm{wr}}^2$. The longitudinal strain in this configuration is

$$\varepsilon_{ss}(s,r)=\frac{\eta^2}{2}\left(r^2-r_{\mathrm{wr}}^2\right)+\frac{1}{4}k^2 f(r)^2-\eta k r f(r)\sin(ks)-\frac{1}{4}k^2 f(r)^2\cos(2ks). \tag{105}$$

Setting $k^2 f(r)^2=2\eta^2(r^2-r_{\mathrm{wr}}^2)$ (Eq. (54)) allows to cancel the s-independent part. However, since this equation implies that the product kf does not vanish in the limit $t\to 0$, we find that the s-dependent terms in the above expression for ε_{ss} remain finite as $t\to 0$, in apparent contradiction to our assumption that the wrinkled state becomes asymptotically strainless in the limit $t,T\to 0$. This shortcoming can be fixed, however, by adding to the deformation (104) a longitudinal displacement term, $u_s(s,r)$,

$$\boldsymbol{X}^{(\mathrm{wr})}(s,r)=\begin{pmatrix}(1-\chi_{\mathrm{FT}})s+u_s(s,r)\\ r\cos(\eta s)-f(r)\cos(ks)\sin(\eta s)\\ r\sin(\eta s)+f(r)\cos(ks)\cos(\eta s)\end{pmatrix}, \tag{106}$$

leading to the longitudinal strain

$$\varepsilon_{ss}(s,r)=-\eta k r f(r)\sin(ks)-\frac{1}{4}k^2 f(r)^2\cos(2ks)+\partial_s u_s(s,r). \tag{107}$$

Setting

$$u_s(s,r)=-\eta r f(r)\cos(ks)+\frac{1}{8}k f(r)^2\sin(2ks), \tag{108}$$

we find that both s-dependent and s-independent terms of the longitudinal strain ε_{ss} in the wrinkled zone vanish for $t\to 0$ (up to higher order terms in η).

The configuration given by Eqs. (106), (108) has also transverse and shear strains, given by

$$\varepsilon_{rr}(s,r)=\frac{1}{2}f'(r)^2\cos(ks)^2, \tag{109}$$

$$\varepsilon_{sr}(s,r)=-\eta f(r)\cos(ks)-\frac{1}{8}k f(r) f'(r)\sin(2ks). \tag{110}$$

However, in contrast to the individual terms in Eq. (105), which are proportional to the product kf (that remains finite as $t\to 0$), all terms on the r.h.s. of Eqs. (109), (110) vanish in the limit of small thickness (since, while $kf(r)$ is finite, the amplitude $f(r)\to 0$ in this limit).

Appendix D: The Stability Analysis of Kirchoff Rod Equations

Here we translate the relevant results of [9], which addressed the stability analysis of a Kirchoff rod with non-symmetric cross section to the terminology of our paper.

The relevant results for us pertain to the linear stability analysis of the helicoidal ("straight") state of the ribbon. This is summarized in Eqs. (58) and (59) that provide the threshold for the two types of instabilities of the centerline ("tapelike" = TL, and "thick" = th). We explain the meaning of the parameters a, b and ρ.

The parameter a (Eq. (9) of [9]) is the ratio between the two principal moments of inertia of the rod ($I_1 < I_2$). The parameter b (also Eq. (9) of [9]) involves also the Poisson ratio (denoted σ in [9], and ν in our manuscript), and the "mixed" moment of inertia J. In the limit $t \ll 1$: $I_1 \sim \mathrm{W}t^3$, $I_2 \sim t\mathrm{W}^3$ and $J \sim t\mathrm{W}^3$, with some numerical coefficients that depend on the exact shape of the cross section. In Eq. (12), both a and b are evaluated for an ellipsoidal cross section, but we assume that the same expressions (i.e., the exact respective ratios between J, I_1, I_2) hold also for a rectangular cross section, from which we can translate to our terminology:

$$a \to t^2, \qquad b \to \frac{2t^2}{1+\nu}, \tag{111}$$

where we assume already the limit $t \ll 1$ and expanded b to lowest order in t.

Now, let us consider the parameter $\rho = F_3^{(0)}/{\kappa_3^{(0)}}^2$ (Eq. (35) of [9]), where $F_3^{(0)}$ is the normalized force exerted along the centerline and $\kappa_3^{(0)}$ is the exerted "torsion" of the centerline. We will show that the translation to our terminology is:

$$\rho \to \frac{t^2 T}{\eta^2}. \tag{112}$$

To see this, first note that $F_3^{(0)}$ and $\kappa_3^{(0)}$ are defined as the tension and the twist density in the sentence after Eq. (14) of [9]. In order to understand the normalization, we need the normalization of lengths and forces, given, respectively in Eqs. (6) and (7b). Note that lengths are measured in units of t (since $I_1 \sim t^3 W$ and $A \sim tW$). The expressions of κ_3 and F_3 in our parameters is therefore: $\kappa_3^{(0)} = (\theta/L)/t = \eta/t$, and $F_3^{(0)} = \text{force}/(EtW) = \text{force}/(YW) = T$. Substituting this expression for $F_3^{(0)}$ and $\kappa_3^{(0)}$ in Eq. (35) of [9], we find the above transformation of the parameter ρ to our parameters. Importantly, ρ of [9] is inversely proportional to the ratio η^2/T, and hence the unstable range of the helicoidal state (gray zones in Fig. 4 of [9]) corresponds to large value of twist/tension (namely η^2/T *above* some threshold).

Let us turn now to Eqs. (58), (59), and express them in our terminology. From Eq. (58) we obtain the threshold for the "tapelike" mode to be:

$$\left(\frac{\eta^2}{T}\right)_{\mathrm{TL}} \approx \frac{1+\nu}{1-\nu}, \tag{113}$$

in the limit $t \ll 1$, which ranges from 1 to 3 as ν ranges from 0 to 1/2. Equation (59) leads the threshold for the "thick" mode,

$$\left(\frac{\eta^2}{T}\right)_{\mathrm{th}} \approx \frac{1+\nu}{2(1+2\nu-2\sqrt{\nu(1+\nu)})}, \tag{114}$$

in the limit $t \ll 1$. This expression ranges from 1/2 to 2.8 as ν ranges from 0 to 1/2 and it is smaller than the first threshold for any value of ν.

Appendix E: Estimating the Clamping-Induced Energy

The transverse displacement u_r must vanish near the clamped edges ($s = \pm L/2$), and is expected to approach $u_r \approx -\nu Tr/2$ beyond a characteristic length ℓ from the clamped edges. In the region $s \in (-L/2+\ell, L/2-\ell)$ the Poisson contraction applies, such that the strain can be approximated as:

$$\varepsilon_{ss} = T, \qquad \varepsilon_{rr} = -\nu T, \qquad \varepsilon_{xy} = 0, \tag{115}$$

and the corresponding energy per length is:

$$\left(1 - \frac{2\ell}{L}\right) T^2. \tag{116}$$

In the near-boundary zones $s \in \pm(L/2-\ell, L/2)$, where u_r is not determined by the Poisson effect, we may express the strain field as:

$$\begin{aligned} \varepsilon_{ss} &= \frac{1}{1-\nu^2} T + \frac{\nu T}{\ell} f_1, \\ \varepsilon_{rr} &= \frac{\nu T}{\ell} f_2, \qquad \varepsilon_{sr} = \frac{\nu T}{\ell} f_3, \end{aligned} \tag{117}$$

where $f_i(s/\ell, r)$ are $O(1)$ functions that characterize the variation of the displacement field from the clamped edge to its bulk value. Note that the ℓ-independent component of ε_{ss} is derived from the Hookean stress-strain relationship by assuming $\sigma^{ss} \approx T$ and $\varepsilon_{rr} \approx 0$. Integrating over the boundary zones $s \in \pm(L/2-\ell, L/2)$, the energy per length associated with the strain field is estimated as:

$$\frac{1}{L}\left(\frac{T^2 \ell}{1-\nu^2} + \frac{\nu^2 T^2}{\ell}\right) \tag{118}$$

(where some unknown numerical constants, which are independent on ℓ and ν, multiply each of the two terms in the above expression). Combining the two energies, Eqs. (116), (118), and minimizing over ℓ, we obtain:

$$\Delta U_{\text{clamp}} \sim \frac{\nu F(\nu) T^2}{L}, \qquad \ell \sim \nu \tag{119}$$

where $F(\nu)$ is some smooth function of ν that satisfies $F(\nu) \to \text{cst}$ for $\nu \to 0$.

References

1. Green, A.E.: Proc. R. Soc. Lond. Ser. A, Math. Phys. Sci. **154**(882), 430 (1936)
2. Green, A.E.: Proc. R. Soc. Lond. Ser. A, Math. Phys. Sci. **161**(905), 197 (1937). http://www.jstor.org/stable/96910
3. Chopin, J., Kudrolli, A.: Phys. Rev. Lett. **111**(17), 174302 (2013). doi:10.1103/PhysRevLett.111.174302
4. Mockensturm, E.M.: J. Appl. Mech. **68**(4), 561 (2000). doi:10.1115/1.1357517
5. Coman, C.D., Bassom, A.P.: Acta Mechanica **200**(1–2), 59 (2008). doi:10.1007/s00707-007-0572-3
6. Sadowsky, M.: Teil II. Verhandl. des 3. Intern. Kongr. f. Techn. Mechanik, 444–451 (1930)
7. Korte, A.P., Starostin, E.L., van der Heijden, G.H.M.: Proc. R. Soc. A, Math. Phys. Eng. Sci. **467**(2125), 285 (2011). doi:10.1098/rspa.2010.0200
8. Cerda, E., Mahadevan, L.: Phys. Rev. Lett. **90**(7), 074302 (2003). doi:10 1103/PhysRevLett.90.074302

9. Goriely, A., Nizette, M., Tabor, M.: J. Nonlinear Sci. **11**(1), 3 (2001). doi:10.1007/s003320010009
10. Champneys, A.R., Thompson, J.M.T.: Proc. R. Soc. Lond. Ser. A, Math. Phys. Eng. Sci. **452**(1954), 2467 (1996). doi:10.1098/rspa.1996.0132
11. van der Heijden, G., Thompson, J.: Phys. D: Nonlinear Phenom. **112**(1–2), 201 (1998). Proceedings of the Workshop on Time-Reversal Symmetry in Dynamical Systems. doi:10.1016/S0167-2789(97)00211-X
12. van der Heijden, G.H.M., Thompson, J.M.T.: Nonlinear Dyn. **21**(1), 71 (2000). doi:10.1023/A:1008310425967
13. Santangelo, C.: Geometric frustration in twisted strips. J. Club Condens. Matter Phys. (2014). http://www.condmatjournalclub.org/?p=2330
14. Davidovitch, B., Schroll, R.D., Vella, D., Adda-Bedia, M., Cerda, E.A.: Proc. Natl. Acad. Sci. **108**(45), 18227 (2011). doi:10.1073/pnas.1108553108
15. King, H., Schroll, R.D., Davidovitch, B., Menon, N.: Proc. Natl. Acad. Sci. **109**(25), 9716 (2012). doi:10.1073/pnas.1201201109
16. Grason, G.M., Davidovitch, B.: Proc. Natl. Acad. Sci. **110**(32), 12893 (2013). doi:10.1073/pnas.1301695110
17. Landau, L.D., Lifchitz, E.M., Kosevitch, A.M., Pitaevski, L.P., Sykes, J.B., Reid, W.: Course of Theoretical Physics: Theory of Elasticity. Butterworth-Heinemann, Stoneham (1986)
18. Cranford, S., Buehler, M.J.: Model. Simul. Mater. Sci. Eng. **19**(5), 054003 (2011). http://stacks.iop.org/0965-0393/19/i=5/a=054003
19. Kit, O.O., Tallinen, T., Mahadevan, L., Timonen, J., Koskinen, P.: Phys. Rev. B **85**(8), 085428 (2012). doi:10.1103/PhysRevB.85.085428
20. Ogden, R.W.: Non-linear Elastic Deformations. Courier Dover Publications, New York (1997)
21. Efrati, E., Sharon, E., Kupferman, R.: J. Mech. Phys. Solids **57**(4), 762 (2009). doi:10.1016/j.jmps.2008.12.004
22. Dias, M.A., Hanna, J.A., Santangelo, C.D.: Phys. Rev. E **84**(3), 036603 (2011). doi:10.1103/PhysRevE.84.036603
23. Hohlfeld, E., Davidovitch, B.: (2014, submitted)
24. Stein, M., Hedgepeth, J.M.: Analysis of Partly Wrinkled Membranes. National Aeronautics and Space Administration, Washington (1961)
25. Pipkin, A.C.: IMA J. Appl. Math. **36**(1), 85 (1986). doi:10.1093/imamat/36.1.85
26. Mansfield, E.H.: The Bending and Stretching of Plates. Cambridge University Press, Cambridge (2005)
27. Davidovitch, B., Schroll, R.D., Cerda, E.: Phys. Rev. E **85**(6), 066115 (2012). doi:10.1103/PhysRevE.85.066115
28. Bella, P., Kohn, R.V.: Commun. Pure Appl. Math. (2013)
29. Audoly, B., Pomeau, Y.: Elasticity and Geometry: From Hair Curls to the Non-linear Response of Shells. Oxford University Press, Oxford (2010)
30. Huang, J., Davidovitch, B., Santangelo, C.D., Russell, T.P., Menon, N.: Phys. Rev. Lett. **105**(3), 038302 (2010). doi:10.1103/PhysRevLett.105.038302
31. Vandeparre, H., Piñeirua, M., Brau, F., Roman, B., Bico, J., Gay, C., Bao, W., Lau, C.N., Reis, P.M., Damman, P.: Phys. Rev. Lett. **106**, 224301 (2011). doi:10.1103/PhysRevLett.106.224301
32. Schroll, R.D., Adda-Bedia, M., Cerda, E., Huang, J., Menon, N., Russell, T.P., Toga, K.B., Vella, D., Davidovitch, B.: Phys. Rev. Lett. **111**(1), 014301 (2013). doi:10.1103/PhysRevLett.111.014301
33. Pogorelov, A.: Extrinsic Geometry of Convex Surfaces
34. Witten, T.A.: Rev. Mod. Phys. **79**, 643 (2007). doi:10.1103/RevModPhys.79.643
35. Sharon, E., Roman, B., Marder, M., Shin, G.S., Swinney, H.L.: Nature **419**(6907), 579 (2002). doi:10.1038/419579a
36. Audoly, B., Boudaoud, A.: Phys. Rev. Lett. **91**, 086105 (2003). doi:10.1103/PhysRevLett.91.086105
37. Klein, Y., Venkataramani, S., Sharon, E.: Phys. Rev. Lett. **106**, 118303 (2011). doi:10.1103/PhysRevLett.106.118303
38. Gemmer, J.A., Venkataramani, S.C.: Nonlinearity **25**(12), 3553 (2012). http://stacks.iop.org/0951-7715/25/i=12/a=3553
39. Giomi, L., Mahadevan, L.: Phys. Rev. Lett. **104**, 238104 (2010). doi:10.1103/PhysRevLett.104.238104
40. Kohn, R.V., Nguyen, H.M.: J. Nonlinear Sci. **23**(3), 343 (2013). doi:10.1007/s00332-012-9154-1
41. Audoly, B., Boudaoud, A.: J. Mech. Phys. Solids **56**(7), 2444 (2008). doi:10.1016/j.jmps.2008.03.001
42. Michell, J.: Messenger of Math. **11**, 181 (1889–1890)
43. Goriely, A.: J. Elast. **84**(3), 281 (2006). doi:10.1007/s10659-006-9055-3
44. Majumdar, A., Prior, C., Goriely, A.: J. Elast. **109**(1), 75 (2012). doi:10.1007/s10659-012-9371-8
45. Love, A.E.H.: A Treatise on the Mathematical Theory of Elasticity. Cambridge University Press, Cambridge (2013)
46. Ashwell, D.G.: Q. J. Mech. Appl. Math. **15**(1), 91 (1962). doi:10.1093/qjmam/15.1.91

47. Nayyar, V., Ravi-Chandar, K., Huang, R.: Int. J. Solids Struct. **48**(25–26), 3471 (2011). doi:10.1016/j.ijsolstr.2011.09.004
48. Healey, T., Li, Q., Cheng, R.B.: J. Nonlinear Sci. **23**(5), 777 (2013). doi:10.1007/s00332-013-9168-3
49. Kim, T.Y., Puntel, E., Fried, E.: Int. J. Solids Struct. **49**(5), 771 (2012). doi:10.1016/j.ijsolstr.2011.11.018
50. Lee, C., Wei, X., Kysar, J.W., Hone, J.: Science **321**(5887), 385 (2008)
51. Novoselov, K.S., Fal[prime]ko, V.I., Colombo, L., Gellert, P.R., Schwab, M.G., Kim, K.: Nature **490**(7419), 192 (2012). doi:10.1038/nature11458
52. Mahadevan, L., Vaziri, A., Das, M.: Europhys. Lett. **77**(4), 40003 (2007). http://stacks.iop.org/0295-5075/77/i=4/a=40003
53. Schroll, R.D., Katifori, E., Davidovitch, B.: Phys. Rev. Lett. **106**, 074301 (2011). doi:10.1103/PhysRevLett.106.074301
54. Klein, Y., Efrati, E., Sharon, E.: Science **315**(5815), 1116 (2007). doi:10.1126/science.1135994
55. Kim, J., Hanna, J.A., Byun, M., Santangelo, C.D., Hayward, R.C.: Science **335**(6073), 1201 (2012). doi:10.1126/science.1215309

DOI 10.1007/978-94-017-7300-3_11
Reprinted from *Journal of Elasticity* Journal, DOI 10.1007/s10659-014-9502-5

The Shrinking Figure Eight and Other Solitons for the Curve Diffusion Flow

Maureen Edwards · Alexander Gerhardt-Bourke · James McCoy · Glen Wheeler · Valentina-Mira Wheeler

Received: 19 January 2014 / Published online: 9 October 2014

Abstract In this article we investigate the dynamics of special solutions to the surface diffusion flow of idealised ribbons. This reduces to studying the curve diffusion flow for the profile curve of the ribbon. We provide: (1) a complete classification of stationary solutions; (2) qualitative results on shrinkers, translators, and rotators; and (3) an explicit parametrisation of a shrinking figure eight curve.

Keywords Curvature flow · Curve diffusion · Self-similar solution

Mathematics Subject Classification (2010) 35J35 · 35K30 · 53C44 · 58J35

1 Introduction

The curve diffusion flow is a particular evolution law governing the changing shape of a curve in time. It is the one-intrinsic-dimension analogue of the surface diffusion flow for evolving surfaces. These flows are gradient flows of a certain energy functional in a certain function space, but not the more common L^2-gradient flow such as the curve shortening

The research of the second author was supported by a summer vacation scholarship of the Australian Mathematical Sciences Institute. The research of the third, fourth and fifth authors was supported by Discovery Project grant DP120100097 of the Australian Research Council.

M. Edwards · A. Gerhardt-Bourke · J. McCoy (✉) · G. Wheeler · V.-M. Wheeler
University of Wollongong, Northfields Avenue, Wollongong, New South Wales 2522, Australia
e-mail: jamesm@uow.edu.au

M. Edwards
e-mail: maureen@uow.edu.au

A. Gerhardt-Bourke
e-mail: agb526@uowmail.edu.au

G. Wheeler
e-mail: glenw@uow.edu.au

V.-M. Wheeler
e-mail: vwheeler@uow.edu.au

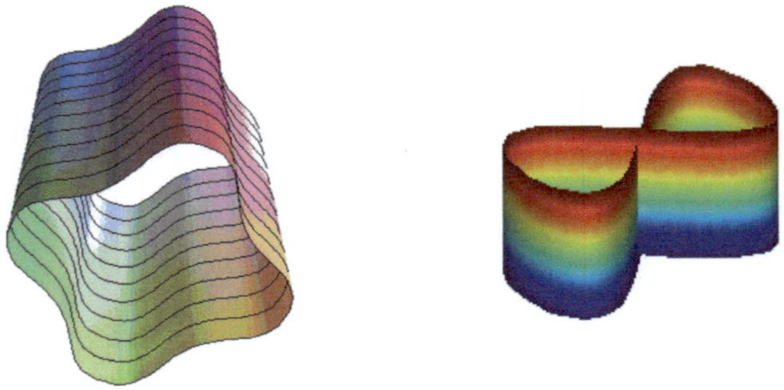

Fig. 1 Pieces of two cylindrical ribbons, immersed via $f(u,z) = (x(u), y(u), z)$ where $(u,d) \in \mathbb{S} \times \mathbb{R}$. The curve $\gamma(u) = (x(u), y(u))$ is called the profile curve. The ribbon on the *left* is embedded, while the ribbon on the *right* is immersed. The profile curve of the ribbon on the *right* is the self-similar figure eight curve we introduce in Sect. 6

flow (corresponding to decreasing the length functional for a curve), the mean curvature flow (decreasing the area functional for a surface), or the Willmore flow (decreasing the elastic energy for a curve or the Willmore energy for a surface).

Surface diffusion flow was first described by W.W. Mullins in 1956 as a model for the development of grooving at grain boundaries of heated crystal structures [19]. The flow is also related to the Cahn-Hilliard equation, which describes phase separation and coarsening in the quenching process of binary alloys [3]. For further analysis and historical remarks on the surface diffusion flow and its constrained variations we refer the interested reader to [8, 18, 23, 24].

The surface diffusion flow is a fourth-order nonlinear parabolic system of equations that, at each time t, moves points p on the solution surface f perpendicularly to the surface with speed $(\Delta H)(p)$. Here Δ is the Laplace-Beltrami operator and H is the mean curvature, which is given by $H = \langle \Delta f, \nu \rangle$. The inner product $\langle \cdot, \cdot \rangle$ is the standard dot product in $\mathbb{R}^3$.

Surface diffusion flow is the geometric analogue of the classical parabolic clamped plate equation, which expresses the dynamics of a flat plate realised as the graph of a function over a domain in $\mathbb{R}^2$. The curve diffusion flow is similarly the geometric analogue of the classical parabolic Euler-Bernoulli model for a clamped rod, modelled as the graph of a function over an interval in $\mathbb{R}$. When the domain is no longer flat, as is the case for a clamped arc segment of a circle in the plane or cylindrical ring in space, the geometry of the problem must enter into the model [5, 22]. The surface diffusion flow or curve diffusion flow can capture the notion of stability for an elastic plate or rod under stress respectively. Indeed, if the flow returns all small perturbations of a given initial configuration to that initial configuration without exception, then this configuration is called stable [20].

If the evolving surfaces are cylindrical, as would be the case for idealised closed ribbons with their width extended to infinity, then this remains so under the flow. In this case the flow equation reduces to a one-dimensional parabolic equation for the profile curve γ. For the surface diffusion flow, this equation is exactly the curve diffusion flow. This reduction is also performed in [22]. Two cylindrical ribbons are shown in Fig. 1.

Let us briefly remark on the above procedure with respect to the particular case of a Möbius band. Extending the width of the Möbius band to infinity must result in self-intersections, and so the profile curve does not project uniquely to a plane curve in $\mathbb{R}^2$. This is an essential characteristic of the Möbius band. One way we could overcome this difficulty is instead to immerse the Möbous band in $\mathbb{R}^4$, then extend the width to infinity, and project to $\mathbb{R}^3$. This would produce a profile curve in $\mathbb{R}^3$, the analysis of the dynamics of which are much more complicated than for the planar curve case.

In this paper we concentrate on the dynamics of solutions to the curve diffusion flow which evolve either by scaling, translation or rotation only. These solutions are known as *solitons*. Solitons are of interest both from a theoretical and practical viewpoint due to their relationship with singularities: given a family of evolving curves, if the family exists for at most a finite amount of time, then it must develop a singularity at the final time. Upon rescaling the flow, effectively 'zooming in on the singularity', we expect to construct a family of curves which approximate one of the solitons for the flow.

The existence of finite-time singularities is typically a difficult question, especially in the case of fourth and higher-order curvature flows. For the curve diffusion flow, this was settled in Polden [21]. The analysis there was later extended by Escher and Ito [6] to more complex initial configurations. In particular, they prove the following theorem.

Theorem 1 (Escher-Ito-Polden) *Let γ_0 be a plane curve with winding number zero. Let T denote the maximal time of existence for the curve diffusion flow $\gamma_t := \gamma(\cdot, t)$ with initial data $\gamma_0 = \gamma(\cdot, 0)$. Then*

(i) *The estimate $T^*(\gamma_0) := \frac{L^4(\gamma_0)}{64\pi^4} \geq T$ holds, where $L(\gamma_0)$ denotes the length of γ_0; and*
(ii) *If $T = T^*(\gamma_0)$ then the solution shrinks to a point as t approaches T.*

This theorem raises two interesting questions:

1. Is there a solution with winding number zero whose maximal existence time satisfies $T = T^*(\gamma)$?
2. Is there a solution with winding number zero that shrinks to a point? (That is, no other kind of singularity occurs.)

The first question investigates the sharpness of the theorem. Its answer is no, since in the proof of Theorem 1, Fenchel's inequality is used, and for plane curves with winding number zero, it is strict. Furthermore, the upper bound from Theorem 1 may be improved by a factor of $\frac{\pi^2}{12}$ by using the following argument.[1]

Let γ_t be a family of curves evolving by curve diffusion flow. Differentiating the length yields

$$L'(\gamma_t) = -\int_\gamma |\partial_s \kappa|^2 d\ell, \tag{1}$$

where $\kappa(s)$ is the curvature of γ_t that is parametrised by arc length. We use ∂_s and $d\ell$ to denote the arc length derivative and arc length element respectively. (We refer the reader to Sect. 2 for more detail on the notation.) For curves with winding number zero, the curvature κ has zero average, and furthermore

$$\left(\int_\gamma |\kappa|\, d\ell\right)^2 \leq \frac{\pi^2}{12} L(\gamma_t) \int_\gamma |\kappa|^2 d\ell$$

[1] We would like to thank one of the anonymous referees for pointing out this improved bound.

(this holds for any function with zero average), so, using the standard sharp Poincaré inequality we compute

$$\left(\int_\gamma |\kappa|\,d\ell\right)^2 \le \frac{\pi^2}{12}L\int_\gamma |\kappa|^2 d\ell \le \frac{\pi^2}{12}L\cdot\frac{L^2}{4\pi^2}\int_\gamma |\partial_s\kappa|^2 d\ell = \frac{L^3}{48}\int_\gamma |\partial_s\kappa|^2 d\ell.$$

In the above we have omitted the argument γ_t from L. Since $\|\kappa\|_1 := \int_\gamma |\kappa|\,d\ell \ge 2\pi$ (by a simple variant of Fenchel's theorem), we combine this with (1) to conclude

$$\left(L^4\right)' = 4L^3L' = -4\|\partial_s\kappa\|_2^2 L^3 \le -4\left(48\|\kappa\|_1^2\right) \le -3\cdot 4^4\pi^2$$

and therefore

$$L^4(\gamma_t) \le -3\cdot 4^4\pi^2 t + L^4(\gamma_0).$$

Consequently we obtain the estimate $T \le \tilde{T}(\gamma_0) := L^4(\gamma_0)/768\pi^2$, which is an improvement on the estimate of Theorem 1.

In the above proof, there are three key estimates used: Fenchel's inequality, the Poincaré inequality, and a sharp Hölder-type inequality for functions with zero average. It is not the case that the same curve satisfies equality for all three of these and therefore it appears clear that the improved bound $T \le \tilde{T}(\gamma_0)$ is again not optimal. The broader question therefore on characterising the optimal upper bound on the maximal time of existence in terms of initial length for curves with winding number zero remains open:

1′. Is there a $\hat{T}$ depending only on initial length $L(\gamma_0)$ such that:
   (a) All curve diffusion flows with winding number zero have maximal time bounded by $\hat{T}$; and
   (b) Does there exist at least one curve diffusion flow with maximal time equal to $\hat{T}$, or, does there exist a sequence of solutions γ_i with initial data $\gamma_i^{(0)}$ and winding number zero whose maximal existence time T_i satisfies $|\hat{T}(\gamma_i^{(0)}) - T_i| \to 0$?

In this paper we shall not answer this question. We do however provide an answer to the above question 2. This question is interesting from the perspective of classifying finite-time singularities. The key issue there is to demonstrate that solutions may become singular by simply becoming small—a rescaling of such solutions will exist for all time and be asymptotic to a smooth model curve. A finite-time singularity of this type is said to be of Type I, in analogy with existing work on the mean curvature flow [14]. Other singularities are called Type II. Type I singularities are expected to be simpler in structure and satisfy natural stability criteria when compared with Type II singularities.

We answer the second question in the affirmative by way of an explicit example: a self-similarly shrinking immersed figure eight. This figure eight is in fact the lemniscate of Bernoulli.[2] Since one requires zero area to have the possibility that the solution shrinks to a point (see Lemma 5), and the maximal existence time is finite (see Theorem 1), a figure eight with one leaf smaller than the other will certainly become singular with non-zero area, ruling out the possibility that it is asymptotic to a point.

Let γ be the shrinking figure eight described in Sect. 6. Upon scaling γ to satisfy $L(\gamma) = 1$, we compute the extinction time of the flow to be

$$T = \frac{1}{3\cdot 2^{11}\cdot \mathcal{K}(-1)^4},$$

[2] The authors would like to thank Hojoo Lee for pointing this out.

where $\mathcal{K}(m)$ is the complete elliptic integral of the first kind with parameter $m = k^2$. In this particular case the maximal time estimate T^* overshoots T by a factor of roughly 2.9, and $\tilde{T}$ overshoots T by a factor of roughly 2.4:

$$\frac{T^*}{T} = 2.9115257845\ldots, \qquad \frac{\tilde{T}}{T} = 2.3946339747\ldots.$$

In a sense, the self-similar figure eight represents a "best-case" scenario. The generic case is that an arbitrary smooth curve with winding number zero will be driven to a curvature singularity and not shrink to a point. Furthermore, such a singularity involves blowup of the curvature and its derivatives, so in particular the speed of the flow is greater in this circumstance. These heuristics lead us to make the following conjecture.

Conjecture The maximal existence time of a curve diffusion flow emanating from a curve with winding number zero is bounded by the existence time of a self-similar figure eight with the same initial length.

Of course, verifying the conjecture has uniqueness of the self-similar figure eight, modulo reparametrisations and rigid motions (compositions of rotations and translations), as a corollary.

Another popular fourth-order surface evolution equation is the Willmore flow, whose one-dimensional counterpart, the elastic flow, also models cylindrical elastic ribbons. The key difference between the two flows is that the elastic flow minimises the integral of the square of the curvature of the curve, while the curve diffusion flow minimises length, keeps enclosed volume fixed, and keeps the quantity

$$\mathcal{Q}(t) = \int_0^t \left(\int_\gamma |\partial_s \kappa|^2 \, d\ell \right) dt$$

uniformly bounded, with a bound dependent only on the initial data. A proof of these statements can be found in [25]. We remark on the similarities and differences between this flow and the curve diffusion flow throughout this article.

The structure and specific contributions of this article are as follows. We begin Sect. 2 by setting up notation, defining key geometric quantities and giving the fourth-order nonlinear parabolic system of partial differential equations for the curve diffusion flow. We then briefly review relevant previous work and the question of local existence before moving on to a classification of stationary solutions—those that remain constant under the flow. We provide a five-parameter family of parametrisations which completely classifies these solutions. In Sect. 3 we record some elementary properties of the flow that are relevant to our investigations. In Sect. 4 we focus on curves that evolve self-similarly, deriving the corresponding fourth-order ordinary differential equation satisfied by such curves. We prove several properties of self-similar solutions. Section 5 is devoted to curves that evolve by translation only. We prove there that the only closed translators are circles, and that the only open translators (open immersions of $\mathbb{R}$) satisfying either of two conditions at infinity are straight lines. In Sect. 6 we provide an explicit parametrisation of a figure eight immersed curve and verify that this curve does indeed evolve self-similarly under the curve diffusion flow, contracting to a point in finite time. In Sect. 7 we consider the case of rotating solutions and prove some non-existence results analogous to those for translators in Sect. 5.

2 The Curve Diffusion Flow Equation

We will parametrise closed curves over the unit circle $\mathbb{S}$ and open curves over the real line $\mathbb{R}$. Let γ denote the curve, with components

$$\gamma(u) = \big(x(u), y(u)\big).$$

The tangent direction to the curve is given by

$$\gamma_u(u) = \big(x_u(u), y_u(u)\big),$$

and therefore a choice of unit normal is

$$\nu(u) = \frac{1}{|\gamma_u(u)|}\big(y_u(u), -x_u(u)\big),$$

where $|\cdot|$ denotes the length of a vector. We will always assume our curves are smooth and regular; the latter means $|\gamma_u| > 0$ for all u. Recall that, in a general parametrisation, the arc length of a curve γ, beginning at point $\gamma(u_0)$, is given by

$$s(u) = \int_{u_0}^{u} \big|\gamma_u(\tilde{u})\big| d\tilde{u}. \tag{2}$$

The arc length induces an arc length element $d\ell = |\gamma_u|\, du$. The curvature vector and curvature of a plane curve are given respectively by

$$\vec{\kappa} = \gamma_{ss} \quad \text{and} \quad \kappa = \langle \nu, \vec{\kappa} \rangle,$$

where we use $\langle \cdot, \cdot \rangle$ to denote the standard inner product between vectors in the plane. For closed embedded curves, we will always choose an inward-pointing unit normal, so that the curvature of a circle is positive. Otherwise, we select either normal—the curve diffusion flow is invariant under changes in orientation.

We say $\gamma_t = \gamma(\cdot, t)$ evolves under the curve diffusion flow if it satisfies the system of fourth-order partial differential equations

$$\frac{\partial \gamma}{\partial t}(u,t) = -\kappa_{ss}(u,t)\nu(u,t) \quad \text{for } (u,t) \in \mathbb{S} \times (0,T) \tag{3}$$

with initial condition

$$\gamma(u,0) = \gamma_0(u),$$

for some given initial curve immersion γ_0. Note that an immersed curve may have intersections, however, it has a well-defined tangent vector for each u. We label the maximal existence time of a solution to (3) as T.

We are interested in the shape of the evolving curves, that is, the image $\gamma(\mathbb{S}, t)$ for closed solutions, or $\gamma(\mathbb{R}, t)$ for open solutions, as a set in $\mathbb{R}^2$, and not in the orbit of a specific point. There is a degeneracy in the flow equation (3): reparametrisation and tangential movement of points in the direction of γ_s do not affect the shape of the image. The degeneracy consists of every diffeomorphism $\varphi : \mathbb{S} \to \mathbb{S}$. This implies that, in terms of the images $\gamma(\mathbb{S}, t)$, the flow (3) is equivalent to

$$\left\langle \frac{\partial \gamma}{\partial t}(u,t), \nu(u,t) \right\rangle = -\kappa_{ss}(u,t) \quad \text{for } (u,t) \in \mathbb{S} \times (0,T). \tag{4}$$

The impact of this observation on our analysis here is that when considering solutions with a particular form, we must also take into account a possible diffeomorphism φ acting on $\mathbb{S}$. This typically forces us to take normal projections in order to remove the influence of φ.

Note that

$$\kappa_{ss} = \partial_s^2 \langle \gamma_{ss}, \nu \rangle = \frac{1}{|\gamma_u|}\left(\frac{1}{|\gamma_u|}\left(\left\langle \frac{1}{|\gamma_u|}\left(\frac{\gamma_u}{|\gamma_u|}\right)_u, \nu \right\rangle\right)_u\right)_u$$

so that the leading order term of $-\kappa_{ss} = \langle \partial_t \gamma, \nu \rangle$ is precisely

$$-\frac{1}{|\gamma_u|^4}\langle \gamma_{uuuu}, \nu \rangle.$$

The normal direction is 'positive' for the flow (3), and so this shows that (3) is a quasilinear parabolic system.[3]

Current understanding of the curve diffusion flow is quite limited, apart from short time existence. There are several methods for obtaining short time existence of solutions to (3) given suitable initial data; we refer the reader to the discussion of these in [25] and the references contained therein. Very little else is known about the behaviour of the flow. Giga and Ito [11] provided an example of a simple, closed, embedded plane curve that develops a self-intersection in finite time under the flow (3). They also gave an example in [12] of a closed, strictly convex plane curve that becomes non-convex in finite time. Elliot and Maier-Paape showed that curves that are initially graphical may evolve under (3) to become non-graphical in finite time [7]. On the other hand, the fourth author recently showed [25] that under the curve diffusion flow, solution curves that are initially close to a circle, in the sense of the normalised integral of the square of the oscillation of curvature, exist for all time and converge exponentially fast to a (possibly different) circle. Further, those admissible curves which are initially embedded remain embedded. A fundamental obstacle in the analysis of such flows and of (3) is that, as a system of higher order partial differential equations, arguments based upon maximum principles are in general not available.

Let us now move on to the classification of stationary curve diffusion flows. We call the family of curves $\gamma(\cdot, t)$ a *closed* solution if γ satisfies (4) and $\gamma(u) = \gamma(u + P)$ for some $P > 0$. The smallest such P is the period of γ.

A solution is called *stationary* if it does not change in time under (4), that is, if $\kappa_{ss}(u, t) \equiv 0$.

Lemma 2 *Suppose* $\gamma : \mathbb{S} \to \mathbb{R}^2$ *is a closed stationary solution to the curve diffusion flow. Then* γ *is a round circle.*

Proof Integrating by parts and using $\kappa_{ss} = 0$ we have

$$0 = -\int_\gamma \kappa \kappa_{ss}\, d\ell = \int_\gamma \kappa_s^2\, d\ell$$

and so $\kappa_s = 0$, that is, κ is constant. Therefore γ is a smooth closed curve with constant curvature, and by the classification of curves in $\mathbb{R}^2$, any curve with constant curvature in the plane is a straight line or round circle. Since γ is closed, γ can not be a straight line, and so it must be a round circle. □

[3] It may be helpful for the reader to recall the biharmonic heat flow $(\partial_t + \Delta^2)v = 0$ for a function v.

The situation is more complicated for open solutions. We define an *open curve* to be a regular immersion $\gamma : \mathbb{R} \to \mathbb{R}^2$ of $\mathbb{R}$ for which there does not exist a pair of intervals I, J such that

$$I \cap J = \emptyset, \qquad |I| = \infty, \qquad |J| < \infty, \quad \text{and} \quad \gamma(I) = \gamma(J). \tag{5}$$

This last condition captures the notion of not allowing γ to 'close' while still allowing infinitely many self-intersections, including possibly intersecting on an arbitrarily large open set.

Open curves may concentrate around a particular point, such as a logarithmic spiral. This concentration behaviour is encapsulated by the notion of *properness* of the curve $\gamma : \mathbb{R} \to \mathbb{R}^2$. The curve γ is *proper* if for any compact set $C \subset \mathbb{R}^2$ the inverse image

$$\gamma^{-1}(C) := \{u \in \mathbb{R} : \gamma(u) \in C\}$$

is a compact set.

Lemma 3 *Suppose $\gamma : \mathbb{R} \to \mathbb{R}^2$ is an open stationary solution to the curve diffusion flow. Then γ is either a straight line, or a similarity transformation of the standard Cornu spiral. Up to translation and rotation such curves satisfy*

$$\kappa(s) = k_2 s + k_1$$

for a pair $(k_1, k_2) \in \mathbb{R}^2$.

Proof If $\kappa_{ss} = 0$, then there is a $k_2 \in \mathbb{R}$ such that

$$\kappa_s(s) = k_2.$$

The curvature κ is therefore linear in s; that is, there exists a $k_1 \in \mathbb{R}$ such that

$$\kappa(s) = k_2 s + k_1.$$

If $k_2 \neq 0$, then after reparametrising by $t = s + \frac{k_1}{k_2}$ we have

$$\kappa(t) = k_2 t.$$

Furthermore, by scaling the curve γ by a factor of $\sqrt{k_2}/\sqrt{\pi}$ we have

$$\hat{\kappa}(t) = \frac{\sqrt{\pi}}{\sqrt{k_2}} \kappa(t) = \sqrt{k_2 \pi} t = \pi \hat{t}.$$

Therefore, either $k_2 = 0$, in which case we have a straight line, or $k_2 \neq 0$ and we have a similarity transformation of the standard Cornu spiral. □

Remark Combining the above two lemmata, we see that up to similarity transformations the only three stationary solutions are the circle, straight line, and the Cornu spiral.

Corollary 4 *Suppose $\gamma : \mathbb{R} \to \mathbb{R}^2$ is an open, stationary, properly immersed solution to the curve diffusion flow. Then γ is a straight line.*

Proof We apply Lemma 3 above to obtain that γ has curvature satisfying

$$\kappa(s) = k_2 s + k_1.$$

If $k_2 = 0$ then κ is a constant and so γ is a straight line, as required.

Suppose on the contrary that $k_2 \neq 0$. We will show that this leads to a contradiction to the properness assumption. Consider the restriction $\hat{\gamma}$ of γ to the interval $(0, \infty)$ if $k_2 > 0$ or to the interval $(-\infty, 0)$ if $k_2 < 0$. The curve $\hat{\gamma}$ has strictly monotone positive curvature. We may therefore apply the classical Tait-Kneser theorem to obtain that the osculating circles of $\hat{\gamma}$ are pairwise disjoint and nested (see [10] and the references therein for a discussion of Tait's original paper and some interesting extensions). Recall that the osculating circle at $\gamma(s_0)$ is the best approximating circle to $\gamma(s_0)$. It is a standard round circle with radius $1/\kappa(s_0)$ and centre $(\gamma(s_0) + \nu(s_0)/\kappa(s_0))$.

Let $\mathcal{C}(s_0)$ denote the osculating circle at $\hat{\gamma}(s_0)$ and $\mathcal{D}(s_0)$ denote the disk with boundary $\mathcal{C}(s_0)$. The theorem implies that

$$\mathcal{D}(s_0) \supset \bigcup_{s\in[s_0,\infty)} \mathcal{C}(s) \supset \bigcup_{s\in[s_0,\infty)} \gamma(s).$$

Clearly $\mathcal{D}(s_0)$ is compact. The above computation shows that

$$\gamma^{-1}\big(\mathcal{D}(s_0)\big) = [s_0, \infty),$$

which is not compact, and so γ is not proper. This is contrary to the hypotheses of this Corollary, and so $k_2 = 0$, as required. □

The two-parameter family of open curves given by Lemma 3 therefore consists of straight lines and a particular family of improperly immersed curves. The improperly immersed curves are Eulerian spirals, with the clothoid or Cornu spiral ($k_2 = \pi$ and $k_1 = 0$) a standard example. The arc length parametrisation of the clothoid is given in terms of the Fresnel C and S integrals:

$$\gamma(s) = \big(C(s), S(s)\big), \quad \text{where } C(s) = \int_0^s \cos\left(\frac{\pi t^2}{2}\right) dt \text{ and}$$

$$S(s) = \int_0^s \sin\left(\frac{\pi t^2}{2}\right) dt.$$

The curvature of the clothoid is $\kappa(s) = \pi s$. In order to produce the whole two-parameter family, let us define the modified Fresnel integrals $C(s, c_1, c_2)$ and $S(s, c_1, c_2)$ by

$$C(s, c_1, c_2) = \int_0^s \cos\big(c_1 t + c_2 t^2\big)\, dt \quad \text{and} \quad S(s, c_1, c_2) = \int_0^s \sin\big(c_1 t + c_2 t^2\big)\, dt.$$

The curve $\gamma : \mathbb{R} \to \mathbb{R}^2$ with arc length parametrisation

$$\gamma(s) = \big(C(s, c_1, c_2), S(s, c_1, c_2)\big)$$

then has curvature $\kappa(s) = 2c_2 s + c_1$.

Note that this includes the case of circles ($c_2 = 0$) and straight lines ($c_1 = c_2 = 0$). The curvature κ is invariant under rigid motions and therefore, in order to account for the entire

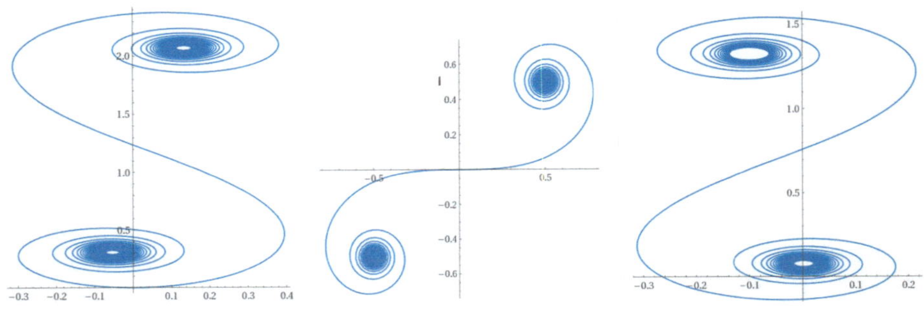

Fig. 2 Three members of $\mathcal{F}$: the *left* with $\kappa(s) = -2s + 3$, the *middle* with $\kappa(s) = s$, and the *right* with $\kappa(s) = 4s + 13$. We choose $(\theta, V_1, V_2) = (0, 0, 0)$ for each

family of stationary solutions, we must include all possible rotations and translations of the model members $\gamma(s) = (C(s, c_1, c_2), S(s, c_1, c_2))$. Therefore the family $\mathcal{F}$ of immersed open or closed stationary solutions to the curve diffusion flow consists precisely of the five parameter family

$$\mathcal{F} = \left\{ \gamma(s) = \left[\begin{pmatrix} \cos\theta & -\sin\theta \\ \sin\theta & \cos\theta \end{pmatrix} \begin{pmatrix} C(s, c_1, c_2) \\ S(s, c_1, c_2) \end{pmatrix} \right]^T + (V_1, V_2) \right. \\ \left. \text{for some } (c_1, c_2, V_1, V_2, \theta) \in \mathbb{R}^4 \times \mathbb{S} \right\},$$

up to reparametrisation. Some members of $\mathcal{F}$ are depicted in Fig. 2.

Remark (Elastic flow) We say γ_t evolves under the elastic flow if $\gamma(\cdot, t) = \gamma_t$ satisfies

$$\frac{\partial \gamma}{\partial t}(u, t) = \left(-\kappa_{ss}(u, t) - \frac{1}{2}\kappa^3(u, t) \right) \nu(u, t).$$

The elastic flow is the steepest descent L^2-gradient flow of $\int_\gamma \kappa^2 \, d\ell$. For global regularity results and further historical references we refer to [4]. Stationary solutions are termed *elasticae* and satisfy

$$\kappa_{ss}(u, t) = -\frac{1}{2}\kappa^3(u, t).$$

A classification of solutions to this equation is not yet known in the sense discussed here, that is, in terms of explicit parametrisations. Indeed, there are no known closed solutions to this equation. Open solutions are yet to be classified. We conjecture that there are no open properly immersed solutions apart from the straight line.

3 Elementary Properties of the Curve Diffusion Flow

In this section we examine the behaviour of the arc length of γ_t, and the area enclosed within γ_t, under the flow (3). Because γ_t may be an immersed curve, care is required in defining the enclosed area, or simply the change in enclosed area; we refer the reader to the discussion in [17] for details (that article concerns evolving immersed surfaces and discusses the definition of the signed enclosed volume, but the issues are analogous). When the curve γ_t is an embedded curve, the calculations below are the one intrinsic dimension analogues of those in [15], for example. The full details for the specific case of the curve diffusion flow of immersed curves are given in [25].

Lemma 5 *Under the flow* (3), *the area enclosed by the curve γ_t remains constant. In particular, if the family of curves γ_t contract to a point, then the initial data γ_0 must have zero signed enclosed area.*

Proof The signed area $A(\gamma_t)$ enclosed by γ_t evolves under (3) according to

$$\frac{d}{dt}A(\gamma_t) = -\int_\gamma \kappa_{ss}\, d\ell = 0.$$

That is, $A(t)$ is constant under the flow.

Since the enclosed area of a point is zero, and the enclosed area is constant under the flow, the signed enclosed area at any time along a family of curves that contract to a point must be zero. In particular, the signed area enclosed by the initial curve γ_0 must be zero. □

Lemma 6 *Under the flow* (3), *the length of the evolving curve γ_t is monotone nonincreasing.*

Proof Under the flow (3), the length of the evolving curve γ evolves according to

$$\frac{d}{dt}L(t) = \int_\gamma \kappa_{ss}\kappa\, d\ell = -\int_\gamma \kappa_s^2\, d\ell \le 0,$$

that is, the length is nonincreasing. □

The above lemmata make the curve diffusion flow interesting from an isoperimetric point of view. In particular, the isoperimetric ratio is monotonically decreasing:

Corollary 7 *The signed isoperimetric ratio*

$$I(\gamma_t) = \frac{L^2(\gamma_t)}{4\pi A(\gamma_t)}$$

is decreasing in absolute value with value given by

$$I(\gamma_t) = I(\gamma_0)e^{-\int_0^t \frac{2}{L(\gamma_t)}\int_\gamma \kappa_s^2\, d\ell\, d\tau}.$$

Proof Differentiate I and apply Lemmata 5 and 6 to obtain

$$\frac{d}{dt}I(\gamma_t) = -\frac{2I(\gamma_t)}{L(\gamma_t)}\int_\gamma \kappa_s^2\, d\ell.$$

Integrating yields the result. □

Remark (Elastic flow) Although the elastic flow reduces $\int_\gamma \kappa^2\,d\ell$ with velocity

$$\frac{d}{dt}\int_\gamma \kappa^2\,d\ell = -2\int_\gamma \left|\kappa_{ss} + \frac{1}{2}\kappa^3\right|^2 d\ell,$$

the behaviour of the length, area, and isoperimetric ratio are, for general initial data, unknown.

4 Curves Evolving Homothetically

A self-similar solution to a curvature flow equation is a solution whose image maintains the same shape as it evolves, that is, it changes in time only by scaling, translation and/or rotation. In this section we focus on the case where γ evolves by scaling, and in the next section we study solutions which evolve by translation. Section 7 studies curves evolving by rotation.

Closed, self-similar solution hypersurfaces of positive mean curvature for the mean curvature flow are known to be spheres [14]; the third author generalised this result for a class of second order fully nonlinear curvature flows in [16]. In the corresponding case of the curve shortening flow, self-similar curves were classified in [1].

Definition We say γ_t is a solution of (3) (or equivalently (4)) evolving *homothetically* if and only if there is a differentiable function of time f and an initial curve γ_0 such that

$$\gamma(u,t) = f(t)\gamma_0(u).$$

This is equivalent to

$$\langle \partial_t\gamma, \nu\rangle = f'\langle \gamma_0, \nu\rangle.$$

In general, homothetic solutions may be expanding, shrinking, stationary, or exhibit more complex breathing behaviour. For the case of curve diffusion flow, Lemmata 5 and 6 show that any solution must decrease its length monotonically (and be asymptotic to a curve with constant curvature unless it exists only for a finite amount of time). We therefore term a homothetic solution γ a *shrinker* or *shrinking soliton*. We take $f(0) = \rho$ where $\rho \in \mathbb{R}$ is an initial value. This freedom will be useful when we estimate the maximal existence time of the shrinking figure eight in Sect. 6.

The simplest (trivial) example of a closed shrinking curve which is a solution to (3) is the circle.

Lemma 8 *Let γ be defined by $\gamma(u,t) = R(\cos u, \sin u)$ for some constant $R > 0$. Then γ is a solution to* (3).

Proof We simply observe

$$\frac{\partial\gamma}{\partial t}(u,t) = \frac{\partial}{\partial t}R(\cos u, \sin u) = 0$$

and since the curvature κ of γ is identically equal to $\frac{1}{R}$,

$$\kappa_{ss}(u,t) \equiv 0.$$

Hence (3) is trivially satisfied. □

If a curve is evolving homothetically under (3), it must satisfy a corresponding ordinary differential equation (ODE); we may write this equation down using 'separation of variables'.

Lemma 9 *Let $\gamma : \mathbb{S} \to \mathbb{R}^2$ be a curve. Then γ is a self-similar solution to* (3) *iff there is a constant K such that*

$$\kappa_{ss}(u) = K\langle \gamma(u), \nu(u)\rangle \tag{6}$$

for all $u \in \mathbb{S}$.

Proof Suppose the curve γ is such that $\gamma(u,t) = f(t)\gamma_0(u,t)$ for some positive differentiable function $f(t)$ with $f(0) = \rho$, that is, at the initial time $t = 0$, $\gamma(u,0) = \rho\,\gamma_0(u)$. Geometric quantities associated to γ are related to those of γ_0 as follows. In the calculations here we often omit the arguments of γ, γ_0 and f. The tangent direction is

$$\gamma_u = f\,\partial_u \gamma_0$$

and the unit normal is

$$\nu = \nu_0.$$

Above and in what follows we will include a zero subscript on geometric quantities associated with γ_0. The curvature of γ is related to that of γ_0 by

$$\kappa = f^{-1}\kappa_0. \tag{7}$$

In view of (2), the arc length of the curves γ and γ_0 are related via

$$s = f s_0$$

and so differentiating (7) twice, we find

$$\kappa_{ss} = f^{-3}\partial^2_{s_0 s_0}\kappa_0.$$

Using the above we find that (3) is satisfied if and only if

$$f'\langle \gamma_0, \nu_0\rangle = -f^{-3}\partial^2_{s_0 s_0}\kappa_0\langle \nu_0, \nu_0\rangle.$$

Following a standard separation of variables argument (note that any zeros of $\langle \gamma_0, \nu_0\rangle$ are isolated unless γ is a straight line passing through the origin—this case is trivial and can be treated separately), there must be a constant K such that

$$\left(f^4\right)' = -4\frac{\partial^2_{s_0 s_0}\kappa_0}{\langle \gamma_0, \nu_0\rangle} = 4K.$$

Simplifying this equation finishes the proof. □

Remarks

1. Solving the ordinary differential equation for f in the above proof, we have

$$f(t) = \sqrt[4]{\rho^4 + 4Kt}.$$

Assuming also that (6) is satisfied, we observe that

- If $K = 0$, then $f(t) \equiv 1$ and the curve γ is static under (3), that is, a stationary solution that exists for all time t.
- If $K < 0$, then the solution to (3) shrinks self-similarly to a point at time $T = -\frac{\rho^4}{4K}$.
- If $K > 0$ then any self-similar solution would have to expand indefinitely. However, for closed curves this is not possible in view of Lemma 6. For open curves this is possible, and there could be open expanding solutions. These could be important models for how the curve diffusion flow smooths out isolated singularities, in the same way that the error function generates a self-similar solution to the heat equation with the Heaviside step function as initial data.

2. Clearly, the shrinker condition (6) is invariant under translation and rotation; that is, applying these operations yields another solution to (6) with the same constant K. Under scaling by a factor ρ, one again obtains a solution to (6), but with a new constant $\hat{K} = K\rho^4$. One may express the ODE (6) in terms of a pair of equations for $(x(u), y(u))$ in the original (non-arclength) parametrisation, and apply a standard Lie symmetry analysis to the system. The result of this analysis is that there are no other possible transformations that one may apply in order to obtain more solutions.

Remark (Elastic flow) One naturally expects *expanders* for the elastic flow, in contrast to the shrinkers of curve diffusion flow. The solution to the elastic flow with initial data equal to a circle expands under the flow for all time, and so the standard round circle is the canonical example of an expander. Indeed, Theorems 3.2 and 3.3 of [4] establish global existence and convergence to an elastica under the condition that the length of the evolving curves remain fixed. Without this constraint, it appears impossible to obtain convergence of the flow in any sense, since there is no way to control the length. There are no other known expanders for the elastic flow.

5 Curves Evolving by Translation

A family of curves $\gamma : \mathbb{S} \times [0, T) \to \mathbb{R}^2$ evolving purely by translation satisfies

$$\frac{\partial \gamma}{\partial t}(u,t) = \gamma_0(u) + Vt + \gamma_s(u,t)\phi(u,t) \tag{8}$$

for some constant vector V and smooth diffeomorphism ϕ. In this case

$$\left\langle \frac{\partial \gamma}{\partial t}, \nu \right\rangle = \langle V, \nu \rangle,$$

and if (3) is to hold then γ must satisfy

$$-\kappa_{ss} \equiv \langle V, \nu \rangle. \tag{9}$$

We call the solution γ a translator. If $V = 0$, then the solution γ is stationary, a trivial translator.

It is common that for a given curvature flow there are no closed non-trivial translators. Nevertheless one typically expects open immersed non-trivial translating solutions. A family of open immersed curves $\gamma : \mathbb{R} \times [0, T) \to \mathbb{R}^2$ evolving purely by translation satisfies (8) and (9) as in the closed setting. Such curves usually arise as blowup limits of a Type 2 singularity [9]. Singularities of this type have engendered much interest in the literature.

In this section we investigate translators for the curve diffusion flow. We prove by a simple argument that there are no smooth closed non-trivial translators. Our proof implies a stronger statement about open curves, which is sharp, by the earlier example of the clothoid.

Proposition 10 *Let $\gamma : \mathbb{S} \to \mathbb{R}^2$ be a smooth closed translator. Then γ is trivial; that is, $\gamma(\mathbb{S})$ is a standard round circle.*

Proof Since γ is closed, integration by parts gives

$$-\int_\gamma \kappa\kappa_{ss}\,d\ell = \int_\gamma \kappa_s^2 d\ell. \tag{10}$$

Also, noting that V is a constant vector and that the curvature vector $\vec{\kappa}$ satisfies $\kappa\nu = \vec{\kappa} = \gamma_{ss}$, we have

$$\int_\gamma \langle V, \vec{\kappa}\rangle\, d\ell = \int_\gamma \partial_s\big(\langle V, \gamma_s\rangle\big)\, d\ell = 0. \tag{11}$$

Multiplying (9) by κ and integrating gives

$$-\int_\gamma \kappa\kappa_{ss}\,d\ell = \int_\gamma \langle V, \vec{\kappa}\rangle\, d\ell,$$

which when combined with (10) and (11) above yields

$$\int_\gamma \kappa_s^2 d\ell = 0.$$

Since γ is smooth, this implies that $\kappa_s \equiv 0$ on γ, that is, that the curvature is constant. Therefore $\kappa_{ss} = 0$, and so $V = 0$; that is, the translator is trivial. By Lemma 2, γ is a round circle. □

Now let us investigate open curves.

Proposition 11 *Suppose $\gamma : \mathbb{R} \to \mathbb{R}^2$ is a smooth open translator satisfying*

$$\lim_{i\to\infty}\big[\big(\kappa(p_i)\kappa_s(p_i) + \langle V, \gamma_s(p_i)\rangle\big) - \big(\kappa(q_i)\kappa_s(q_i) + \langle V, \gamma_s(q_i)\rangle\big)\big] = 0 \tag{12}$$

for arbitrary sequences $\{p_i\}$ and $\{q_i\}$ approaching $-\infty$ and $+\infty$ respectively. Then $\gamma(\mathbb{R})$ is a straight line.

Proof Let us follow the idea of Proposition 10. We integrate (9) against κ to find

$$\begin{aligned}\int_{p_i}^{q_i} \kappa_s^2 d\ell &= -\int_{p_i}^{q_i} \kappa\kappa_{ss}\,d\ell + \kappa\kappa_s|_{p_i}^{q_i} = \int_{p_i}^{q_i} \langle V, \vec{\kappa}\rangle\, d\ell + \kappa\kappa_s|_{p_i}^{q_i}\\ &= \int_{p_i}^{q_i} \partial_s\big(\langle V, \gamma_s\rangle\big)\, d\ell + \kappa\kappa_s|_{p_i}^{q_i} = \big[\langle V, \gamma_s\rangle + \kappa\kappa_s\big]_{p_i}^{q_i}.\end{aligned}$$

Taking $i \to \infty$ yields the result. Note that circles are excluded by the open hypothesis. □

Remark The result of Proposition 11 is false without condition (12). To see this, consider the clothoid: a smooth open curve with $\kappa(s) = s$ where s is the arc length parameter. Clearly this curve satisfies $\kappa_{ss} = 0$ and so is a stationary translator with $V = 0$ in (9). The clothoid however does not satisfy condition (12), since $\kappa(s)\kappa_s(s) = s$ is an odd function, and so the limit does not exist.

Although the above remark shows in some sense the sharpness of Proposition 11, it leaves open the possibility of replacing the decay condition (12) with something different.

The following proposition explores this and converts (12) from a pointwise decay condition on the curvature to a natural growth condition on κ in L^2 and V in L^1.

Proposition 12 *Suppose $\gamma : \mathbb{R} \to \mathbb{R}^2$ is a smooth open proper translator satisfying*

$$\lim_{\rho\to\infty}\left(\frac{1}{\rho^2}\int_{\gamma^{-1}(B_\rho(0))}\kappa^2\,d\ell + \frac{1}{\rho}\int_{\gamma^{-1}(B_{2\rho}(0))}\left|\langle V,\gamma_s\rangle\right|d\ell\right) = 0, \tag{13}$$

where $B_\rho(0)$ denotes the closed ball of radius ρ centred at the origin. Then $\gamma(\mathbb{R})$ is a straight line.

Proof We use again a similar idea, however this time we localise the estimate in the plane using a cutoff function $\eta_\rho = \tilde{\eta}_\rho \circ \gamma$ where $\tilde{\eta}_\rho : \mathbb{R}^2 \to [0, 1]$ satisfies

$$\tilde{\eta}_\rho(x) = 1 \quad \text{on } B_{\frac{\rho}{2}}(0),$$

$$\tilde{\eta}_\rho(x) = 0 \quad \text{on } B_\rho(0),$$

$$\eta \text{ is of class } C^\infty,\ \operatorname{supp}\eta \text{ is compact, and}$$

$$\text{there exist constants } c_k \text{ such that } \partial_s^k\eta \le \frac{c_k}{\rho^k}.$$

The existence of such a function is straightforward. A constructive proof can be found in [26]. Let us denote by $\gamma^{-1}(S)$ the inverse image of a set $S \subset \mathbb{R}^2$. First, note that since γ is proper, the integral

$$\int_{\gamma^{-1}(B_\rho(0))}\kappa_s^2\,\eta^4\,d\ell$$

is well-defined, irrespective of ρ. We compute

$$\begin{aligned}
\int_{\gamma^{-1}(B_\rho(0))}\kappa_s^2\,\eta^4\,d\ell &= -\int_{\gamma^{-1}(B_\rho(0))}\kappa\kappa_{ss}\,\eta^4\,d\ell - 4\int_{\gamma^{-1}(B_\rho(0))}\kappa\kappa_s\,\eta_s\,\eta^3\,d\ell\\
&= \int_{\gamma^{-1}(B_\rho(0))}\langle V,\vec{\kappa}\rangle\,\eta^4\,d\ell - 4\int_{\gamma^{-1}(B_\rho(0))}\kappa\kappa_s\,\eta_s\,\eta^3\,d\ell\\
&= \int_{\gamma^{-1}(B_\rho(0))}\partial_s\big(\langle V,\gamma_s\rangle\big)\,\eta^4\,d\ell - 4\int_{\gamma^{-1}(B_\rho(0))}\kappa\kappa_s\,\eta_s\,\eta^3\,d\ell\\
&= -4\int_{\gamma^{-1}(B_\rho(0))}\langle V,\gamma_s\rangle\,\eta_s\,\eta^3\,d\ell - 4\int_{\gamma^{-1}(B_\rho(0))}\kappa\kappa_s\,\eta_s\,\eta^3\,d\ell
\end{aligned}$$

$$\leq \varepsilon \int_{\gamma^{-1}(B_\rho(0))} \kappa_s^2 \eta^4 \, d\ell + 4 \int_{\gamma^{-1}(B_\rho(0))} \big|\langle V, \gamma_s\rangle\big| \, |\eta_s| \, \eta^3 \, d\ell$$

$$+ \frac{4}{\varepsilon} \int_{\gamma^{-1}(B_\rho(0))} \kappa^2 \eta_s^2 \eta^2 \, d\ell,$$

for any $\varepsilon > 0$. Choosing $\varepsilon = \frac{1}{4}$ and absorbing the first term on the left, we find

$$\int_{\gamma^{-1}(B_\rho(0))} \kappa_s^2 \eta^4 \, d\ell \leq \frac{c}{\rho^2} \int_{\gamma^{-1}(B_{2\rho}(0))} \kappa^2 \, d\ell + \frac{c}{\rho} \int_{\gamma^{-1}(B_{2\rho}(0))} \big|\langle V, \gamma_s\rangle\big| \, d\ell.$$

Taking $\rho \to \infty$ in this estimate yields $\kappa_s = 0$ and hence the result. □

Remark It is an open problem to determine if either the properness condition or growth condition (13) are necessary in Proposition 12. One advantage of the integral growth condition over the pointwise decay condition is that the rigidity statement holds for a much wider class of weak solutions, which may not even have continuous tangent vector. We will not explore notions of weak solution here.

6 The Shrinking Figure Eight

Proposition 13 *Let* $\gamma : \mathbb{S} \to \mathbb{R}^2$ *be defined by*

$$\gamma(u) = \frac{1}{(1 + \sin^2 u)} \left(\cos u, \frac{1}{2} \sin 2u \right). \tag{14}$$

Then γ *evolves self-similarly under* (3).

Proof The tangent direction to γ is given by

$$\gamma_u(u) = \frac{1}{(2 - \cos^2 u)^2} \big(-(2 + \cos^2 u) \sin u, -2 + 3\cos^2 u \big)$$

and the unit normal is

$$\nu(u) = \left(\frac{3\sin^2 u - 1}{(1 + \sin^2 u)^{\frac{3}{2}}}, \frac{-\sin u (2 + \cos^2 u)}{(1 + \sin^2 u)^{\frac{3}{2}}} \right).$$

Therefore, we have

$$\big\langle \gamma(u), \nu(u) \big\rangle = \frac{\cos u (\sin^2 u - 1)}{(1 + \sin^2 u)^{\frac{3}{2}}}. \tag{15}$$

By further straightforward calculations, the curvature of γ is given by

$$\kappa(u) = \frac{3\cos(u)}{(1 + \sin^2 u)^{\frac{1}{2}}}.$$

We calculate using (2) and the chain rule that the derivatives or curvature with respect to arc length are

$$\kappa_s(u) = \frac{-6 \sin u}{1 + \sin^2 u} \quad \text{and} \quad \kappa_{ss}(u) = \frac{-6\cos^3 u}{(1 + \sin^2 u)^{\frac{3}{2}}}.$$

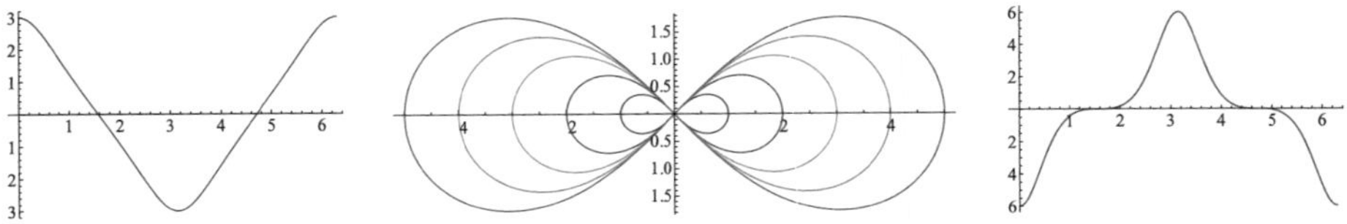

Fig. 3 The shrinking figure eight (*middle*), along with plots of its curvature (*left*) and second arc length derivative of curvature (*right*). The horizontal axis represents the u variable and the vertical axis represents $\kappa(u)$ and $\kappa_{ss}(u)$ in the *left* and *right* figures respectively

Comparing with (15), we see that

$$6\langle\gamma(u), \nu(u)\rangle = \kappa_{ss}(u),$$

so γ given by (14) satisfies (6) with $K = -6$. □

Figure 3 above depicts a shrinking figure eight.

Remarks

1. Since $K < 0$ above, in view of the Remark at the end of Sect. 4, the figure eight given by (14) shrinks to a point at time $T = \frac{1}{24}$. This is in fact consistent with Lemma 5, since the symmetry of the figure eight implies that the signed area is equal to zero and remains constant under (3).
2. The length of the parametrisation of the figure eight above is

$$L(\gamma) = 4\mathcal{K}(-1).$$

Therefore, choosing $\rho = \frac{1}{4\mathcal{K}(-1)}$ allows us to obtain the maximal existence time of the figure eight with unit length as claimed in the introduction.

7 Curves Evolving by Rotation

A family of curves $\gamma : \mathbb{S} \times [0, T) \to \mathbb{R}^2$ evolving purely by rotation satisfies

$$-\kappa_{ss} \equiv 2S(t)\langle\gamma_s, \gamma\rangle. \tag{16}$$

In particular, if (16) is satisfied for a curve γ_0, then the curve diffusion flow evolving from that curve evolves purely by translation. We call the solution γ_t a rotator.

Rotators have been conjectured to exist for the curve shortening flow for some time, with Altschuler making a specific conjecture in [2]. A rigorous classification of these rotators has only recently appeared in the literature however [13]. In this section we prove some classification results for rotators. Our method is to multiply (16) by the curvature and integrate.

Proposition 14 *Let $\gamma : \mathbb{S} \to \mathbb{R}^2$ be a smooth closed rotator. Then γ is a standard round circle.*

Proof The unit normal vector satisfies

$$0 = \frac{1}{2}\partial_s|\nu|^2 = \langle\nu, \nu_s\rangle.$$

Using the Frenet-Serret frame equations,

$$\int_\gamma \kappa_s^2\,d\ell = -\int_\gamma \kappa\,\kappa_{ss}\,d\ell = -2S\int_\gamma \langle k\gamma_s,\gamma\rangle\,d\ell = 2S\int_\gamma \langle \nu_s,\gamma\rangle\,d\ell = -2S\int_\gamma \langle \nu,\gamma_s\rangle\,d\ell = 0,$$

irrespective of S. The curvature of γ is therefore constant and its image must be a round circle. □

Let us now turn to open curves.

Proposition 15 *Suppose $\gamma:\mathbb{R}\to\mathbb{R}^2$ is a smooth open rotator satisfying*

$$\lim_{i\to\infty}\big[\big(\kappa(p_i)\kappa_s(p_i) - 2S\langle\nu(p_i),\gamma(p_i)\rangle\big) - \big(\kappa(q_i)\kappa_s(q_i) - 2S\langle\nu(q_i),\gamma(q_i)\rangle\big)\big] = 0 \qquad (17)$$

for arbitrary sequences $\{p_i\}$ and $\{q_i\}$ approaching $-\infty$ and $+\infty$ respectively. Then $\gamma(\mathbb{R})$ is a straight line.

Proof Multiplying (16) by κ and integrating, we find

$$\int_{p_i}^{q_i}\kappa_s^2 d\ell = -\int_{p_i}^{q_i}\kappa\kappa_{ss}\,d\ell + \kappa\kappa_s|_{p_i}^{q_i} = -2S\int_{p_i}^{q_i}\langle\nu_s,\gamma\rangle\,d\ell + \kappa\kappa_s|_{p_i}^{q_i} = \big[-2S\langle\nu,\gamma\rangle + \kappa\kappa_s\big]_{p_i}^{q_i}.$$

Taking $i\to\infty$ yields the result. □

Remark The result of Proposition 15 is false without condition (17), for the same reason that Proposition 11 is false without condition (12).

As with translators earlier, we may convert (17) from a pointwise decay condition on the curvature to a natural growth condition on κ in L^2 and $S\langle\nu,\gamma\rangle$ in L^1.

Proposition 16 *Suppose $\gamma:\mathbb{R}\to\mathbb{R}^2$ is a smooth open proper rotator satisfying*

$$\lim_{\rho\to\infty}\left(\frac{1}{\rho^2}\int_{\gamma^{-1}(B_\rho(0))}\kappa^2\,d\ell + \frac{|S|}{\rho}\int_{\gamma^{-1}(B_{2\rho}(0))}\big|\langle\nu,\gamma\rangle\big|\,d\ell\right) = 0, \qquad (18)$$

where $B_\rho(0)$ denotes the closed ball of radius ρ centred at the origin. Then $\gamma(\mathbb{R})$ is a straight line.

Proof Let $\eta_\rho = \tilde\eta_\rho\circ\gamma$ be the cutoff function used in the proof of Proposition 12. We compute

$$\begin{aligned}\int_{\gamma^{-1}(B_\rho(0))}\kappa_s^2\,\eta^4\,d\ell &= -\int_{\gamma^{-1}(B_\rho(0))}\kappa\kappa_{ss}\,\eta^4\,d\ell - 4\int_{\gamma^{-1}(B_\rho(0))}\kappa\kappa_s\,\eta_s\,\eta^3\,d\ell\\ &= 2S\int_{\gamma^{-1}(B_\rho(0))}\langle\nu_s,\gamma\rangle\,\eta^4\,d\ell - 4\int_{\gamma^{-1}(B_\rho(0))}\kappa\kappa_s\,\eta_s\,\eta^3\,d\ell\\ &= -8S\int_{\gamma^{-1}(B_\rho(0))}\langle\nu,\gamma\rangle\,\eta_s\,\eta^3\,d\ell - 4\int_{\gamma^{-1}(B_\rho(0))}\kappa\kappa_s\,\eta_s\,\eta^3\,d\ell\end{aligned}$$

$$\leq \varepsilon \int_{\gamma^{-1}(B_\rho(0))} \kappa_s^2 \eta^4 \, d\ell + 8|S| \int_{\gamma^{-1}(B_\rho(0))} \big|\langle \nu, \gamma\rangle\big| \, |\eta_s| \, \eta^3 \, d\ell$$

$$+ \frac{4}{\varepsilon} \int_{\gamma^{-1}(B_\rho(0))} \kappa^2 \eta_s^2 \eta^2 \, d\ell,$$

for any $\varepsilon > 0$. Choosing $\varepsilon = \frac{1}{4}$ and absorbing the first term on the left, we find

$$\int_{\gamma^{-1}(B_\rho(0))} \kappa_s^2 \eta^4 \, d\ell \leq \frac{c}{\rho^2} \int_{\gamma^{-1}(B_{2\rho}(0))} \kappa^2 \, d\ell + \frac{c|S|}{\rho} \int_{\gamma^{-1}(B_{2\rho}(0))} \big|\langle \nu, \gamma\rangle\big| \, d\ell.$$

Taking $\rho \to \infty$ in this estimate yields $\kappa_s = 0$ and hence the result. □

Acknowledgements The research of the second author was supported by a summer vacation scholarship of the Australian Mathematical Sciences Institute. The research of the third, fourth and fifth authors was supported by Discovery Project grant DP120100097 of the Australian Research Council.

The authors would like to thank the anonymous referees for their careful reading and comments that have led to improvements in the article. The authors would also like to thank Hojoo Lee for enlightening discussions related to this work.

References

1. Abresch, U., Langer, J.: The normalized curve shortening flow and homothetic solutions. J. Differ. Geom. **23**(2), 175–196 (1986)
2. Altschuler, A.: Singularities for the curve shrinking flow for space curves. J. Differ. Geom. **34**(2), 491–514 (1991)
3. Cahn, J.W., Elliott, C.M., Novick-Cohen, A.: The Cahn-Hilliard equation with a concentration dependent mobility: motion by minus the Laplacian of the mean curvature. Eur. J. Appl. Math. **7**(3), 287–301 (1996)
4. Dziuk, G., Kuwert, E., Schätzle, R.: Evolution of elastic curves in $\mathbb{R}^n$: Existence and computation. SIAM J. Math. Anal. **33**(5), 1228–1245 (2002)
5. Efrati, E., Sharon, E., Kupferman, R.: Elastic theory of unconstrained non-Euclidean plates. J. Mech. Phys. Solids **57**(4), 762–775 (2009)
6. Escher, J., Ito, K.: Some dynamic properties of volume preserving curvature driven flows. Math. Ann. **333**(1), 213–230 (2005)
7. Escher, J., Maier-Paape, S.: Losing a graph with surface diffusion. Hokkaido Math. J. **30**, 297–305 (2001)
8. Escher, J., Mayer, U., Simonett, G.: On the surface diffusion flow. In: Navier-Stokes Equations and Related Nonlinear Problems (Palanga, 1997). VSP, Utrecht (1998)
9. Gage, M., Hamilton, R.: The heat equation shrinking convex plane curves. J. Differ. Geom. **23**(1), 69–96 (1986)
10. Ghys, E., Tabachnikov, S., Timorin, V.: Osculating curves: Around the Tait-Kneser theorem. Math. Intell. **35**(1), 61–66 (2013)
11. Giga, Y., Ito, K.: On pinching of curves moved by surface diffusion. Commun. Appl. Anal. **2**(3), 393–406 (1998)
12. Giga, Y., Ito, K.: Loss of convexity of simple closed curves moved by surface diffusion. In: Escher, J., Simonett, G. (eds.) Topics in Nonlinear Analysis, the Herbert Amann Anniversary Volume. Progress in nonlinear differential equations and their applications, vol. 35. Birkhäuser, Basel (1999)
13. Halldorsson, H.: Self-similar solutions to the curve shortening flow. Trans. Am. Math. Soc. **364**, 5285–5309 (2012)
14. Huisken, G.: Asymptotic behaviour for singularities of the mean curvature flow. J. Differ. Geom. **31**, 285–299 (1990)
15. McCoy, J.: The surface area preserving mean curvature flow. Asian J. Math. **7**(1), 7–30 (2003)
16. McCoy, J.: Self-similar solutions of fully nonlinear curvature flows. Ann. Sc. Norm. Super. Pisa Cl. Sci. (5) **10**, 317–333 (2011)
17. McCoy, J., Wheeler, G.: A classification theorem for Helfrich surfaces. Math. Ann. **357**, 1485–1508 (2013)

18. McCoy, J., Wheeler, G., Williams, G.: Lifespan theorem for constrained surface diffusion flows. Math. Z. **269**(1–2), 147–178 (2011)
19. Mullins, W.W.: Theory of thermal grooving. J. Appl. Phys. **28**, 333–339 (1957)
20. Novick-Cohen, A.: The Cahn-Hilliard equation. In: Handbook of Differential Equations: Evolutionary Equations, vol. 4, pp. 201–228 (2008)
21. Polden, A.: Curves and surfaces of least total curvature and fourth-order flows. Ph.D. Thesis (1996)
22. Sharon, E., Efrati, E.: The mechanics of non-Euclidean plates. Soft Matter **6**, 5693–5704 (2010)
23. Wheeler, G.: Lifespan theorem for simple constrained surface diffusion flows. J. Math. Anal. Appl. **375**(2), 685–698 (2011)
24. Wheeler, G.: Surface diffusion flow near spheres. Calc. Var. Partial Differ. Equ. **44**(1–2), 131–151 (2012)
25. Wheeler, G.: On the curve diffusion flow of closed plane curves. Ann. Mat. **192**, 931–950 (2013)
26. Willmore, T.: Riemannian Geometry, vol. 33. Clarendon Press, Oxford (1993)

DOI 10.1007/978-94-017-7300-3_12
Reprinted from *Journal of Elasticity* Journal, DOI 10.1007/s10659-014-9504-3

Kinematical Aspects of Levi-Civita Transport of Vectors and Tensors Along a Surface Curve

James Casey

Received: 7 August 2014 / Published online: 12 November 2014

Abstract The concept of parallelism along a surface curve, which was introduced by Levi-Civita in the context of n-dimensional Riemannian manifolds, is re-examined from a kinematical viewpoint. A special type of frame, whose angular velocity is determined by the rate at which the tangent plane turns as one moves along a surface curve, is defined and is called a Levi-Civita frame. The surface may be orientable or not. Vectors and tensors fixed on Levi-Civita frames are parallel transported. Covariant differentiation of vectors and tensors along a surface curve can be expressed in terms of the corresponding corotational rates measured on Levi-Civita frames. Relevant results on ruled surfaces are also included.

Keywords Levi-Civita parallelism · Parallel transport · Differential geometry of surfaces · Adapted frames · Covariant differentiation · Ruled surfaces · Developable surfaces · Möbius band

Mathematics Subject Classification 53A05 · 53A17 · 53A45

1 Introduction

As the 20th century unfolded, the methods of tensor calculus combined with differential geometry were increasingly found to be useful in physics, Lagrangian and Hamiltonian dynamics, shell theory, nonlinear elasticity, and other areas. Continuum mechanics easily lends itself to geometrical interpretations, even if in the end, boundary value problems generally require highly analytical or computational methods of solution. Indeed, the most basic concept in continuum mechanics, that of a *body*, is fundamentally a differentiable manifold having additional physical properties. As observed in the many seminal works that began to appear around 1950, if curvilinear coordinates are employed, the square of the line element of a deformable body is given by the fundamental quadratic form of a Riemannian

J. Casey (✉)
Department of Mechanical Engineering, University of California, Berkeley, CA, USA
e-mail: jimcasey@berkeley.edu

manifold.[1] Further, the compatibility conditions for finite strain are best understood as the vanishing of the corresponding Riemann-Christoffel tensor in Euclidean space.[2] It is also worth mentioning that the description of the transport of various sets and fields by a deforming body is facilitated by the use of modern geometrical concepts, such as Lie dragging, or push-forward and pull-back operations, and the Lie derivative itself.[3]

The concept of parallelism lies at the heart of geometry. For plane geometry, the ancient Greek mathematicians were able to prove that for any point A not belonging to a given straight line EF, there exists a straight line BAC which is parallel to EF. Many believed that the uniqueness of the parallel BAC should also be provable, but there were hidden flaws in the available "proofs." It was Euclid alone who realized that no proof was possible, and he boldly added the epochal Postulate 5 (which was far from being self-evident) to the beginning of his *Elements*.[4]

The differential geometry of curved surfaces was initiated in the 18th century, principally by Leonhard Euler, and was greatly augmented by Gaspard Monge and his pupils. But, it is Carl Friedrich Gauss's ground-breaking paper [11], presented in Latin to the Royal Society of Göttingen on 8 October 1827, that is justly regarded as "the birth certificate of differential geometry."[5] While it was well known before Gauss that portions of surfaces such as cylinders and cones, viewed as flexible, inextensible membranes (or films or sheets) could be rolled out, or "developed," on a plane without any extension or contraction of lines (i.e., isometrically), Gauss was the first to realize that surfaces have certain inner, or *intrinsic*, properties that are preserved under bending deformations, and which are consequently independent of the shape that the surface has in space. Thus, surfaces have both an intrinsic geometry which is determined by measurements made along the surface itself and also have extrinsic properties which depend on the manner in which they occupy space. In particular, as proved by Gauss in his "Theorema Egregium" (remarkable theorem), the Gaussian curvature K is not changed by bending: it is an intrinsic property of the surface.

Gauss's work strongly influenced his sublimely gifted pupil Bernhard Riemann, whose revolutionary paper of 1854 on the geometry of n-dimensional manifolds, read in the presence of the aging Gauss, was eventually to change geometrical and physical ideas forever.[6] Following Riemann, other mathematicians, including Beltrami, Christoffel, Lipschitz, and

---

[1] A foundational memoir on nonlinear continuum mechanics was completed in 1949 by Clifford Truesdell [1]. In it, he mentions [1, p. 129]: "The very real advantage of tensor methods in continuum mechanics I first learned in the courses of Professor Michal." Michal's book [2] was published in 1947, and he acknowledges the influence of a number of earlier writers. See also Green and Zerna [3].

[2] This topic is treated by Truesdell and Toupin [4, Sec. 34] and Fosdick [5].

[3] An important contribution to the promulgation of geometrical methods in continuum mechanics is the monograph by Marsden and Hughes [6]. Additionally, see [7, 8]. A discussion of material transport may be found in [9].

[4] The commentary by Sir Thomas Heath [10] on Euclid's Definition 23 (of parallel lines) and Postulate 5 is fascinating. Heath cannot contain his admiration for Euclid: "When we consider the countless successive attempts made through more than twenty centuries to prove the Postulate, many of them by geometers of ability, we cannot but admire the genius of the man who concluded that such a hypothesis, which he found necessary to the validity of his whole system of geometry, was really indemonstrable." It was not until 1829 that Lobachevsky–and somewhat later Bolyai–independently established that a logically consistent geometry could be created by negating Postulate 5 and retaining Euclid's other assumptions. In this non-Euclidean plane, there is a fan of straight lines through A that never meet EF.

[5] The quotation is from the renowned geometer S.-S. Chern [12]. For an analysis of Gauss's paper and an accompanying translation of a major part of it, see Spivak [13, Vol. 2, Chap. 3].

[6] A translation and analysis of Riemann's paper may be found in [13, Vol. 2, Chap. 4]. The paper was not published until 1868 (which was after Riemann's untimely death). Spivak's translation is reprinted in the

Helmholtz, vigorously pursued the new geometrical ideas. The process of covariant differentiation was discovered and the scalar Gaussian curvature blossomed into the celebrated Riemann-Christoffel tensor. At the hands of Gregorio Ricci-Curbastro, the "absolute differential calculus," ideally suited to the expression of invariant and covariant properties of manifolds, was skillfully developed. The rich fruits of 19th century research in tensor calculus and differential geometry were laid out in an invited memoir [17] by Ricci and his pupil Tullio Levi-Civita in 1900. It was here that Einstein was to find the perfect instrument for expressing his profound ideas on general relativity, and the differential geometry of curved manifolds received a totally unanticipated impetus from physics and cosmology.[7]

In 1917, Levi-Civita discovered a new concept of parallelism for n-dimensional Riemannian manifolds [19]. In a Euclidean plane, this concept reduces to parallelism in the ordinary sense, but it already exhibits a striking contrast with the latter in the case of curved surfaces ($n = 2$). For curved surfaces embedded in 3-dimensional space, Levi-Civita appealed to properties of developable surfaces to explain his idea ([18, Chap. V] and [20]), and I will return to this momentarily. A more immediate explanation can be given in terms of geodesics.[8] The geodesics on a surface are the counterparts of the straight lines in a Euclidean plane. Now, if parallel lines are drawn through any number of points along a straight line, each of the parallels will make a constant angle with the straight line. Let AB be any portion of a geodesic on a surface and let AM be any line in the tangent plane at A. Let α be the angle between AM and the geodesic. Then, we can always find a straight line, BN, that lies in the tangent plane at B and makes the same angle α with the geodesic. If we do this for every point along the geodesic, we say that AM has been *parallel transported* or (*parallel propagated*) along it. A little reflection will reveal that, in general, AM and BN will not be parallel in the usual sense, and more surprisingly perhaps, that the propagation process is path-dependent.[9] It follows that parallel propagation around a closed curve consisting of geodesic segments will generally lead to a line AN which is not coincident with AM.[10]

The concept of parallel propagation can be extended to an arbitrary surface curve APB [18]. Let $\mathcal{D}$ be a developable surface that touches a given surface $\mathcal{S}$ along APB (Fig. 1(a)).[11]

splendid collection of papers [14]. An insightful history of Riemannian geometry and tensor calculus was written by Weatherburn [15]; excerpts may be found in [16].

[7] Levi-Civita elaborated further on the new tensor calculus in his monograph [18]. (The word "tensor" came into widespread use only after the publication of Einstein's paper on general relativity in 1916. It had been employed previously by Gibbs in the early 1880's and by Voigt in 1898. It was originally introduced into mathematics, but with a different meaning, by Hamilton in 1846.)

[8] See e.g., Graustein [21, Sec. 66]. A geodesic curve can be defined in a number of different ways, the most fundamental of which is actually in terms of Levi-Civita parallelism (see Sect. 3). However, for now, let us characterize a geodesic on a surface by the property that every sufficiently small arc PQ of it is the shortest arc joining P and Q.

[9] To repeat a well-known example, consider a unit sphere centered at O. Let N be its north pole. Take an arc NC of a great circle that joins N to any point C on the equator. Let ND be a second such arc with the angle $COD = \pi/2$. Take a line tangent to NC at N and propagate it along the geodesic NC. At C, it will be parallel to NO in space. Next, propagate it from C along the equator to D (where it will still be parallel to NO in space), and finally propagate it along the geodesic DN. At N, the resulting line will form an angle of $\pi/2$ with its initial direction. (Incidentally, the geodesic triangle NCD has an angle-sum of three right angles, in contrast to the Euclidean case.)

[10] This geometrical phenomenon is referred to as *holonomy* in the mathematical literature, but should properly be called *anholonomy*, in keeping with the established usages of holonomic and nonholonomic in analytical mechanics.

[11] The mathematical details of the process being discussed verbally here will be provided in Sect. 5. The surface $\mathcal{S}$ itself need not be developable.

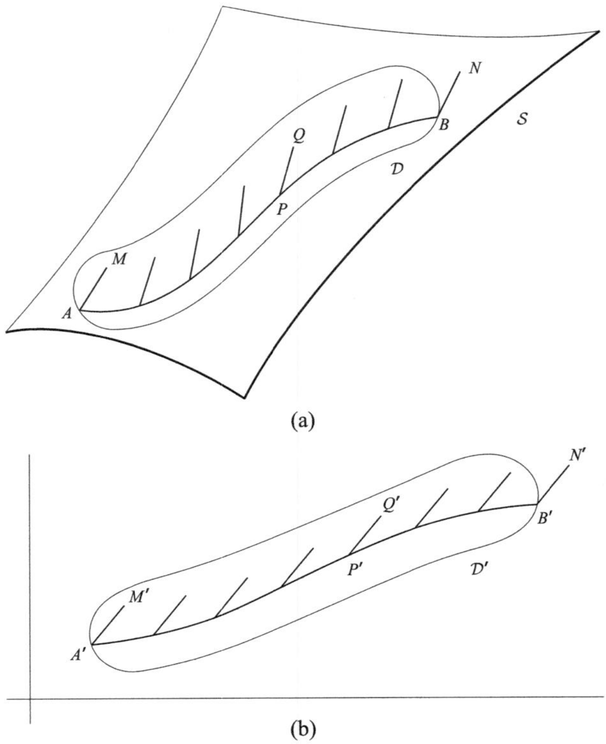

Fig. 1 $\mathcal{D}$ is a developable surface that shares the same tangent planes with the surface $\mathcal{S}$ along the curve AB. $\mathcal{D}'$ is the developed image of $\mathcal{D}$ on a Euclidean plane. Ordinary parallelism along $A'P'B'$ becomes Levi-Civita parallelism along APB

Choose an arbitrary line AM in the tangent plane to $\mathcal{D}$ at A. Develop the surface $\mathcal{D}$ on a Euclidean plane, producing $\mathcal{D}'$ (Fig. 1(b)), and draw ordinary parallels to $A'M'$ all along the developed image $A'P'B'$ of the curve APB. Wrap the developed surface $\mathcal{D}'$ back into its original configuration along APB. The images of the Euclidean parallels PQ, etc., are defined to be the Levi-Civita parallels along the surface curve APB.[12]

The present paper is concerned with kinematical aspects of parallel transport. By examining how various orthonormal bases (or "frames") move along a surface curve by rigid body motions, we will be able to view parallel transport and covariant differentiation in a new light.

To set the stage for the kinematical developments, it may be instructive to describe a simple geometrical experiment, in which two different frames are transported around a closed line $P_0QRSTUP_0$ on a cardboard box, as indicated in Figs. 2 and 3. Each frame is modeled by a small stiff paper square $ABCD$ that can be slid along the surface of the box in a manner which will now be specified. The side of the square may be taken as the unit of length.

In Fig. 2, two unit vectors $\boldsymbol{t}$ and $\boldsymbol{e}$ are inscribed on the square, in directions perpendicular to its sides. The square is slid along the line $P_0QRSTUP_0$ in such a way that $\boldsymbol{t}$ is tangent to each of the line segments P_0Q, QR, RS, ST, TU, and UP_0 in succession. The square tilts over the edges at Q, S, and U, and rotates through right angles at R, T, and finally P_0,

[12] A satisfactory physical model of this operation is obtained as follows: (1) Draw short parallel lines along an arbitrary curve on a page; (2) Cut out a narrow strip of paper containing these parallels; (3) Tape one end of the strip on a curved surface and run a finger along the strip to lay it on the surface. Several examples are given in [22]. (It may also be remarked that if the original planar curve is a straight line, it will be bent into a geodesic on the surface.)

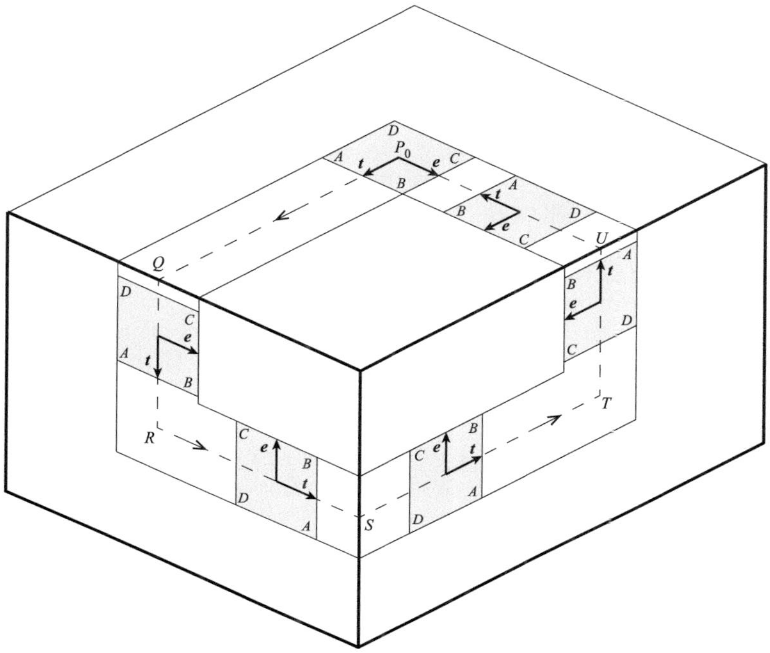

Fig. 2 Transport of a Darboux frame around a closed path. Upon completion of the circuit, the frame regains its original position

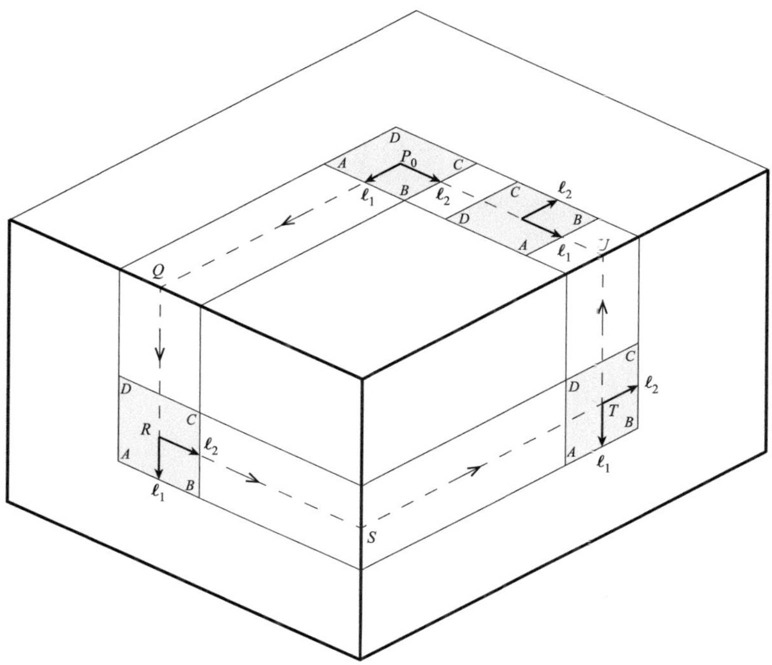

Fig. 3 Transport of a Levi-Civita frame around a closed path. Upon completion of the circuit, the frame has rotated through a right angle about the surface normal

where its original position is restored. The vectors $\boldsymbol{t}$, $\boldsymbol{e}$, and the outward unit surface normal form a right-handed triad, called the *Darboux frame*.[13]

In Fig. 3, two vectors $\boldsymbol{\ell}_1$ and $\boldsymbol{\ell}_2$ are similarly inscribed on another paper square. At the initial point P_0, $\boldsymbol{\ell}_1$ may be conveniently chosen to be tangent to P_0Q. However, this square will follow the tangent plane to the box along $P_0QRSTUP_0$ rather than following the tangents to the line itself. Thus, it tilts at the edges at Q, S, and U, but it just translates whenever it is away from the edges. Upon returning to P_0, the square is found to have undergone a net positive rotation of 90° in the tangent plane at P_0. When a frame is moved along a surface curve in such a manner that it senses only the change in the orientation of the tangent plane, we will call it a *Levi-Civita frame* (Sect. 3). Examination of Fig. 3 reveals that each of the unit vectors $\boldsymbol{\ell}_1$ and $\boldsymbol{\ell}_2$ undergoes parallel propagation around the circuit $P_0QRSTUP_0$.[14] An anholonomy of 90° of rotation has taken place. It is due to the Gaussian curvature of the part of the box's surface that is enclosed by the circuit; this curvature is concentrated at the corner.

On non-orientable surfaces, we may also define Darboux and Levi-Civita frames along a curve, choosing either one of the two available unit normals at the starting point of the curve. Thus, a paper model of a Möbius band is everywhere locally developable and its centerline is a geodesic. Let us slide a paper square $ABCD$ along the centerline (in the same manner as in Fig. 3). Upon completion of one circuit, the sides AD and BC will be seen to have switched places and the vector $\boldsymbol{\ell}_2$ will have become $-\boldsymbol{\ell}_2$. A reflection in the centerline has taken place.[15] Also, of course, the unit normal has been reversed. The initial Levi-Civita frame is rotated into a frame having the same positive orientation. The anholonomy of the (inextensible) Möbius band is characterized by a reflection rather than a rotation, and derives from a global property (non-orientability) rather than surface curvature.

The contents of the paper are as follows. Background material on frames, finite rotations, corotational rates of vectors and second-order tensors, and the elements of surface geometry are reviewed in Sect. 2.[16] An analysis of how the tangent plane changes its spatial orientation along a surface curve is presented in Sect. 3 and a geometrically important frame, denoted by $\{\boldsymbol{u}, \boldsymbol{g}, \boldsymbol{n}\}$, is identified. Its angular velocity $\boldsymbol{\omega}_G$ has two components (Eq. (46)). One component measures the angular velocity $\boldsymbol{\omega}_T$ (Eq. (54)) of the tangent plane, corresponding to motion along a surface curve; the other is a spin about the surface normal. We formally define a Levi-Civita frame to be any adapted frame whose angular velocity vector is equal to $\boldsymbol{\omega}_T$. Parallel transport of a vector will then correspond to the vector being fixed on a Levi-Civita frame, and its corresponding corotational rate will be zero; similarly for tensors.

Covariant differentiation is treated in Sect. 4, where it is shown that the covariant derivative of a vector $\boldsymbol{w}$ along a surface curve is equal to the corotational rate of that vector relative to a Levi-Civita frame (which is denoted by $\overset{\square}{\boldsymbol{w}}$), and likewise for tensors (Eqs. (129) and (150)).

[13] Any frame that consists of the unit surface normal and two unit vectors orthogonal to it is said to be *adapted* to the surface [23].

[14] For the box in Fig. 2, the band along which the paper squares slide can be covered by series of rectangular strips of paper that can be successively attached to one another, except that the end should not be joined to the beginning (the band is everywhere locally developable, but not globally so). When laid out on a plane, these strips have a horizontal bottom side and two vertical sides.

[15] For an illustration, see Weeks [24, Fig. 4.2].

[16] This material is based on standard works on differential geometry, tensor calculus, and continuum mechanics, especially [3, 13, 18, 23, 25–37].

As mentioned earlier, in [18, 20] Levi-Civita explained his concept of parallel propagation using the device of a developable surface. In Sect. 5, the envelope of the tangent planes along a surface curve is discussed. It is shown how this envelope and the initial tangent plane are applicable to one another, i.e., one can be bent onto the other by a one-parameter family of isometric mappings. The rotation tensor involved is that of the Levi-Civita frame.

Relevant material on ruled surfaces is collected in the Appendix.

2 Background Material

2.1 Preliminaries

Let $\mathcal{E}$ denote a 3-dimensional Euclidean point space and let $\mathcal{E}^*$ be the associated vector space. The inner product of any pair of vectors $\boldsymbol{a}, \boldsymbol{b} \in \mathcal{E}^*$ is denoted by $\boldsymbol{a} \cdot \boldsymbol{b}$. Choose an arbitrary origin $O \in \mathcal{E}$, a fixed right-handed orthonormal basis $\{\boldsymbol{E}_i\}$ for $\mathcal{E}^*$, and rectangular Cartesian coordinates x_i. In this paper, lower-case Latin indices will have a range $1, 2, 3$, while Greek indices will have a range $1, 2$, and summation over repeated indices will always be understood. For components on an orthonormal basis, only subscripts will be used. For general bases, superscripts will be employed for curvilinear coordinates and for contravariant objects, while subscripts will indicate covariant objects. For brevity, we shall refer to any orthonormal basis in $\mathcal{E}^*$ as a *frame*.

Any linear mapping $\boldsymbol{A}$ from $\mathcal{E}^*$ into $\mathcal{E}^*$ that is invariant under every change of coordinates will be called a (second-order) tensor. The "bun" product of $\boldsymbol{a}, \boldsymbol{b} \in \mathcal{E}^*$ is the linear mapping $\boldsymbol{a} \otimes \boldsymbol{b}$ defined by $(\boldsymbol{a} \otimes \boldsymbol{b})\boldsymbol{u} = (\boldsymbol{b} \cdot \boldsymbol{u})\boldsymbol{a}$ for all $\boldsymbol{u} \in \mathcal{E}^*$.[17] The identity tensor $\boldsymbol{I}$ on $\mathcal{E}^*$ has a useful representation: $\boldsymbol{I} = \boldsymbol{E}_i \otimes \boldsymbol{E}_i$. We recall that $(\boldsymbol{a} \otimes \boldsymbol{b})(\boldsymbol{c} \otimes \boldsymbol{d}) = (\boldsymbol{b} \cdot \boldsymbol{c})\boldsymbol{a} \otimes \boldsymbol{d}$, $\boldsymbol{A}(\boldsymbol{u} \otimes \boldsymbol{v}) = (\boldsymbol{A}\boldsymbol{u}) \otimes \boldsymbol{v}$, and $(\boldsymbol{u} \otimes \boldsymbol{v})\boldsymbol{A} = \boldsymbol{u} \otimes (\boldsymbol{A}^T \boldsymbol{v})$, where $\boldsymbol{A}^T$ is the transpose of the tensor $\boldsymbol{A}$. Also $(\boldsymbol{a} \otimes \boldsymbol{b})^T = \boldsymbol{b} \otimes \boldsymbol{a}$. Further, it is convenient to have available the products $\boldsymbol{a} \otimes \boldsymbol{b} \otimes \boldsymbol{c}$ which act linearly as follows: $(\boldsymbol{a} \otimes \boldsymbol{b} \otimes \boldsymbol{c})\boldsymbol{u} = (\boldsymbol{c} \cdot \boldsymbol{u})\boldsymbol{a} \otimes \boldsymbol{b}$, $(\boldsymbol{a} \otimes \boldsymbol{b} \otimes \boldsymbol{c})[\boldsymbol{u} \otimes \boldsymbol{v}] = (\boldsymbol{b} \cdot \boldsymbol{u})(\boldsymbol{c} \cdot \boldsymbol{v})\boldsymbol{a}$. An inner product for tensors is furnished by $\boldsymbol{A} \cdot \boldsymbol{B} = \mathrm{tr}(\boldsymbol{A}^T \boldsymbol{B})$.

Let $\boldsymbol{R}$ be a proper orthogonal tensor, i.e.,

$$\boldsymbol{R}\boldsymbol{R}^T = \boldsymbol{R}^T\boldsymbol{R} = \boldsymbol{I}, \qquad \det \boldsymbol{R} = 1. \tag{1}$$

Such tensors represent rotations. Define three vectors $\boldsymbol{e}_i$ by

$$\boldsymbol{e}_i = \boldsymbol{R}\boldsymbol{E}_i. \tag{2}$$

Clearly, $\boldsymbol{e}_i \cdot \boldsymbol{e}_j = \delta_{ij}$, where the Kronecker delta δ_{ij} satisfies $\delta_{ij} = 0$ if $i \neq j$, and $\delta_{ij} = 1$ if $i = j$ (no sum); thus, $\boldsymbol{R}$ maps an orthonormal basis into an orthonormal basis. Moreover, the scalar triple product $[\boldsymbol{e}_1, \boldsymbol{e}_2, \boldsymbol{e}_3] = (\det \boldsymbol{R})[\boldsymbol{E}_1, \boldsymbol{E}_2, \boldsymbol{E}_3] = 1$ and the frame $\{\boldsymbol{e}_i\}$ is right-handed. We note that

$$\boldsymbol{R} = \boldsymbol{R}\boldsymbol{I} = (\boldsymbol{R}\boldsymbol{E}_i) \otimes \boldsymbol{E}_i = \boldsymbol{e}_i \otimes \boldsymbol{E}_i. \tag{3}$$

[17] It is common in the literature to regard any linear mapping on a vector space as being a tensor, but linearity alone is not sufficient. For example, the bun product of the tangent vectors to two coordinate curves at a point is a linear mapping, but it changes under coordinate transformations and is therefore not a tensor. When linear transformations are defined using a coordinate system, a check must be made to find out whether they are tensors or not.

Thus, if two right-handed frames $\{\boldsymbol{E}_i\}$ and $\{\boldsymbol{e}_i\}$ are specified, then a corresponding rotation tensor $\boldsymbol{R}$ is uniquely defined. $\boldsymbol{R}$ also has the standard representations

$$\boldsymbol{R} = R_{ij}\boldsymbol{E}_i \otimes \boldsymbol{E}_j = R'_{ij}\boldsymbol{e}_i \otimes \boldsymbol{e}_j, \tag{4}$$

with $R_{ij} = \boldsymbol{E}_i \cdot \boldsymbol{R}\boldsymbol{E}_j = \boldsymbol{E}_i \cdot \boldsymbol{e}_j$, $R'_{ij} = \boldsymbol{e}_i \cdot \boldsymbol{R}\boldsymbol{e}_j = (\boldsymbol{R}^T\boldsymbol{e}_i)\cdot \boldsymbol{e}_j = \boldsymbol{E}_i \cdot \boldsymbol{e}_j = R_{ij}$. In general, the nine components R_{ij} depend on three independent parameters (such as Euler angles).

Consider a differentiable one-parameter family of rotations $\boldsymbol{R}(t)$, where t represents time, and rotate the frame $\{\boldsymbol{E}_i\}$ by $\boldsymbol{R}(t)$ to obtain a one-parameter family of frames $\{\boldsymbol{e}_i(t)\}$. It is suggestive to think of $\{\boldsymbol{e}_i\}$ as representing a massless 3-dimensional rigid body. Differentiation of $\boldsymbol{R}(t)\boldsymbol{R}^T(t)$ in $(1)_1$ yields

$$\boldsymbol{\Omega} + \boldsymbol{\Omega}^T = \boldsymbol{0}, \qquad \boldsymbol{\Omega} = \dot{\boldsymbol{R}}\boldsymbol{R}^T, \tag{5}$$

where the overdot represents d/dt. The skew-symmetric tensor $\boldsymbol{\Omega}$ is the angular velocity tensor of the frame $\{\boldsymbol{e}_i(t)\}$. Associated with it is the angular velocity vector $\boldsymbol{\omega}$, such that

$$\boldsymbol{\Omega}\boldsymbol{c} = \boldsymbol{\omega} \times \boldsymbol{c} \tag{6}$$

for every $\boldsymbol{c} \in \mathcal{E}^*$; $\boldsymbol{\omega}$ and $\boldsymbol{\Omega}$ are related explicitly through the formulas

$$\boldsymbol{\omega} = -\frac{1}{2}\boldsymbol{\varepsilon}[\boldsymbol{\Omega}], \qquad \boldsymbol{\Omega} = -\boldsymbol{\varepsilon}\boldsymbol{\omega}, \tag{7}$$

where the alternator $\boldsymbol{\varepsilon}$ is given on the right-handed orthonormal basis $\{\boldsymbol{e}_i\}$ by $\boldsymbol{\varepsilon} = e_{ijk}\boldsymbol{e}_i \otimes \boldsymbol{e}_j \otimes \boldsymbol{e}_k$, with $e_{ijk} = 1$ if ijk is an even permutation of $1, 2, 3$, $e_{ijk} = -1$ if ijk is an odd permutation of $1, 2, 3$, and $e_{ijk} = 0$ otherwise.

If a vector $\boldsymbol{\omega}$ is specified as a continuously differentiable function of time t, a corresponding skew-symmetric tensor $\boldsymbol{\Omega}$ can be found from $(7)_2$. Consider then the initial value problem

$$\dot{\boldsymbol{R}} = \boldsymbol{\Omega}\boldsymbol{R}, \qquad \boldsymbol{R}(0) = \boldsymbol{I}. \tag{8}$$

This possesses a unique solution $\boldsymbol{R}(t)$ for some interval of time, and by virtue of the skew-symmetry of $\boldsymbol{\Omega}$,

$$\dot{\overline{\boldsymbol{R}\boldsymbol{R}^T}} = \boldsymbol{0}. \tag{9}$$

Hence, $\boldsymbol{R}\boldsymbol{R}^T$ is a constant tensor, which in view of $(8)_2$ must be $\boldsymbol{I}$. Therefore, the solution $\boldsymbol{R}(t)$ to $(8)_{1,2}$ is a rotation.

From (2), $(5)_2$, and (6), it is obvious that

$$\dot{\boldsymbol{e}}_i = \boldsymbol{\Omega}\boldsymbol{e}_i = \boldsymbol{\omega} \times \boldsymbol{e}_i. \tag{10}$$

Representing $\boldsymbol{\Omega}$ and $\boldsymbol{\omega}$ as $\boldsymbol{\Omega} = \Omega_{ij}\boldsymbol{e}_i \otimes \boldsymbol{e}_j$, $\boldsymbol{\omega} = \omega_i\boldsymbol{e}_i$, we see that

$$\Omega_{ij} = \boldsymbol{e}_i \cdot \boldsymbol{\Omega}\boldsymbol{e}_j = \boldsymbol{e}_i \cdot \dot{\boldsymbol{e}}_j. \tag{11}$$

Also, Eqs. $(7)_{1,2}$ yield $\omega_i = -\frac{1}{2}e_{ijk}\Omega_{jk}$ and $\Omega_{ij} = -e_{ijk}\omega_k$. Hence, $\Omega_{32} = \omega_1$, $\Omega_{13} = \omega_2$, $\Omega_{21} = \omega_3$.

Fig. 4 A surface $\mathcal{S}$ embedded in 3-dimensional Euclidean space and the image of its unit normal $\boldsymbol{n}$ on a unit sphere

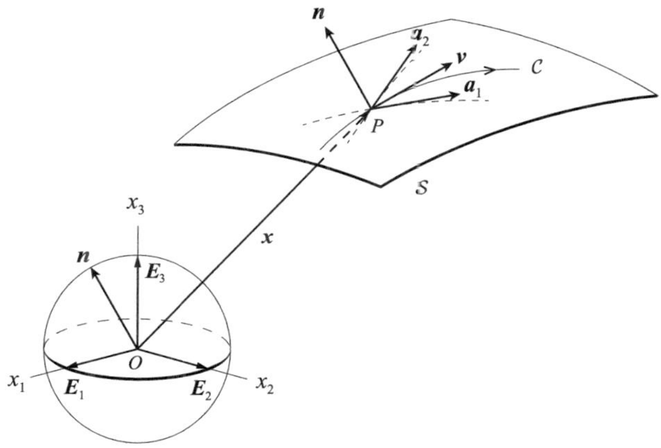

For any differentiable vector-valued function $\boldsymbol{u}(t)$, we write $\boldsymbol{u} = u_i \boldsymbol{e}_i$ and use (10) to obtain

$$\dot{\boldsymbol{u}} = \overset{\circ}{\boldsymbol{u}} + \boldsymbol{\Omega} \boldsymbol{u} = \overset{\circ}{\boldsymbol{u}} + \omega \times \boldsymbol{u}, \tag{12}$$

where $\overset{\circ}{\boldsymbol{u}} = \dot{u}_i \boldsymbol{e}_i$ is called the *corotational rate* of $\boldsymbol{u}$ with respect to the rotating frame $\{\boldsymbol{e}_i\}$. Clearly, $\overset{\circ}{\boldsymbol{u}} = \boldsymbol{0}$ if and only if the components u_i are constants. Also, note that $\overset{\circ}{\boldsymbol{e}}_i = \boldsymbol{0}$ and $\dot{\boldsymbol{\omega}} = \overset{\circ}{\boldsymbol{\omega}}$. For any differentiable tensor-valued function $\boldsymbol{U}(t)$, we similarly obtain

$$\dot{\boldsymbol{U}} = \frac{d}{dt}(U_{ij} \boldsymbol{e}_i \otimes \boldsymbol{e}_j) = \overset{\circ}{\boldsymbol{U}} + \boldsymbol{\Omega} \boldsymbol{U} - \boldsymbol{U} \boldsymbol{\Omega}, \tag{13}$$

where $\overset{\circ}{\boldsymbol{U}} = \dot{U}_{ij} \boldsymbol{e}_i \otimes \boldsymbol{e}_j$ is the corotational rate of $\boldsymbol{U}$.

2.2 Elements of Surface Geometry

Consider a fixed surface $\mathcal{S}$ embedded in $\mathcal{E}$; $\mathcal{S}$ is regarded as a 2-dimensional, differentiable, and connected manifold. Let P be any point on $\mathcal{S}$, with position vector $\boldsymbol{x} = x_i \boldsymbol{E}_i$ (Fig. 4). Let u^α be local Gaussian coordinates (i.e., in a neighborhood of P): $\boldsymbol{x} = \boldsymbol{x}(u^\alpha)$. The tangent vectors at P to the two coordinate curves on $\mathcal{S}$ are given by

$$\boldsymbol{a}_\alpha = \frac{\partial \boldsymbol{x}}{\partial u^\alpha} = \boldsymbol{x}_{,\alpha}, \tag{14}$$

with $\boldsymbol{a}_1 \times \boldsymbol{a}_2 \neq \boldsymbol{0}$. The tangent plane $\mathcal{T}$ to $\mathcal{S}$ at P consists of all straight lines in $\mathcal{E}$ that are tangent to curves on $\mathcal{S}$ at P. The geometry of $\mathcal{T}$ is induced from that of the containing space $\mathcal{E}$. Let $\mathcal{T}^* \subset \mathcal{E}^*$ be the 2-dimensional vector space associated with $\mathcal{T}$; we call $\mathcal{T}^*$ the *tangent space* at P. The set of vectors $\{\boldsymbol{a}_\alpha\}$ form a basis for $\mathcal{T}^*$, the "coordinate basis." Let

$$a_{\alpha\beta} = \boldsymbol{a}_\alpha \cdot \boldsymbol{a}_\beta \ (= a_{\beta\alpha}). \tag{15}$$

Linear mappings from $\mathcal{T}^*$ to $\mathcal{T}^*$ that are invariant under every change of Gaussian coordinates are called surface tensors. Let $\boldsymbol{I}_{\mathcal{T}}$ denote the identity surface tensor on $\mathcal{T}^*$.

The unit vector

$$\boldsymbol{n} = \frac{\boldsymbol{a}_1 \times \boldsymbol{a}_2}{\|\boldsymbol{a}_1 \times \boldsymbol{a}_2\|} \quad (\boldsymbol{a}_1 \times \boldsymbol{a}_2 \neq \boldsymbol{0}) \tag{16}$$

is orthogonal to every vector belonging to $\mathcal{T}^*$. It is convenient to think of $\boldsymbol{n}$ as the position vector of a point on the unit sphere (i.e., the Gauss map, see Fig. 4.). Clearly, $\boldsymbol{n}$ determines the orientation (or attitude) of the tangent plane in space, and two parameters suffice to specify $\boldsymbol{n}$.

Let $\mathcal{C}$ be any differentiable curve lying on $\mathcal{S}$, and parametrized by $t : \boldsymbol{x}(t) = \boldsymbol{x}(u^\alpha(t))$; let the positive orientation of $\mathcal{C}$ be defined by increasing values of t. Also, let s denote the arc length of $\mathcal{C}$, with $\dot{s} = ds/dt > 0$. The velocity vector $\boldsymbol{v} \in \mathcal{T}^*$, and the unit tangent vector $\boldsymbol{t} \in \mathcal{T}^*$, to $\mathcal{C}$ at P, satisfy the relations

$$\begin{aligned} \boldsymbol{v} = \dot{\boldsymbol{x}} = \frac{d\boldsymbol{x}}{dt} = \frac{\partial \boldsymbol{x}}{\partial u^\alpha}\dot{u}^\alpha = \dot{u}^\alpha \boldsymbol{a}_\alpha, \\ \boldsymbol{t} = \frac{d\boldsymbol{x}}{ds} = \frac{du^\alpha}{ds}\boldsymbol{a}_\alpha, \qquad \boldsymbol{v} = \dot{s}\boldsymbol{t}, \qquad \boldsymbol{t}\cdot\boldsymbol{t} = 1. \end{aligned} \tag{17}$$

We also use the notation $v^\alpha = \dot{u}^\alpha$, $t^\alpha = du^\alpha/ds$ for these components of $\boldsymbol{v}$ and $\boldsymbol{t}$.

Corresponding to the coordinate basis $\{\boldsymbol{a}_\alpha\}$, a reciprocal basis $\{\boldsymbol{a}^\alpha\}$ for $\mathcal{T}^*$ is given by

$$\boldsymbol{a}^1 = \frac{\boldsymbol{a}_2 \times \boldsymbol{n}}{\|\boldsymbol{a}_1 \times \boldsymbol{a}_2\|}, \qquad \boldsymbol{a}^2 = \frac{\boldsymbol{n} \times \boldsymbol{a}_1}{\|\boldsymbol{a}_1 \times \boldsymbol{a}_2\|}. \tag{18}$$

Evidently,

$$\boldsymbol{a}^\alpha \cdot \boldsymbol{a}_\beta = \delta^\alpha_\beta, \tag{19}$$

where $\delta^1_1 = \delta^2_2 = 1$, $\delta^1_2 = \delta^2_1 = 0$. Let

$$a^{\alpha\beta} = \boldsymbol{a}^\alpha \cdot \boldsymbol{a}^\beta \; (= a^{\beta\alpha}). \tag{20}$$

For any vector $\boldsymbol{w} \in \mathcal{T}^*$, we have the pair of representations

$$\boldsymbol{w} = w^\alpha \boldsymbol{a}_\alpha = w_\alpha \boldsymbol{a}^\alpha, \tag{21}$$

with

$$w^\alpha = \boldsymbol{w} \cdot \boldsymbol{a}^\alpha, \qquad w_\alpha = \boldsymbol{w} \cdot \boldsymbol{a}_\alpha. \tag{22}$$

The components w^α are called contravariant, while w_α are called covariant.[18] Utilizing representations of the form (21) for the basis vectors themselves, we find that

$$\boldsymbol{a}_\alpha = (\boldsymbol{a}_\alpha \cdot \boldsymbol{a}_\beta)\boldsymbol{a}^\beta = a_{\alpha\beta}\boldsymbol{a}^\beta, \qquad \boldsymbol{a}^\alpha = a^{\alpha\beta}\boldsymbol{a}_\beta. \tag{23}$$

It then follows that $w^\alpha = \boldsymbol{w} \cdot (a^{\alpha\beta}\boldsymbol{a}_\beta) = a^{\alpha\beta} w_\beta$ and $w_\alpha = a_{\alpha\beta} w^\beta$.

The identity surface tensor $\boldsymbol{I}_T$ for $\mathcal{T}^*$ has the property that $\boldsymbol{I}_T\boldsymbol{w} = \boldsymbol{w}$ for every $\boldsymbol{w} \in \mathcal{T}^*$. But, $(\boldsymbol{a}_\alpha \otimes \boldsymbol{a}^\alpha)\boldsymbol{w} = (\boldsymbol{w} \cdot \boldsymbol{a}^\alpha)\boldsymbol{a}_\alpha = \boldsymbol{w}$. Therefore, $\boldsymbol{I}_T$ has the representations $\boldsymbol{I}_T = \boldsymbol{a}_\alpha \otimes \boldsymbol{a}^\alpha = \delta^\alpha_\beta \boldsymbol{a}_\alpha \otimes \boldsymbol{a}^\beta$. Likewise, $\boldsymbol{I}_T = \boldsymbol{a}^\alpha \otimes \boldsymbol{a}_\alpha$. Note that, in general, $\boldsymbol{I}_T$ varies across the surface $\mathcal{S}$. In view of $(23)_{1,2}$, $\boldsymbol{I}_T$ can also be expressed as $\boldsymbol{I}_T = a_{\alpha\beta}\boldsymbol{a}^\alpha \otimes \boldsymbol{a}^\beta = a^{\alpha\beta}\boldsymbol{a}_\alpha \otimes \boldsymbol{a}_\beta$. If any orthonormal basis $\{\bar{\boldsymbol{e}}_\alpha\}$ is chosen for $\mathcal{T}^*$, then $\boldsymbol{I}_T$ may be written as $\boldsymbol{I}_T = \bar{\boldsymbol{e}}_1 \otimes \bar{\boldsymbol{e}}_1 + \bar{\boldsymbol{e}}_2 \otimes \bar{\boldsymbol{e}}_2$,

[18] The contravariant and covariant components of a vector transform in definite, different ways under a general change of coordinates, whilst the two bases in (22) also transform.

and it follows that $\boldsymbol{I}_T = \boldsymbol{I} - \boldsymbol{n} \otimes \boldsymbol{n}$.[19] Clearly, $\boldsymbol{I}_T \cdot \boldsymbol{n} \otimes \boldsymbol{n} = 0$. If $d\boldsymbol{x} = du^\alpha \boldsymbol{a}_\alpha$, we note that $\|d\boldsymbol{x}\|^2 = d\boldsymbol{x} \cdot d\boldsymbol{x} = a_{\alpha\beta} du^\alpha du^\beta$, which is traditionally called the first fundamental form. In keeping with this terminology, $\boldsymbol{I}_T$ may be referred to as the first fundamental tensor of $\mathcal{S}$.

Since $\mathcal{S}$ is a differentiable manifold, it follows from (14) that

$$\boldsymbol{a}_{\alpha},_{\beta} = \frac{\partial}{\partial u^\beta}\left(\frac{\partial \boldsymbol{x}}{\partial u^\alpha}\right) = \frac{\partial}{\partial u^\alpha}\left(\frac{\partial \boldsymbol{x}}{\partial u^\beta}\right) = \boldsymbol{a}_{\beta},_{\alpha}. \tag{24}$$

The Christoffel symbols of the first and second kinds for $\mathcal{S}$ are defined by

$$\Gamma_{\alpha\beta\gamma} = \frac{1}{2}(a_{\beta\gamma},_\alpha + a_{\alpha\gamma},_\beta - a_{\alpha\beta},_\gamma), \qquad \Gamma^\gamma_{\alpha\beta} = a^{\gamma\lambda}\Gamma_{\alpha\beta\lambda} \tag{25}$$

respectively, and possess the symmetries $\Gamma_{\alpha\beta\gamma} = \Gamma_{\beta\alpha\gamma}$, $\Gamma^\gamma_{\beta\alpha} = \Gamma^\gamma_{\alpha\beta}$. It follows from $(25)_{1,2}$, (15), (24), and $(23)_2$, that

$$\Gamma_{\alpha\beta\gamma} = \boldsymbol{a}_\alpha,_\beta \cdot \boldsymbol{a}_\gamma, \qquad \Gamma^\gamma_{\alpha\beta} = \boldsymbol{a}_\alpha,_\beta \cdot \boldsymbol{a}^\gamma. \tag{26}$$

The partial derivatives $\boldsymbol{a}_\alpha,_\beta$, which are vectors belonging to $\mathcal{E}^*$, may therefore be expressed in the form

$$\begin{aligned}\boldsymbol{a}_\alpha,_\beta &= (\boldsymbol{a}_\alpha,_\beta \cdot \boldsymbol{a}^\gamma)\boldsymbol{a}_\gamma + (\boldsymbol{a}_\alpha,_\beta \cdot \boldsymbol{n})\boldsymbol{n} \\ &= \Gamma^\gamma_{\alpha\beta}\boldsymbol{a}_\gamma + b_{\alpha\beta}\boldsymbol{n} = \Gamma_{\alpha\beta\gamma}\boldsymbol{a}^\gamma + b_{\alpha\beta}\boldsymbol{n},\end{aligned} \tag{27}$$

where

$$b_{\alpha\beta} = \boldsymbol{a}_\alpha,_\beta \cdot \boldsymbol{n} = b_{\beta\alpha} \tag{28}$$

and the symmetry condition in (24) has been invoked. Equations (27) are known as the formulas of Gauss.

For the partial derivatives of $\boldsymbol{n}$, we have

$$\boldsymbol{n},_\alpha = (\boldsymbol{n},_\alpha \cdot \boldsymbol{a}_\beta)\boldsymbol{a}^\beta + (\boldsymbol{n},_\alpha \cdot \boldsymbol{n})\boldsymbol{n}. \tag{29}$$

But,

$$\boldsymbol{n} \cdot \boldsymbol{n} = 1, \qquad \boldsymbol{n} \cdot \boldsymbol{a}_\alpha = 0, \tag{30}$$

and hence

$$\boldsymbol{n},_\alpha \cdot \boldsymbol{n} = 0, \qquad \boldsymbol{n},_\beta \cdot \boldsymbol{a}_\alpha = -b_{\alpha\beta}, \tag{31}$$

where (28) has also been used. Therefore, Eq. (29) reduces to

$$\boldsymbol{n},_\alpha = -b_{\beta\alpha}\boldsymbol{a}^\beta = -b^\gamma_\alpha \boldsymbol{a}_\gamma, \tag{32}$$

where $b^\gamma_\alpha = a^{\beta\gamma} b_{\beta\alpha} = a^{\gamma\beta} b_{\beta\alpha}$, and $(23)_2$ has been utilized. Equations (32) are known as Weingarten's formulas and indicate that the partial derivatives of $\boldsymbol{n}$, with respect to the

[19] Strictly speaking, it is the *restriction* of the mapping $\boldsymbol{I} - \boldsymbol{n} \otimes \boldsymbol{n}$ to the domain $\mathcal{T}^*$ that coincides with $\boldsymbol{I}_T$. ($\boldsymbol{I} - \boldsymbol{n} \otimes \boldsymbol{n}$ acts on vectors in the 3-dimensional vector space $\mathcal{E}^*$ and projects them onto $\mathcal{T}^*$.)

Gaussian coordinates u^α, belong to the vector space $\mathcal{T}^*$. Moreover, for a surface curve $\mathcal{C}$ that passes through P, it is clear from (32) that

$$\dot{\boldsymbol{n}} = \boldsymbol{n}_{,\alpha}\, \dot{u}^\alpha = -b_{\beta\alpha} v^\alpha \boldsymbol{a}^\beta = -b^\beta_\alpha v^\alpha \boldsymbol{a}_\beta. \tag{33}$$

Now, for each choice of $\boldsymbol{v}$, corresponding to each of the different curves that pass through P, the vector $\dot{\boldsymbol{n}}$ is a uniquely defined element of $\mathcal{T}^*$; hence, $b_{\beta\alpha} v^\alpha \boldsymbol{a}^\beta$ is independent of the choice of surface coordinates (even though $\boldsymbol{a}^\beta$, v^α, and $b_{\beta\alpha}$ individually depend on the choice of coordinates in definite ways); thus, $b_{\beta\alpha} v^\alpha \boldsymbol{a}^\beta$ is an invariant. Let us define a linear mapping on $\mathcal{T}^*$ by

$$\boldsymbol{b} = b_{\alpha\beta} \boldsymbol{a}^\alpha \otimes \boldsymbol{a}^\beta = b^\alpha_\beta \boldsymbol{a}_\alpha \otimes \boldsymbol{a}^\beta. \tag{34}$$

It follows from (34), (33), and $(17)_3$ that

$$\dot{\boldsymbol{n}} = -\boldsymbol{b}\boldsymbol{v}, \qquad \frac{d\boldsymbol{n}}{ds} = -\boldsymbol{b}\boldsymbol{t}. \tag{35}$$

Even though $\boldsymbol{b}$ is defined in (34) using the coordinates u^α, it is clear from either relation in (35) that $\boldsymbol{b}$ is independent of coordinates. For, if we take any other coordinate system $\bar{u}^\alpha$ and set $\bar{\boldsymbol{b}} = \bar{b}_{\alpha\beta} \bar{\boldsymbol{a}}^\alpha \otimes \bar{\boldsymbol{a}}^\beta$, then $(35)_1$ would imply that $\bar{\boldsymbol{b}}\boldsymbol{v} = \boldsymbol{b}\boldsymbol{v}$ for all $\boldsymbol{v} \in \mathcal{T}^*$, and hence the mappings $\bar{\boldsymbol{b}}$ and $\boldsymbol{b}$ are equal to one another. Thus, $\boldsymbol{b}$ is a surface tensor.[20] For each $\boldsymbol{t} \in \mathcal{T}^*$, $\boldsymbol{b}$ maps $\boldsymbol{v}$ into a vector $d\boldsymbol{n}/ds \in \mathcal{T}^*$, which is equal to the arc rate at which the attitude of the tangent plane $\mathcal{T}$ changes corresponding to travel through P in the direction $\boldsymbol{t}$. (Recall the Gauss map in Fig. 4.) Along the coordinate curves, it is evident from (32) and (34) that $\boldsymbol{n}_{,\alpha} = -\boldsymbol{b}\boldsymbol{a}_\alpha$. The surface tensor $\boldsymbol{b}$, which is symmetric ($\boldsymbol{b}^T = \boldsymbol{b}$), may be called the second fundamental tensor of $\mathcal{S}$.

The curvature vector of $\mathcal{C}$ is $\frac{d\boldsymbol{t}}{ds}$, and it is obvious that

$$\frac{d\boldsymbol{t}}{ds} \cdot \boldsymbol{t} = 0. \tag{36}$$

Further, with the aid of $(17)_1$ and (27), it is clear that

$$\begin{aligned} \frac{d\boldsymbol{t}}{ds} &= \frac{dt^\gamma}{ds}\boldsymbol{a}_\gamma + \boldsymbol{a}_{\alpha,\beta} t^\alpha t^\beta \\ &= \left(\frac{dt^\gamma}{ds} + \Gamma^\gamma_{\alpha\beta} t^\alpha t^\beta\right)\boldsymbol{a}_\gamma + b_{\alpha\beta} t^\alpha t^\beta \boldsymbol{n}. \end{aligned} \tag{37}$$

Thus, the curvature vector has one component which lies in the tangent space $\mathcal{T}^*$ and another which lies along $\boldsymbol{n}$. Note that all curves on $\mathcal{S}$ that share the same tangent vector $\boldsymbol{t}$ at P will have the same normal component of their curvature vectors. Consider the intersection of $\mathcal{S}$ with a plane containing $\boldsymbol{n}$ at P: this will produce a (plane) curve (called a normal section of $\mathcal{S}$) whose curvature at P is $(\boldsymbol{t} \cdot \boldsymbol{b}\boldsymbol{t})\boldsymbol{n} = \kappa_n \boldsymbol{n}$, say. The scalar κ_n is called the curvature of the normal section corresponding to $\boldsymbol{t}$, or simply the *normal curvature*.

[20] It may also be mentioned that it can be deduced from (28) that under a change of surface coordinates, $b_{\alpha\beta}$ transform covariantly. In contrast, the Christoffel symbols do *not* transform tensorially (see, e.g.: [34, Sec. 32]). However, as will be seen in Sect. 4, a suitable combination of partial derivatives and Christoffel terms furnish a derivative which is tensorial.

Substituting (37) into (35), and remembering that $\boldsymbol{t}\cdot\boldsymbol{n}=0$, we find that

$$\boldsymbol{t}\cdot\left(\frac{dt^{\gamma}}{ds}+\Gamma^{\gamma}_{\alpha\beta}t^{\alpha}t^{\beta}\right)\boldsymbol{a}_{\gamma}=0, \tag{38}$$

where use has also been made of (36). Consequently, if we introduce a unit vector $\boldsymbol{e}=\boldsymbol{n}\times\boldsymbol{t}$ to form the Darboux frame, we may write

$$\left(\frac{dt^{\gamma}}{ds}+\Gamma^{\gamma}_{\alpha\beta}t^{\alpha}t^{\beta}\right)\boldsymbol{a}_{\gamma}=\kappa_{g}\boldsymbol{e}. \tag{39}$$

The scalar κ_g is the *geodesic curvature* of $\mathcal{C}$ at P. It thus transpires that the curvature vector of $\mathcal{C}$ may be expressed as

$$\frac{d\boldsymbol{t}}{ds}=\kappa_{g}\boldsymbol{e}+\kappa_{n}\boldsymbol{n}. \tag{40}$$

Since the second fundamental tensor $\boldsymbol{b}$ is symmetric, it possesses two real eigenvalues (which need not be distinct) and corresponding mutually orthogonal eigenvectors. Thus, $\boldsymbol{b}$ has a spectral decomposition $\boldsymbol{b}=\kappa_1\boldsymbol{t}_1\otimes\boldsymbol{t}_1+\kappa_2\boldsymbol{t}_2\otimes\boldsymbol{t}_2$, where the unit eigenvectors $\boldsymbol{t}_\alpha\in\mathcal{T}^*$. The eigenvalue κ_1 is the curvature of a normal section corresponding to $\boldsymbol{t}_1$ (since $\kappa_1=\boldsymbol{t}_1\cdot\boldsymbol{b}\boldsymbol{t}_1$); similarly for κ_2 and $\boldsymbol{t}_2$. These eigenvalues are the principal curvatures of $\mathcal{S}$ at P. Also, the mean curvature H and the Gaussian curvature K are given by $2H=\kappa_1+\kappa_2=\operatorname{tr}\boldsymbol{b}$, $K=\kappa_1\kappa_2=\det\boldsymbol{b}$.

3 Tilting Velocity of Tangent Plane along a Surface Curve. Levi-Civita Frames

In this section, we will examine how adapted frames move along a surface curve and we will show that they all share one component of angular velocity, which is related to the rate at which the tangent plane is sensed to turn as one travels along the curve.

It is clear from (34) and $(35)_{1,2}$ that the vectors $\frac{d\boldsymbol{n}}{ds}$ and $\dot{\boldsymbol{n}}=\dot{s}\frac{d\boldsymbol{n}}{ds}$ always lie in the tangent space $\mathcal{T}^*$. Let

$$\beta=\left\|\frac{d\boldsymbol{n}}{ds}\right\|. \tag{41}$$

Suppose that along the surface curve $\mathcal{C}$ in Fig. 4, $\beta\neq 0$. (We will return to the case $\dot{\boldsymbol{n}}=\boldsymbol{0}$ presently.) Then, we may define a unique unit vector $\boldsymbol{u}\in\mathcal{T}^*$ such that

$$\frac{d\boldsymbol{n}}{ds}=\beta\boldsymbol{u},\qquad \dot{\boldsymbol{n}}=\beta\dot{s}\boldsymbol{u}=\|\dot{\boldsymbol{n}}\|\boldsymbol{u}. \tag{42}$$

Furthermore, the vector $\boldsymbol{g}$, defined by $\boldsymbol{g}=\boldsymbol{n}\times\boldsymbol{u}$, also lies in $\mathcal{T}^*$. The frame $\{\boldsymbol{u},\boldsymbol{g},\boldsymbol{n}\}$ is an adapted frame on $\mathcal{S}$.

Let us denote the angular velocity tensor of the frame $\{\boldsymbol{u},\boldsymbol{g},\boldsymbol{n}\}$ by $\boldsymbol{\Omega}_G$. In accordance with (11), and taking account of the skew-symmetry of $\boldsymbol{\Omega}_G$, for the corresponding components Ω_{Gij} of $\boldsymbol{\Omega}_G$ we have

$$\begin{aligned}
\Omega_{G12}&=\boldsymbol{u}\cdot\dot{\boldsymbol{g}}, &\quad \Omega_{G21}&=\boldsymbol{g}\cdot\dot{\boldsymbol{u}}=-\Omega_{G12},\\
\Omega_{G13}&=\boldsymbol{u}\cdot\dot{\boldsymbol{n}}=\beta\dot{s}, &\quad \Omega_{G31}&=-\beta\dot{s},\\
\Omega_{G23}&=\boldsymbol{g}\cdot\dot{\boldsymbol{n}}=0, &\quad \Omega_{G32}&=\boldsymbol{n}\cdot\dot{\boldsymbol{g}}=0,
\end{aligned} \tag{43}$$

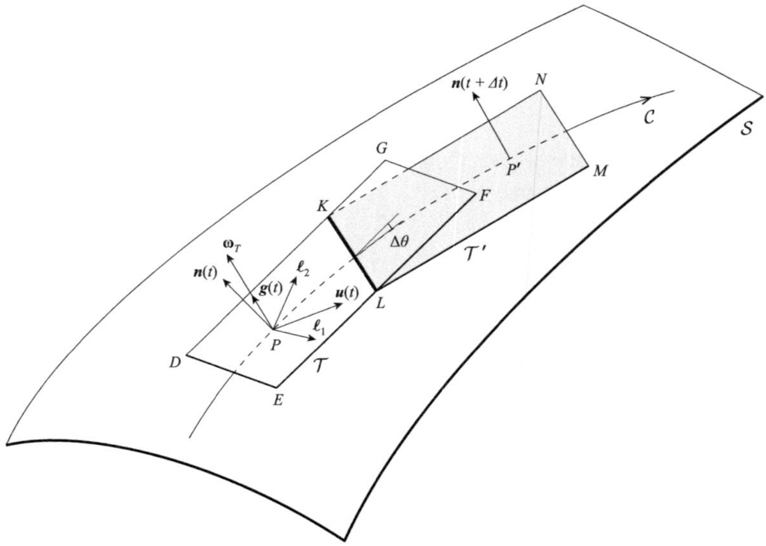

Fig. 5 Two neighboring tangent planes, $\mathcal{T}$ and $\mathcal{T}'$, to a surface $\mathcal{S}$ intersect in a line LK. The limiting position of LK lies along the unit vector $\boldsymbol{g}(t)$

where use has been made of (42). Setting

$$\frac{d\boldsymbol{u}}{ds}\cdot\boldsymbol{g}=\gamma, \tag{44}$$

we then obtain the representation

$$\boldsymbol{\Omega}_G=\left[-\gamma(\boldsymbol{u}\otimes\boldsymbol{g}-\boldsymbol{g}\otimes\boldsymbol{u})+\beta(\boldsymbol{u}\otimes\boldsymbol{n}-\boldsymbol{n}\otimes\boldsymbol{u})\right]\dot{s}. \tag{45}$$

The corresponding angular velocity vector is

$$\boldsymbol{\omega}_G=\Omega_{32}\boldsymbol{u}+\Omega_{13}\boldsymbol{g}+\Omega_{21}\boldsymbol{n}=(\beta\boldsymbol{g}+\gamma\boldsymbol{n})\dot{s}. \tag{46}$$

In view of (10), we then have

$$\begin{aligned}\dot{\boldsymbol{u}}&=\boldsymbol{\Omega}_G\boldsymbol{u}=\boldsymbol{\omega}_G\times\boldsymbol{u}=(\gamma\boldsymbol{g}-\beta\boldsymbol{n})\dot{s},\\ \dot{\boldsymbol{g}}&=\boldsymbol{\Omega}_G\boldsymbol{g}=\boldsymbol{\omega}_G\times\boldsymbol{g}=-\gamma\dot{s}\boldsymbol{u},\\ \dot{\boldsymbol{n}}&=\boldsymbol{\Omega}_G\boldsymbol{n}=\boldsymbol{\omega}_G\times\boldsymbol{n}=\beta\dot{s}\boldsymbol{u},\end{aligned} \tag{47}$$

or in matrix form

$$\begin{pmatrix}\dot{\boldsymbol{u}}\\ \dot{\boldsymbol{g}}\\ \dot{\boldsymbol{n}}\end{pmatrix}=\dot{s}\begin{pmatrix}0 & \gamma & -\beta\\ -\gamma & 0 & 0\\ \beta & 0 & 0\end{pmatrix}\begin{pmatrix}\boldsymbol{u}\\ \boldsymbol{g}\\ \boldsymbol{n}\end{pmatrix}. \tag{48}$$

The geometrical role that the frame $\{\boldsymbol{u},\boldsymbol{g},\boldsymbol{n}\}$ plays is very interesting. To elaborate, consider the tangent planes to $\mathcal{S}$ at P and at a neighboring point P' (Fig. 5). Let $\mathcal{C}$ be a curve passing through P and P', with $\dot{\boldsymbol{n}}\neq\boldsymbol{0}$ at P. The two tangent planes $\mathcal{T}$ and $\mathcal{T}'$ will intersect in a straight line LK which is perpendicular to both $\boldsymbol{n}(t)$ and the unit normal $\boldsymbol{n}(t+\Delta t)$

to $\mathcal{T}'$. The direction of LK is therefore along the vector $\boldsymbol{n}(t) \times \boldsymbol{n}(t+\Delta t)$. The limiting direction of LK as $\Delta t \to 0$ will be parallel to the vector[21]

$$\begin{aligned}\lim_{\Delta t\to 0}\frac{\boldsymbol{n}(t)\times\boldsymbol{n}(t+\Delta t)}{\Delta t} &= \boldsymbol{n}(t)\times\lim_{\Delta t\to 0}\frac{[\boldsymbol{n}(t+\Delta t)-\boldsymbol{n}(t)]}{\Delta t}\\ &= \boldsymbol{n}(t)\times\dot{\boldsymbol{n}}(t)\\ &= \beta\dot{s}\boldsymbol{g},\end{aligned} \tag{49}$$

where $(42)_2$ has been invoked. Thus, the unit vector $\boldsymbol{g}$ lies along the limiting position of the intersection of $\mathcal{T}$ and $\mathcal{T}'$ as P' tends towards P along the curve $\mathcal{C}$.

We observe that the tangent plane $\mathcal{T}$ can be made to coincide with $\mathcal{T}'$ by rotating it about LK in the sense of the right hand rule through a small dihedral angle $\Delta\theta$. We may therefore define an angular velocity $\omega_T = \lim_{\Delta t\to 0}\frac{\Delta\theta}{\Delta t}$, and an angular velocity vector $\boldsymbol{\omega}_T = \omega_T\boldsymbol{g}$, which describes the rate of rotation of the tangent plane at P about KL as one travels along the curve $\mathcal{C}$. We call $\boldsymbol{\omega}_T$ the *tilting velocity* of the tangent plane, and we call $\boldsymbol{g}$ the *spin axis* of the tangent plane.[22]

Given the vector function $\boldsymbol{\omega}_T(t)$ along the curve $\mathcal{C}$ and the corresponding skew-symmetric tensor $\boldsymbol{\Omega}_T$ $(= -\boldsymbol{\varepsilon}\boldsymbol{\omega}_T)$, we may set the initial value problem

$$\dot{\boldsymbol{R}}_T = \boldsymbol{\Omega}_T\boldsymbol{R}_T, \qquad \boldsymbol{R}_T(0) = \boldsymbol{I} \tag{50}$$

for the rotation tensor $\boldsymbol{R}_T$ that generates $\boldsymbol{\omega}_T$ as its angular velocity. Let $\mathcal{T}_0$ be the tangent plane to $\mathcal{S}$ at an arbitrary point P_0, where both t and s are taken to be zero (see Fig. 6). Choose any adapted frame $\{\boldsymbol{\ell}_{01}, \boldsymbol{\ell}_{02}, \boldsymbol{n}_0\}$ at P_0, where $\boldsymbol{n}_0 = \boldsymbol{n}(0)$, and construct a special family of adapted frames to $\mathcal{S}$ along $\mathcal{C}$ by

$$\boldsymbol{\ell}_\alpha = \boldsymbol{R}_T\boldsymbol{\ell}_{\alpha 0}, \qquad \boldsymbol{n} = \boldsymbol{R}_T\boldsymbol{n}_0. \tag{51}$$

It follows from $(51)_{1,2}$, $(50)_1$, and (6) that

$$\dot{\boldsymbol{\ell}}_\alpha = \boldsymbol{\Omega}_T\boldsymbol{\ell}_\alpha = \boldsymbol{\omega}_T \times \boldsymbol{\ell}_\alpha, \qquad \dot{\boldsymbol{n}} = \boldsymbol{\Omega}_T\boldsymbol{n} = \boldsymbol{\omega}_T \times \boldsymbol{n} = \omega_T\boldsymbol{u}. \tag{52}$$

Thus, the adapted frame $\{\boldsymbol{\ell}_1, \boldsymbol{\ell}_2, \boldsymbol{n}\}$ spins about LK, which lies along the vector $\boldsymbol{g}$, at the same rate as the tangent plane tilts about LK.[23] We call $\{\boldsymbol{\ell}_1, \boldsymbol{\ell}_2, \boldsymbol{n}\}$ a *Levi-Civita frame*. For any Levi-Civita frame, it follows from $(42)_2$ and $(52)_2$ that

$$\omega_T = \beta\dot{s}, \tag{53}$$

and hence the angular velocity of every Levi-Civita frame along $\mathcal{C}$ is given by

$$\boldsymbol{\omega}_T = \beta\dot{s}\boldsymbol{g} = \|\dot{\boldsymbol{n}}\|\boldsymbol{g}. \tag{54}$$

[21] This particular step was suggested by do Carmo's treatment of the envelope of a family of tangent planes along a surface curve [25, pp. 195–197]. The envelope itself will be discussed in Sect. 5. Note that if the opposite surface orientation is chosen, the limit in (49) is unaltered.

[22] One way to sense the tilting of the tangent plane is to slide an inverted bottle cap with one finger along the surface of a bottle or bowl. It is worth mentioning that the frame $\{\boldsymbol{u}, \boldsymbol{g}, \boldsymbol{n}\}$ arises naturally in the spatial treatment of areal velocity in particle mechanics [38].

[23] The vector $\boldsymbol{g}$ is an eigenvector of $\boldsymbol{\Omega}_T$ corresponding to the eigenvalue 0. Further, if $\boldsymbol{h}$ is the eigenvector corresponding to the eigenvalue 0, of the relative rotation tensor of the frame $\{\boldsymbol{\ell}_1, \boldsymbol{\ell}_2, \boldsymbol{n}\}$ between time t and time $t + \Delta t$, then $\lim_{\Delta t\to 0}\boldsymbol{h}(t+\Delta t) = \boldsymbol{g}(t)$ [39].

Fig. 6 Rigid motion of a Levi-Civita frame along a surface curve

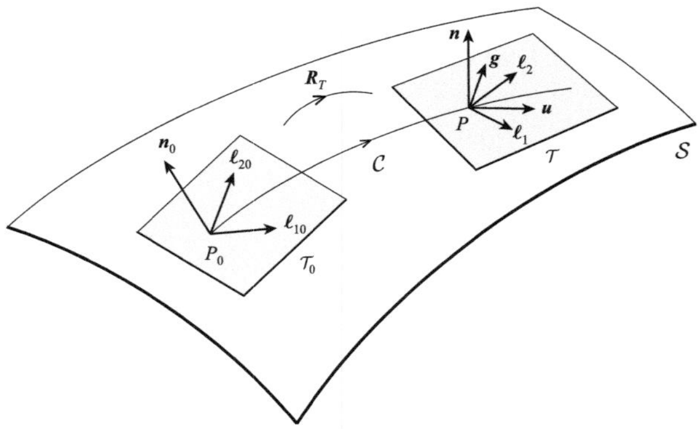

Further,

$$\boldsymbol{\Omega}_T = \|\dot{\boldsymbol{n}}\|(\boldsymbol{u}\otimes\boldsymbol{n} - \boldsymbol{n}\otimes\boldsymbol{u}) = \dot{\boldsymbol{n}}\otimes\boldsymbol{n} - \boldsymbol{n}\otimes\dot{\boldsymbol{n}}$$
$$= -(\boldsymbol{b}\boldsymbol{v})\otimes\boldsymbol{n} + \boldsymbol{n}\otimes(\boldsymbol{b}\boldsymbol{v})$$
$$= -\boldsymbol{b}(\boldsymbol{v}\otimes\boldsymbol{n}) + (\boldsymbol{n}\otimes\boldsymbol{v})\boldsymbol{b}, \tag{55}$$

where $(42)_2$ and $(35)_1$ have been used, along with the symmetry of $\boldsymbol{b}$. If $\dot{\boldsymbol{n}}$ vanishes, we may take $\boldsymbol{\omega}_T$ to be zero.[24]

If we write $\boldsymbol{\omega}_T$ as[25]

$$\boldsymbol{\omega}_T = \omega_{T\alpha}\boldsymbol{\ell}_\alpha, \tag{56}$$

where $\omega_{T\alpha} = \boldsymbol{\omega}_T\cdot\boldsymbol{\ell}_\alpha$, it is obvious that

$$\dot{\boldsymbol{\ell}}_1 = \boldsymbol{\omega}_T\times\boldsymbol{\ell}_1 = -\omega_{T2}\boldsymbol{n}, \qquad \dot{\boldsymbol{\ell}}_2 = \boldsymbol{\omega}_T\times\boldsymbol{\ell}_2 = \omega_{T1}\boldsymbol{n}. \tag{57}$$

Let "□" signify a corotational rate measured on a Levi-Civita frame. Thus, in accordance with the general relation (12), for a Levi-Civita frame we have

$$\dot{\boldsymbol{u}} = \overset{\square}{\boldsymbol{u}} + \boldsymbol{\Omega}_T\boldsymbol{u} = \overset{\square}{\boldsymbol{u}} + \boldsymbol{\omega}_T\times\boldsymbol{u}. \tag{58}$$

We note that

$$\dot{\boldsymbol{\omega}}_T = \overset{\square}{\boldsymbol{\omega}}_T. \tag{59}$$

Moreover, for any vector $\boldsymbol{w}$ belonging to the tangent space $\mathcal{T}^*$,

$$\dot{\boldsymbol{w}} = \overset{\square}{\boldsymbol{w}} + \boldsymbol{\Omega}_T\boldsymbol{w} = \overset{\square}{\boldsymbol{w}} + \boldsymbol{\omega}_T\times\boldsymbol{w}$$
$$= \dot{w}_\alpha\boldsymbol{\ell}_\alpha + \boldsymbol{\omega}_T\times\boldsymbol{w}$$

[24] If the surface $\mathcal{S}$ is a plane, then $\dot{\boldsymbol{n}}$ is zero everywhere. Also, $\dot{\boldsymbol{n}}$ may vanish along certain curves on a surface, or at some points of surface curves. Whenever $\dot{\boldsymbol{n}} = \boldsymbol{0}$, the vectors $\boldsymbol{u}$ and $\boldsymbol{g}$ can be chosen from any adapted frame, as convenient.

[25] Correspondingly, on Levi-Civita frames, $\boldsymbol{\Omega}_T$ may be expressed as $\boldsymbol{\Omega}_T = \omega_{T1}(\boldsymbol{n}\otimes\boldsymbol{\ell}_2 - \boldsymbol{\ell}_2\otimes\boldsymbol{n}) + \omega_{T2}(\boldsymbol{\ell}_1\otimes\boldsymbol{n} - \boldsymbol{n}\otimes\boldsymbol{\ell}_1)$.

$$
\begin{aligned}
&= \overset{\square}{\boldsymbol{w}} + (\omega_{T1} w_2 - \omega_{T2} w_1)\boldsymbol{n} \\
&= \overset{\square}{\boldsymbol{w}} - (\boldsymbol{w} \cdot \dot{\boldsymbol{n}})\boldsymbol{n},
\end{aligned} \tag{60}
$$

where (56) and $(52)_2$ have been employed. Thus, for such a vector $\boldsymbol{w}$, $\overset{\square}{\boldsymbol{w}} \in \mathcal{T}^*$, while the other component lies along $\boldsymbol{n}$. It should be observed that $\overset{\square}{\boldsymbol{w}}$ is invariant under every change of surface coordinates.

If $\boldsymbol{w} \in \mathcal{T}^*$ is fixed on a (moving) Levi-Civita frame, then $\boldsymbol{w}$ is said to be *Levi-Civita transported*, or *parallel transported*, along the surface curve $\mathcal{C}$. A necessary and sufficient condition for Levi-Civita transport (or Levi-Civita parallelism) of a vector is

$$
\overset{\square}{\boldsymbol{w}} = \boldsymbol{0}. \tag{61}
$$

By virtue of (60) and $(35)_1$, the condition (61) is equivalent to

$$
\dot{\boldsymbol{w}} = -(\boldsymbol{w} \cdot \dot{\boldsymbol{n}})\boldsymbol{n} = (\boldsymbol{w} \cdot \boldsymbol{b}\boldsymbol{v})\boldsymbol{n}, \tag{62}
$$

which implies that $\dot{\boldsymbol{w}}$ lies along the surface normal.[26] Clearly, the basis vectors $\boldsymbol{\ell}_1$ and $\boldsymbol{\ell}_2$ themselves are parallel transported. If a vector $\boldsymbol{w} \in \mathcal{T}^*$ is Levi-Civita transported along a curve, it is evident that its magnitude $(\sqrt{w_1^2 + w_2^2})$ remains constant. Moreover, if two vectors $\boldsymbol{w}, \boldsymbol{w}' \in \mathcal{T}^*$ are parallel transported along a curve, the inner product $\boldsymbol{w} \cdot \boldsymbol{w}' = w_1 w_1' + w_2 w_2'$ remains constant, and so also does the angle between them. If $\mathcal{S}$ is a plane, $\boldsymbol{\omega}_T = \boldsymbol{0}$, and $\boldsymbol{\ell}_1$ and $\boldsymbol{\ell}_2$ may be chosen to be any fixed basis in the plane $\mathcal{S}$, and Levi-Civita parallelism reduces to ordinary Euclidean parallelism.

For any adapted frame $\{\boldsymbol{e}_1, \boldsymbol{e}_2, \boldsymbol{n}\}$, we may introduce an angle ψ in the tangent plane $\mathcal{T}$ such that

$$
\begin{pmatrix} \boldsymbol{e}_1 \\ \boldsymbol{e}_2 \end{pmatrix} = \begin{pmatrix} \cos\psi & \sin\psi \\ -\sin\psi & \cos\psi \end{pmatrix} \begin{pmatrix} \boldsymbol{\ell}_1 \\ \boldsymbol{\ell}_2 \end{pmatrix}. \tag{63}
$$

Hence, with the aid of (60), we obtain

$$
\dot{\boldsymbol{e}}_1 = \dot{\psi}\boldsymbol{e}_2 + \boldsymbol{\omega}_T \times \boldsymbol{e}_1, \qquad \dot{\boldsymbol{e}}_2 = -\dot{\psi}\boldsymbol{e}_1 + \boldsymbol{\omega}_T \times \boldsymbol{e}_2. \tag{64}
$$

The angular velocity of the adapted frame $\{\boldsymbol{e}_1, \boldsymbol{e}_2, \boldsymbol{n}\}$ relative to a Levi-Civita frame is denoted by $\hat{\boldsymbol{\omega}}$ and is given by [40]

$$
\hat{\boldsymbol{\omega}} = \dot{\psi}\boldsymbol{n}. \tag{65}
$$

Then,

$$
\dot{\boldsymbol{e}}_\alpha = (\hat{\boldsymbol{\omega}} + \boldsymbol{\omega}_T) \times \boldsymbol{e}_\alpha, \qquad \dot{\boldsymbol{n}} = (\hat{\boldsymbol{\omega}} + \boldsymbol{\omega}_T) \times \boldsymbol{n}, \tag{66}
$$

where $(52)_2$ has also been invoked. Thus, the angular velocity $\boldsymbol{\omega}$ of any adapted frame $\{\boldsymbol{e}_1, \boldsymbol{e}_2, \boldsymbol{n}\}$ may be expressed in the form

$$
\boldsymbol{\omega} = \boldsymbol{\omega}_T + \hat{\boldsymbol{\omega}} = \|\dot{\boldsymbol{n}}\|\boldsymbol{g} + \dot{\psi}\boldsymbol{n}, \tag{67}
$$

[26] Since $\boldsymbol{w} \cdot \boldsymbol{n} = 0$, it follows that $\dot{\boldsymbol{w}} \cdot \boldsymbol{n} = -\boldsymbol{w} \cdot \dot{\boldsymbol{n}}$, and the condition (62) may be written as $\dot{\boldsymbol{w}} = (\dot{\boldsymbol{w}} \cdot \boldsymbol{n})\boldsymbol{n}$. Hence, if $\dot{\boldsymbol{w}}$ lies along $\boldsymbol{n}$, then $\boldsymbol{w}$ will be parallel transported.

where (54) has also been utilized. Equation (67) indicates that the angular velocity of any adapted frame is equal to the tilting velocity $\boldsymbol{\omega}_T$ of the tangent plane plus the relative angular velocity $\hat{\boldsymbol{\omega}}$, which is a spin about the surface normal $\boldsymbol{n}$. All adapted frames have the same component of angular velocity $\boldsymbol{\omega}_T$ in the tangent space $\mathcal{T}^*$, but they can have different spins about $\boldsymbol{n}$.

The corotational rates of $\boldsymbol{e}_1$, $\boldsymbol{e}_2$, and $\boldsymbol{n}$ with respect to a Levi-Civita frame are

$$\overset{\square}{\boldsymbol{e}}_1 = \dot{\boldsymbol{e}}_1 - \boldsymbol{\omega}_T \times \boldsymbol{e}_1 = \dot{\psi}\boldsymbol{e}_2 = \hat{\boldsymbol{\omega}} \times \boldsymbol{e}_1, \qquad \overset{\square}{\boldsymbol{e}}_2 = -\dot{\psi}\boldsymbol{e}_1 = \hat{\boldsymbol{\omega}} \times \boldsymbol{e}_2, \qquad \overset{\square}{\boldsymbol{n}} = \boldsymbol{0}, \tag{68}$$

where use has been made of (58), $(64)_{1,2}$, (65), and $(52)_2$.

If $\boldsymbol{W}$ is any surface tensor, then

$$\begin{aligned}\dot{\boldsymbol{W}} &= \dot{W}_{\alpha\beta}\boldsymbol{\ell}_\alpha \otimes \boldsymbol{\ell}_\beta + W_{\alpha\beta}(\boldsymbol{\Omega}_T\boldsymbol{\ell}_\alpha) \otimes \boldsymbol{\ell}_\beta + W_{\alpha\beta}\boldsymbol{\ell}_\alpha \otimes (\boldsymbol{\Omega}_T\boldsymbol{\ell}_\beta) \\ &= \overset{\square}{\boldsymbol{W}} + \boldsymbol{\Omega}_T\boldsymbol{W} - \boldsymbol{W}\boldsymbol{\Omega}_T, \\ &= \overset{\square}{\boldsymbol{W}} - (\boldsymbol{n} \otimes \dot{\boldsymbol{n}})\boldsymbol{W} - \boldsymbol{W}(\dot{\boldsymbol{n}} \otimes \boldsymbol{n}) \\ &= \overset{\square}{\boldsymbol{W}} - (\boldsymbol{W}\dot{\boldsymbol{n}}) \otimes \boldsymbol{n} - \boldsymbol{n} \otimes \left(\boldsymbol{W}^T\dot{\boldsymbol{n}}\right),\end{aligned} \tag{69}$$

where $(52)_1$ and (55) have been employed, as well as the fact that $(W_{\alpha\beta}\boldsymbol{\ell}_\alpha \otimes \boldsymbol{\ell}_\beta)(\boldsymbol{n} \otimes \dot{\boldsymbol{n}}) = 0$ and $W_{\alpha\beta}(\dot{\boldsymbol{n}} \otimes \boldsymbol{n})(\boldsymbol{\ell}_\alpha \otimes \boldsymbol{\ell}_\beta) = 0$. We observe that $\overset{\square}{\boldsymbol{W}}$ is a surface tensor.

A surface tensor $\boldsymbol{W}$ is said to be Levi-Civita transported if and only if its components on a Levi-Civita frame remain constant. A necessary and sufficient condition for Levi-Civita transport of $\boldsymbol{W}$ is that

$$\overset{\square}{\boldsymbol{W}} = \boldsymbol{0}. \tag{70}$$

Since the identity tensor $\boldsymbol{I}_T$ on $\mathcal{T}^*$ has a representation $\boldsymbol{I}_T = \delta_{\alpha\beta}\boldsymbol{\ell}_\alpha \otimes \boldsymbol{\ell}_\beta$, it is obvious that

$$\overset{\square}{\boldsymbol{I}}_T = \boldsymbol{0}. \tag{71}$$

Thus, $\boldsymbol{I}_T$ is Levi-Civita transported along every curve $\mathcal{C}$ on $\mathcal{S}$.

In the case of the frame $\{\boldsymbol{u}, \boldsymbol{g}, \boldsymbol{n}\}$, it is clear from (46) and (54) that the angular velocity of $\{\boldsymbol{u}, \boldsymbol{g}, \boldsymbol{n}\}$ with respect to a Levi-Civita frame $\{\boldsymbol{\ell}_1, \boldsymbol{\ell}_2, \boldsymbol{n}\}$ is given by

$$\hat{\boldsymbol{\omega}}_G = \dot{\psi}_G\boldsymbol{n} = \gamma\dot{s}\boldsymbol{n}. \tag{72}$$

In general, $\boldsymbol{g}$ and $\boldsymbol{u}$ change their directions relative to a Levi-Civita frame as one travels along the curve $\mathcal{C}$. Indeed, it is evident from $(68)_{1,2}$ and (72) that

$$\overset{\square}{\boldsymbol{u}} = \hat{\boldsymbol{\omega}}_G \times \boldsymbol{u} = \gamma\dot{s}\boldsymbol{g}, \qquad \overset{\square}{\boldsymbol{g}} = \hat{\boldsymbol{\omega}}_G \times \boldsymbol{g} = -\gamma\dot{s}\boldsymbol{u}. \tag{73}$$

Clearly, $\{\boldsymbol{u}, \boldsymbol{g}, \boldsymbol{n}\}$ is not a Levi-Civita frame unless $\gamma = 0$.

Consider again the Darboux frame $\{\boldsymbol{t}, \boldsymbol{e}, \boldsymbol{n}\}$, and let $\boldsymbol{\Omega}_D$ and $\boldsymbol{\omega}_D$ be its angular velocity tensor and vector. In view of (11) and (40), the corresponding components of $\boldsymbol{\Omega}_D$ are

$$\begin{aligned}\Omega_{D21} &= \boldsymbol{e} \cdot \dot{\boldsymbol{t}} = \kappa_g\dot{s} = -\Omega_{D12}, \\ \Omega_{D31} &= \boldsymbol{n} \cdot \dot{\boldsymbol{t}} = \kappa_n\dot{s} = -\Omega_{D13}, \\ \Omega_{D23} &= \boldsymbol{e} \cdot \dot{\boldsymbol{n}} = -\Omega_{D32}.\end{aligned} \tag{74}$$

Let[27]

$$\tau = -\frac{d\boldsymbol{n}}{ds} \cdot \boldsymbol{e} (= \boldsymbol{bt} \cdot \boldsymbol{e}). \tag{75}$$

The angular velocity vector of the Darboux frame is given by

$$\begin{aligned} \boldsymbol{\omega}_D &= \omega_{D1}\boldsymbol{t} + \omega_{D2}\boldsymbol{e} + \omega_{D3}\boldsymbol{n} \\ &= \Omega_{D32}\boldsymbol{t} + \Omega_{D13}\boldsymbol{e} + \Omega_{D21}\boldsymbol{n} \\ &= \dot{s}(\tau\boldsymbol{t} - \kappa_n\boldsymbol{e} + \kappa_g\boldsymbol{n}). \end{aligned} \tag{76}$$

Hence,

$$\begin{aligned} \dot{\boldsymbol{t}} &= \boldsymbol{\omega}_D \times \boldsymbol{t} = \dot{s}(\kappa_g\boldsymbol{e} + \kappa_n\boldsymbol{n}), \\ \dot{\boldsymbol{e}} &= \boldsymbol{\omega}_D \times \boldsymbol{e} = \dot{s}(-\kappa_g\boldsymbol{t} + \tau\boldsymbol{n}), \\ \dot{\boldsymbol{n}} &= \boldsymbol{\omega}_D \times \boldsymbol{n} = \dot{s}(-\kappa_n\boldsymbol{t} - \tau\boldsymbol{e}), \end{aligned} \tag{77}$$

or, in matrix form,

$$\begin{pmatrix} \dot{\boldsymbol{t}} \\ \dot{\boldsymbol{e}} \\ \dot{\boldsymbol{n}} \end{pmatrix} = \dot{s} \begin{pmatrix} 0 & \kappa_g & \kappa_n \\ -\kappa_g & 0 & \tau \\ -\kappa_n & -\tau & 0 \end{pmatrix} \begin{pmatrix} \boldsymbol{t} \\ \boldsymbol{e} \\ \boldsymbol{n} \end{pmatrix}. \tag{78}$$

Applying the decomposition (67) to the Darboux frame, we may write

$$\boldsymbol{\omega}_D = \boldsymbol{\omega}_T + \hat{\boldsymbol{\omega}}_D = \|\dot{\boldsymbol{n}}\|\boldsymbol{g} + \dot{\psi}_D\boldsymbol{n}, \tag{79}$$

where $\hat{\boldsymbol{\omega}}_D$ is the angular velocity of the Darboux frame relative to a Levi-Civita frame, and ψ_D is the angle between the vectors $\boldsymbol{t}$ and $\boldsymbol{\ell}_1$. From (76) and (79), it follows that

$$\boldsymbol{\omega}_D \cdot \boldsymbol{n} = \dot{\psi}_D = \kappa_g\dot{s}, \qquad \frac{d\psi_D}{ds} = \kappa_g, \qquad \hat{\boldsymbol{\omega}}_D = \kappa_g\dot{s}\boldsymbol{n}, \tag{80}$$

and

$$\boldsymbol{\omega}_T = \|\dot{\boldsymbol{n}}\|\boldsymbol{g} = \dot{s}(\tau\boldsymbol{t} - \kappa_n\boldsymbol{e}). \tag{81}$$

The Darboux frame is not a Levi-Civita frame unless κ_g vanishes along C.

It follows from (42) and $(77)_3$ that

$$\dot{\boldsymbol{n}} = \|\dot{\boldsymbol{n}}\|\boldsymbol{u} = \beta\dot{s}\boldsymbol{u} = -\dot{s}(\kappa_n\boldsymbol{t} + \tau\boldsymbol{e}). \tag{82}$$

[27] Some authors call τ the *geodesic torsion* of the surface curve, while others take the opposite sign (see, e.g., [13, Vol. 3], [25], and [28]).

Hence,[28]

$$\|\dot{\boldsymbol{n}}\|^2 = \dot{s}^2\left(\kappa_n^2 + \tau^2\right), \qquad \beta^2 = \kappa_n^2 + \tau^2. \tag{83}$$

The Levi-Civita corotational rates of $\boldsymbol{t}$ and $\boldsymbol{e}$ may be found from $(68)_{1,2}$ and $(80)_3$:

$$\overset{\square}{\boldsymbol{t}} = \hat{\boldsymbol{\omega}}_D \times \boldsymbol{t} = \kappa_g \dot{s} \boldsymbol{e}, \qquad \overset{\square}{\boldsymbol{e}} = \hat{\boldsymbol{\omega}}_D \times \boldsymbol{e} = -\kappa_g \dot{s} \boldsymbol{t}. \tag{84}$$

The results $(80)_2$ and $(84)_{1,2}$ indicate that the arc rate of rotation of the Darboux frame relative to the Levi-Civita frame and the corotational rates $\overset{\square}{\boldsymbol{t}}$ and $\overset{\square}{\boldsymbol{e}}$ are determined by the geodesic curvature κ_g of $\mathcal{C}$.[29] Geodesics on $\mathcal{S}$ may be characterized by the condition that $\kappa_g = 0$ along them; equivalently,

$$\overset{\square}{\boldsymbol{t}} = \boldsymbol{0} \tag{85}$$

along a geodesic. Thus, the unit tangent vector to a geodesic is Levi-Civita transported; this is called "auto-parallelism." The Darboux frame along a geodesic maintains a constant angle ψ_D relative to a Levi-Civita frame and is itself a Levi-Civita frame.

Example 3.1 The case of parallel transport around a line of latitude on a fixed sphere is of particular physical interest. Introducing spherical coordinates $\{r, \phi, \theta\}$, we have

$$\boldsymbol{x} = r(\sin\phi\cos\theta\boldsymbol{E}_1 + \sin\phi\sin\theta\boldsymbol{E}_2 + \cos\phi\boldsymbol{E}_3) = r\boldsymbol{e}_r, \tag{86}$$

where $\boldsymbol{e}_r$ is a radial unit vector. A right-handed orthonormal basis $\{\boldsymbol{e}_r, \boldsymbol{e}_\phi, \boldsymbol{e}_\theta\}$ is completed by

$$\begin{aligned} \boldsymbol{e}_\phi &= \cos\phi\cos\theta\boldsymbol{E}_1 + \cos\phi\sin\theta\boldsymbol{E}_2 - \sin\phi\boldsymbol{E}_3, \\ \boldsymbol{e}_\theta &= -\sin\theta\boldsymbol{E}_1 + \cos\theta\boldsymbol{E}_2. \end{aligned} \tag{87}$$

The velocities of the basis vectors are given by

$$\dot{\boldsymbol{e}}_r = \boldsymbol{\omega} \times \boldsymbol{e}_r, \qquad \dot{\boldsymbol{e}}_\phi = \boldsymbol{\omega} \times \boldsymbol{e}_\phi, \qquad \dot{\boldsymbol{e}}_\theta = \boldsymbol{\omega} \times \boldsymbol{e}_\theta, \tag{88}$$

where

$$\boldsymbol{\omega} = \dot{\theta}\boldsymbol{E}_3 + \dot{\phi}\boldsymbol{e}_\theta = \dot{\theta}\cos\phi\boldsymbol{e}_r - \dot{\theta}\sin\phi\boldsymbol{e}_\phi + \dot{\phi}\boldsymbol{e}_\theta \tag{89}$$

is the angular velocity of the frame.

The frame $\{\boldsymbol{e}_r, \boldsymbol{e}_\phi, \boldsymbol{e}_\theta\}$ is an adapted frame on the sphere ($\boldsymbol{n} = \boldsymbol{e}_r$). Applying the relations (67) and (65) to (89), we obtain

$$\hat{\boldsymbol{\omega}} = \dot{\theta}\cos\phi\boldsymbol{e}_r, \qquad \boldsymbol{\omega}_T = -\dot{\theta}\sin\phi\boldsymbol{e}_\phi + \dot{\phi}\boldsymbol{e}_\theta. \tag{90}$$

---

[28] A referee kindly brought to my attention a paper by Beetle [41], in which an elaborated version of the formula $(83)_2$ is derived. For the left-hand side of $(83)_2$, Beetle gives the expression $2H\kappa_n - K$ $[= (\kappa_1 + \kappa_2)\kappa_n - \kappa_1\kappa_2$, where κ_1 and κ_2 are the principal curvatures]. But $\kappa_n = \boldsymbol{t} \cdot \boldsymbol{bt} = \kappa_1 t_1^2 + \kappa_2 t_2^2$, where $t_1 = \boldsymbol{t} \cdot \boldsymbol{t}_1$, $t_2 = \boldsymbol{t} \cdot \boldsymbol{t}_2$ and $\boldsymbol{t}_1$, $\boldsymbol{t}_2$ are unit eigenvectors of $\boldsymbol{b}$ (see the last paragraph in Sect. 2). Also, in view of (41) and $(35)_2$, $\beta^2 = \frac{d\boldsymbol{n}}{ds} \cdot \frac{d\boldsymbol{n}}{ds} = \boldsymbol{bt} \cdot \boldsymbol{bt} = \kappa_1^2 t_1^2 + \kappa_2^2 t_2^2$. Noting that $1 = \boldsymbol{t} \cdot \boldsymbol{t} = t_1^2 + t_2^2$, it is then seen that $2H\kappa_n - K = \beta^2$. Beetle [41] deduces a number of interesting geometrical results from his formula.

[29] The relations $(84)_{1,2}$ are the surface counterparts of the Serret-Frenet formulas for a plane curve.

Along a line of latitude, $\phi = \text{const.}$, and $(90)_{1,2}$ reduce to[30]

$$\hat{\boldsymbol{\omega}} = \dot{\theta} \cos\phi \boldsymbol{e}_r, \qquad \boldsymbol{\omega}_T = -\dot{\theta} \sin\phi \boldsymbol{e}_\phi. \tag{91}$$

It follows from (53), (54) and $(91)_2$ that

$$\omega_T = \dot{\theta} \sin\phi, \qquad \boldsymbol{g} = -\boldsymbol{e}_\theta. \tag{92}$$

Thus, along a line of latitude, the tangent plane turns about the unit vector $\boldsymbol{g} = -\boldsymbol{e}_\theta$ with angular velocity $\omega_T = \dot{\theta} \sin\phi$. The frame $\{\boldsymbol{e}_r, \boldsymbol{e}_\phi, \boldsymbol{e}_\theta\}$ spins with an angular velocity $\hat{\boldsymbol{\omega}} = \dot{\theta} \cos\theta \boldsymbol{e}_r$ relative to a Levi-Civita frame, and is not parallel-transported (except trivially at the poles).

Consider a Foucault pendulum located at latitude λ and performing small oscillations. As a result of the earth's angular velocity Ω, the elliptical trace of the bob's motion on a local horizontal plane spins about the local vertical with angular velocity $-\Omega \sin\lambda \boldsymbol{e}_r$ [42, 43]. But, the angular velocity of a Levi-Civita frame relative to the $\{\boldsymbol{e}_r, \boldsymbol{e}_\phi, \boldsymbol{e}_\theta\}$ frame is $-\hat{\boldsymbol{\omega}} = -\dot{\theta} \cos\phi \boldsymbol{e}_r = -\Omega \sin\lambda \boldsymbol{e}_r$. Thus, the elliptical trace of the Foucault pendulum is parallel transported around the line of latitude on the fixed sphere.[31] The Levi-Civita frame completes one revolution relative to the $\{\boldsymbol{e}_r, \boldsymbol{e}_\phi, \boldsymbol{e}_\theta\}$ frame in a period $2\pi/\Omega \sin\lambda$, or in (1 sidereal day)/ $\sin\lambda$.

Example 3.2 The standard parametrization of the Möbius band (Fig. 7) is[32]

$$\begin{aligned} \boldsymbol{x}(t, \eta) &= \boldsymbol{e}_R + \eta \boldsymbol{f}, \\ \boldsymbol{e}_R &= \cos t \boldsymbol{E}_1 + \sin t \boldsymbol{E}_2, \\ \boldsymbol{f} &= \sin \frac{t}{2} \boldsymbol{e}_R + \cos \frac{t}{2} \boldsymbol{E}_3. \end{aligned} \tag{93}$$

Here, we have taken a polar coordinate $\theta = t$, and since $\boldsymbol{e}_R(t)$ is a unit vector, $s = t$ also, and hence $\dot{s} = 1$. The coordinate η $(-\frac{1}{2} < \eta < \frac{1}{2})$ measures signed distance along a ruling.

Clearly,

$$\dot{\boldsymbol{e}}_R = \boldsymbol{t}. \qquad \boldsymbol{t} \cdot \boldsymbol{f} = 0, \qquad \dot{\boldsymbol{f}} = \sin \frac{t}{2} \boldsymbol{t} + \frac{1}{2} \cos \frac{t}{2} \boldsymbol{e}_R - \frac{1}{2} \sin \frac{t}{2} \boldsymbol{E}_3. \tag{94}$$

The basis vectors $\boldsymbol{a}_1$ and $\boldsymbol{a}_2$ on the band are

$$\boldsymbol{a}_1 = \frac{\partial \boldsymbol{x}}{\partial t} = \boldsymbol{t} + \eta \dot{\boldsymbol{f}}, \qquad \boldsymbol{a}_2 = \frac{\partial \boldsymbol{x}}{\partial \eta} = \boldsymbol{f}. \tag{95}$$

Along the circle, $\eta = 0$ and $\boldsymbol{a}_1$ reduces to

$$\boldsymbol{a}_1 = \boldsymbol{t}. \tag{96}$$

[30] The Darboux frame along a line of latitude is $\{\boldsymbol{e}_\theta, -\boldsymbol{e}_\phi, \boldsymbol{e}_r\}$.

[31] See [44, 45].

[32] See e.g.: [13, Vol. 3], [25]. This embedding of the Möbius band has negative Gaussian curvature [25], and is not the same band as one makes from a paper strip. Equation $(93)_1$ describes the Möbius band as a ruled surface, i.e., one that can be swept out by the motion of a straight line (the *ruling*) along a curve (the *directrix*), in this case a unit circle. The unit vector $\boldsymbol{f}(t)$ may be referred to as the ruling vector.

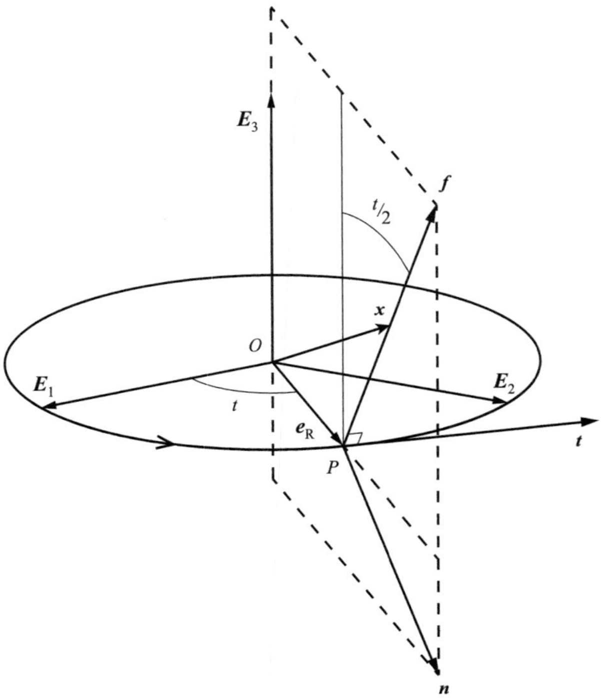

Fig. 7 A Möbius band can be formed by sweeping a line around a unit circle

Hence, along the circle, a local unit normal to the band is provided by[33]

$$\boldsymbol{n} = \boldsymbol{t} \times \boldsymbol{f}. \tag{97}$$

Consequently, the right-handed orthonormal frame $\{\boldsymbol{t}, \boldsymbol{f}, \boldsymbol{n}\}$ is actually a Darboux frame, with

$$\boldsymbol{e} = \boldsymbol{f}. \tag{98}$$

It is obvious that

$$\begin{pmatrix} \boldsymbol{e} \\ \boldsymbol{n} \end{pmatrix} = \begin{pmatrix} \sin\frac{t}{2} & \cos\frac{t}{2} \\ \cos\frac{t}{2} & -\sin\frac{t}{2} \end{pmatrix} \begin{pmatrix} \boldsymbol{e}_R \\ \boldsymbol{E}_3 \end{pmatrix}. \tag{99}$$

Further,

$$\begin{aligned} \dot{\boldsymbol{t}} &= -\boldsymbol{e}_R = -\left(\sin\frac{t}{2}\boldsymbol{e} + \cos\frac{t}{2}\boldsymbol{n} \right), \\ \dot{\boldsymbol{e}} &= \sin\frac{t}{2}\boldsymbol{t} + \frac{1}{2}\boldsymbol{n}, \\ \dot{\boldsymbol{n}} &= \cos\frac{t}{2}\boldsymbol{t} - \frac{1}{2}\boldsymbol{e}. \end{aligned} \tag{100}$$

[33] Of course, $\boldsymbol{e}$ changes from $\boldsymbol{E}_3$ to $-\boldsymbol{E}_3$ and $\boldsymbol{n}$ changes from $\boldsymbol{E}_1$ to $-\boldsymbol{E}_1$ when t goes from 0 to 2π.

It follows from $(74)_{1,2,3}$, (76), and $(100)_{1,2,3}$ that the angular velocity of the Darboux frame is

$$\boldsymbol{\omega}_D = \frac{1}{2}\boldsymbol{t} + \cos\frac{t}{2}\boldsymbol{e} - \sin\frac{t}{2}\boldsymbol{n} \tag{101}$$

and

$$\tau = \frac{1}{2}, \qquad \kappa_n = -\cos\frac{t}{2}, \qquad \kappa_g = -\sin\frac{t}{2}. \tag{102}$$

By virtue of the decomposition (79), it is clear from (101) that the angular velocity of the Darboux frame relative to a Levi-Civita frame $\{\boldsymbol{\ell}_1, \boldsymbol{\ell}_2, \boldsymbol{n}\}$ is now

$$\hat{\boldsymbol{\omega}}_D = \dot{\psi}_D\boldsymbol{n} = -\sin\frac{t}{2}\boldsymbol{n}, \tag{103}$$

and that the angular velocity of every Levi-Civita frame along the circle of the Möbius band is

$$\boldsymbol{\omega}_T = \beta\boldsymbol{g} = \frac{1}{2}\boldsymbol{t} + \cos\frac{t}{2}\boldsymbol{e}, \tag{104}$$

where (54) has been noted. It follows from (103) that

$$\psi_D = 2\cos\frac{t}{2} + \text{const.} \tag{105}$$

We may construct a Levi-Civita frame along the circle by taking

$$\boldsymbol{\ell}_{10} = \boldsymbol{t}(0) = \boldsymbol{E}_2, \qquad \boldsymbol{\ell}_{20} = \boldsymbol{e}(0) = \boldsymbol{E}_3, \tag{106}$$

so that the constant of integration in (105) is -2 and

$$\psi_D = 2\left(\cos\frac{t}{2} - 1\right). \tag{107}$$

Then, by utilizing (63) for the Darboux frame, we obtain

$$\begin{pmatrix} \boldsymbol{\ell}_1 \\ \boldsymbol{\ell}_2 \end{pmatrix} = \begin{pmatrix} \cos\psi_D & -\sin\psi_D \\ \sin\psi_D & \cos\psi_D \end{pmatrix} \begin{pmatrix} \boldsymbol{t} \\ \boldsymbol{e} \end{pmatrix}. \tag{108}$$

In this example, κ_g is not identically zero, and the Darboux frame is not parallel transported around the circle.

At $t = 2\pi$, the Levi-Civita frame is

$$\begin{aligned} \boldsymbol{\ell}_1(2\pi) &= \cos(-4)\boldsymbol{E}_2 + \sin(-4)\boldsymbol{E}_3 = \boldsymbol{R}_T(2\pi)\boldsymbol{E}_2, \\ \boldsymbol{\ell}_2(2\pi) &= \sin(-4)\boldsymbol{E}_2 - \cos(-4)\boldsymbol{E}_3 = \boldsymbol{R}_T(2\pi)\boldsymbol{E}_3, \\ \boldsymbol{n}(2\pi) &= -\boldsymbol{E}_1 = \boldsymbol{R}_T(2\pi)\boldsymbol{E}_1. \end{aligned} \tag{109}$$

Hence, by the relation (3),

$$\boldsymbol{R}_T(2\pi) = -\boldsymbol{E}_1 \otimes \boldsymbol{E}_1 + \cos(-4)(\boldsymbol{E}_2 \otimes \boldsymbol{E}_2 - \boldsymbol{E}_3 \otimes \boldsymbol{E}_3) + \sin(-4)(\boldsymbol{E}_2 \otimes \boldsymbol{E}_3 + \boldsymbol{E}_3 \otimes \boldsymbol{E}_2). \tag{110}$$

This rotation tensor is equal to the product $\boldsymbol{R}_2\boldsymbol{R}_1$ of a rotation

$$\boldsymbol{R}_1 = \boldsymbol{E}_1 \otimes \boldsymbol{E}_1 + \cos(-4)(\boldsymbol{E}_2 \otimes \boldsymbol{E}_2 + \boldsymbol{E}_3 \otimes \boldsymbol{E}_3) + \sin(-4)(\boldsymbol{E}_2 \otimes \boldsymbol{E}_3 - \boldsymbol{E}_3 \otimes \boldsymbol{E}_2) \quad (111)$$

and a rotation

$$\boldsymbol{R}_2 = -\boldsymbol{E}_1 \otimes \boldsymbol{E}_1 + \boldsymbol{E}_2 \otimes \boldsymbol{E}_2 - \boldsymbol{E}_3 \otimes \boldsymbol{E}_3. \quad (112)$$

Obviously,

$$\boldsymbol{R}_2\boldsymbol{E}_2 = \boldsymbol{E}_2, \qquad \boldsymbol{R}_2\boldsymbol{E}_3 = -\boldsymbol{E}_3, \quad (113)$$

so that the plane perpendicular to $\boldsymbol{E}_1$ is reflected in $\boldsymbol{E}_2$. The anholonomy from the tangent plane perpendicular to $\boldsymbol{E}_1$ into itself is therefore the composition of a reflection in $\boldsymbol{E}_2$ and a rotation about $\boldsymbol{E}_1$. Both Gaussian curvature and nonorientability contribute.

Let us next find the $\{\boldsymbol{g}, \boldsymbol{u}, \boldsymbol{n}\}$ frame. Thus, Eqs. (41) and (104) yield

$$\|\dot{\boldsymbol{n}}\| = \beta = \sqrt{\frac{1}{4} + \cos^2\frac{t}{2}}, \qquad \boldsymbol{g} = \frac{1}{\beta}\left(\frac{1}{2}\boldsymbol{t} + \cos\frac{t}{2}\boldsymbol{e}\right). \quad (114)$$

Notice that the spin axis $\boldsymbol{g}$ is not the same as the ruling $\boldsymbol{e}$.[34] The unit vector $\boldsymbol{u}$ is given by

$$\boldsymbol{u} = \boldsymbol{g} \times \boldsymbol{n} = \frac{1}{\beta}\left(\cos\frac{t}{2}\boldsymbol{t} - \frac{1}{2}\boldsymbol{e}\right). \quad (115)$$

It is clear from $(114)_2$ and $(100)_{1,2}$ that

$$\dot{\beta}\boldsymbol{g} + \beta\dot{\boldsymbol{g}} = \sin\frac{t}{2}\left(\cos\frac{t}{2}\boldsymbol{t} - \boldsymbol{e}\right). \quad (116)$$

But, in view of $(114)_2$ and (115),

$$\begin{pmatrix}\boldsymbol{t}\\ \boldsymbol{e}\end{pmatrix} = \frac{1}{\beta}\begin{pmatrix}\cos\frac{t}{2} & \frac{1}{2}\\ -\frac{1}{2} & \cos\frac{t}{2}\end{pmatrix}\begin{pmatrix}\boldsymbol{u}\\ \boldsymbol{g}\end{pmatrix}. \quad (117)$$

Substituting for $\boldsymbol{t}$ and $\boldsymbol{e}$ in (116), and recalling $(47)_2$, we see that

$$\beta\dot{\beta}\boldsymbol{g} - \beta^2\gamma\boldsymbol{u} = \sin\frac{t}{2}\left[\left(\frac{1}{2} + \cos^2\frac{t}{2}\right)\boldsymbol{u} - \frac{1}{2}\cos\frac{t}{2}\boldsymbol{g}\right]. \quad (118)$$

Consequently,

$$\beta\dot{\beta} = -\frac{1}{2}\sin\frac{t}{2}\cos\frac{t}{2}, \qquad \beta^2\gamma = -\left(\frac{1}{2} + \cos^2\frac{t}{2}\right)\sin\frac{t}{2}. \quad (119)$$

The angular velocity of the $\{\boldsymbol{u}, \boldsymbol{g}, \boldsymbol{n}\}$ frame relative to a Levi-Civita frame is given by (72). In the present case, it is not identically zero, and hence $\{\boldsymbol{u}, \boldsymbol{g}, \boldsymbol{n}\}$ is not parallel transported around the circle in Fig. 7.

[34] See the discussion in the Appendix.

4 Covariant Differentiation on a Surface

In this section, it will be shown that the covariant derivative of a vector along a surface curve may be interpreted as the corotational rate of that vector on a moving Levi-Civita frame, and similarly for a tensor.

Let $\boldsymbol{w}$ be a differentiable vector-valued field on the surface $\mathcal{S}$ ($\boldsymbol{w} \in \mathcal{T}^*$), and represent $\boldsymbol{w}$ both on the coordinate basis $\boldsymbol{a}_\alpha$ and on a Levi-Civita basis $\boldsymbol{\ell}_\alpha$:

$$\boldsymbol{w} = w^\alpha \boldsymbol{a}_\alpha = \bar{w}_\alpha \boldsymbol{\ell}_\alpha. \tag{120}$$

The derivative of $\boldsymbol{w}$ along a differentiable curve $\mathcal{C}$ on $\mathcal{S}$ may be calculated using each of these representations. In terms of the Levi-Civita representation, it is clear from (60) and $(35)_1$ that

$$\dot{\boldsymbol{w}} = \overset{\square}{\boldsymbol{w}} - \boldsymbol{w} \cdot \dot{\boldsymbol{n}}\boldsymbol{n} = \overset{\square}{\boldsymbol{w}} + \boldsymbol{w} \cdot \boldsymbol{b}\boldsymbol{v}\boldsymbol{n}. \tag{121}$$

As noted previously, the corotational component, $\overset{\square}{\boldsymbol{w}}$, belongs to the tangent space $\mathcal{T}^*$.

In terms of the other representation in (120), we may use (27) to obtain

$$\begin{aligned}\dot{\boldsymbol{w}} &= \dot{w}^\alpha \boldsymbol{a}_\alpha + w^\alpha \dot{\boldsymbol{a}}_\alpha \\ &= \dot{w}^\alpha \boldsymbol{a}_\alpha + w^\alpha \boldsymbol{a}_{\alpha},_{\beta}\, \dot{u}^\beta \\ &= \big(w^\alpha{},_\beta\, \boldsymbol{a}_\alpha + w^\alpha \Gamma^\gamma_{\alpha\beta} \boldsymbol{a}_\gamma + w^\alpha b_{\alpha\beta} \boldsymbol{n}\big) v^\beta.\end{aligned} \tag{122}$$

Setting

$$w^\gamma|_\beta = w^\gamma_{,\beta} + \Gamma^\gamma_{\alpha\beta} w^\alpha, \tag{123}$$

and recalling (34), we may rewrite (122) in the form

$$\dot{\boldsymbol{w}} = w^\gamma|_\beta v^\beta \boldsymbol{a}_\gamma + \boldsymbol{w} \cdot \boldsymbol{b}\boldsymbol{v}\boldsymbol{n}. \tag{124}$$

It is important to observe that the term $w^\gamma|_\beta v^\beta \boldsymbol{a}_\gamma$ is independent of the choice of surface coordinates u^α, since it is equal to the invariant quantity $\overset{\square}{\boldsymbol{w}}$.[35]

Let

$$\nabla \boldsymbol{w} = w^\gamma|_\beta \boldsymbol{a}_\gamma \otimes \boldsymbol{a}^\beta. \tag{125}$$

Then, associated with a differentiable curve having velocity vector $\boldsymbol{v} \in \mathcal{T}^*$, there is a unique vector

$$(\nabla \boldsymbol{w})\boldsymbol{v} = w^\gamma|_\beta \big(\boldsymbol{a}^\beta \cdot \boldsymbol{v}\big)\boldsymbol{a}_\gamma. \tag{126}$$

Clearly, $\nabla \boldsymbol{w}$ is a linear mapping $\mathcal{T}^* \to \mathcal{T}^*$. Thus, $\nabla \boldsymbol{w}$ is a surface tensor; it is called the *covariant derivative* of $\boldsymbol{w}$. From (124) and (126), it follows that

$$\dot{\boldsymbol{w}} = (\nabla \boldsymbol{w})\boldsymbol{v} + \boldsymbol{w} \cdot \boldsymbol{b}\boldsymbol{v}\boldsymbol{n}. \tag{127}$$

[35] A more complicated classical argument, based on changes of coordinates, can also be applied here. The linear mappings $w^\alpha{},_\beta\, \boldsymbol{a}_\gamma \otimes \boldsymbol{a}^\beta$ and $\Gamma^\gamma_{\alpha\beta} w^\alpha \boldsymbol{a}_\gamma \otimes \boldsymbol{a}^\beta$ separately depend on the choice of surface coordinates–they are not tensors–but their sum does not and hence it is a surface tensor.

The vector $(\nabla \boldsymbol{w})\boldsymbol{v} \in \mathcal{T}^*$ is called the *covariant derivative of* $\boldsymbol{w}$ *along the curve* $\mathcal{C}$.[36] If the arc length s is employed as a parameter, then (127) and $(17)_3$ yield

$$\frac{d\boldsymbol{w}}{ds} = (\nabla \boldsymbol{w})\boldsymbol{t} + \boldsymbol{w} \cdot \boldsymbol{btn}. \tag{128}$$

Comparing the two expressions (121) and (127) for $\dot{\boldsymbol{w}}$, we conclude that

$$(\nabla \boldsymbol{w})\boldsymbol{v} = \overset{\square}{\boldsymbol{w}}, \tag{129}$$

i.e., the covariant derivative of $\boldsymbol{w}$ along $\mathcal{C}$ is equal to the Levi-Civita corotational rate of $\boldsymbol{w}$ along $\mathcal{C}$.[37]

It follows immediately from (61) and (129) that a vector $\boldsymbol{w} \in \mathcal{T}^*$ is Levi-Civita transported along $\mathcal{C}$ if and only if

$$(\nabla \boldsymbol{w})\boldsymbol{v} = \boldsymbol{0}, \tag{130}$$

or equivalently

$$(\nabla \boldsymbol{w})\boldsymbol{t} = \boldsymbol{0}, \tag{131}$$

where $(17)_3$ has been invoked.

For any adapted frame, it is clear from $(68)_{1,2}$ and (129) that

$$(\nabla \boldsymbol{e}_1)\boldsymbol{v} = \dot{\psi}\boldsymbol{e}_2, \qquad (\nabla \boldsymbol{e}_2)\boldsymbol{v} = -\dot{\psi}\boldsymbol{e}_1. \tag{132}$$

In particular, for the $\{\boldsymbol{u}, \boldsymbol{g}, \boldsymbol{n}\}$ frame, Eqs. $(73)_{1,2}$ and (129) yield

$$(\nabla \boldsymbol{u})\boldsymbol{v} = \gamma\dot{s}\boldsymbol{g}, \qquad (\nabla \boldsymbol{g})\boldsymbol{v} = -\gamma\dot{s}\boldsymbol{u}. \tag{133}$$

Likewise, for the Darboux frame, Eqs. $(84)_{1,2}$ and (129) furnish

$$(\nabla \boldsymbol{t})\boldsymbol{v} = \kappa_g\dot{s}\boldsymbol{e}, \qquad (\nabla \boldsymbol{e})\boldsymbol{v} = -\kappa_g\dot{s}\boldsymbol{t}. \tag{134}$$

Along a geodesic, $\kappa_g = 0$, and

$$(\nabla \boldsymbol{t})\boldsymbol{v} = \boldsymbol{0}, \qquad (\nabla \boldsymbol{t})\boldsymbol{t} = \boldsymbol{0}. \tag{135}$$

Let us calculate the covariant derivatives of the coordinate basis vectors $\boldsymbol{a}_\alpha$ along a curve. Thus, in view of (27) and (127), along a curve $\mathcal{C}$,

$$\dot{\boldsymbol{a}}_\alpha = \Gamma^\gamma_{\alpha\beta} v^\beta \boldsymbol{a}_\gamma + b_{\alpha\beta} v^\beta \boldsymbol{n}$$

[36] Other common notations for $(\nabla \boldsymbol{w})\boldsymbol{v}$ are $\nabla_{\boldsymbol{v}}\boldsymbol{w}$, $\frac{D\boldsymbol{w}}{dt}$ (which may look strange but is suggestive), and $\frac{\delta \boldsymbol{w}}{\delta t}$. Also, the components in (123) are often written as $\nabla_\beta w^\gamma$.

[37] The rules for covariant differentiation follow at once from the usual rules for the corotational rate $\overset{\square}{\boldsymbol{w}}$. With regard to the formula (129), we may interpret the mapping ∇ as assigning a surface vector $\overset{\square}{\boldsymbol{w}}$ to a given pair $\boldsymbol{w}$ and $\boldsymbol{v}$ of surface vectors while satisfying differentiation rules. It is to be emphasized, however, that while $\nabla \boldsymbol{w}$ acts linearly on $\boldsymbol{v}$ in (129) to produce a vector, in contrast, the mapping ∇ obeys a product rule when it acts on $f\boldsymbol{w}$, where f is a scalar-valued function. The function ∇ is called a *Koszul connection* on the surface (see e.g., [13, Vol. 2, Chap. 6] or [33]).

$$
\begin{aligned}
&= (\nabla \boldsymbol{a}_\alpha)\boldsymbol{v} + \boldsymbol{a}_\alpha \cdot \boldsymbol{b}\boldsymbol{v}\boldsymbol{n} \\
&= (\nabla \boldsymbol{a}_\alpha)\boldsymbol{v} + b_{\alpha\beta} v^\beta \boldsymbol{n}. \qquad (136)
\end{aligned}
$$

Hence,

$$
(\nabla \boldsymbol{a}_\alpha)\boldsymbol{v} = \Gamma^\gamma_{\alpha\beta} v^\beta \boldsymbol{a}_\gamma. \qquad (137)
$$

If we now choose C to be the u^1-coordinate curve, we have $\boldsymbol{v} = v^1 \boldsymbol{a}_1$ and it follows from (137) that

$$
(\nabla \boldsymbol{a}_\alpha)\boldsymbol{a}_1 = \Gamma^\gamma_{\alpha 1} \boldsymbol{a}_\gamma. \qquad (138)
$$

Likewise, taking $\boldsymbol{v} = v^2 \boldsymbol{a}_2$, we get

$$
(\nabla \boldsymbol{a}_\alpha)\boldsymbol{a}_2 = \Gamma^\gamma_{\alpha 2} \boldsymbol{a}_\gamma. \qquad (139)
$$

Thus,

$$
(\nabla \boldsymbol{a}_\alpha)\boldsymbol{a}_\beta = \Gamma^\gamma_{\alpha\beta} \boldsymbol{a}_\gamma. \qquad (140)
$$

Consequently,

$$
\boldsymbol{a}^\lambda \cdot (\nabla \boldsymbol{a}_\alpha)\boldsymbol{a}_\beta = \Gamma^\lambda_{\alpha\beta}, \qquad (141)
$$

which should be compared with $(26)_2$.

Turning now to surface tensors, recall the formulas (69) which express the relationship between the time derivative $\dot{\boldsymbol{W}}$ of a surface tensor $\boldsymbol{W}$ and its Levi-Civita corotational rate $\overset{\square}{\boldsymbol{W}}$. Using a coordinate basis, we may proceed to calculate $\dot{\boldsymbol{W}}$ as follows:

$$
\begin{aligned}
\dot{\boldsymbol{W}} &= \dot{W}^{\alpha\beta} \boldsymbol{a}_\alpha \otimes \boldsymbol{a}_\beta + W^{\alpha\beta}(\dot{\boldsymbol{a}}_\alpha \otimes \boldsymbol{a}_\beta + \boldsymbol{a}_\alpha \otimes \dot{\boldsymbol{a}}_\beta) \\
&= \left[W^{\alpha\beta},_\gamma \boldsymbol{a}_\alpha \otimes \boldsymbol{a}_\beta + W^{\alpha\beta}(\boldsymbol{a}_\alpha,_\gamma \otimes \boldsymbol{a}_\beta + \boldsymbol{a}_\alpha \otimes \boldsymbol{a}_{\beta},_\gamma) \right] \dot{u}^\gamma \\
&= \left[W^{\alpha\beta},_\gamma \boldsymbol{a}_\alpha \otimes \boldsymbol{a}_\beta + W^{\alpha\beta}\left(\Gamma^\lambda_{\alpha\gamma} \boldsymbol{a}_\lambda \otimes \boldsymbol{a}_\beta + \Gamma^\lambda_{\beta\gamma} \boldsymbol{a}_\alpha \otimes \boldsymbol{a}_\lambda\right) \right. \\
&\quad \left. + W^{\alpha\beta}(b_{\alpha\gamma} \boldsymbol{n} \otimes \boldsymbol{a}_\beta + b_{\beta\gamma} \boldsymbol{a}_\alpha \otimes \boldsymbol{n}) \right] v^\gamma, \qquad (142)
\end{aligned}
$$

where use has been made of (27). Re-labeling indices in the terms involving Christoffel symbols, we may rewrite (142) as

$$
\begin{aligned}
\dot{\boldsymbol{W}} &= \left\{ W^{\alpha\beta},_\gamma + \Gamma^\alpha_{\lambda\gamma} W^{\lambda\beta} + \Gamma^\beta_{\lambda\gamma} W^{\alpha\lambda} \right\} v^\gamma \boldsymbol{a}_\alpha \otimes \boldsymbol{a}_\beta \\
&\quad + \boldsymbol{n} \otimes \left(W^{\alpha\beta} b_{\alpha\gamma} v^\gamma\right)\boldsymbol{a}_\beta + \left(W^{\alpha\beta} b_{\beta\gamma} v^\gamma \boldsymbol{a}_\alpha\right) \otimes \boldsymbol{n}. \qquad (143)
\end{aligned}
$$

But, in view of (33) and (19),

$$
\begin{aligned}
-\boldsymbol{W}\dot{\boldsymbol{n}} &= \left(W^{\alpha\beta} \boldsymbol{a}_\alpha \otimes \boldsymbol{a}_\beta\right)\left(b_{\lambda\gamma} v^\gamma \boldsymbol{a}^\lambda\right) = W^{\alpha\beta} b_{\beta\gamma} v^\gamma \boldsymbol{a}_\alpha \\
-\boldsymbol{W}^T \dot{\boldsymbol{n}} &= W^{\alpha\beta} b_{\alpha\gamma} v^\gamma \boldsymbol{a}_\beta.
\end{aligned} \qquad (144)
$$

Therefore, Eq. (143) becomes

$$
\dot{\boldsymbol{W}} = \left\{ W^{\alpha\beta},_\gamma + \Gamma^\alpha_{\lambda\gamma} W^{\lambda\beta} + \Gamma^\beta_{\lambda\gamma} W^{\alpha\lambda} \right\} v^\gamma \boldsymbol{a}_\alpha \otimes \boldsymbol{a}_\beta - (\boldsymbol{W}\dot{\boldsymbol{n}}) \otimes \boldsymbol{n} - \boldsymbol{n} \otimes \left(\boldsymbol{W}^T \boldsymbol{n}\right). \qquad (145)
$$

It follows from (69) and (145) that the Levi-Civita corotational rate, $\overset{\square}{\boldsymbol{W}}$, which we know to be a surface tensor, is given by

$$\overset{\square}{\boldsymbol{W}} = \left\{ W^{\alpha\beta},_{\gamma} + \Gamma^{\alpha}_{\lambda\gamma} W^{\lambda\beta} + \Gamma^{\beta}_{\lambda\gamma} W^{\alpha\lambda} \right\} v^{\gamma} \boldsymbol{a}_{\alpha} \otimes \boldsymbol{a}_{\beta}. \tag{146}$$

Setting

$$W^{\alpha\beta}|_{\gamma} = W^{\alpha\beta},_{\gamma} + \Gamma^{\alpha}_{\lambda\gamma} W^{\lambda\beta} + \Gamma^{\beta}_{\lambda\gamma} W^{\alpha\lambda}, \tag{147}$$

we obtain

$$\overset{\square}{\boldsymbol{W}} = W^{\alpha\beta}|_{\gamma} v^{\gamma} \boldsymbol{a}_{\alpha} \otimes \boldsymbol{a}_{\beta}. \tag{148}$$

Let

$$\nabla \boldsymbol{W} = W^{\alpha\beta}|_{\gamma} \boldsymbol{a}_{\alpha} \otimes \boldsymbol{a}_{\beta} \otimes \boldsymbol{a}^{\gamma}. \tag{149}$$

From (148) and (149), it is evident that

$$(\nabla \boldsymbol{W})\boldsymbol{v} = \overset{\square}{\boldsymbol{W}}. \tag{150}$$

$\nabla \boldsymbol{W}$ maps each $\boldsymbol{v} \in \mathcal{T}^*$ into a unique surface tensor $\overset{\square}{\boldsymbol{W}}$, and its action is linear. It is therefore a surface tensor. It is called the *covariant derivative* of $\boldsymbol{W}$, and the vector $(\nabla \boldsymbol{W})\boldsymbol{v}$ is called the *covariant derivative of* $\boldsymbol{W}$ *along the curve* $\mathcal{C}$.

It is obvious from (70) and (150) that the surface tensor $\boldsymbol{W}$ is Levi-Civita transported along a curve $\mathcal{C}$ if and only if

$$(\nabla \boldsymbol{W})\boldsymbol{v} = \boldsymbol{0}. \tag{151}$$

In particular, for the identity tensor $\boldsymbol{I}_T$ on $\mathcal{T}^*$, it is clear from (71) and (150) that

$$(\nabla \boldsymbol{I}_T)\boldsymbol{v} = \boldsymbol{0} \tag{152}$$

for all $\boldsymbol{v} \in \mathcal{T}^*$, and hence

$$\nabla \boldsymbol{I}_T = \boldsymbol{0} \tag{153}$$

on $\mathcal{S}$, which is known as Ricci's theorem (or Ricci's lemma). Thus, the first fundamental tensor of a surface is parallel-transported along every surface curve; it is carried corotationally by every Levi-Civita frame.[38]

5 The Envelope of the Tangent Planes along a Surface Curve

In this section, we discuss the properties of the envelope of a family of tangent planes along a surface curve, and indicate the way in which Levi-Civita explained his concept of parallel transport in [18, 20].

[38]In terms of components of $\boldsymbol{I}_T$, the covariant derivatives $a_{\alpha\beta}|_{\gamma}$, $a^{\alpha\beta}|_{\gamma}$, $\delta^{\alpha}_{\beta}|_{\gamma}$ all vanish, and they may therefore be moved across the sign for covariant differentiation. It follows that the operation of covariant differentiation commutes with that of raising or lowering an index on a tensor component.

Consider the curve $\mathcal{C}$ ($\boldsymbol{x} = \boldsymbol{x}(s)$) on $\mathcal{S}$ together with the unit vector $\boldsymbol{g}(s)$ defined along it (see Fig. 6). Taking ξ and η as Gaussian coordinates, we may construct a special differentiable ruled surface $\tilde{\mathcal{S}}$ by means of the equation

$$\tilde{\boldsymbol{x}}(\xi, \eta) = \boldsymbol{x}(\xi) + \eta \boldsymbol{g}(\xi), \tag{154}$$

in which the parameter η is taken to be signed distance along the straight lines $\xi = \text{const.}$ (the rulings); also, along the directrix $\mathcal{C}$ ($\eta = 0$), the parameter ξ may be conveniently chosen to be the arc length of $\mathcal{C}$.[39] Then, along the ruling through P, ξ is numerically equal to s. Also,

$$\tilde{\boldsymbol{x}}(\xi, 0) = \boldsymbol{x}(\xi) = \boldsymbol{x}(s), \qquad \boldsymbol{g}(\xi) = \boldsymbol{g}(s). \tag{155}$$

The coordinate basis on $\tilde{\mathcal{S}}$ is given by

$$\begin{aligned} \tilde{\boldsymbol{a}}_1(\xi, \eta) &= \frac{\partial \tilde{\boldsymbol{x}}}{\partial \xi} = \boldsymbol{t}(\xi) - \eta\gamma \boldsymbol{u}(\xi) = \boldsymbol{t}(s) - \eta\gamma \boldsymbol{u}(s), \\ \tilde{\boldsymbol{a}}_2(\xi, \eta) &= \frac{\partial \tilde{\boldsymbol{x}}}{\partial \eta} = \boldsymbol{g}(\xi) = \boldsymbol{g}(s), \end{aligned} \tag{156}$$

where $\boldsymbol{t} = d\boldsymbol{x}/ds = d\boldsymbol{x}/d\xi$ is the unit tangent vector to $\mathcal{C}$ and $(47)_2$ has been utilized.[40] Clearly,

$$\tilde{\boldsymbol{a}}_1(\xi, 0) = \boldsymbol{t}(s), \qquad \tilde{\boldsymbol{a}}_2(\xi, 0) = \boldsymbol{g}(s), \tag{157}$$

and hence the unit surface normal $\tilde{\boldsymbol{n}}(s, 0)$ to $\tilde{\mathcal{S}}$ along $\mathcal{C}$ is perpendicular to the tangent plane $\mathcal{T}$ to $\mathcal{S}$. Since the surfaces $\mathcal{S}$ and $\tilde{\mathcal{S}}$ share the same tangent plane at each point of $\mathcal{C}$, $\tilde{\mathcal{S}}$ is the *envelope* of the family of tangent planes to $\mathcal{S}$ along $\mathcal{C}$.

It follows from $(156)_{1,2}$ that

$$\tilde{\boldsymbol{a}}_1(\xi, \eta) \times \tilde{\boldsymbol{a}}_2(\xi, \eta) = \boldsymbol{t}(s) \times \boldsymbol{g}(s) - \eta\gamma \boldsymbol{n}(s). \tag{158}$$

Consequently, along each ruling, the unit surface normal $\tilde{\boldsymbol{n}}(\xi, \eta)$ to $\tilde{\mathcal{S}}$ is parallel to $\boldsymbol{n}(s)$, the unit normal to $\mathcal{S}$ at P. Hence,

$$\frac{\partial \tilde{\boldsymbol{n}}}{\partial \eta}(\xi, \eta) = \boldsymbol{0}, \tag{159}$$

and

$$\tilde{\boldsymbol{n}}(\xi, \eta) = \tilde{\boldsymbol{n}}(\xi, 0) = \boldsymbol{n}(s). \tag{160}$$

In other words, the tangent plane $\tilde{\mathcal{T}}(\xi, \eta)$ to $\tilde{\mathcal{S}}$ does not vary with η and coincides with $\tilde{\mathcal{T}}(\xi, 0) = \mathcal{T}$.[41] Also, the arc rate of change of $\tilde{\boldsymbol{n}}$ along $\mathcal{C}$ is given by $(42)_1$.

[39] While ξ does measure arc length along $\mathcal{C}$, it merely serves as a parameter along other curves in the family. We could use s instead of ξ in (154) as a surface coordinate, but shortly we will employ s to parametrize a family of surfaces having (ξ, η) as Gaussian coordinates.

[40] For a general ruled surface, the condition $(47)_2$ does not hold for the ruling: see Example 3.2 and the Appendix.

[41] This may be taken as the defining condition for a ruled surface to be developable [27, Sec. 58]. The term "applicable" is then employed for the older meaning of "developable" (see below).

Let $\tilde{\boldsymbol{b}}(\xi, \eta)$ be the second fundamental tensor of the envelope $\tilde{\mathcal{S}}$. Applying the result $(35)_2$ to $\tilde{\mathcal{S}}$, we see that along each ruling

$$\mathbf{0} = \frac{\partial \tilde{\boldsymbol{n}}}{\partial \eta} = -\tilde{\boldsymbol{b}}(\xi, \eta)\boldsymbol{g}(s). \tag{161}$$

Resolving $\tilde{\boldsymbol{b}}(\xi, \eta)$ on the orthonormal basis $\{\boldsymbol{u}(s), \boldsymbol{g}(s)\}$ for the tangent space $\tilde{\mathcal{T}}(\xi, \eta)(= \tilde{\mathcal{T}}(s, 0))$, we have

$$\begin{aligned}\tilde{\boldsymbol{b}}(\xi, \eta) = {} & \tilde{b}_{11}(\xi, \eta)\boldsymbol{u}(s) \otimes \boldsymbol{u}(s) + \tilde{b}_{12}(\xi, \eta)\big(\boldsymbol{u}(s) \otimes \boldsymbol{g}(s) + \boldsymbol{g}(s) \otimes \boldsymbol{u}(s)\big) \\ & + \tilde{b}_{22}(\xi, \eta)\boldsymbol{g}(s) \otimes \boldsymbol{g}(s).\end{aligned} \tag{162}$$

It follows from (161) and (162) that

$$\tilde{b}_{12}(\xi, \eta)\boldsymbol{u}(s) + \tilde{b}_{22}(\xi, \eta)\boldsymbol{g}(s) = \mathbf{0}, \tag{163}$$

and hence

$$\tilde{b}_{12}(\xi, \eta) = 0, \qquad \tilde{b}_{22}(\xi, \eta) = 0. \tag{164}$$

The tensor $\tilde{\boldsymbol{b}}(\xi, \eta)$ can now be expressed as

$$\tilde{\boldsymbol{b}}(\xi, \eta) = \tilde{b}_{11}(\xi, \eta)\boldsymbol{u}(s) \otimes \boldsymbol{u}(s). \tag{165}$$

The Gaussian curvature $\tilde{K}$ of the envelope $\tilde{\mathcal{S}}$ is therefore

$$\tilde{K} = \det \tilde{\boldsymbol{b}} = 0. \tag{166}$$

Now, the Gaussian curvature of any plane is also zero, and consequently, any sufficiently small region of the envelope $\tilde{\mathcal{S}}$ can be mapped isometrically into a plane, with the preservation of distances and angles. Let us take the local isometry Ψ that maps a region of $\tilde{\mathcal{S}}$ into the tangent plane $\mathcal{T}_0$ to the surface $\mathcal{S}$ at P_0 (Fig. 8). Recall that the unit normal vector to $\mathcal{S}$ at P_0 is $\boldsymbol{n}_0$. Let $\boldsymbol{X}$ be the position vector of points in $\mathcal{T}_0$. The η-coordinate curves ($\xi = \text{const.} = s$) on the envelope $\tilde{\mathcal{S}}$ will be mapped into straight lines in $\mathcal{T}_0$, while the ξ-coordinate curves ($\eta = \text{const.}$) will be mapped into curves in $\mathcal{T}_0$. The curve $\mathcal{C}$ will be mapped into a curve $\bar{\mathcal{C}}$ in $\mathcal{T}_0$ ($\eta = 0$), having the same arc length s. Representing $\bar{\mathcal{C}}$ by $\boldsymbol{X} = \boldsymbol{X}(s)$, for the unit tangent vector to $\bar{\mathcal{C}}$, we have

$$\bar{\boldsymbol{t}}(s) = \frac{\partial \boldsymbol{X}}{\partial s}. \tag{167}$$

$\bar{\mathcal{C}}$ will have curvature

$$\kappa(s) = \kappa_g(s), \tag{168}$$

the geodesic curvature of $\mathcal{C}$.[42] Let $\bar{\boldsymbol{e}}(s)$ be the image of the unit vector $\boldsymbol{e}(s)$ under the isometry Ψ, i.e.,

$$\bar{\boldsymbol{e}}(s) = \Psi\big(\boldsymbol{e}(s)\big). \tag{169}$$

Trivially, $\{\bar{\boldsymbol{t}}, \bar{\boldsymbol{e}}, \boldsymbol{n}_0\}$ is a Darboux frame on $\mathcal{T}_0$ along $\bar{\mathcal{C}}$.

[42] Here, it is convenient to allow $\kappa(s)$ to take on the same sign as $\kappa_g(s)$, rather than being always non-negative.

Fig. 8 Isometric mapping Ψ of the envelope $\tilde{\mathcal{S}}$ into the initial tangent plane $\mathcal{T}_0$

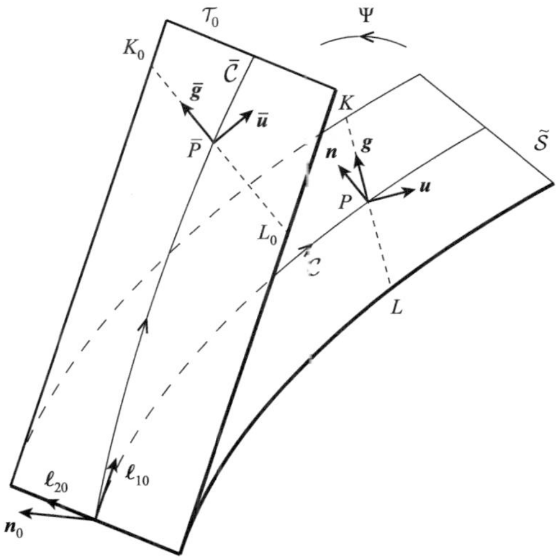

Let $\bar{g}(s)$ be a unit vector along each coordinate line $\xi = \text{const}$. Also, let

$$\bar{u}(s) = \bar{g}(s) \times n_0. \tag{170}$$

Clearly,

$$\bar{g}(s) = \Psi\big(g(s)\big), \qquad \bar{u}(s) = \Psi\big(u(s)\big). \tag{171}$$

The angle that $\bar{g}(s)$ makes with $\bar{t}(s)$ will be equal to the angle which $g(s)$ makes with $t(s)$. The tangent plane $\mathcal{T}_0$ may be described by the equation

$$X(\xi, \eta) = X(\xi) + \eta \bar{g}(\xi), \tag{172}$$

which has the form of a (trivial) ruled surface. The corresponding basis is

$$A_1(\xi, \eta) = \frac{\partial X}{\partial \xi}(\xi, \eta) = t(\xi) + \eta \frac{\partial \bar{g}}{\partial \xi}(\xi, \eta), \qquad A_2(\xi, \eta) = \frac{\partial X}{\partial \eta}(\xi, \eta) = \bar{g}(\xi). \tag{173}$$

Along the curve $\bar{\mathcal{C}}$,

$$A_1(\xi, 0) = \bar{t}(s), \qquad A_2(\xi, 0) = \bar{g}(s). \tag{174}$$

Let us choose a Levi-Civita frame $\{\ell_{01}, \ell_{02}, n_0\}$ on $\tilde{\mathcal{S}}$ at P_0 such that

$$\ell_{01} = \bar{t}(0), \qquad \ell_{02} = \bar{e}(0). \tag{175}$$

Let $\psi(s)$ be the angle that $\bar{t}(s)$ makes with $\bar{t}(0)$. The curvature of $\bar{\mathcal{C}}$ may be expressed as

$$\kappa(s) = \frac{d\psi}{ds}. \tag{176}$$

It follows from $(80)_2$, (168), and (176) that

$$\psi(s) = \psi_D(s), \tag{177}$$

which is the angle between the unit tangent vector $\boldsymbol{t}(s)$ to $\mathcal{C}$ and $\boldsymbol{\ell}_1(s)$. It follows that the Levi-Civita frame at each point of $\mathcal{C}$ is mapped by the isometry Ψ into the fixed frame $\{\boldsymbol{\ell}_{01}, \boldsymbol{\ell}_{02}, \boldsymbol{n}_0\}$, a copy of which we may place at each point of $\bar{\mathcal{C}}$. Thus, the Levi-Civita frames along $\mathcal{C}$ are mapped into a set of Euclidean parallel frames along $\bar{\mathcal{C}}$ in the tangent plane $\mathcal{T}_0$.

It is clear from $(51)_1$ that

$$\boldsymbol{\ell}_\alpha(s) = \boldsymbol{R}_T(s)\boldsymbol{\ell}_{0\alpha}. \tag{178}$$

Since $\boldsymbol{R}_T$ preserves both lengths and angles, it is evident that

$$\boldsymbol{R}_T(s)\bar{\boldsymbol{g}}(s) = \boldsymbol{g}(s), \qquad \boldsymbol{R}_T(s)\bar{\boldsymbol{u}}(s) = \boldsymbol{u}(s). \tag{179}$$

The frame $\{\bar{\boldsymbol{u}}, \bar{\boldsymbol{g}}, \boldsymbol{n}_0\}$ for the tangent plane $\mathcal{T}_0$ satisfies equations of the form $(47)_{1,2,3}$, with

$$\bar{\beta} = \left\| \frac{d\boldsymbol{n}_0}{ds} \right\| = 0, \tag{180}$$

where (41) has been recalled. Thus,

$$\frac{d\bar{\boldsymbol{u}}}{ds}(s) = \bar{\gamma}\bar{\boldsymbol{g}}(s), \qquad \frac{d\bar{\boldsymbol{g}}}{ds}(s) = -\bar{\gamma}\bar{\boldsymbol{u}}(s). \tag{181}$$

Also, in view of (46) and (180), the angular velocity $\bar{\boldsymbol{\omega}}_G$ of this frame is

$$\bar{\boldsymbol{\omega}}_G = \bar{\gamma}\dot{s}\boldsymbol{n}_0. \tag{182}$$

This is also equal to the relative angular velocity given by (72), since the Levi-Civita frames are fixed along $\bar{\mathcal{C}}$. Consequently,

$$\bar{\gamma} = \frac{d\bar{\psi}_G}{ds}(s), \tag{183}$$

where $\bar{\psi}_G(s)$ is the angle that $\bar{\boldsymbol{u}}(s)$ makes with $\boldsymbol{\ell}_{01}$. Now, it follows from $(179)_1$, $(8)_1$, (6), and (54) that

$$\dot{\boldsymbol{g}} = \boldsymbol{\omega}_T \times \boldsymbol{g} + \boldsymbol{R}_T\dot{\bar{\boldsymbol{g}}} = \boldsymbol{R}_T\dot{\bar{\boldsymbol{g}}}. \tag{184}$$

With the aid of $(47)_2$, $(181)_2$ and $(179)_2$, (184) becomes

$$\gamma\boldsymbol{u} = \bar{\gamma}\boldsymbol{u}, \tag{185}$$

and hence,

$$\gamma = \bar{\gamma} = \frac{d\bar{\psi}_G}{ds}. \tag{186}$$

Thus, the angular velocity component $\gamma\dot{s}$ in (46) is equal to the time rate of change of the angle $\bar{\psi}_G$.

Next, we argue that the envelope $\tilde{\mathcal{S}}$ and the tangent plane $\mathcal{T}_0$ are *applicable* to one another, i.e., that there exists a continuous one-parameter family of isometries by means of which $\mathcal{T}_0$ can be (locally) deformed into $\tilde{\mathcal{S}}$; such a family of mappings—or a *motion*, in the sense of continuum mechanics–is called a *bending*. To this end, for any fixed value of s, consider an isometry Ψ_s that takes the tangent plane $\mathcal{T}_0$ into the part of the envelope $\tilde{\mathcal{S}}$ given by $0 \leq s$,

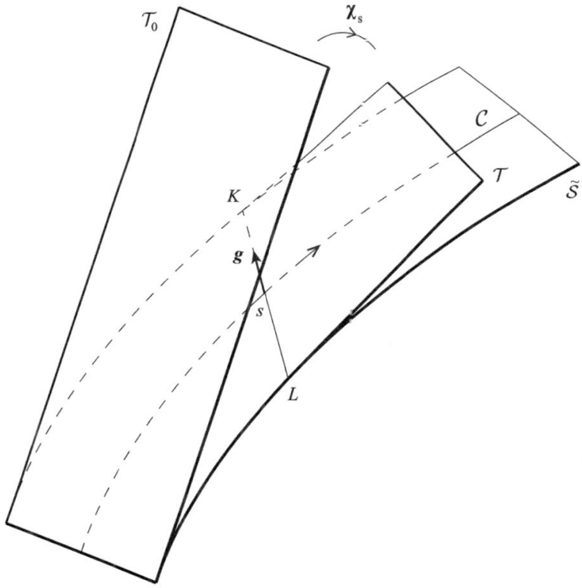

Fig. 9 The initial tangent plane $\mathcal{T}_0$ can be continuously bent onto the envelope $\tilde{\mathcal{S}}$

joined to the continuation of the tangent plane $\mathcal{T}$ to $\tilde{\mathcal{S}}$ at s as illustrated in Fig. 9. Employing s as a parameter, we then have a family $\boldsymbol{\chi}_s$ of isometric mappings given by

$$\boldsymbol{\chi}_s(\xi, \eta) = \begin{cases} \boldsymbol{x}(\xi) + \eta \boldsymbol{g}(\xi), & \xi \leq s \\ \boldsymbol{x}(s) + \eta \boldsymbol{g}(s) + (\xi - s)\boldsymbol{u}(s), & \xi \geq s. \end{cases} \tag{187}$$

These mappings bend the tangent plane $\mathcal{T}_0$ onto the envelope $\tilde{\mathcal{S}}$. In view of $(179)_{1,2}$, the bending in (187) involves a one-parameter family of rotations $\boldsymbol{R}_{\mathcal{T}}(s)$; the vector $\boldsymbol{g}(s)$ is the spin axis of the tangent plane common to $\mathcal{S}$ and $\tilde{\mathcal{S}}$ and the corresponding angular velocity is $\boldsymbol{\omega}_T$, the tilting velocity of the tangent plane.

In Sect. 3, we defined Levi-Civita transport of a vector $\boldsymbol{w}$ along a curve by the condition that the components of the vector remain constant in a Levi-Civita frame. As noted above, the continuous development of the envelope $\tilde{\mathcal{S}}$ results in a Levi-Civita frame that is constant along $\bar{\mathcal{C}}$ in the plane $\mathcal{T}_0$. Therefore, the images $\bar{\boldsymbol{w}}(s) = \Psi(\boldsymbol{w}(s))$ will form a Euclidean parallel field along $\bar{\mathcal{C}}$ in $\mathcal{T}_0$. This is essentially how Levi-Civita described his concept of parallelism in [18, 20].

Example 5.1 Returning to the Möbius band in Example 3.2, recall that the spin axis $\boldsymbol{g}$ is given by $(114)_{1,2}$. In accordance with (154), around the circle in Fig. 7, the envelope to the band is defined by

$$\tilde{\boldsymbol{x}}(t, \eta) = \boldsymbol{e}_R(t) + \eta \boldsymbol{g}(t) = \boldsymbol{e}_R(t) + \eta \frac{\frac{1}{2}\boldsymbol{t} + \cos\frac{t}{2}\boldsymbol{e}}{\sqrt{\frac{1}{4} + \cos^2\frac{t}{2}}}. \tag{188}$$

The ruling vector for the band is the vector $\boldsymbol{f}(=\boldsymbol{e})$, while the ruling vector for the envelope is the spin axis $\boldsymbol{g}$. The derivative of $\boldsymbol{g}$ satisfies $(47)_2$, and the envelope is a developable surface. The derivative of $\boldsymbol{e}$ in $(100)_2$ has a nonzero component along $\boldsymbol{n}$ and the ruled surface described by $(93)_{1,2,3}$ is therefore not developable.[43]

Acknowledgements I should like to thank the editors, Professor Fosdick and Professor Fried and an anonymous reviewer for their suggestions. I am also grateful to my colleague Professor O. O'Reilly and Berkeley Mechanical Engineering graduate students G. de Vera, S. Kshirsagar, and C. Diamond for their comments.

Appendix: Ruled Surfaces

We wish to consider conditions under which a ruled surface is developable or not.

Consider a general ruled surface $\bar{\mathcal{S}}$ given by an equation of the form

$$\boldsymbol{x}(s,\eta) = \boldsymbol{x}(s) + \eta\boldsymbol{r}(s), \tag{A.1}$$

where $\boldsymbol{r}(s)$ is a unit vector (the ruling vector) and $\boldsymbol{x}(s)$ describes a curve $\bar{\mathcal{C}}$ (the directrix). The unit tangent vector to $\bar{\mathcal{C}}$ is

$$\boldsymbol{t}(s) = \frac{d\boldsymbol{x}}{ds}(s) \tag{A.2}$$

and we assume that $\boldsymbol{t}(s) \times \boldsymbol{r}(s) \neq \boldsymbol{0}$. The coordinate basis for $\bar{\mathcal{S}}$ is

$$\boldsymbol{a}_1(s,\eta) = \frac{\partial \boldsymbol{x}}{\partial s} = \boldsymbol{t}(s) + \eta\frac{d\boldsymbol{r}}{ds}(s), \qquad \boldsymbol{a}_2(s,\eta) = \frac{\partial \boldsymbol{x}}{\partial \eta} = \boldsymbol{r}(s). \tag{A.3}$$

Then,

$$\boldsymbol{a}_1(s,\eta) \times \boldsymbol{a}_2(s,\eta) = \boldsymbol{t}(s) \times \boldsymbol{r}(s) + \eta\frac{d\boldsymbol{r}}{ds}(s) \times \boldsymbol{r}(s), \tag{A.4}$$

which will be nonzero for some range of values of the parameter η, and will furnish a unit normal vector

$$\boldsymbol{n}(s,\eta) = \frac{\boldsymbol{a}_1(s,\eta) \times \boldsymbol{a}_2(s,\eta)}{\|\boldsymbol{a}_1(s,\eta) \times \boldsymbol{a}_2(s,\eta)\|}. \tag{A.5}$$

Along the directrix $\bar{\mathcal{C}}$, this becomes

$$\boldsymbol{n}(s,0) = \frac{\boldsymbol{t}(s) \times \boldsymbol{r}(s)}{\|\boldsymbol{t}(s) \times \boldsymbol{r}(s)\|}. \tag{A.6}$$

By definition, a ruled surface $\bar{\mathcal{S}}$ is *developable* if and only if [27, Sec. 58]

$$\boldsymbol{n}(s,\eta) = \boldsymbol{n}(s,0), \tag{A.7}$$

i.e., along each ruling, the tangent plane to $\bar{\mathcal{S}}$ is the same plane. Suppose that $\bar{\mathcal{S}}$ is developable. Then, the basis vector $\boldsymbol{a}_1(s,\eta)$ will be perpendicular to $\boldsymbol{n}(s,0)$ for all admissible values of η. Consequently, we may express it as a linear combination of the vectors $\boldsymbol{t}(s)$ and $\boldsymbol{r}(s)$:

$$\boldsymbol{t}(s) + \eta\frac{d\boldsymbol{r}}{ds}(s) = \mu_1(s,\eta)\boldsymbol{t}(s) + \mu_2(s,\eta)\boldsymbol{r}(s). \tag{A.8}$$

[43] See the condition (A.10) in the Appendix.

It follows immediately that

$$\eta \frac{d\boldsymbol{r}}{ds}(s) \cdot \boldsymbol{n}(s,0) = 0 \tag{A.9}$$

for all admissible values of η. Hence,

$$\frac{d\boldsymbol{r}}{ds}(s) \cdot \boldsymbol{n}(s,0) = 0. \tag{A.10}$$

This is a necessary condition for a ruled surface to be developable.[44] It is also a sufficient condition: If (A.10) holds, then in view of $(A.3)_1$,

$$\boldsymbol{a}_1(s,\eta) \cdot \boldsymbol{n}(0,\eta) = \boldsymbol{t}(s) \cdot \boldsymbol{n}(s,0) = 0. \tag{A.11}$$

But, by virtue of $(A.3)_2$,

$$\boldsymbol{a}_2(s,\eta) \cdot \boldsymbol{n}(0,\eta) = 0 \tag{A.12}$$

also. The tangent plane to $\bar{S}$ at (s,η) is spanned by $\boldsymbol{a}_1(s,\eta)$ and $\boldsymbol{a}_2(s,\eta)$. It must therefore be the case then that for each value of s, the tangent plane to $\bar{S}$ is the same for all admissible values of η.

In Sect. 3, it was shown that along a surface curve the tangent plane turns with angular velocity

$$\boldsymbol{\omega}_T = \beta \dot{s} \boldsymbol{g}(t), \qquad \beta = \left\| \frac{d\boldsymbol{n}}{ds} \right\|. \tag{A.13}$$

Also,[45]

$$\frac{d\boldsymbol{n}}{ds} = \beta \boldsymbol{u}, \qquad \frac{d\boldsymbol{g}}{ds} = -\gamma \boldsymbol{u}, \qquad \boldsymbol{u} \times \boldsymbol{g} = \boldsymbol{n}. \tag{A.14}$$

Let us now apply these results to the ruled surface $\bar{S}$.

Consider the term $\frac{d\boldsymbol{r}}{ds}(s)$. Since $\boldsymbol{r}$ is a unit vector

$$\frac{d\boldsymbol{r}}{ds}(s) \cdot \boldsymbol{r}(s) = 0. \tag{A.15}$$

Further, since

$$\boldsymbol{r}(s) \cdot \boldsymbol{n}(s,0) = 0, \tag{A.16}$$

it follows that

$$\frac{d\boldsymbol{r}}{ds}(s) \cdot \boldsymbol{n}(s,0) + \boldsymbol{r}(s) \cdot \frac{d\boldsymbol{n}}{ds}(s,0) = 0. \tag{A.17}$$

Therefore, by virtue of $(A.14)_1$,

$$\frac{d\boldsymbol{r}}{ds}(s) \cdot \boldsymbol{n}(s,0) = -\beta \boldsymbol{r}(s) \cdot \boldsymbol{u}(s). \tag{A.18}$$

[44] See, e.g., [27, Sec. 58], where it is written in the form $\boldsymbol{t}(s) \times \boldsymbol{r}(s) \cdot \frac{d\boldsymbol{r}}{ds}(s) = 0$.

[45] See $(42)_1$ and $(47)_2$.

If the ruling vector $\boldsymbol{r}(s)$ is equal to $\pm\boldsymbol{g}(s)$, then the right-hand side of (A.18) vanishes, and the condition (A.10) results. The ruled surface $\bar{\mathcal{S}}$ is then developable. Thus, if the ruling is always parallel to the spin axis, the ruled surface is developable.

Conversely, suppose that the ruled surface $\bar{\mathcal{S}}$ is developable. Then, the condition (A.10) holds, and it follows from (A.18) that

$$\beta \boldsymbol{r}(s) \cdot \boldsymbol{u}(s) = 0. \tag{A.19}$$

If for some value s^* of s, $\beta(s^*) \neq 0$, then β will be nonzero in some open interval containing s^* and

$$\boldsymbol{r}(s) \cdot \boldsymbol{u}(s) = 0 \tag{A.20}$$

in this interval. Consequently, the ruling vector $\boldsymbol{r}(s)$ will then be parallel to the spin axis $\boldsymbol{g}(s)$.

References

1. Truesdell, C.: The mechanical foundations of elasticity and continuum mechanics. J. Ration. Mech. Anal. **1**, 125–300 (1952)
2. Michal, A.D.: Matrix and Tensor Calculus with Applications to Mechanics, Elasticity, and Aeronautics. Wiley, New York (1947)
3. Green, A.E., Zerna, W.: Theoretical Elasticity. Clarendon Press, Oxford (1954)
4. Truesdell, C., Toupin, R.: The classical field theories. In: Flügge, C. (ed.) Handbuch der Physik, vol. III, pp. 226–858. Springer, Berlin (1960)
5. Fosdick, R.L.: Remarks on compatibility. In: Eskinazi, S. (ed.) Modern Developments in the Mechanics of Continua, pp. 109–127. Academic Press, New York (1966)
6. Marsden, J.E., Hughes, T.J.R.: Mathematical Foundations of Elasticity. Prentice Hall, Englewood Cliffs, New Jersey (1983). Reprinted in Dover, New York (1994)
7. Hughes, T.J.R., Marsden, J.E.: Some applications of geometry in continuum mechanics. Rep. Math. Phys. **12**, 35–44 (1977)
8. Stumpf, H., Hoppe, U.: The application of tensor algebra on manifolds to nonlinear continuum mechanics—invited survey article. Z. Angew. Math. Mech. **5**, 327–339 (1997)
9. Casey, J., Papadopoulos, P.: Material transport of sets and fields. Math. Mech. Solids **7**, 647–676 (2002)
10. Heath, S.T.L. (ed.): The Thirteen Books of Euclid's Elements, 3 Vols., 2nd edn. Cambridge University Press, Cambridge (1926). Reprinted in Dover, New York (1956)
11. Gauss, K.F.: General Investigations of Curved Surfaces of 1827 and 1825. Princeton University Press, Princeton (1902). Translated by: Morehead, J.C., Hiltebeitel, A.M. (See also General Investigations of Curved Surfaces. Raven Press, Hewlett (1965). Translated by: Hiltebeitel, A.M., Morehead, J.C.)
12. Chern, S.-S.: From triangles to manifolds. Am. Math. Mon. **86**, 339–349 (1979)
13. Spivak, M.: A Comprehensive Introduction to Differential Geometry, 5 Vols., 3rd edn. Publish or Perish, Houston (2005)
14. Pesic, P.: Beyond Geometry: Classic Papers from Riemann to Einstein. Dover, New York (2007)
15. Weatherburn, C.E.: The development of multidimensional differential geometry. In: Report of 21st Meeting of Australian and New Zealand Association for the Advancement of Science, Sydney 1932, pp. 12–28. A.J. Kent, Government Printer, Sydney (1933)
16. Weatherburn, C.E.: An Introduction to Riemannian Geometry and the Tensor Calculus. Cambridge University Press, Cambridge (1938)
17. Ricci, G., Levi-Civita, T.: Méthodes de calcul differential absolu et leurs applications. Math. Ann. **54**, 125–201 (1900). Reprinted in Levi-Civita, T.: Opere Matematiche, vol. 1, pp. 479–559. Nicola Zanichelli Editore, Bologna (1954)
18. Levi-Civita, T.: The Absolute Differential Calculus. Blackie & Son, London (1926). Reprinted in Dover, New York (1977)
19. Levi-Civita, T.: Nozione di parallelismo in una varietà qualunque e conseguente specificazione geometrica della curvature Riemanniana. Rend. Circ. Mat. Palermo **42**, 173–205 (1917). Reprinted in Levi-Civita, T.: Opere Matematiche, vol. 4, pp. 1–39. Nicola Zanichelli Editore, Bologna (1960)

20. Levi-Civita, T.: Parallelismo e curvature in una varietà qualunque. Questioni di Meccanica classica e relativista pp. 95–143. Zanichelli, Bologna (1924). Reprinted in Levi-Civita, T.: Opere Matematiche, vol. 6, pp. 169–197. Nicola Zanichelli Editore, Bologna (1960)
21. Graustein, W.C.: Differential Geometry. Macmillan Co., New York (1935). Reprinted in Dover, New York (1966) and (2006)
22. Casey, J.: Exploring Curvature. Vieweg, Wiesbaden (1996)
23. O'Neill, B.: Elementary Differential Geometry, 2nd edn. Academic Press, San Diego (1997)
24. Weeks, J.R.: The Shape of Space. Dekker, New York (1985)
25. do Carmo, M.P.: Differential Geometry of Curves and Surfaces. Prentice Hall, New York (1976)
26. Henderson, D.W.: Differential Geometry: A Geometric Introduction. Prentice Hall, New York (1998)
27. Kreyszig, E.: Differential Geometry. University of Toronto Press, Toronto (1959). Reprinted in Dover, New York (1991)
28. Struik, D.J.: Lectures on Classical Differential Geometry, 2nd edn. Addison-Wesley, Reading (1961). Reprinted in Dover, New York (1988)
29. Weatherburn, C.E.: Differential Geometry of Three Dimensions, vol. II. Cambridge University Press, Cambridge (1930)
30. Stoker, J.J.: Differential Geometry. Wiley, New York (1969)
31. Naghdi, P.M.: Geometry of a surface and related results. In: Truesdell, C. (ed.) The Theory of Shells and Plates. Handbuch der Physik, vol. VIa/2, pp. 615–633. Springer, Berlin (1972). Appendix
32. Willmore, T.J.: An Introduction to Differential Geometry. Clarendon Press, Oxford (1959)
33. Willmore, T.J.: Riemannian Geometry. Clarendon Press, Oxford (1993)
34. Sokolnikoff, I.S.: Tensor Analysis: Theory and Applications to Geometry and Mechanics of Continua, 2nd edn. Wiley, New York (1964)
35. Schutz, B.: Geometrical Methods of Mathematical Physics. Cambridge University Press, Cambridge (1980)
36. Chadwick, P.: Continuum Mechanics. Allen & Unwin, London (1976). Reprinted in Dover, New York (1999)
37. Gurtin, M.E., Fried, E., Anand, L.: The Mechanics and Thermodynamics of Continua. Cambridge University Press, Cambridge (2010)
38. Casey, J.: Areal velocity and angular momentum for non-planar problems in particle mechanics. Am. J. Phys. **75**, 677–685 (2007)
39. Casey, J.: A remark on the definition of angular velocity. Z. Angew. Math. Mech. **89**, 922–930 (2009)
40. Casey, J., Lam, V.C.: On the relative angular velocity tensor. J. Mech. Transm. Autom. Des. **108**, 399–400 (1986)
41. Beetle, R.D.: A formula in the theory of surfaces. Ann. Math. **15**, 179–183 (1913–1914)
42. Synge, J.L., Griffith, B.A.: Principles of Mechanics, 3rd edn. McGraw-Hill, New York (1959)
43. Meirovitch, L.: Methods of Analytical Dynamics. McGraw-Hill, New York (1970)
44. Hart, J.B., Miller, R.E., Mills, R.L.: A simple geometric model for visualizing the motion of a Foucault pendulum. Am. J. Phys. **55**, 67–70 (1987)
45. Berry, M.: Anticipations of the geometric phase. Phys. Today **43**, 34–40 (1990)

DOI 10.1007/978-94-017-7300-3_13
Reprinted from *Journal of Elasticity* Journal, DOI 10.1007/s10659-014-9509-y

Non-Euclidean Ribbons

Generalized Sadowsky Functional for Residually-Stressed Thin and Narrow Bodies

Efi Efrati

Received: 14 April 2014 / Published online: 31 December 2014

Abstract The classical theory of ribbons as developed by Sadowsky and Wunderlich has received much attention in recent years. It concerns the equilibrium conformations of thin and narrow ribbons whose intrinsic structure favors a rectangular and flat state. However, the intrinsic structure of naturally formed ribbons will often be more complicated; Spatial variations in the in-plane distance metric can give rise to both geodesic curvature and Gaussian curvature, curving the ribbon in and out of its plane. Moreover, metric variation across the thickness of the ribbon may result in nontrivial reference normal curvatures. The resulting geometric structure is likely to have no zero-energy (stress-free) realizations in Euclidean space.

This paper presents a generalization of the Sadowsky functional, which measures the bending energy of narrow ribbons, for the case of incompatible ribbons (having no stress-free configuration). Specific solutions to special cases where the reference normal curvatures vanish, and for a naturally curved developable ribbon are presented and the resulting twist-stretch relations are discussed.

Keywords Non-Euclidean plates · Residual stress · Ribbons · Sadowsky

Mathematics Subject Classification (2010) 74K20 · 74K25 · 53A05

1 Introduction

The term ribbons is used to describe elastic bodies in which the thickness, the width and the length are all very different, $t \ll w \ll L$. As their length, L, is much larger than all other

E. Efrati (✉)
Department of Physics of Complex Systems, Weizmann Institute of Science, PO Box 26, Rehovot 76100, Israel
e-mail: efi.efrati@weizmann.ac.il

E. Efrati
James Franck Institute, The University of Chicago, 929 E. 57 st., Chicago, IL 60637, USA

dimensions one would like to describe these objects and their elastic response in terms of a one dimensional theory using, for example, the ribbon's center line, as done for rod-like objects. However, in contrast to rod-like elastic structures, which display two independent soft bending directions of deformations, and much like thin sheets, ribbons have only one independent soft bending direction.

There are two approaches to derive the elastic theory for ribbons. The first follows by endowing a rod-like theory with additional internal structure as is done for example in [1]; (this approach is to be distinguished from the elastic rod approximation where one of the bending stiffnesses is sent to infinity, which was shown in [2] to not appropriately capture the Mobius band solution). Alternatively, one could reduce a thin sheet theory to an effective one dimensional description as done for example in [3, 4]. This work follows the latter approach, however, in contrast with the standard derivations of [4] it starts from the theory of non-Euclidean thin sheets [5], which is aimed at describing residually-stressed thin elastic bodies.

The elastic theory of thin non-Euclidean ribbons is expected to bear relevance not only to ribbon-like continuous bodies, but also to the much broader class of slender inextensible filamentous bodies possessing internal structure resulting in only one soft direction of deformation, such as supramolecular assemblies [6].

The present work joins the recent revived interest in elastic ribbons (e.g., [1, 2, 6–8]) and incorporates into these systems the notion of incompatibility [5, 9]. In particular it demonstrates how the results presented in [3] can be understood through a straightforward generalization of the Sadowsky functional. This works also broadens these results to the case of general incompatibility that in particular allows for arbitrary reference normal curvature. The remainder of this paper is organized as follows: Sect. 2 discusses the elastic energy of residually-stressed thin sheets focusing on the thin limit. In Sect. 3 the curvature and torsion of the mid-curve of a ribbon are used to construct a second fundamental form using the compatibility condition. Section 4 presents an application of the obtained reduced elastic energy for the case of a ribbon with negative Gaussian curvature and vanishing reference curvatures. Section 5 presents an application to the case of a ribbon with non-vanishing reference normal curvature, and discusses the twist-stretch couplings that arise in the different cases.

2 The Vanishing Thickness Limit of Non-Euclidean Plates and Shells

When the thickness of an elastic body is much smaller than its other spatial dimensions it is often possible to describe the thin three dimensional body using a two dimensional surface such as its mid-surface, $\mathbf{r}$. This results in a dimensionally reduced elastic energy, E, which is expressed in terms of surface properties representing the mid-surface; namely the first and second fundamental forms (also known as the metric tensor and the normal curvature form). In component form these are given by

$$a_{\alpha\beta} = \frac{\partial \mathbf{r}}{\partial x^\alpha} \cdot \frac{\partial \mathbf{r}}{\partial x^\beta}, \quad \text{and} \quad b_{\alpha\beta} = \frac{(\partial)^2 \mathbf{r}}{\partial x^\alpha \partial x^b} \cdot \hat{\mathbf{N}},$$

where $\hat{\mathbf{N}}$ is the unit normal to the surface, x^α are the components of two dimensional curvilinear coordinates on the surface and $\alpha, \beta = \{1, 2\}$. To distinguish the components of contravariant vectors and tensors from exponentiation, the latter will be bracketed, i.e., $(x)^y$ will imply x to the power y, whereas x^y will denote the yth component of the vector x.

The total energy is decomposed into two distinct terms each penalizing a different type of deformation; stretching deformations and bending deformations, ϵ_S and e_B:

$$\mathcal{E} = Y_S e_S + Y_B e_B. \tag{1}$$

For a continuous body of thickness t, Poisson ratio ν and Young's modulus E, the stretching stiffness and bending stiffness, Y_S and Y_B, are given by

$$Y_S = \frac{Et}{8(1-(\nu)^2)}, \qquad Y_B = \frac{E(t)^3}{24(1-(\nu)^2)}. \tag{2}$$

The stretching content, e_S, measures length mismatches in the local tangent plane expressed as differences between the realized metric, a, and the reference metric, $\bar{a}$:

$$e_S = \iint A^{\alpha\beta\gamma\delta}(a_{\alpha\beta} - \bar{a}_{\alpha\beta})(a_{\gamma\delta} - \bar{a}_{\gamma\delta})dS. \tag{3}$$

The bending content measures the normal curvature mismatch between the realized second fundamental form, b, and the reference second fundamental form, $\bar{b}$:

$$e_B = \iint A^{\alpha\beta\gamma\delta}(b_{\alpha\beta} - \bar{b}_{\alpha\beta})(b_{\gamma\delta} - \bar{b}_{\gamma\delta})dS, \tag{4}$$

where for a material with isotropic response the elasticity tensor is given by

$$A^{\alpha\beta\gamma\delta} = \left(\frac{1-\nu}{2}(\bar{a}^{\alpha\gamma}\bar{a}^{\beta\delta} + \bar{a}^{\alpha\delta}\bar{a}^{\beta\gamma}) + \nu\bar{a}^{\alpha\beta}\bar{a}^{\gamma\delta}\right),$$

and $dS = \sqrt{\det(\bar{a})}dx^1dx^2$ is the area element with respect to the reference metric. For a detailed derivation see [5] and [10].

Not every two symmetric quadratic forms represent a surface embedded in the Euclidean three dimensional space. In order for the forms a and b to represent such a realizable embedding, they need satisfy a set of three non-linear partial differential equations called the Gauss–Peterson–Mainardi–Codazzi compatibility conditions (GPMC) [11]. Thus, finding an elastic equilibrium with respect to the above energy amounts to finding the first and second fundamental forms, a and b, that satisfy the GPMC compatibility conditions and minimize the total energy (1) among all such compatible forms. The reference metric and curvature, $\bar{a}$ and $\bar{b}$, however, need not satisfy the GPMC conditions; whenever they do not, the ribbon's intrinsic geometry is said to be incompatible and allows for no stress free configuration. In this case the zero energy level sets of the stretching content and the bending content become distinct; one cannot find admissible first and second fundamental forms which will make both the stretching and bending energies vanish. Therefore, for every thickness, equilibrium is achieved by balancing the competing energetic costs of stretching and bending.

The limit of vanishing thickness corresponds to the case where the bending stiffness is infinitely smaller than the in plane stretching stiffness. In [5] it was shown for the case of non-Euclidean plates that if a vanishing-thickness limit equilibrium configuration exists, it corresponds to a minimizer of the bending content, e_B, among all isometric embeddings of the reference metric, $a = \bar{a}$. Later, in [12] it was shown that if $\bar{a}$ admits an isometric embedding with finite bending content then such a limit exists.

Thus, for sufficiently thin ribbons the equilibrium configuration is obtained by finding the bending content minimizing isometry, or equivalently finding a second fundamental form, b, that satisfies the GPMC compatibility conditions with respect to the given metric $a = \bar{a}$, and minimizes (4) among all such compatible second forms. The latter approach is next expanded.

3 The Compatibility Conditions Expansion in Semi-geodesic Form

The compatibility conditions can be derived by considering an embedded surface $\mathbf{r}(x^1, x^2)$ and equating the different mixed third derivatives. They can be written compactly as

$$|b| = K|a|, \qquad \nabla_1 b_{12} = \nabla_2 b_{11}, \qquad \nabla_1 b_{22} = \nabla_2 b_{12},$$

where K denotes the Gaussian curvature as calculated from the metric alone, $|\cdot|$ denotes the determinant, and ∇ is to be understood as the covariant derivative with respect to the metric a.

A particularly convenient parametrization is the linear semi-geodesic parametrization. In this parametrization the $x^1 = 0$ parametric line corresponds to the mid-curve parametrized by the arc-length parameter x^2, and the $x^2 = const$ parametric lines are geodesics that perpendicularly intersect the mid-curve. In this case the metric can be expressed in terms of a single scalar field ϕ [11]:

$$a = \begin{pmatrix} 1 & 0 \\ 0 & (\phi)^2 \end{pmatrix}.$$

With respect to this parametrization the compatibility conditions take the following form:

$$\partial_1 b_{12} = \partial_2 b_{11} - b_{12}\partial_1\phi/\phi, \tag{5a}$$

$$\partial_1 b_{22} = \partial_2 b_{12} + b_{11}\phi\,\partial_1\phi - b_{12}\partial_2\phi/\phi + b_{22}\partial_1\phi/\phi, \tag{5b}$$

$$b_{11} = \big((b_{12})^2 - \phi\partial_1\partial_1\phi\big)/b_{22}. \tag{5c}$$

When $b_{22} \neq 0$ Gauss' equation (5c) reduces the number of variables to only two. The metric function ϕ and the remaining variables, b_{11} and b_{12}, may then be expanded in orders of x^1:

$$\begin{aligned} \phi &= \overset{0}{\phi} + \overset{1}{\phi}(x^1) + \frac{1}{2}\overset{2}{\phi}(x^1)^2 + \frac{1}{6}\overset{3}{\phi}(x^1)^3 + \mathcal{O}(x^1)^4 \\ &\equiv 1 - k_g x^1 - \frac{1}{2}K(x^1)^2 + \frac{1}{6}(K\,k_g - G)(x^1)^3 + \mathcal{O}(x^1)^4, \\ b_{ij} &= \overset{0}{b}_{ij} + \overset{1}{b}_{ij}(x^1) + \frac{1}{2}\overset{2}{b}_{ij}(x^1)^2 + \mathcal{O}(x^1)^3. \end{aligned}$$

where all the functions above may depend on the variable x^2 and correspond to the mid-curve ($x^1 = 0$) values of the relevant surface properties; namely the geodesic curvature (k_g), the Gaussian curvature (K) and the transverse derivative of the Gaussian curvature, $\partial_1 K (= G)$. Note that ϕ and all the parameters derived from it (K, k_g and G) are considered to be given functions.

Substituting this expansion back into Eqs. (5a) and (5b) and collecting orders in x^1 gives a hierarchy of equations in which the nth order curvature components $\overset{n}{b}_{11}$ and $\overset{n}{b}_{12}$ are expressed in terms of the known function ϕ and lower order contributions to the curvature components, $\overset{m}{b}_{ij}$, where $m < n$.

The zeroth order contribution to the curvature tensor, as explicitly derived in the next section, is expressible in terms of mid-curve properties, i.e., the curvature and torsion, κ and τ. It follows that the expansion of the second fundamental form for every finite order is determined by the curvature and torsion functions characterizing the mid-curve, and their derivatives.

3.1 Leading Order Contributions to the Mid-Curve Curvature Form

The parameter x^2 serves as an arc-length parameter of the mid-curve (defined by $x^1 = 0$). The tangent, normal and binormal Serret–Frenet directions $\hat{\mathbf{t}}$, $\hat{\mathbf{n}}$ and $\hat{\mathbf{b}}$, may thus be related to the x^2 derivatives of the surface $\mathbf{r}$ evaluated along the mid-curve through

$$
\begin{aligned}
\partial_2 \mathbf{r} &= \hat{\mathbf{t}}, \\
\partial_2\partial_2 \mathbf{r} &= k_g \partial_1 \mathbf{r} + \overset{0}{b}_{22} \hat{\mathbf{N}} = \kappa \hat{\mathbf{n}}, \\
\partial_2\partial_2\partial_2 \mathbf{r} &= \big(k_g' - \overset{0}{b}_{22}\overset{0}{b}_{12}\big)\partial_1 \mathbf{r} - k_g\big(k_g + (\overset{0}{b}_{22})^2\big)\partial_2 \mathbf{r} + (k_g \overset{0}{b}_{12} + \partial_2 \overset{0}{b}_{22})\hat{\mathbf{N}}, \\
&= \kappa' \hat{\mathbf{n}} - (\kappa)^2 \hat{\mathbf{t}} + \kappa\tau \hat{\mathbf{b}},
\end{aligned}
\tag{6}
$$

where, as before, k_g corresponds to the geodesic curvature of the mid-curve and prime denotes differentiation along the x^2 direction. Combining the above yields

$$
\begin{aligned}
(\partial_2 \mathbf{r} \times \partial_2\partial_2 \mathbf{r}) \cdot \partial_2\partial_2\partial_2 \mathbf{r} = (\kappa)^2 \tau &= b_{22}\big(k_g' - \overset{0}{b}_{22}\overset{0}{b}_{12}\big) - k_g(k_g \overset{0}{b}_{12} + \partial_2 \overset{0}{b}_{22}) \\
&= -\overset{0}{b}_{12}\big((k_g)^2 + (\overset{0}{b}_{22})^2\big) + (\overset{0}{b}_{22})^2 (k_g/\overset{0}{b}_{22})'.
\end{aligned}
\tag{7}
$$

From (6) it is easy to recover that the curvature of the mid-curve satisfies $\kappa = \sqrt{(\overset{0}{b}_{22})^2 + (k_g)^2}$. Defining θ to measure the angle between the Serret–Frenet normal to the mid-curve and the surface's normal gives

$$
\hat{n} \cdot \hat{\mathbf{N}} = \cos(\theta) = \overset{0}{b}_{22}/\kappa, \quad \text{and} \quad \sin(\theta) = k_g/\kappa, \tag{8}
$$

which in turn reduce Eq. (7) to $b_{12} = \theta' - \tau$. The leading-order contribution to the second fundamental form can thus be compactly written in the following form:

$$
\overset{0}{b} = \begin{pmatrix} \frac{(\theta'-\tau)^2 + K}{\sqrt{(\kappa)^2 - (k_g)^2}} & \theta' - \tau \\ \theta' - \tau & \sqrt{(\kappa)^2 - (k_g)^2} \end{pmatrix}. \tag{9}
$$

4 Non-Euclidean Plate-Like Ribbons

Non-Euclidean plates are a subset of frustrated elastic thin sheets in which $\bar{b} = 0$, but the reference metric, $\bar{a}$, is associated with a non-vanishing reference Gaussian curvature, $\bar{K}$. Their bending energy thus favors flat configurations. However, as discussed above, in the limit of vanishing thickness $a = \bar{a}$, thus precluding the possibility of flat embeddings, where $b_{ij} = 0$ (and therefore also $K = 0$). Their bending content (4) reduces to

$$
e_B = \iint \big(4(H)^2 - 2(1-\nu)K\big) dS, \tag{10}
$$

where $H = \frac{1}{2}\,\mathrm{Tr}(a^{-1}b)$ and $K = |b|/|a|$ are the mean and Gaussian curvatures of the surface respectively. For isometric transformations the Gaussian curvature is fixed. Therefore in order to determine the vanishing thickness configuration of the thin sheet it suffices to minimize the Willmore functional, $\iint (H)^2 dS$, with respect to all isometries of the reference metric [5, 13]. However, the values that the mean curvature may assume are not independent of the metric and must satisfy the GPMC compatibility conditions, as discussed in the next section.

4.1 Generalized Sadowsky Functional for Non-Euclidean Plate-Like Ribbons and Its Trivial Minimizers

Substituting the expressions for the second fundamental form in terms of the mid-curve properties (9) into (10) gives the generalized Sadowsky functional:

$$\iint 4(H)^2 dS \approx w \int \frac{((\tau - \theta')^2 + K + (\kappa)^2 - (k_g)^2)^2}{((\kappa)^2 - (k_g)^2)} ds. \tag{11}$$

When $k_g = K = 0$ (which implies $\theta' = 0$) the familiar form of the Sadowsky functional is recovered, $w \int \frac{((\tau)^2 + (\kappa)^2)^2}{(\kappa)^2} ds$, [14]. However, the form (11) enables considering ribbons with additional intrinsic structure, namely Gaussian curvature and geodesic curvature.

In the general case, ribbons may be forced or constrained by boundary conditions. Such constraints may render isometric configurations inaccessible or lead to curvature localization invalidating the reduced energy (11), much in the same way the Sadowsky reduced energy was recently shown to be invalid under strong stretch and twist conditions [15]. However, in some cases, such as those considered next, not only does Eq. (11) predict the equilibrium configuration, but it is also the case that these configurations are obtained by point-wise minimization of the integrand. The integrand in (11) is non-negative, thus if there exists a solution for which it vanishes identically, then such a solution will be the minimal energy solution in the absence of external constraints. These specific solutions capture some of the variety of solutions one may obtain by allowing ribbons to assume non-trivial intrinsic geometries.

For the case $K = 0$, (recalling that by definition $(\kappa)^2 \geq (k_g)^2$) the only zero energy solution of (11) satisfies $\kappa = k_g$ (leading to $\theta' = 0$) and thus also $\tau = 0$. If k_g is constant this corresponds to a section of a planar annulus. Considering the case $K > 0$, minimal energy is obtained for $\tau = 0$ and $(\kappa)^2 = K + (k_g)^2$. Up to the choice of sign for the curvature, here again the configuration is unique. These two results can be easily understood geometrically: for a flat sheet only a single $H = 0$ solution exists. For the positive Gaussian curvature case the minimal value of (the absolute value of) the mean curvature is obtained when the principal curvatures are identical $k_1 = k_2 = \sqrt{K}$, and as in the flat case this solution is unique.

However, when considering the hyperbolic case, where $K < 0$, the minimal energy configuration is not unique. Setting $k_1 = -k_2 = \sqrt{K}$ gives $H = 0$, yet there are infinitely many such configurations differing by the relative orientation of the principal curvatures and the mid-curve. Within the framework of (11) one sees this by observing that there is a one-parameter family of solutions that set (11) to zero identically. This family of solutions is studied next.

4.2 Unconstrained Hyperbolic Ribbons

A simple example of the one parameter family of solutions obtained for $K < 0$ follows by considering the case $k_g = 0$. In this case every curve that satisfies $(\kappa)^2 + (\tau)^2 = -K$

corresponds to a vanishing leading order contribution to the mean curvature (when $\kappa = 0$ then τ is ill defined and the assumption $b_{22} \neq 0$ does not hold. However, in this case a zero energy solution still exists, and satisfies $(b_{12})^2 = -K$). This results in a one dimensional minimal energy level set with the parameter κ:

$$\mathbf{r}(s) = \big(\kappa/(\alpha)^2 \cos(\alpha s), \kappa/(\alpha)^2 \sin(\alpha s), s\sqrt{1-(\kappa/\alpha)^2}\big), \quad \alpha = \sqrt{(\kappa)^2 + (\tau)^2}, \tag{12}$$

where $s = x^2$ is the arc-length parameter along the mid-curve. Varying κ changes the end-to-end distance of the ribbon, yet as $\alpha = \sqrt{-K} = const$ this variation carries no end rotation (additional winding) of the mid-curve of the ribbon.

The above results may now be used to calculate the leading contribution to the mean curvature:

$$H = \frac{1}{2} a^{\alpha\beta} b_{\alpha\beta} = \frac{1}{2}\big(b_{11} + (\phi)^{-2} b_{22}\big) = \frac{(\overset{0}{b}_{12})^2 + K + (\overset{0}{b}_{22})^2}{2\overset{0}{b}_{22}} + \mathcal{O}\big(x^1\big).$$

For the case $K < 0$ (hyperbolic ribbons), as described in the previous section, there are infinitely many configurations for which the leading order bending energy term (11) vanishes identically. These configurations all satisfy

$$\big(\tau - \theta'\big)^2 + K + (\kappa)^2 - (k_g)^2 = 0.$$

With the help of this relation the torsion is eliminated from the curvature form (9) resulting in $\overset{0}{b}_{22} = -\overset{0}{b}_{11} = \sqrt{\kappa^2 - k_g^2}$ and $\overset{0}{b}_{12} = \sqrt{-K - (\kappa)^2 + (k_g)^2}$. The next order correction to the generalized Sadowsky functional in this case reads

$$\iint 4(H)^2 dS = (w)^3 \int \big(2\partial_2 \overset{0}{b}_{12} - 2\overset{0}{b}_{12}\partial_2 \overset{0}{b}_{22}/\overset{0}{b}_{22} + (G - 4Kk_g)/\overset{0}{b}_{22}\big)^2 ds + \mathcal{O}\big((w)^5\big).$$

Here again the integrand can be set identically to zero resulting in the equation

$$\frac{\partial_2 f}{1 + (f)^2} = G/2K - 2k_g, \quad \text{where } f = \overset{0}{b}_{12}/\overset{0}{b}_{22}.$$

The solution to the equation yields

$$\overset{0}{b} = \sqrt{-K}\begin{pmatrix} -\cos(\beta(x^2)) & \sin(\beta(x^2)) \\ \sin(\beta(x^2)) & \cos(\beta(x^2)) \end{pmatrix}, \quad \beta\big(x^2\big) = \int (G/2K - 2k_g)\,dx^2.$$

This result was first derived (in a slightly different fashion) in ref [3]. where further study of these solutions and their properties can be found. Figure 1 depicts typical solutions for the case $K = -1$ and varying values of k_g.

5 Non-Euclidean Ribbons with Non-vanishing Reference Curvature

The intrinsic structure of plate-like ribbons is symmetric under reflections about the mid-plane of the ribbon. Therefore, naturally handed structures cannot be appropriately described as plate-like ribbons. Including the effect of rest-length variation across the thickness of the

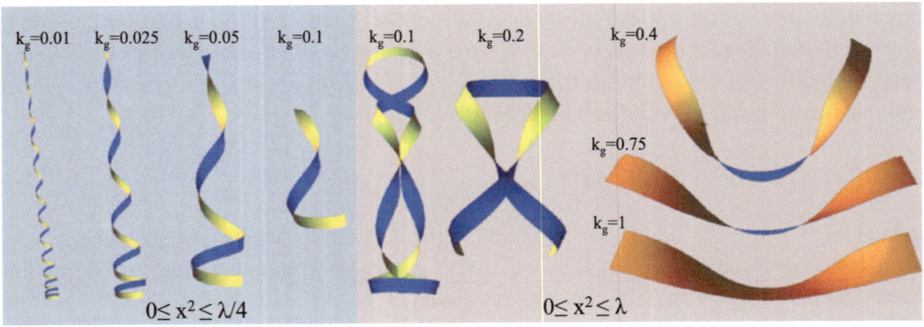

Fig. 1 *Non-Euclidean plate-like ribbons with negative Gaussian curvature.* Minimal energy configurations of ribbons with Gaussian curvature $K=-1$, width $w=0.5$ and varying geodesic curvature. For visualization purposes for the lower values of the geodesic curvature only a quarter of the typical solution period λ is plotted (*left*), whereas for the higher values of k_g a full period is plotted (*right*)

ribbon breaks the symmetry for reflection about the mid-plane, and suffices to generate handed structures [7]. When the ribbon is associated with a non-vanishing reference second fundamental form $\bar{b}$, which gives rise to the reference mean curvature $\bar{H}=\frac{1}{2}\bar{b}^{\alpha}_{\alpha}$, the generalized Willmore functional can be written compactly as

$$e_B=\iint\big(4(H-\nu\bar{H})^2-2(1-\nu)\bar{b}^{\gamma\delta}b_{\gamma\delta}\big)dS. \tag{13}$$

Note that the form above, while compact, may be somewhat misleading for interpretation; the first term is obtained by completing the square for $4(H-\bar{H})^2+8(1-\nu)H\bar{H}$, and two constant terms, $4(1-\nu^2)(\bar{H})^2-4(1-\nu)\bar{K}$, were omitted. From here on forward, the bending energy e_B should be interpreted as the bending energy up to additive constants (independent of the configuration of the ribbon). The next section describes a specific example in which the functional (13) is expressed in terms of mid-curve properties.

5.1 Uniaxial Reference Curvature: A Cylindrical Helical Ribbon

For a flat reference metric and a uniaxial reference curvature a stress free configuration for the ribbon exists and gives rise to a cylindrical helical ribbon. Setting the principal curvature direction at 45° to the mid-line of the ribbon gives rise to the following parameters:

$$\bar{a}=\begin{pmatrix}1&0\\0&1\end{pmatrix},\qquad \bar{b}=\begin{pmatrix}\frac{1}{\sqrt{2}}&\frac{1}{\sqrt{2}}\\-\frac{1}{\sqrt{2}}&\frac{1}{\sqrt{2}}\end{pmatrix}\begin{pmatrix}1&0\\0&0\end{pmatrix}\begin{pmatrix}\frac{1}{\sqrt{2}}&-\frac{1}{\sqrt{2}}\\\frac{1}{\sqrt{2}}&\frac{1}{\sqrt{2}}\end{pmatrix}=\frac{1}{2}\begin{pmatrix}1&-1\\-1&1\end{pmatrix},$$

$$\bar{H}=\frac{1}{2}.$$

These, in turn, give rise to the reduced energy

$$e_B=\iint\big(4(H-\nu/2)^2-2(1-\nu)H+(1-\nu)2b_{12}\big)dS. \tag{14}$$

Assuming the mid-curve adopts a helical configuration of the form (12) the above energy may be rewritten as

$$\int\left(\frac{(\alpha)^2}{(1-(\zeta)^2)}-2(1-\nu)\alpha\zeta-\frac{(1+\nu)\alpha}{\sqrt{1-(\zeta)^2}}\right)dx^2,$$

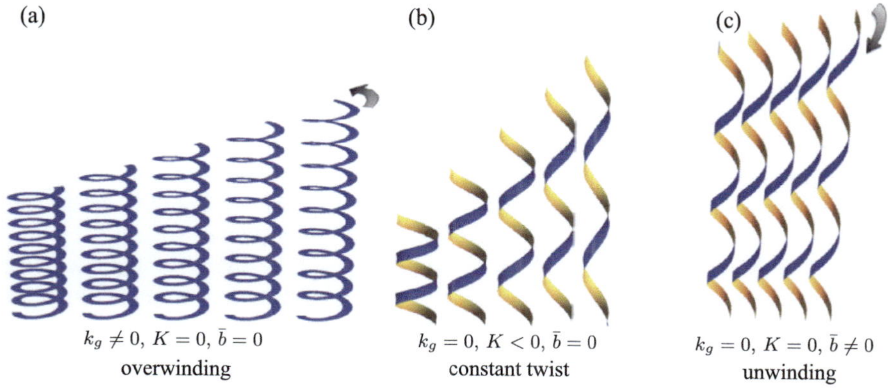

Fig. 2 *Twist stretch response of different ribbons.* By tuning the intrinsic structure of a ribbon one can generate all possible twist-stretch coupling. (**a**) A slinky-like ribbon with geodesic curvature but no Gaussian curvature and no reference normal curvature shows overwinding when stretched. The *arrow* denotes the rotation induced by extension. Similar behavior is observed for DNA molecules [16] and is also typical for simple (rod-like) elastic springs under small displacements [17]. (**b**) For negative Gaussian curvature, vanishing geodesic curvature and vanishing normal reference curvatures, stretching results in no twist. A well-known geometric example exhibiting this behavior is the continuous isometric transformation between the catenoid and the helicoid. Every surface in this family is minimal (has vanishing mean curvature everywhere), thus constitute a minimizer of the plate-like elastic bending energy with the prescribed end displacements. (**c**) For a cylindrical ribbon (vanishing Gaussian curvature and vanishing geodesic curvature) infinitesimal stretching leads to unwinding. The *arrow* again denotes the rotation induced by extension

where $\zeta = \sqrt{1-(\kappa/\alpha)^2} = \tau/\alpha$ relates the axial height of a helix, ΔZ, and its total length L via $\zeta L = \Delta Z$, and $\alpha = \sqrt{(\kappa)^2 + (\tau)^2}$ measures the total angular winding through $\theta_{max} = \alpha L$, (see Eq. (12)). For every given end displacement value ζL there is a corresponding energy minimizing winding rate α^*. Calculating the quantity $\frac{d\alpha^*}{a\zeta}$ gives the torque free twist-stretch coupling. In the present case this quantity is negative, reading $\frac{d\alpha^*}{d\zeta}| = -1$ at the natural length $\zeta^* = \frac{1}{\sqrt{2}}$. This result implies that the cylindrical helical ribbon unwinds as it stretches, (see Fig. 2). This is in contrast with the plate-like hyperbolic ribbon (12), which exhibited zero twist stretch coupling, and also in contrast with a slinky-like ribbon ($\bar{b} = 0$, $k_g \neq 0$) which, similarly to simple springs [17] and DNA molecules [16], overwinds when stretched.

6 Discussion

This paper describes a generalization of the Sadowsky functional for ribbons with arbitrary Gaussian curvature, geodesic curvature and reference normal curvatures. The analysis presented is expected to appropriately describe only thin and narrow ribbons. For an elastic strip to be considered narrow its width should be smaller not only compared to its length, but also compared to the local geometric scale as embodied in the Gaussian curvature. However, thin sheets are known to display a boundary layer in which finite thickness effects become significant and in plane strains arise [18]. Therefore to be considered thin, the width of the ribbon needs to be much larger than the width of the boundary layer which scales as $t^{1/2}K^{-1/4}$. Thus, for the above theory to be valid the width of the ribbon needs satisfy

$$t \ll w \ll L, \quad \text{and} \quad t^{1/2}K^{-1/4} \ll w \ll K^{-1/2}.$$

The present work only treats the infinitely thin limit and no finite thickness corrections are considered. Even in this isometric limit, there are subtle issues regarding the uniqueness of the resulting embedding of the ribbons. These issues are not captured by the formal hierarchy of equations for the second fundamental form (5a)–(5c) which predicts the configuration for narrow ribbons. For example, for the case of ribbons of negative Gaussian curvature where the directions of principal curvature rotate along the mid-curve, it was shown that there are regions whose configuration is not uniquely determined by the configuration of the mid-curve. The reader is referred to [3] for further discussion of these subtle issues.

Molecular assemblies can display ribbon-like properties. If a polymer cannot freely twist about its mid-curve, as is the case for double stranded DNA, then impenetrability gives rise to non-trivial topological constraints on its allowed motions [19]. However, the theory presented here does not necessarily hold for every structure exhibiting ribbon-like properties. The mechanical response properties of DNA, for example, have been shown to be appropriately captured by rod-like models [17, 20], or bi-rod models [21] rendering it a "thick" ribbon, and thus outside the scope of the present work.

There are molecular structures, however, ranging from graphene nano-ribbons [22] to lipid crystals grown from solution [23], to metastable states of cholesterol in bile [24] that display sufficient anisotropic response properties to allow an elastic description as thin non-Euclidean ribbons. The internal structure of such molecular assemblies, e.g., spatial variation in interatomic bond-length and bond-angle, can be interpreted as providing a reference metric and reference normal curvatures that give rise to non-trivial unconstrained configuration. In such cases, the reduced energy presented here allows a simple and direct relation between the underlying geometric properties of the structure and its mechanical response.

Acknowledgements I would like to thank D. Biron, O. Feinerman, M. Moshe and T.A. Witten for helpful discussions and M. Dias and B. Audoly for providing me with a copy of their manuscript ahead of print and for helpful comments.

References

1. Dias, M.A., Audoly, B.: arXiv preprint (2014). arXiv:1403.2094. doi:10.1007/s10659-014-9487-0
2. Starostin, E.L., Van der Heijden, G.H.M.: Nat. Mater. **6**(8), 563 (2007)
3. Efrati, E., Sharon, E., Kupferman, R.: Phys. Rev. E **83**(4), 046602 (2011)
4. Wunderlich, W.: Monatshefte Math. **66**(3), 276 (1962)
5. Efrati, E., Sharon, E., Kupferman, R.: J. Mech. Phys. Solids **57**(4), 762 (2009)
6. Armon, S., Aharoni, H., Moshe, M., Sharon, E.: Soft Matter **10**(16), 2733 (2014)
7. Armon, S., Efrati, E., Kupferman, R., Sharon, E.: Science **333**(6050), 1726 (2011)
8. Chopin, J., Kudrolli, A.: Phys. Rev. Lett. **111**(17), 174302 (2013)
9. Efrati, E., Sharon, E., Kupferman, R.: Soft Matter **9(34)**(8187), 00002 (2013)
10. Ciarlet, P.G.: An Introduction to Differential Geometry with Applications to Elasticity, 2005th edn. Springer, Berlin (2006)
11. Struik, D.J.: Lectures on Classical Differential Geometry, 2nd edn. Dover, New York (1988)
12. Lewicka, M., Reza Pakzad, M.: ESAIM Control Optim. Calc. Var. **17**(4), 1158 (2011)
13. Willmore, T.J.: An Introduction to Differential Geometry, 1st edn. Clarendon, Oxford (1959)
14. Sadowsky, M.: Sitz.ber. Preuss. Akad. Wiss. Berl. Philos.-Hist. Kl. **22**, 412–415 (1930)
15. Chopin, J., Démery, V., Davidovitch, B.: J. Elast., 1–53 (2014). doi:10.1007/s10659-014-9498-x
16. Gore, J., Bryant, Z., Nöllmann, M., Le, M.U., Cozzarelli, N.R., Bustamante, C.: Nature **442**(7104), 836 (2006)
17. Đuričković, B., Goriely, A., Maddocks, J.H.: Phys. Rev. Lett. **111**, 108103 (2013)
18. Efrati, E., Sharon, E., Kupferman, R.: Phys. Rev. E **80**(1), 016602 (2009)
19. Plewa, J.S., Witten, T.A.: J. Chem. Phys. **112**(22), 10042 (2000)
20. Kamien, R.D., Lubensky, T.C., Nelson, P., O'Hern, C.S.: Europhys. Lett. **38**(3), 237 (1997)
21. Moakher, M., Maddocks, J.H.: Arch. Ration. Mech. Anal. **177**(1), 53 (2005)

22. Bets, K.V., Yakobson, B.I.: Nano Res. **2**(2), 161 (2009)
23. Thomas, B.N., Lindemann, C.M., Corcoran, R.C., Cotant, C.L., Kirsch, J.E., Persichini, P.J.: J. Am. Chem. Soc. **124**(7), 1227 (2002)
24. Chung, D.S., Benedek, G.B., Konikoff, F.M., Donovan, J.M.: Proc. Natl. Acad. Sci. USA **90**(23), 11341 (1993)

DOI 10.1007/978-94-017-7300-3_14
Reprinted from *Journal of Elasticity* Journal, DOI 10.1007/s10659-015-9518-5

The Second-Order L^2-Flow of Inextensible Elastic Curves with Hinged Ends in the Plane

Chun-Chi Lin[1,2] · **Yang-Kai Lue**[1] · **Hartmut R. Schwetlick**[3]

Received: 14 January 2014 / Published online: 12 February 2015

Abstract In this article, we study the evolution of *open* inextensible planar curves with hinged ends. We obtain long time existence of C^∞-smooth solutions during the evolution, given the initial curves that are only C^2-smooth with vanishing curvature at the boundary. Moreover, the asymptotic limits of this flow are inextensible elasticae. Our method and result extend the work by Wen (Duke Math. J. 70(3):683–698, 1993).

Keywords Geometric flow · Second-order parabolic equation · Hinged boundary conditions · Elastic energy · Willmore functional

Mathematics Subject Classification (2010) 35B65 · 35K51 · 53A04 · 53C44

1 Introduction

The Euler–Kirchhoff theory of elastic rods is one of the simplest mathematical models in elasticity [2] or [15]; meanwhile a different but related mechanical model is the Möbius strip [23]. These simple mechanical models provide many challenging mathematical problems in the calculus of variations and related dynamics. On the other hand, there has been a lot of recent progress in geometric gradient flows, leading to the development of new techniques in mathematical analysis. In this article, we combine these two areas and study an

✉ C.-C. Lin
chunlin@math.ntnu.edu.tw

Y.-K. Lue
luf961@yahoo.com.tw

H.R. Schwetlick
schwetlick@maths.bath.ac.uk

1 Department of Mathematics, National Taiwan Normal University, 116 Taipei, Taiwan

2 2F of Astronomy-Mathematics Building, National Taiwan University, No. 1, Sec. 4, Roosevelt Road, 106 Taipei, Taiwan

3 Department of Mathematical Sciences, University of Bath, Bath, UK

evolution equation for elastic curves using a geometric formulation which can be viewed as an over-damped dynamics of the physical system.

Geometric flows of curves associated with energy functionals of higher-order derivatives have been motivated in various settings, for example, higher-order variational problems in differential geometry [3, 26], interpolation problems of curves by splines in computer-aided geometric design [4, 11, 18], and mechanical modeling of DNA molecules [12, 20, 21]. These energy functionals are often related to certain Sobolev norms of first-order derivatives (e.g., stretching energy) and second-order derivatives (e.g., bending energy) of curves.

The geometric flow of elastic curves in a parabolic PDE has been investigated for the case of *closed* curves [9, 13, 22, 25]. The study of open curves in the PDE setting only started recently: for example, the 4th-order flow of elastic curves with a positive stretching coefficient and clamped ends in [16]; the 4th-order flow for elastic curves of Willmore–Helfrich type with a positive stretching coefficient and hinged ends in [6]; the 4th-order flow of elastic curves with fixed length and with hinged or clamped ends in [7] or [8]; the 4th-order flow of elastic inextensible planar curves with hinged ends or infinite length in [19]. Notice that literature on the so-called curve-straightening flow, in which the min-max method [14] is used, does not rely on a PDE setting. However, the results hold for the cases of either open or closed curves.

In this article, we investigate the initial-boundary value problem of a one-dimensional *second-order* parabolic PDE for $T : \bar{I} \times [0, t_0) \to \mathbb{S}^1(1) \subset \mathbb{R}^2$, $t_0 > 0$, $I := (0, L)$,

$$\partial_t T = \nabla_s^2 T - \langle \vec{\lambda}, \vec{n} \rangle \vec{n}, \quad \text{in } I \times (0, t_0), \tag{1.1}$$

$$(\partial_s T)_{\lfloor \partial I \times (0, t_0)} = 0, \tag{1.2}$$

$$T_{\lfloor \bar{I} \times \{t=0\}} = T_0, \tag{1.3}$$

where ∇_s is the covariant differentiation defined in (1.6) or (1.7) below, $\vec{n}$ is the unit normal vector in $\mathbb{R}^2$ derived from a counterclockwise rotation of T by 90 degrees, and $\vec{\lambda} = \vec{\lambda}(t)$ is defined by

$$\vec{\lambda} := \left(\int_I |\partial_s T|^2 T ds \right) \cdot A_T^{-1}, \tag{1.4}$$

where

$$A_T := \int_I (\vec{n})^t \vec{n} ds \tag{1.5}$$

is a 2×2 matrix. Thus, if $\det A_T = 0$, the vector-valued Lagrange multiplier $\vec{\lambda}$ is not well-defined. Note that (1.2) will be called the *hinged* boundary condition in this paper. From the fundamental theorem of differential geometry of planar curves, a unit $\mathbb{R}^2$-valued function T can represent a planar curve γ up to rotation and translation with T being the unit tangent vector field of γ. Thus, in the remainder of this article, T will be called a tangent indicatrix. Equation (1.1) has been investigated by Wen [24] for the case where the domain I is topologically a circle $\mathbb{S}^1$, i.e., $\partial I = \emptyset$. Our result is an extension of Wen's work, as our approach to treating open intervals can also be applied to the problem considered by Wen. The solutions of (1.1) give a way of evolving planar inextensible curves (see the next section for details). Notice that (1.1) for the indicatrix T is a second-order nonlinear parabolic partial differential equation, instead of fourth-order parabolic equation for the position vector of the curve [9, 22, 25]. One remarkable property of solutions of (1.1) is that its smooth solutions demonstrate nice geometric properties during evolution: e.g., convexity-preserving, non-increasing of the number of inflection points [1, 24]. These geometric properties are analogous to those of curve-shortening flow [10] and could be useful in applications. In contrast, Linnér [17] gave an example showing the failure of convexity-preserving during the so-called curve-straightening flow of planar curves. To the best of our knowledge, it is not clear whether these geometric properties could hold for the fourth-order L^2-flow of elastic curves considered in [9, 16, 22, 25].

The long time existence of smooth solutions for the L^2-flow, (1.1), (1.2), (1.3), can be derived by applying Gagliardo–Nirenberg type inequalities to the estimates of L^2-norms of high-order derivatives of curvature. The main difficulty in extending the work of Wen [24] to the case of open curves comes from the extra terms associated to the boundary. It does not appear to be possible to deal with these extra terms by invoking Gagliardo–Nirenberg type inequalities in the L^2-estimates. In [16], we found that good boundary conditions provide certain algebraic relationships to derive "higher-order energy inequalities" which allow us to obtain the uniform bounds of higher-order derivatives of curvature. Namely, as replacing the spatial derivative estimates of $\|\nabla_s^m \vec{\kappa}\|_{L^2}$ by the time-like derivative $\|\nabla_t^m f\|_{L^2}$, where f is the position vector of curves and the covariant differentiation ∇_t is defined in (1.7), we avoid the estimates of boundary terms in the case of clamped ends. However, in this article, we find that the approach by estimating $\|\nabla_s^m \vec{\kappa}\|_{L^2}$ still works in the case of hinged end boundary conditions.

Below we give the main result of this article and use the notations, $\mathbb{N}_0 := \mathbb{N} \cup \{0\}$ and T is always the tangent indicatrix. For a vector field $g : I = (0, L) \to \mathbb{R}^N$, let $g^{(k)} := \frac{d^k}{ds^k} g$ for $k \in \mathbb{N}_0$. Moreover, we use the notation $g \in C^\infty(\bar{I})$ to mean that g and all of its derivatives, $g^{(k)}$, are continuous up to the boundary ∂I. Let

$$\nabla_s g := \partial_s g - \langle \partial_s g, T \rangle T \tag{1.6}$$

denote the covariant derivative of the vector field g. Similarly, for any vector field $h : (0, t_0) \times I \to \mathbb{R}^2$, we let

$$\nabla_t h := \partial_t h - \langle \partial_t h, T \rangle T; \qquad \nabla_s h := \partial_s h - \langle \partial_s h, T \rangle T. \tag{1.7}$$

For a normal vector field defined over I, i.e., vector fields perpendicular to T, we put an arrow on top, for example, $\vec{\eta}$.

Theorem 1 *Let the initial data $T_0 : \bar{I} = [0, L] \to \mathbb{S}^1(1) \subset \mathbb{R}^2$ belong to the class $C^1(\bar{I})$ fulfilling $T_0'(0) = T_0'(L) = 0$. Then, subject to the initial-boundary conditions* (1.2) *and* (1.3), *there exists a global solution $T : \bar{I} \times [0, \infty) \to \mathbb{S}^1(1)$ of the L^2-flow* (1.1), *where*

$$T, \partial_s T \in C^0\big(\bar{I} \times [0, \infty)\big) \cap C^\infty\big(\bar{I} \times (0, \infty)\big).$$

Moreover, there exists a convergent subsequence, $\{T_i(\cdot) := T(\cdot, t_i)\}$ for a sequence $t_i \to \infty$, such that $T_\infty = \lim_{i\to\infty} T_i$ and T_∞ is a stationary solution, i.e., T_∞ is a critical point of the functional (2.6).

The following corollary is a simple consequence of Theorem 1 by applying the fundamental theorem of planar curves.

Corollary 1 *Let $f_0 : \bar{I} = [0, L] \to \mathbb{R}^2$ be a planar curve and belong to the class $C^2(\bar{I})$ with vanishing curvature at boundary. Then, subject to the boundary conditions of vanishing curvature and fixed end points $f_0(0)$, $f_0(L)$ (i.e., the hinged ends), there exists an inextensible* *evolution of planar curves* $f : \bar{I} \times [0, \infty) \to \mathbb{R}^2$ *such that*

$$f, \partial_s f, \partial_s^2 f \in C^0\big(\bar{I} \times [0, \infty)\big) \cap C^\infty\big(\bar{I} \times (0, \infty)\big).$$

Moreover, the inextensible evolution of curve f_0 *subconverges to an inextensible elastica, i.e., an equilibrium configuration of the functional* $\mathcal{E}$ *in the class of inextensible curves* $\mathcal{D}_{f_0}$ *(see* (2.2)*) subject to the hinged ends.*

The rest of this article is organized as follows: In Sect. 2, we derive the Euler–Lagrange equation and set up the nonlinear second-order parabolic flow, (1.1). In Sect. 3, we prove Theorem 1, the main result of this article.

2 The L^2-Flow Equation

Let $I := (0, L)$ and $f : [0, L] = \bar{I} \to \mathbb{R}^2$ be an open (or *non-closed*) regular planar curve, $L > 0$ represent the total length of f, and $s \in \bar{I}$ be the arclength parameter of f. The curve is still said to be open even if $f(0) = f(L)$, since we don't treat this point $f(0) = f(L)$ as an interior point of the curve f. Denote by T the tangent indicatrix of f and by $\vec{\kappa}$ the curvature vector of f. The bending energy of planar curves is defined by

$$\mathcal{E}[f] = \int_I \frac{1}{2} |\vec{\kappa}|^2 ds. \tag{2.1}$$

When a smooth and regular planar curve $f_0 : \bar{I} \to \mathbb{R}^2$ is a critical point of the bending energy in (2.1), it is said to be inextensible if its perturbation class is restricted to

$$\mathcal{D}_{f_0} = \big\{ f \in C^\infty\big(I \times (-1, 1)\big) : f(s, 0) = f_0, \big|\partial_s f(s, \varepsilon)\big| = 1, \ \forall s \in I, \ \forall \varepsilon \in (-1, 1)\big\}. \tag{2.2}$$

Suppose $f(0) = p_-$, $f(L) = p_+$ and $L > |p_+ - p_-|$. The family of inextensible planar curves with fixed length L and fixed end points p_-, p_+ can be equivalently described by a family of tangent indicatrices. Namely, we first let

$$\mathcal{A}_L := \big\{ T : T \in C^\infty\big(\bar{I}, \mathbb{S}^1(1)\big)\big\} \tag{2.3}$$

and

$$\mathcal{A}_{L,0} := \big\{ T : T \in \mathcal{A}_L, T'_{\lfloor \partial I} = 0 \big\}. \tag{2.4}$$

Observe that if the family of tangent indicatrices $\{T(\cdot, \varepsilon) : \varepsilon \in (-1, 1)\} \subset \mathcal{A}_L$ fulfills the constraint

$$\int_I T(s, \varepsilon) ds = p_+ - p_- =: \Delta p, \quad \forall \text{ fixed } \varepsilon \in (-1, 1), \tag{2.5}$$

then the fundamental theorem of planar curves allows us to construct a family of inextensible planar curves with fixed length L and fixed end points p_-, p_+ from $\{T(\cdot, \varepsilon) : \varepsilon \in (-1, 1)\}$.

Consider the functionals $\mathcal{F}_L$ and $\mathcal{F}_{L,\Delta p}$: $\mathcal{A}_L \to \mathbb{R}$ defined according to

$$\mathcal{F}_L[T] := \int_I \frac{1}{2} |\partial_s T|^2 ds \tag{2.6}$$

and

$$\mathcal{F}_{L,\Delta p}[T] := \int_I \frac{1}{2}|\partial_s T|^2 ds + \vec{\lambda} \cdot \left[\int_I T ds - \Delta p\right] \tag{2.7}$$

where $\vec{\lambda} = (\lambda_1, \lambda_2)$ is the $\mathbb{R}^2$-valued Lagrange multiplier. Applying integration by parts to the first variation of $\mathcal{F}_{L,\Delta p}$ in the class $\mathcal{A}_L$ gives

$$\partial_\varepsilon \mathcal{F}_{L,\Delta p}\big[T(\cdot,\varepsilon)\big]_{|\varepsilon=0} = \big\langle \vec{\kappa}, (\partial_\varepsilon T)_{|\varepsilon=0}\big\rangle_{|\partial I} - \int_I \big\langle \partial_s \vec{\kappa} - \vec{\lambda}, (\partial_\varepsilon T)_{|\varepsilon=0}\big\rangle ds, \tag{2.8}$$

where $\vec{\kappa} = \partial_s T$ is the curvature vector of a planar curve f. The boundary condition (1.2) implies the vanishing of the boundary term in (2.8) and, thus, is called a natural boundary condition. Moreover, since $|T(\cdot,\varepsilon)| \equiv 1$ implies that $\langle \partial_\varepsilon T, T\rangle \equiv 0$, i.e., that $\partial_\varepsilon T$ is parallel to $\vec{n}$, we may rewrite (2.8) as

$$\partial_\varepsilon \mathcal{F}_{L,\Delta p}\big[T(\cdot,\varepsilon)\big]_{|\varepsilon=0} = -\int_I \big\langle \nabla_s \vec{\kappa} - \langle\vec{\lambda}, \vec{n}\rangle\vec{n}, (\partial_\varepsilon T)_{|\varepsilon=0}\big\rangle ds. \tag{2.9}$$

From (2.9), we obtain the Euler–Lagrange equation for $\mathcal{F}_{L,\Delta p}$:

$$\nabla_s^2 T - \langle\vec{\lambda}, \vec{n}\rangle\vec{n} = 0. \tag{2.10}$$

If $T \in \mathcal{A}_{L,0}$ satisfies (2.10), it can thus be concluded from (2.8) that $T \in \mathcal{A}_L$ is a critical point of $\mathcal{F}_{L,\Delta p}$.

Now we consider the flow (1.1) with initial and boundary conditions (1.2) and (1.3). If there is a smooth solution of this flow in short-time $[0, t_0)$, then $\{T(\cdot,t)\}$ is a family of tangent indicatrices in $\mathcal{A}_{L,0}$. Moreover,

$$0 = \frac{d}{dt}\int_I T(\cdot,t) ds, \tag{2.11}$$

which implies $\frac{d}{dt}\mathcal{F}_{L,\Delta p}[T(\cdot,t)] = \frac{d}{dt}\mathcal{F}_L[T(\cdot,t)]$ during the flow (1.1). To see why (2.11) holds, first note that in (1.1) one may write

$$\langle\vec{\lambda}, \vec{n}\rangle\vec{n} = \vec{\lambda} \cdot \big[(\vec{n})^t \vec{n}\big],$$

where $[(\vec{n})^t\vec{n}] \in \mathbb{M}_{2\times 2}$ (the set of 2×2 matrices). From integration over I, (1.5), and the definition of $\vec{\lambda}$ in (1.4), we have

$$\int_I \langle\vec{\lambda}, \vec{n}\rangle\vec{n} ds = \vec{\lambda} \cdot A_T = \int_I |\vec{\kappa}|^2 T ds.$$

On the other hand, by applying integration by parts, the hinged boundary condition (1.2) and the product rule of differentiation, we have $\langle \partial_s\vec{\kappa}, T\rangle = -|\vec{\kappa}|^2$ and

$$\int_I \nabla_s^2 T ds = \int_I \nabla_s \vec{\kappa} ds = \int_I \partial_s \vec{\kappa} - \langle \partial_s\vec{\kappa}, T\rangle T ds = \int_I |\vec{\kappa}|^2 T ds.$$

Thus, from (1.1) and above,

$$\frac{d}{dt}\int_I T ds = \int_I \partial_t T ds = \int_I \big(\nabla_s^2 T - \langle\vec{\lambda}, \vec{n}\rangle\big) ds = 0,$$

and therefore the constraint in (2.5) is fulfilled during the flow, if the initial data T_0 fulfills it, too.

Furthermore, one may apply the formula (2.9) and the flow equations (1.1) and (1.2) to obtain the energy identity

$$\frac{d}{dt}\mathcal{F}_{L,\Delta p}\big[T(\cdot,t)\big] = -\int_I |\partial_t T|^2 ds \le 0 \tag{2.12}$$

for any fixed t. It implies the non-increasing property of $\mathcal{F}_{L,\Delta p}[T(\cdot,t)]$ w.r.t. t, and thus that

$$\mathcal{F}_{L,\Delta p}\big[T(\cdot,t)\big] \le \mathcal{F}_{L,\Delta p}[T_0], \quad \forall t \in [0,t_0) \tag{2.13}$$

if the smooth solutions of (1.1) exist, $\forall t \in [0,t_0)$.

From the evolution equation of the tangent indicatrix in (1.1) and the property $\nabla_s \vec{n} = 0$, we obtain

$$\nabla_t \vec{\kappa} = (\partial_t \vec{\kappa})^\perp = (\partial_t \partial_s T)^\perp = (\partial_s \partial_t T)^\perp = \big(\partial_s\big(\nabla_s \vec{\kappa} - \langle \vec{\lambda}, \vec{n}\rangle \vec{n}\big)\big)^\perp,$$

which gives

$$\nabla_t \vec{\kappa} = \nabla_s^2 \vec{\kappa} + \langle \vec{\lambda}, T\rangle \vec{\kappa}. \tag{2.14}$$

By letting

$$T = T_\varphi = (\cos\varphi, \sin\varphi), \tag{2.15}$$

where $\varphi : \bar{I} \times [0,t_1) \to \mathbb{R}$, we can convert (1.1) to an evolution equation for the scalar function φ:

$$\partial_t \varphi = \partial_s^2 \varphi + \lambda_1 \sin\varphi - \lambda_2 \cos\varphi = \partial_s^2 \varphi - \langle \vec{\lambda}, \vec{n}\rangle. \tag{2.16}$$

Similarly, the evolution equation for the curvature vector $\vec{\kappa}$ in (2.14) can be written in terms of the signed curvature $k = \partial_s \varphi$ as

$$\partial_t k = \partial_s^2 k + \langle \vec{\lambda}, T\rangle k. \tag{2.17}$$

Using the expression of tangent indicatrix T in (2.15), (1.5) yields

$$A_T = \begin{pmatrix} \int_I \sin^2\varphi ds & -\int_I \sin\varphi\cos\varphi ds \\ -\int_I \sin\varphi\cos\varphi ds & \int_I \cos^2\varphi ds \end{pmatrix}, \tag{2.18}$$

therefore

$$A_T^{-1} = \frac{1}{\det A_T}\begin{pmatrix} \int_I \cos^2\varphi ds & \int_I \sin\varphi\cos\varphi ds \\ \int_I \sin\varphi\cos\varphi ds & \int_I \sin^2\varphi ds \end{pmatrix}, \tag{2.19}$$

and

$$\det A_T = \left(\int_I \cos^2\varphi ds\right)\cdot\left(\int_I \sin^2\varphi ds\right) - \left(\int_I \cos\varphi\sin\varphi ds\right)^2, \tag{2.20}$$

which is non-negative by Cauchy–Schwarz inequality. In fact, one can write

$$\det A_T = \frac{1}{2}\iint_{I\times I} \sin^2\big[\varphi(\sigma) - \varphi(s)\big] d\sigma ds, \tag{2.21}$$

which is useful in the estimates of $|\vec{\lambda}|$ later. The formula (2.21) is simply obtained by

$$\begin{aligned}
&2 \cdot \det A_T \\
&= \left(\int_I \cos^2\varphi(s)ds\right) \cdot \left(\int_I \sin^2\varphi(\sigma)d\sigma\right) \\
&\quad - 2\left(\int_I \cos\varphi(s)\sin\varphi(s)ds\right) \cdot \left(\int_I \cos\varphi(\sigma)\sin\varphi(\sigma)d\sigma\right) \\
&\quad + \left(\int_I \cos^2\varphi(\sigma)d\sigma\right) \cdot \left(\int_I \sin^2\varphi(s)ds\right) \\
&= \iint_{I\times I} \big[\cos^2\varphi(s)\sin^2\varphi(\sigma) - 2\cos\varphi(s)\sin\varphi(s)\cos\varphi(\sigma)\sin\varphi(\sigma) \\
&\quad + \cos^2\varphi(\sigma)\sin^2\varphi(s)\big]d\sigma ds \\
&= \iint_{I\times I} \big[\cos\varphi(s)\sin\varphi(\sigma) - \cos\varphi(\sigma)\sin\varphi(s)\big]^2 d\sigma ds \\
&= \iint_{I\times I} \sin^2\big[\varphi(\sigma) - \varphi(s)\big]d\sigma ds.
\end{aligned}$$

Moreover, from (1.4), the Lagrange multipliers in the case of the hinged boundary condition can be expressed as

$$\begin{cases}
\lambda_1 = (\det A_T)^{-1} \cdot \Big[\left(\int_I (\partial_s\varphi)^2 \cos\varphi ds\right) \cdot \left(\int_I \cos^2\varphi ds\right) \\
\qquad + \left(\int_I (\partial_s\varphi)^2 \sin\varphi ds\right) \cdot \left(\int_I \sin\varphi\cos\varphi ds\right)\Big], \\
\lambda_2 = (\det A_T)^{-1} \cdot \Big[\left(\int_I (\partial_s\varphi)^2 \cos\varphi ds\right) \cdot \left(\int_I \sin\varphi\cos\varphi ds\right) \\
\qquad + \left(\int_I (\partial_s\varphi)^2 \sin\varphi ds\right) \cdot \left(\int_I \sin^2\varphi ds\right)\Big].
\end{cases} \tag{2.22}$$

3 The Existence of Global Smooth Solutions

In this section, we prove long time existence of smooth solutions. The key step is to show that $\|\partial_s^m \vec{\kappa}\|_{L^2}$ remains uniformly bounded during the geometric flow (1.1) for any $m \in \mathbb{N}_0$.

3.1 Some Preliminary Estimates

The following lemma gives a crucial estimate on the Lagrange multipliers. We follow the notation used in the previous sections.

Lemma 1 *Let $I = (0, L)$ and $\varphi : \bar{I} \to \mathbb{R}$ be a continuous function with positive oscillation d_0, i.e.,*

$$\operatorname*{osc}_{\bar{I}} \varphi \geq d_0 > 0. \tag{3.1}$$

Suppose $\omega : [0, \infty] \to [0, \infty]$ is a continuous monotonic function and is the modulus of continuity of φ, i.e.,

$$\big|\varphi(s) - \varphi(\sigma)\big| \le \omega\big(|s - \sigma|\big), \quad \forall s, \sigma \in \bar{I}, \tag{3.2}$$

and $\omega(0) = 0$. Then, we have the following estimates.

(i) There exists a positive constant

$$C = \sin^2 \frac{\delta_0}{4} \cdot \min\big\{\omega^{-1}(\delta_0/4), L/2\big\},$$

where $\delta_0 = \min\{d_0, \pi\}$, such that

$$C \le \int_I \sin^2\big(\varphi(s) + \varphi_*\big) ds,$$
$$C \le \int_I \cos^2\big(\varphi(s) + \varphi_*\big) ds,$$

for any arbitrary constant φ_*.

(ii) From (2.18), the matrix A_{T_φ} defined by φ satisfies

$$\det A_{T_\varphi} \ge \frac{L}{2} C.$$

Proof (*i*) Let φ_* be an arbitrarily given constant. The continuity of φ and the oscillation of φ in (3.1) implies that there exists $s_0 \in \bar{I}$ such that

$$\mathrm{dist}_{\mathcal{H}}\big(\varphi(s_0) + \varphi_*, S\big) \ge \min\{d_0/2, \pi/2\} =: \delta_0/2 \in (0, \pi/2],$$

where $\mathrm{dist}_{\mathcal{H}}(S_1, S_2)$ represents the Hausdorff distance between two subsets S_1 and S_2 in $\mathbb{R}$, and

$$S := \{n\pi : n \in \mathbb{Z}\}.$$

From (3.2), and letting

$$r_0 := \omega^{-1}(\delta_0/4),$$
$$J_{r_0}(s_0) := \big\{\sigma \in \mathbb{R} : |\sigma - s_0| \le r_0\big\} \cap I,$$

we have

$$\varphi\big(J_{r_0}(s_0)\big) \subset \left[\varphi(s_0) - \frac{\delta_0}{4}, \varphi(s_0) + \frac{\delta_0}{4}\right].$$

Thus, for any $s \in J_{r_0}(s_0)$,

$$\mathrm{dist}_{\mathcal{H}}\big(\varphi(s) + \varphi_*, S\big) \ge \delta_0/4,$$

from which it follows that

$$\big|\sin\big(\varphi(s) + \varphi_*\big)\big| \ge \sin \frac{\delta_0}{4} \in (0, 1/\sqrt{2}].$$

Note that since $|J_{r_0}(s_0)| \geq \min\{\omega^{-1}(\delta_0/4), L/2\}$, we conclude

$$\int_I \sin^2\big(\varphi(s)+\varphi_*\big)ds \geq \int_{J_{r_0}(s_0)} \sin^2\big(\varphi(s)+\varphi_*\big)ds$$
$$\geq \sin^2\frac{\delta_0}{4}\cdot\min\big\{\omega^{-1}(\delta_0/4), L/2\big\} =: C.$$

Similarly, the inequality for $\int_I \cos^2(\varphi(s)+\varphi_*)ds$ can be obtained on replacing the set S by

$$\left\{\frac{\pi}{2}+n\pi : n\in\mathbb{Z}\right\}$$

and then applying the same argument.

(ii) From (2.20), (2.21), and (i) above, we find that

$$\det A_{T_\varphi} = \int_0^L \cos^2\psi ds\cdot\int_0^L \sin^2\psi ds - \left(\int_0^L \cos\psi\sin\psi ds\right)^2$$
$$= \frac{1}{2}\int_I\left(\int_I \sin^2\big[\psi(s)-\psi(\sigma)\big]ds\right)d\sigma \geq \frac{L}{2}C.$$

□

To simplify the presentation, we introduce some additional notation. We denote by $\vec{\phi} : \bar{I}\to\mathbb{R}^2$ a normal vector field of a planar curve f if $\langle\vec{\phi}, \partial_s f\rangle \equiv 0$. For normal vector fields $\vec{\phi}_1,\dots,\vec{\phi}_\ell$ along a curve f, we denote by $\vec{\phi}_1 *** \vec{\phi}_\ell$ a term of the type

$$\vec{\phi}_1 *** \vec{\phi}_\ell = \begin{cases} \langle\vec{\phi}_{i_1},\vec{\phi}_{i_2}\rangle\cdots\langle\vec{\phi}_{i_{\ell-1}},\vec{\phi}_{i_\ell}\rangle, & \text{for } \ell \text{ even},\\ \langle\vec{\phi}_{i_1},\vec{\phi}_{i_2}\rangle\cdots\langle\vec{\phi}_{i_{\ell-2}},\vec{\phi}_{i_{\ell-1}}\rangle\cdot\vec{\phi}_{i_\ell}, & \text{for } \ell \text{ odd},\end{cases}$$

where $(i_1,\dots,i_\ell)$ is any permutation of $(1,\dots,\ell)$. Somewhat more generally, we allow some of the $\vec{\phi}_i$ to be functions, in which case the $*$-product reduces to a multiplication. Denote by $P_b^{a,c}(\vec{\phi})$ any linear combination of terms of the type

$$\nabla_s^{i_1}\vec{\phi} * \cdots * \nabla_s^{i_b}\vec{\phi}, \quad i_1+\cdots+i_b = a \text{ with } \max i_j \leq c,$$

where all coefficients are bounded from above and below by some universal constants, depending only on a and b. Moreover, let

$$\sum_{\substack{[\![a,b]\!]\leq[\![A,B]\!]\\ c\leq C}} P_b^{a,c}(\vec{\kappa}) := \sum_{a=0}^{A}\sum_{b=1}^{2A+B-2a}\sum_{c=0}^{C} P_b^{a,c}(\vec{\kappa}), \tag{3.3}$$

where $[\![a,b]\!] := 2a+b$.

Although the formulae in the lemma below as known [6, 9], for convenience we give the statement and proof here.

Lemma 2 *Suppose $T : \bar{I}\times[0,t_1)\to\mathbb{S}^1(1)\subset\mathbb{R}^2$ is a smooth solution of (1.1). Let $\vec{\phi}_\ell := \nabla_s^\ell\vec{\kappa}$. Denote by $\vec{\psi} : \bar{I}\times[0,t_1)\to\mathbb{R}^2$ a smooth normal vector field along the planar curve, i.e., $\vec{\psi}(s,t)\perp T(s,t)$, for all $(s,t)\in\bar{I}\times[0,t_1)$. Then, for any integer $\ell\geq 2$ and*

$k, m \in \mathbb{N}$, we have the following formulae,

$$\nabla_t \nabla_s \vec{\psi} = \nabla_s \nabla_t \vec{\psi}, \tag{3.4}$$

$$\partial_s \vec{\psi} = \nabla_s \vec{\psi} - \langle \vec{\psi}, \vec{\kappa} \rangle T, \tag{3.5}$$

$$\partial_s^\ell \vec{\kappa} - \nabla_s^\ell \vec{\kappa} = \Bigg[\sum_{\substack{[\![a,b]\!] \le [\![\ell-1,2]\!] \\ c \le \ell-1}} P_b^{a,c}(\vec{\kappa}) \Bigg] T + \sum_{\substack{[\![a,b]\!] \le [\![\ell-2,3]\!] \\ c \le \ell-2}} P_b^{a,c}(\vec{\kappa}), \tag{3.6}$$

where the coefficient of T in (3.6), i.e. the one denoted by $[\cdots]$, is a sum of terms like $\langle \vec{\phi}_{i_1}, \vec{\phi}_{i_2} \rangle \cdots \langle \vec{\phi}_{i_{j-1}}, \vec{\phi}_{i_j} \rangle$; while the last term in (3.6) is a sum of terms of the form $\langle \vec{\phi}_{i_1}, \vec{\phi}_{i_2} \rangle \cdots \langle \vec{\phi}_{i_{k-2}}, \vec{\phi}_{i_{k-1}} \rangle \cdot \vec{\phi}_{i_k}$.

Proof (*i*) We obtain (3.4) by applying the property: if $\vec{\psi} = \varphi \cdot \vec{n}$, then $\nabla_s \vec{\psi} = \partial_s \varphi \cdot \vec{n}$ and $\nabla_t \vec{\psi} = \partial_t \varphi \cdot \vec{n}$.

(*ii*) We derive (3.5) from

$$\partial_s \vec{\psi} = \nabla_s \vec{\psi} + \langle \partial_s \vec{\psi}, T \rangle T = \nabla_s \vec{\psi} - \langle \vec{\psi}, \vec{\kappa} \rangle T.$$

(*iii*) The proof of (3.6) is an induction argument. From (3.5), we have

$$\partial_s \vec{\kappa} = \nabla_s \vec{\kappa} - |\vec{\kappa}|^2 T,$$

which gives the case of $\ell = 1$ in (3.6). By induction, assume (3.6) holds for any $\ell \in \{1, 2, \ldots, m\}$. Then, we can keep applying (3.5) to derive

$$\begin{aligned}
\partial_s^{m+1} \vec{\kappa} &= \partial_s \Bigg(\nabla_s^m \vec{\kappa} + \Bigg[\sum_{\substack{[\![a,b]\!] \le [\![m-1,2]\!] \\ c \le m-1}} P_b^{a,c}(\vec{\kappa}) \Bigg] T + \sum_{\substack{[\![a,b]\!] \le [\![m-2,3]\!] \\ c \le m-2}} P_b^{a,c}(\vec{\kappa}) \Bigg) \\
&= \nabla_s^{m+1} \vec{\kappa} + \Bigg[\sum_{\substack{[\![a,b]\!] \le [\![m,2]\!] \\ c \le m}} P_b^{a,c}(\vec{\kappa}) \Bigg] T + \Bigg(\sum_{\substack{[\![a,b]\!] \le [\![m-1,2]\!] \\ c \le m-1}} P_b^{a,c}(\vec{\kappa}) \Bigg) \vec{\kappa} \\
&\quad + \partial_s \Bigg(\sum_{k=0}^{m-2} \Bigg[\sum_{\substack{[\![a,b]\!] \le [\![m-2-k,2]\!] \\ c \le m-2-k}} P_b^{a,c}(\vec{\kappa}) \Bigg] \nabla_s^k \kappa \Bigg) \\
&= \nabla_s^{m+1} \vec{\kappa} + \Bigg[\sum_{\substack{[\![a,b]\!] \le [\![m,2]\!] \\ c \le m}} P_b^{a,c}(\vec{\kappa}) \Bigg] T + \sum_{\substack{[\![a,b]\!] \le [\![m-1,3]\!] \\ c \le m-1}} P_b^{a,c}(\vec{\kappa}),
\end{aligned}$$

which proves the case of $\ell = m + 1$ in (3.6). The proof is now finished by induction. □

Below is a formula of integration by parts for the covariant derivative ∇_s and normal vector fields.

Lemma 3 *Let $\vec{\eta}_1, \vec{\eta}_2 : \bar{I} \to \mathbb{R}^2$ be two normal vector fields of curve f, i.e., $\langle \vec{\eta}_i, \vec{n} \rangle \equiv 0$, $\forall i \in \{1, 2\}$. If $\vec{\eta}_i = 0$ at the boundary for some i, then*

$$\int_I \langle \nabla_s \vec{\eta}_1, \vec{\eta}_2 \rangle ds = - \int_I \langle \vec{\eta}_1, \nabla_s \vec{\eta}_2 \rangle ds.$$

Proof The proof is derived by applying integration by parts and the definition of ∇_s. Specifically

$$\int_I \langle \nabla_s \vec{\eta}_1, \vec{\eta}_2 \rangle ds = \int_I \langle \partial_s \vec{\eta}_1, \vec{\eta}_2 \rangle ds = -\int_I \langle \vec{\eta}_1, \partial_s \vec{\eta}_2 \rangle ds = -\int_I \langle \vec{\eta}_1, \nabla_s \vec{\eta}_2 \rangle ds. \qquad \square$$

We recall the Gagliardo–Nirenberg type interpolation inequalities from §4 in [6].

Lemma 4 *Let $f : \bar{I} \to \mathbb{R}^n$ be a regular curve, s be its arclength parameter, and $\vec{\kappa}$ be its curvature vector. Assume that the integers b and c satisfy $b \geq 2$, $0 \leq c \leq m-1$. Then,*

$$\int_I \left| P_b^{a,c}(\vec{\kappa}) \right| ds \leq C \cdot \mathcal{L}[f]^{1-a-b} \|\vec{\kappa}\|_2^{b-\gamma} \|\vec{\kappa}\|_{m,2}^{\gamma}, \tag{3.7}$$

where $\gamma = (a + \frac{b}{2} - 1)/m$, $C = C(n, m, a, b)$, $\mathcal{L}[f]$ is the length of f, and

$$\|\vec{\kappa}\|_{m,p} := \sum_{i=0}^{m} \left\| \nabla_s^i \vec{\kappa} \right\|_p, \quad \left\| \nabla_s^i \vec{\kappa} \right\|_p := \mathcal{L}[f]^{i+1-1/p} \left(\int_I |\nabla_s^i \vec{\kappa}|^p ds \right)^{1/p}.$$

Moreover, if $a + \frac{b}{2} < 2m + 1$, then $\gamma < 2$ and we have for any $\varepsilon > 0$

$$\int_I \left| P_b^{a,c}(\vec{\kappa}) \right| ds \leq \varepsilon \int_I \left| \nabla_s^m \vec{\kappa} \right|^2 ds + C \varepsilon^{\frac{-\gamma}{2-\gamma}} \left(\int_I |\vec{\kappa}|^2 ds \right)^{\frac{b-\gamma}{2-\gamma}} + C \left(\int_I |\vec{\kappa}|^2 ds \right)^{a+b-1}. \tag{3.8}$$

The lemma below shows that the hinged boundary conditions endow the derivatives of the curvature vector at the boundary with a certain nice property. This phenomenon also appears in the fourth-order flow with Navier boundary conditions in [6].

Lemma 5 *Assume for each fixed $t \in [0, t_0)$ the curvature vector $\vec{\kappa}$ remains smooth up to the boundary $\partial I \times \{t\}$ during the L^2-flow (1.1) and subject to (1.2), the null condition of the curvature at boundary. Then for any fixed $\ell \in \mathbb{N}_0$,*

$$\partial_s^{2\ell} \vec{\kappa}(b, t) = 0 = \nabla_s^{2\ell} \vec{\kappa}(b, t), \quad \forall b \in \partial I, \ \forall t \in [0, t_0).$$

Proof (*i*) to show $\partial_s^{2\ell} \vec{\kappa}(b, \cdot) = 0$:

From a direct computation, we have

$$\partial_s^2 \vec{\kappa} = \nabla_s^2 \vec{\kappa} - 3 \langle \nabla_s \vec{\kappa}, \vec{\kappa} \rangle T - |\vec{\kappa}|^2 \vec{\kappa}.$$

Together with the hinged boundary condition (1.2), we derive $\partial_s^2 \vec{\kappa}(b, t) = 0$, $\forall t \in [0, t_0)$. To show that $\partial_s^{2\ell} \vec{\kappa}(b, t) = 0$, $\forall t \in [0, t_0)$, $\forall \ell \geq 2$, we argue inductively on the basis of the identity

$$\begin{aligned}
\partial_t \partial_s^{2\ell-2} \vec{\kappa} &= \partial_s^{2\ell-2} \partial_t \vec{\kappa} = \partial_s^{2\ell-2} \big(\partial_s^2 \vec{\kappa} + \langle \vec{\lambda}, T \rangle \vec{\kappa} + |\vec{\kappa}|^2 \vec{\kappa} + 2 \langle \partial_s \vec{\kappa}, \vec{\kappa} \rangle T + \langle \vec{\lambda}, \vec{\kappa} \rangle T \big) \\
&= \partial_s^{2\ell} \vec{\kappa} + \partial_s^{2\ell-2} \big(\langle \vec{\lambda}, T \rangle \vec{\kappa} + |\vec{\kappa}|^2 \vec{\kappa} + 2 \langle \partial_s \vec{\kappa}, \vec{\kappa} \rangle \cdot T + \langle \vec{\lambda}, \vec{\kappa} \rangle \cdot T \big) \\
&= \partial_s^{2\ell} \vec{\kappa} + \sum_{i+j=2\ell-2} C_1(i, j) \big\langle \vec{\lambda}, \partial_s^i T \big\rangle \cdot \partial_s^j \vec{\kappa} + \sum_{i+j+k=2\ell-2} C_2(i, j, k) \big\langle \partial_s^i \vec{\kappa}, \partial_s^j \vec{\kappa} \big\rangle \cdot \partial_s^k \vec{\kappa} \\
&\quad + \sum_{i+j+k=2\ell-2} C_3(i, j, k) \big\langle \partial_s^{i+1} \vec{\kappa}, \partial_s^j \vec{\kappa} \big\rangle \cdot \partial_s^k T + \sum_{i+j=2\ell-2} C_4(i, j) \big\langle \vec{\lambda}, \partial_s^i \vec{\kappa} \big\rangle \cdot \partial_s^j T,
\end{aligned}$$

where (1.1) is applied to derive the second equality with constants $C_1(i,j)$, $C_2(i,j,k)$, $C_3(i,j,k)$, $C_4(i,j)$. On applying the assumption

$$\partial_s^{2i}\vec{\kappa}(b,t)=0, \quad \forall i\in\{0,1,\dots,\ell-1\},\ \forall b\in\partial I,\ \forall t\in[0,t_0),$$

the term $\partial_t\partial_s^{2\ell-2}\vec{\kappa}$ and the last four terms on the R.H.S. all vanish. This gives $\partial_s^{2\ell}\vec{\kappa}(b,\cdot)=0$, $\forall b\in\partial I$ and thus completes the proof of the first part.

(*ii*) to show that $\nabla_s^{2\ell}\vec{\kappa}(b,\cdot)=0$:

Applying the hinged boundary condition (1.2) to (2.14) gives

$$\nabla_s^2\vec{\kappa}(b,\cdot)=0, \quad \forall b\in\partial I.$$

To show $\nabla_s^{2\ell}\vec{\kappa}(b,\cdot)=0$, $\forall \ell\geq 2$, we argue inductively on the basis of the result $\partial_s^{2\ell}\vec{\kappa}(b,\cdot)=\partial_s^{2\ell+1}T(b,\cdot)=0$ from part 1° and the identity

$$\begin{aligned}\nabla_t\nabla_s^{2\ell-2}\vec{\kappa}&=\nabla_s^{2\ell-2}\nabla_t\vec{\kappa}=\nabla_s^{2\ell-2}\big(\nabla_s^2\vec{\kappa}+\langle\vec{\lambda},T\rangle\vec{\kappa}\big)=\nabla_s^{2\ell}\vec{\kappa}+\nabla_s^{2\ell-2}\big(\langle\vec{\lambda},T\rangle\vec{\kappa}\big)\\&=\nabla_s^{2\ell}\vec{\kappa}+\sum_{i+j=2\ell-2}C_5(i,j)\big\langle\vec{\lambda},\partial_s^iT\big\rangle\cdot\nabla_s^j\vec{\kappa},\end{aligned}$$

where $C_5(i,j)$ is a constant. □

3.2 Short-Time Existence

In this part, we give a proof of the short-time existence of smooth solutions for (3.10) below, i.e., (2.16) with hinged boundary condition. We follow the proof of short-time existence of classical solutions for the semilinear parabolic equations provided in [5], where a contraction map on a suitable function space plays the key role.

Theorem 2 (The short-time existence for hinged boundary condition) *For a given $\varphi_0\in C^1(\bar{I})$ with*

$$\varphi_0'(0)=\varphi_0'(L)=0, \tag{3.9}$$

there exist a positive time $t_0>0$ and a unique continuous solution φ with the regularity, φ, $\partial_s\varphi\in C^0(\bar{I}\times[0,t_0))\cap C^\infty(\bar{I}\times(0,t_0))$, satisfying

$$\begin{cases}\partial_t\varphi=\partial_s^2\varphi+\lambda_1\sin\varphi-\lambda_2\cos\varphi & \text{in } (0,L)\times(0,t_0),\\ \varphi(s,0)=\varphi_0(s) & \forall 0\leq s\leq L,\\ \partial_s\varphi(0,t)=\partial_s\varphi(L,t)=0 & \forall 0<t<t_0,\end{cases} \tag{3.10}$$

where λ_1 and λ_2 are defined in (2.22).

Instead of proving existence of smooth solutions of (3.10) directly, we first work with the equation for a function $\varphi:\mathbb{R}\times(0,t_0)\to\mathbb{R}$ on the entire domain,

$$\begin{cases}\partial_t\varphi=\partial_s^2\varphi+\lambda_1\sin\varphi-\lambda_2\cos\varphi & \text{in } \mathbb{R}\times(0,t_0),\\ \varphi(s,0)=\varphi_0(s) & \forall s\in\mathbb{R},\end{cases} \tag{3.11}$$

where $\lambda_1=\lambda_1(t)$ and $\lambda_2=\lambda_2(t)$ are determined by the integrals over the finite domain $[0,L]$ as shown in (2.20), (2.22).

Theorem 3 (The short-time existence over entire domain) *Suppose φ_0 is a real-valued function belonging to the class $C^1(\mathbb{R}) \cap L^\infty(\mathbb{R})$. Then there exists a positive time $t_0 > 0$ and a unique continuous solution of* (3.11) *with the regularity*

$$\begin{cases} \varphi, \partial_s\varphi \in C^0(\overline{D_0^{t_0}}) \cap L^\infty(\overline{D_0^{t_0}}), \\ \varphi \in C^\infty(D_0^{t_0}). \end{cases}$$

In the following, we let

$$K(x,t) := \frac{1}{\sqrt{4\pi t}} e^{-\frac{x^2}{4t}}. \tag{3.12}$$

For the proof of Theorem 3, we cite the following lemma from [5] with a modification in notation.

Lemma 6 (Lemma 19.2.1 of [5]) *Assume $f : \mathbb{R} \times \mathbb{R}_+ \to \mathbb{R}$ is bounded and $f(\cdot,t)$ is uniformly Hölder continuous with exponent $\alpha \in (0,1)$ for any fixed t. Then, the potential*

$$z(x,t) = \int_0^t \int_{-\infty}^{\infty} K(x-\xi, t-\tau) f(\xi,\tau) d\xi d\tau$$

possesses the following properties:

(1) z, z_x, z_t, *and* z_{xx} *are continuous*;
(2) $z_t = z_{xx} + f(x,t)$, $-\infty < x < \infty$, $0 < t$;
(3) $|z(x,t)| \le t \cdot |f|_t$, $-\infty < x < \infty$, $0 \le t$, *where*

$$|f|_t := \sup_{x \in \mathbb{R}; \tau \in [0,t]} \big|f(x,\tau)\big|;$$

(4) $|z_x(x,t)| \le 2\pi^{-1/2} |f|_t \cdot t^{1/2}$, $-\infty < x < \infty$, $0 \le t$;
(5) $|z_{xx}(x,t)| \le C[f]_\alpha \cdot t^{\alpha/2}$, $-\infty < x < \infty$, $0 \le t$, *where C is a positive number and*

$$[f]_\alpha := \sup_{x \in \mathbb{R}; 0 \le t; \delta > 0} \left\{ \frac{|f(x+\delta,t) - f(x,t)|}{\delta^\alpha} \right\};$$

(6) $|z_t(x,t)| \le |f|_t + C[f]_\alpha \cdot t^{\alpha/2}$, $-\infty < x < \infty$, $0 \le t$.

We first introduce some notation for the proof of Theorem 3. For any $a > \delta \ge 0$, define the following subsets in $\mathbb{R}^2$:

$$\begin{cases} D_\delta^a := \{(s,t) : (s,t) \in \mathbb{R} \times (\delta, a)\}, \\ D_0^a(L) := \{(s,t) : s \in [0,L], t \in (0,a)\}. \end{cases}$$

Let $\overline{D_0^a}$ denote the closure of D_0^a and define

$$\mathfrak{B}_a := \left\{\psi : \overline{D_0^a} \to \mathbb{R} : \psi, \partial_s\psi \in C^0\left(\overline{D_0^a}\right) \text{ with } \|\psi\|_a < \infty\right\},$$

where

$$\|\psi\|_a := \sup_{(s,t) \in \overline{D_0^a}} \left\{ \big|\psi(s,t)\big| + \big|\partial_s\psi(s,t)\big| \right\}.$$

Furthermore, to estimate the Lagrange multipliers, define the space

$$\mathfrak{B}_a^L(d,M) := \Big\{\psi \in \mathfrak{B}_a : \inf_{t\in[0,a]} \operatorname*{osc}_{[0,L]\times\{t\}} \psi \ge d, \sup_{\overline{D_0^a}} |\partial_s \psi| \le M\Big\}.$$

For a given $\varphi_0 \in C^1(\mathbb{R}) \cap L^\infty(\mathbb{R})$ with $\sup_{[0,L]} |\varphi_0'| \le M$ and $\operatorname{osc}_{[0,L]} \varphi_0 \ge d$, we define the map $\mathcal{G}_{\varphi_0}$ from $\{\psi \in \mathfrak{B}_a^L(d,M) : \psi(s,0) = \varphi_0\} \subset \mathfrak{B}_a$ to $\mathfrak{B}_a$ by

$$\mathcal{G}_{\varphi_0}(\psi) := U_{\varphi_0} + H_0^\psi =: G_{\varphi_0}^\psi, \tag{3.13}$$

where U_{φ_0} is the solution of

$$\begin{cases} \partial_t U_{\varphi_0} = \partial_s^2 U_{\varphi_0} & \text{in } D_0^a, \\ U_{\varphi_0}(s,0) = \varphi_0(s), & \forall s \in \mathbb{R}, \end{cases} \tag{3.14}$$

H_0^ψ is the solution of

$$\begin{cases} \partial_t H_0^\psi = \partial_s^2 H_0^\psi + h\circ\psi & \text{in } D_0^a, \\ H_0^\psi(s,0) = 0, & \forall s \in \mathbb{R}, \\ h\circ\psi := \lambda_1 \sin\psi - \lambda_2\cos\psi, \end{cases} \tag{3.15}$$

and $\lambda_1(t)$, $\lambda_2(t)$ are determined by the formulae (2.20), (2.22). Notice that $\{\psi \in \mathfrak{B}_a^L(d,M) : \psi(s,0) = \varphi_0\}$ is a closed subset of the Banach space $(\mathfrak{B}_a, \|\cdot\|_a)$.

Note that for a given $\psi \in \{\phi \in \mathfrak{B}_a^L(d,M) : \phi(s,0) = \varphi_0\}$, $\lambda_1(t)$ and $\lambda_2(t)$ determined by ψ are continuous and bounded functions. Thus, both (3.14) and (3.15) are linear parabolic equations with sufficiently regular coefficients such that we may apply Lemma 6 to obtain the existence of classical C^2-smooth solutions of (3.14) and (3.15) with the following expression

$$U_{\varphi_0}(s,t) = \int_{-\infty}^{\infty} K(s-\xi,t)\cdot\varphi_0(\xi)d\xi, \quad (s,t) \in \mathbb{R}\times[0,\infty), \tag{3.16}$$

$$H_0^\psi(s,t) = \int_0^t \int_{-\infty}^{\infty} K(s-\xi,t-\tau)\cdot(h\circ\psi)(\xi,\tau)d\xi d\tau, \quad (s,t)\in\mathbb{R}\times[0,\infty). \tag{3.17}$$

Therefore, the map $\mathcal{G}_{\varphi_0} : \{\psi \in \mathfrak{B}_a^L(d,M) : \psi(s,0) = \varphi_0\} \to \mathfrak{B}_a$ is well-defined for any $a > 0$.

Lemma 7 *Let $d \in (0,\pi/2)$ be a positive constant. Then for any $\psi \in \mathfrak{B}_a^L(d,M)$, there exist positive numbers C_1, C_2, and C_3, depending only on M and d, such that*

$$C_1 \le \int_0^L \cos^2\big[\psi(s,\cdot)+\theta_*\big]ds, \qquad C_1 \le \int_0^L \sin^2\big[\psi(s,\cdot)+\theta_*\big]ds, \tag{3.18}$$

for any constant $\theta_* \in \mathbb{R}$,

$$C_2 \le \int_0^L \cos^2\psi(s,\cdot)ds \cdot \int_0^L \sin^2\psi(s,\cdot)ds - \left(\int_0^L \cos\psi(s,\cdot)\sin\psi(s,\cdot)ds\right)^2, \tag{3.19}$$

$$\begin{cases} \|\lambda_i\|_{L^\infty(D_0^a)} \le C_3, & i \in \{1,2\}, \\ \|h\circ\psi\|_{L^\infty(D_0^a)} \le 2C_3. \end{cases} \tag{3.20}$$

Proof The assumption, $\psi \in \mathfrak{B}_a^L(d, M)$, implies that

$$\inf_{t\in[0,a]} \operatorname*{osc}_{[0,L]\times\{t\}} \psi \geq d > 0,$$

and for any fixed $t \in [0, a]$ the modulus of continuity of $\psi(\cdot, t)$ can be chosen as $\omega(x) = M \cdot x$. Thus, from Lemma 1, we derive (3.18) by letting

$$C_1 = \sin^2(d/4) \cdot \min\{d/(4M), L/2\}$$

and (3.19) by letting $C_2 = \frac{L}{2} C_1$.

The inequalities in (3.20) are simple consequences of applying (3.19) and the assumption on the Lipschitz norm of $\psi(\cdot, t)$, $\forall t \in [0, a]$, to (2.20), (2.22), (3.15). In fact, we may let

$$C_3 = \frac{2L^2M^2}{C_2}.$$

□

Lemma 8

(i) *For any* $\psi_0, \psi_1 \in \mathfrak{B}_a^L(d, M)$, *we have*

$$\|h \circ \psi_1 - h \circ \psi_0\|_{L^\infty} \leq C_4 \cdot \|\psi_1 - \psi_0\|_a, \tag{3.21}$$

where h *is given in* (3.15) *and* $C_4 = C_4(d, M)$ *is a constant.*

(ii) *Assume* $\varphi \in \mathfrak{B}_a^L(d, M)$ *fulfills* $\partial_s^j \varphi \in C^0(D_0^a)$, $\forall j \in \{0, 1, \ldots, m\}$. *Then, for any* $j \in \{0, 1, \ldots, m\}$, *the composite function* $h \circ \varphi$ *satisfies*

$$\partial_s^j (h \circ \varphi) \in C^0\big(D_0^a\big).$$

Proof (i) From (3.15) and (1.4), we write

$$h \circ \psi_\alpha = -\left\langle \left(\int_I (\partial_s \psi_\alpha)^2 T_{\psi_\alpha} ds \right) A_{T_{\psi_\alpha}}^{-1}, \vec{n}_{\psi_\alpha} \right\rangle,$$

where $\vec{n}_{\psi_\alpha} := (-\sin\psi_\alpha, \cos\psi_\alpha)$. Note that $h : \mathfrak{B}_a^L(d, M) \to \mathbb{R}$ is a well-defined function as $|\det A_{T_\psi}| \neq 0$, $\forall \psi \in \mathfrak{B}_a^L(d, M)$. This condition can be ensured from Lemma 7. Observe that

$$\begin{aligned}
&|h \circ \psi_1 - h \circ \psi_0| \\
&\leq \left| \left\langle \left(\int_I (\partial_s \psi_0)^2 T_{\psi_0} ds \right) A_{T_{\psi_0}}^{-1}, \vec{n}_{\psi_1} - \vec{n}_{\psi_0} \right\rangle \right| \\
&\quad + \left| \left\langle \left(\int_I (\partial_s \psi_0)^2 T_{\psi_0} ds \right) \left(A_{T_{\psi_1}}^{-1} - A_{T_{\psi_0}}^{-1}\right), \vec{n}_{\psi_1} \right\rangle \right| \\
&\quad + \left| \left\langle \left(\int_I (\partial_s \psi_0)^2 \cdot (T_{\psi_1} - T_{\psi_0}) ds \right) A_{T_{\psi_1}}^{-1}, \vec{n}_{\psi_1} \right\rangle \right| \\
&\quad + \left| \left\langle \left(\int_I \left[(\partial_s \psi_1)^2 - (\partial_s \psi_0)^2\right] \cdot T_{\psi_1} ds \right) A_{T_{\psi_1}}^{-1}, \vec{n}_{\psi_1} \right\rangle \right|.
\end{aligned}$$

The Lipschitz continuity of h with global Lipschitz constant, given in (3.21), can be derived from applying (2.15), (2.18), (2.19), (2.20), and Lemma 7. For the sake of conciseness, we leave the verification to the reader.

(*ii*) Note that

$$\partial_s^j (h \circ \varphi)(s,t) = -\langle \vec{\lambda}(t), \partial_s^j \vec{n}_\varphi(s,t)\rangle,$$

where $\vec{n}_\varphi := (-\sin\varphi, \cos\varphi)$. From the definition of $\vec{\lambda}$ and the assumption $\varphi \in \mathfrak{B}_a^L(d, M)$, it is easy to see that $\vec{\lambda} \in C^0(D_0^a)$. The property $\partial_s^j \vec{n}_\varphi \in C^0(D_0^a)$ is derived from the assumption $\partial_s^j \varphi \in C^0(D_0^a)$, $\forall j \in \{0, 1, \ldots, m\}$. The conclusion (*ii*) now follows. □

The following lemma will allow us to obtain local solutions in Theorem 3 within the class of $\mathfrak{B}_a$ by contraction mapping theorem.

Lemma 9 *Assume* $\varphi_0 \in C^1(\mathbb{R}) \cap L^\infty(\mathbb{R})$ *with* $\sup_{\mathbb{R}} |\varphi_0'| = M_0/2 < \infty$ *and* $\operatorname{osc}_{[0,L]} \varphi_0 = 2d_0 > 0$.

(*i*) *There exists a positive number* $b_0 \in (0, a]$ *so that the map given in* (3.13),

$$\mathcal{G}_{\varphi_0} : \{\psi \in \mathfrak{B}_b^L(d_0, M_0) : \psi(s,0) = \varphi_0\} \to \{\psi \in \mathfrak{B}_b^L(d_0, M_0) : \psi(s,0) = \varphi_0\},$$

is well-defined for all $0 < b \le b_0$.

(*ii*) *There exists a positive time* t_0 *such that*

$$\mathcal{G}_{\varphi_0} : \{\psi \in \mathfrak{B}_{t_0}^L(d_0, M_0) : \psi(s,0) = \varphi_0\} \to \{\psi \in \mathfrak{B}_{t_0}^L(d_0, M_0) : \psi(s,0) = \varphi_0\}$$

is a contraction map.

Proof To prove (*i*), we need to verify that there exists a constant $b_0 > 0$ such that, for all $b \in (0, b_0]$ and $\psi \in \{\varphi \in \mathfrak{B}_b^L(d_0, M_0) : \varphi(s,0) = \varphi_0\}$, the function $G_{\varphi_0}^\psi$ defined in (3.13) satisfies

$$\left|\partial_s G_{\varphi_0}^\psi\right| \le M_0 \quad \text{in } \overline{D_0^b}, \qquad \inf_{t\in[0,b]} \operatorname*{osc}_{[0,L]\times\{t\}} G_{\varphi_0}^\psi \ge d_0.$$

Since φ_0 is bounded and continuous on $\mathbb{R}$, from the standard argument on the initial value problem for the heat equation on $\mathbb{R}$ (Chap. 3 of [5]), the function U_{φ_0}, given in (3.16), belongs to the class of $C^\infty(\mathbb{R} \times (0,\infty)) \cap C^0(\mathbb{R} \times [0,\infty))$. Observe that

$$\begin{aligned} \partial_s U_{\varphi_0}(s,t) &= \int_{-\infty}^{\infty} \partial_s K(s-\xi, t) \cdot \varphi_0(\xi) d\xi = \int_{-\infty}^{\infty} -\partial_\xi K(s-\xi,t) \cdot \varphi_0(\xi) d\xi \\ &= \int_{-\infty}^{\infty} K(s-\xi, t) \cdot \varphi_0'(\xi) d\xi, \end{aligned} \tag{3.22}$$

where the assumption $\varphi_0 \in C^1(\mathbb{R}) \cap L^\infty(\mathbb{R})$ and the decay speed of $K(\cdot, t)$ at infinity (as $t \ne 0$) are applied in deriving the last equality. Thus, using the same argument for $\partial_s U_{\varphi_0}$, $\partial_s U_{\varphi_0}$ also belongs to the class $C^\infty(\mathbb{R} \times (0,\infty)) \cap C^0(\mathbb{R} \times [0,\infty))$. From (3.22) and the assumption $\sup_{\mathbb{R}} \|\varphi_0'\| \le M_0/2$ and the identity

$$\int_{-\infty}^{\infty} K(x,t) dx = 1, \tag{3.23}$$

we conclude that

$$\left|\partial_s U_{\varphi_0}(s,t)\right| \le \frac{M_0}{2} \tag{3.24}$$

for all $s \in \mathbb{R}$, $t \geq 0$. In fact, by applying this argument step by step, we obtain

$$\lim_{t \to +0} \partial_s^j U_{\varphi_0}(s,t) = \varphi_0^{(j)}(s) \tag{3.25}$$

for all $s \in \mathbb{R}$, and any $j \in \mathbb{N}_0$.

To show the continuity of H_0^ψ and $\partial_s H_0^\psi$ up to the parabolic boundary $\mathbb{R} \times \{0\}$, we apply Lemma 7 and (3.23) to obtain

$$\left|H_0^\psi(s,t)\right| \leq \int_0^t \int_{-\infty}^\infty K(s-\xi, t-\tau) \cdot \left|(h \circ \psi)(\xi,\tau)\right| d\xi d\tau \leq t \cdot C_3 \tag{3.26}$$

for all fixed $s \in \mathbb{R}$. By applying the identity

$$\int_{-\infty}^\infty \left|\partial_x K(x,t)\right| dx = \frac{1}{\sqrt{\pi t}}, \tag{3.27}$$

we obtain

$$\left|\partial_s H_0^\psi(s,t)\right| \leq \int_0^t \int_{-\infty}^\infty \left|\partial_s K(s-\xi, t-\tau)\right| \cdot \left|(h \circ \psi)(\xi,\tau)\right| d\xi d\tau \leq 2\frac{\sqrt{t}}{\sqrt{\pi}} \cdot C_3. \tag{3.28}$$

Thus, we conclude that for all $s \in \mathbb{R}$

$$\lim_{t \to +0} H_0^\psi(s,t) = 0 \quad \text{and} \quad \lim_{t \to +0} \partial_s H_0^\psi(s,t) = 0. \tag{3.29}$$

Hence by (3.24), (3.26), (3.28), (3.29), $G_{\varphi_0}^\psi$ and $\partial_s G_{\varphi_0}^\psi$ are continuous up to $\mathbb{R} \times \{0\}$. Therefore, we may choose a sufficiently small $\alpha > 0$ such that

$$\left|\partial_s G_{\varphi_0}^\psi(s,t)\right| \leq M_0, \quad \forall (s,t) \in D_0^\alpha.$$

Moreover, by the continuity of $G_{\varphi_0}^\psi$ over the compact set $\overline{D_0^t(L)}$, the property $\operatorname{osc}_{[0,L]} \varphi_0 \geq 2d_0$, and the fact $\lim_{t \to +0} G_{\varphi_0}^\psi(s,t) = \varphi_0(s)$, we may choose a sufficiently small $\beta > 0$ such that

$$\inf_{t \in [0,\beta]} \operatorname*{osc}_{[0,L] \times \{t\}} G_{\varphi_0}^\psi \geq d_0.$$

The conclusion (i) is now obtained by letting $b_0 := \min\{\alpha, \beta\}$.

(ii) From (i), we let $t_0 \in (0, b_0]$ below. By applying Lemma 8, (3.23), and (3.27), we obtain that for all $(s,t) \in \mathbb{R} \times (0, t_0]$

$$\begin{aligned}
&\left|G_{\varphi_0}^{\psi_i}(s,t) - G_{\varphi_0}^{\psi_j}(s,t)\right| \\
&\leq \int_0^{t_0} \int_{-\infty}^\infty K(s-\xi, t-\tau) \cdot \left|(h \circ \psi_i)(\xi,\tau) - (h \circ \psi_j)(\xi,\tau)\right| d\xi d\tau \\
&\leq C_4 \cdot t_0 \|\psi_i - \psi_j\|_{t_0}
\end{aligned} \tag{3.30}$$

and

$$\begin{aligned}&\left|\partial_s G_{\varphi_0}^{\psi_i}(s,t) - \partial_s G_{\varphi_0}^{\psi_j}(s,t)\right| \\ &\quad \le \int_0^{t_0}\int_{-\infty}^{\infty} \left|\partial_s K(s-\xi, t-\tau)\right| \cdot \left|(h\circ\psi_i)(\xi,\tau) - (h\circ\psi_j)(\xi,\tau)\right| d\xi d\tau \\ &\quad \le 2C_4 \cdot \frac{\sqrt{t_0}}{\sqrt{\pi}} \|\psi_i - \psi_j\|_{t_0}. \end{aligned} \tag{3.31}$$

From (3.30) and (3.31), we deduce that

$$\left\|G_{\varphi_0}^{\psi_i} - G_{\varphi_0}^{\psi_j}\right\|_{t_0} \le C_4 \cdot \left(t_0 + 2\frac{\sqrt{t_0}}{\sqrt{\pi}}\right)\|\psi_i - \psi_j\|_{t_0}.$$

Thus, conclusion *(ii)* is obtained by choosing a small $t_0 \le b_0$ such that

$$C_4 \cdot \left(t_0 + 2\frac{\sqrt{t_0}}{\sqrt{\pi}}\right) < 1.$$

□

The following lemma will allow us to improve interior smoothness of solutions $\varphi \in \mathfrak{B}_{t_0}$.

Lemma 10 *Assume that $b > 0$, $\varphi : \mathbb{R} \times [0,b] \to \mathbb{R}$ is bounded, and $\varphi(\cdot,t)$ is continuous for any fixed $t \in [0,b]$. Suppose, $\forall \ell \in \{1,2,\dots,m\}$ and $\forall \delta \in (0,b)$,*

$$\left|\partial_s^\ell \varphi(s,t)\right| \le C_\ell(b,\delta), \quad \forall (s,t) \in D_\delta^b, \tag{3.32}$$

$$\left|\partial_s^\ell \varphi(s_2,t) - \partial_s^\ell \varphi(s_1,t)\right| \le C_\ell(b,\delta) \cdot |s_2 - s_1|, \quad \forall (s_1,t),(s_2,t) \in D_\delta^b, \tag{3.33}$$

$$\partial_s^\ell \varphi \in C^0\left(D_0^b\right), \tag{3.34}$$

for some sequence of positive numbers $\{C_\ell(b,\delta)\}$.

Then, there exists a constant $C'_{m+1}(b,\delta)$ such that

$$\left|\partial_s^{m+1} H_0^\varphi(s,t)\right| \le C'_{m+1}(b,\delta), \quad \forall (s,t) \in D_\delta^b, \tag{3.35}$$

$$\left|\partial_s^{m+1} H_0^\varphi(s_2,t) - \partial_s^{m+1} H_0^\varphi(s_1,t)\right| \le C'_{m+1}(b,\delta)|s_2 - s_1|, \quad \forall (s_1,t),(s_2,t) \in D_\delta^b, \tag{3.36}$$

$$\partial_s^{m+1} H_0^\varphi \in C^0\left(D_0^b\right). \tag{3.37}$$

Moreover, if φ is a fixed point of the map $\mathcal{G}_{\varphi_0}$ between the space $\mathfrak{B}_b$ as defined in (3.13), *i.e., $\varphi = \mathcal{G}_{\varphi_0}(\varphi)$, there exists a constant $C_{m+1}(b,\delta)$ such that*

$$\left|\partial_s^{m+1} \varphi(s,t)\right| \le C_{m+1}(b,\delta), \quad \forall (s,t) \in D_\delta^b, \tag{3.38}$$

$$\left|\partial_s^{m+1} \varphi(s_2,t) - \partial_s^{m+1} \varphi(s_1,t)\right| \le C_{m+1}(b,\delta) \cdot |s_2 - s_1|, \quad \forall (s_1,t),(s_2,t) \in D_\delta^b, \tag{3.39}$$

$$\partial_s^{m+1} \varphi \in C^0\left(D_0^b\right). \tag{3.40}$$

We leave the proof of this lemma to the Appendix.

Proof of Theorem 3 Below, we will use the contraction map constructed in Lemma 9 to find the solutions.

Define the sequence of functions $\{\psi_n : n \in \mathbb{N}_0\}$ by

$$\psi_n = \begin{cases} \varphi_0 & n = 0, \\ \mathcal{G}_{\varphi_0}(\psi_{n-1}) & n \geq 1. \end{cases} \tag{3.41}$$

From Lemma 9, the sequence $\{\psi_n\}$ is a Cauchy sequence in $\mathfrak{B}^L_{t_0}(d_0, M_0)$, which is a closed subset in the Banach space $\mathfrak{B}_{t_0}$. Thus, there exists a unique function, $\varphi \in \mathfrak{B}^L_{t_0}(d_0, M_0)$, such that $\psi_n \to \varphi$ in the topological space $(\mathfrak{B}_{t_0}, \|\cdot\|_{t_0})$. On the other hand, observe from (3.14)–(3.17) that for each $n \geq 0$, the function ψ_{n+1} satisfies

$$\begin{cases} \partial_t \psi_{n+1} = \partial_s^2 \psi_{n+1} + h \circ \psi_n & \text{in } D_0^{t_0}, \\ \psi_{n+1}(\cdot, 0) = \varphi_0(\cdot) & \text{in } \mathbb{R}. \end{cases}$$

Therefore, as $n \to \infty$, the function $\varphi \in \mathfrak{B}_{t_0}$ satisfies Eq. (3.11) and can be written as

$$\begin{aligned} \varphi(s,t) &= U_{\varphi_0}(s,t) + H_0^{\varphi}(s,t) \\ &= U_{\varphi_0}(s,t) + \int_0^t \int_{-\infty}^{\infty} K(s-\xi, t-\tau) \cdot (h \circ \varphi)(\xi, \tau) d\xi d\tau. \end{aligned} \tag{3.42}$$

Since $\varphi \in \mathfrak{B}_{t_0}$, we conclude that $\varphi, \partial_s \varphi \in C^0(\overline{D_0^{t_0}}) \cap L^\infty(\overline{D_0^{t_0}})$.

Moreover, by using Lemma 10 and a bootstrap argument, we obtain

$$\partial_s^m \varphi \in C^0\big(D_0^{t_0}\big), \quad \forall m \in \mathbb{N}. \tag{3.43}$$

Note that for any $k \in \mathbb{N}$,

$$\partial_t^k \varphi = \partial_s^{2k} \varphi + \text{lower-order terms}, \tag{3.44}$$

where "lower-order terms" are terms whose highest-order derivatives w.r.t. space variable s are of order less than $2k$. Thus, we conclude that $\varphi \in C^\infty(D_0^{t_0})$. □

Proof of Theorem 2 We may extend φ_0 as an even function from $[0, L]$ to $[-L, L]$. Then we extend φ_0 as a periodic function with period $2L$ from $[-L, L]$ to the whole real line $\mathbb{R}$. Denote this extended function by $\widetilde{\varphi}_0$. From Theorem 3, there is a positive number t_0, and a continuous and bounded function φ satisfying

$$\begin{cases} \partial_t \varphi = \partial_s^2 \varphi + \lambda_1 \sin \varphi - \lambda_2 \cos \varphi & \text{in } D_0^{t_0}, \\ \varphi(\cdot, 0) = \widetilde{\varphi}_0(\cdot) & \text{on } \mathbb{R} \times \{0\}, \end{cases} \tag{3.45}$$

with the regularity $\varphi, \partial_s \varphi \in C^0(\overline{D_0^{t_0}}) \cap L^\infty(\overline{D_0^{t_0}}) \cap C^\infty(D_0^{t_0})$. Thus, $\varphi_{\lfloor [0,L]\times[0,t_0]}$ fulfills the first two equations in (3.10). To finish the proof, we only need to verify that $\varphi_{\lfloor [0,L]\times[0,t_0]}$ also fulfills the boundary conditions in (3.10), i.e.,

$$\partial_s \varphi(b, t) = k(b, t) = 0, \quad \forall b \in \partial I, \ t \in [0, t_0]. \tag{3.46}$$

To prove (3.46), it remains to show that the solution φ of (3.45) satisfies the periodicity conditions

$$\begin{cases} \varphi(s,t) = \varphi(-s,t), \\ \varphi(s,t) = \varphi(s+2L,t), \end{cases} \tag{3.47}$$

for all $s \in \mathbb{R}$ and $t \in [0, t_0]$. This is because that the regularity $\varphi, \partial_s \varphi \in C^0(\overline{D_0^{t_0}})$ and periodicity properties (3.47), (3.48) imply the condition (3.46). Notice that (3.47) implies that

$$\varphi(L+s,t) = \varphi(L-s,t), \quad \forall s \in \mathbb{R},\ t \in [0, t_0]. \tag{3.48}$$

With the properties (3.47) and (3.48), φ would satisfy the boundary conditions (3.46). To show (3.47), we use an induction argument. Namely, assume the sequence of functions $\{\psi_n : n \in \mathbb{N}_0\}$ defined in (3.41) satisfies (3.47) for all $n \in \{0, 1, \dots, \ell\}$. Then, we show that $\psi_{\ell+1} := U_{\varphi_0} + H_0^{\psi_\ell}$ also satisfies the periodicity conditions in (3.47). This can be achieved by showing that U_{φ_0} and $H_0^{\psi_\ell}$ also fulfills the periodicity conditions in (3.47). This follows straightforward from the explicit integral formulae of U_{φ_0} and $H_0^{\psi_\ell}$. Note that since $\{\psi_n\}_{n\in\mathbb{N}_0}$ is a Cauchy sequence, which converges to φ in the Banach space $(\mathfrak{B}_{t_0}, \|\cdot\|_{t_0})$, the topology of $(\mathfrak{B}_{t_0}, \|\cdot\|_{t_0})$ guarantees that the limit φ also satisfies (3.47). □

3.3 Long-Time Existence and Asymptotics

The long-time existence of C^∞-smooth solution of (1.1) is obtained by a contradiction argument. Namely, assume the solution fails to be C^∞-smooth at the first time $t = t_1 > 0$. Then, we show that the L^2-norm of the derivatives of the curvature of any order remains uniformly bounded for any $t < t_1$. This would imply the required contradiction. Finally, we show the asymptotic behavior for a convergent subsequence of the solutions up to rotation of T.

The following lemma shows that there is a constant $C > 0$ such that $|\vec{\lambda}(t)| \le C$ for all $t \in [0, t_1)$.

Lemma 11 *For fixed $t \in [0, t_1)$, let $\varphi_t(\cdot) = \varphi(\cdot, t)$. Assume $\varphi_t \in W^{1,2}(\bar{I})$ with $0 < \|\varphi_t\|_{W^{1,2}} \le E_0$ and $T_t = (\cos\varphi_t, \sin\varphi_t)$ fulfills $\int_I T_t = \Delta p$, where $\Delta L := L - |\Delta p| > 0$ and Δp are constant. Then, there exists a positive number $C = C(\Delta L, L, E_0)$ such that $|\vec{\lambda}(t)| \le C$, for all $t \in [0, t_1)$.*

Proof The proof is achieved by applying Lemma 1. Thus, we need to find a modulus of continuity of φ_t and positive lower bound of $\operatorname{osc}_{\bar{I}} \varphi_t$, which are independent of $t \in [0, t_1)$.

Observe that, for any $x, y \in \bar{I}$,

$$\bigl|\varphi_t(y) - \varphi_t(x)\bigr| = \left|\int_x^y \partial_s \varphi_t ds\right| \le \left(\int_I |\partial_s \varphi_t|^2 ds\right)^{1/2} \cdot |x-y|^{1/2},$$

by Hölder's inequality. The modulus of continuity of φ_t could be chosen as

$$\omega(x) = E_0 \cdot x^{1/2}. \tag{3.49}$$

Since the integral of the tangent indicatrix T_t, $\int_I T_t ds$, represents the vector in $\mathbb{R}^2$ connecting the end points of the planar curve, we may assume without loss of generality that

$$\int_I \bigl(\cos\varphi_t(s), \sin\varphi_t(s)\bigr) ds = \bigl(|\Delta p|, 0\bigr) \tag{3.50}$$

by a rotation of the coordinates of $\mathbb{R}^2$ (i.e., by considering a new angle function $\phi_t = \varphi_t + \varphi_*$ for some constant φ_*). Note that (3.50) implies that

$$\int_I \bigl[1 - \cos\varphi_t(s)\bigr] ds = \Delta L > 0. \tag{3.51}$$

From (3.50), (3.51), and the mean value theorem, there exist $s_-, s_+ \in \bar{I}$ such that $\sin\varphi_t(s_-) = 0$ and $1 - \cos\varphi_t(s_+) = \frac{\Delta L}{L} \in (0, 1)$. Let

$$\begin{cases} \varphi_-(t) := \varphi_t(s_-) \in \{n\pi : n \in \mathbb{Z}\}, \\ \varphi_+(t) := \varphi_t(s_+) = \arccos(1 - \Delta L/L) \notin \{n\pi : n \in \mathbb{Z}\}, \end{cases}$$

and

$$|\Delta\varphi| := \big|\varphi_+(t) - \varphi_-(t)\big| \geq \big|\arccos(1 - \Delta L/L)\big| \in (0, \pi/2).$$

Thus, for any fixed $t \in [0, t_1)$,

$$\operatorname*{osc}_{\bar{I}} \varphi_t \geq |\Delta\varphi| > 0. \tag{3.52}$$

From (3.49) and (3.52), we may apply Lemma 1 to derive that, for any fixed $t \in [0, t_1)$,

$$\int_I \sin^2\big[\varphi_t(s) + \varphi_*\big] ds \geq \sin^2\left(\frac{\delta_0}{4}\right) \cdot \min\big\{\omega^{-1}(\delta_0/4), L/2\big\} =: C_1,$$

where $\delta_0 = \min\{|\Delta\varphi|, \pi\}$, and

$$\det A_{T_{\varphi_t}} \geq \frac{L}{2} C_1 =: C_2.$$

Then, together with the assumption, $0 < \|\varphi_t\|_{W^{1,2}} \leq E_0$, $\forall t \in [0, t_1)$, and (2.22), we finish the proof by choosing

$$C := \frac{2\sqrt{2} L E_0}{C_2}.$$

□

Lemma 12 *Assume the curvature vector $\vec{\kappa}$ remains smooth during the L^2-flow* (1.1) *up to $t_1 > 0$. Then, $\forall t \in (0, t_1)$ and $\forall m \in \mathbb{N}$,*

$$\frac{d}{dt}\frac{1}{2}\int_I \big|\nabla_s^m \vec{\kappa}\big|^2 ds + \int_I \big|\nabla_s^{m+1}\vec{\kappa}\big|^2 ds = -\int_I \big\langle \nabla_s^{m+1}\vec{\kappa}, \nabla_s^{m-1}\big(\langle \vec{\lambda}, T\rangle \cdot \vec{\kappa}\big)\big\rangle ds.$$

Proof From Lemma 5, we have $\nabla_s^{2\ell}\vec{\kappa} = 0$ and $\nabla_t \nabla_s^{2\ell}\vec{\kappa} = 0$ for any $\ell \in \mathbb{N}_0$ at the boundary. Together with applying (3.4), (2.14), and the formula of integration by parts in Lemma 3, we have

$$\begin{aligned} \frac{d}{dt}\frac{1}{2}\int_I \big|\nabla_s^m\vec{\kappa}\big|^2 ds &= \int_I \big\langle \nabla_s^m\vec{\kappa}, \nabla_t\nabla_s^m\vec{\kappa}\big\rangle ds = \int_I \big\langle \nabla_s^m\vec{\kappa}, \nabla_s\nabla_t\nabla_s^{m-1}\vec{\kappa}\big\rangle ds \\ &= -\int_I \big\langle \nabla_s^{m+1}\vec{\kappa}, \nabla_t\nabla_s^{m-1}\vec{\kappa}\big\rangle ds = -\int_I \big\langle \nabla_s^{m+1}\vec{\kappa}, \nabla_s^{m-1}\nabla_t\vec{\kappa}\big\rangle ds \\ &= -\int_I \big\langle \nabla_s^{m+1}\vec{\kappa}, \nabla_s^{m-1}\big(\nabla_s^2\vec{\kappa} + \langle \vec{\lambda}, T\rangle \cdot \vec{\kappa}\big)\big\rangle ds \\ &= -\int_I \big|\nabla_s^{m+1}\vec{\kappa}\big|^2 ds - \int_I \big\langle \nabla_s^{m+1}\vec{\kappa}, \nabla_s^{m-1}\big(\langle \vec{\lambda}, T\rangle \cdot \vec{\kappa}\big)\big\rangle ds. \end{aligned}$$

□

Proof of Theorem 1 For any $m \in \mathbb{N}$, we will obtain the long-time existence by deriving the following estimates for curvature integrals,

$$\frac{d}{dt}\int_I \big|\nabla_s^m\vec{\kappa}\big|^2 ds + \int_I \big|\nabla_s^m\vec{\kappa}\big|^2 ds \leq C\big(\|\vec{\kappa}_0\|_{L^2}, L, |\Delta p|, m\big). \tag{3.53}$$

By Gronwall inequality, (3.53) implies

$$\int_I \left|\nabla_s^m \vec{\kappa}\right|^2 ds \leq C\big(\|\vec{\kappa}_0\|_{L^2}, L, |\Delta p|, m\big), \quad \forall t \in (0, t_1), \tag{3.54}$$

where $\vec{\kappa}_0 = \vec{\kappa}(s, 0)$ is the curvature vector of the initial curve f_0.

Note that (2.12) gives the non-increasing property of the energy $\mathcal{F}_{L,\Delta p}[T]$. Thus, as long as the smooth solutions of the L^2-flow exist for all $t \in (0, t_1)$, one has $\mathcal{F}_{L,\Delta p}(T) \leq \mathcal{F}_{L,\Delta p}(T_0)$, where T_0 is the initial tangent indicatrix. Therefore, for each $t \in (0, t_1)$, $\|\vec{\kappa}(\cdot, t)\|_{L^2}^2 \leq 2\mathcal{F}_{L,\Delta p}(T_0)$.

From Lemma 12, we have

$$\begin{aligned} \frac{d}{dt}\frac{1}{2}\int_I \left|\nabla_s^m \vec{\kappa}\right|^2 ds + \int_I \left|\nabla_s^{m+1}\vec{\kappa}\right|^2 ds &= -\int_I \big\langle \nabla_s^{m+1}\vec{\kappa}, \nabla_s^{m-1}\big(\langle \vec{\lambda}, T\rangle \cdot \vec{\kappa}\big)\big\rangle ds \\ &\leq \left(\int_I \left|\nabla_s^{m+1}\vec{\kappa}\right|^2 ds\right)^{1/2} \cdot \left(\int_I \left|\nabla_s^{m-1}\big(\langle \vec{\lambda}, T\rangle \cdot \vec{\kappa}\big)\right|^2 ds\right)^{1/2}. \end{aligned} \tag{3.55}$$

Observe that

$$\nabla_s^{m-1}\big(\langle \vec{\lambda}, T\rangle \cdot \vec{\kappa}\big) = \langle \vec{\lambda}, T\rangle \cdot \nabla_s^{m-1}\vec{\kappa} + \sum_{i+j=m-2} C(i,j) \cdot \big\langle \vec{\lambda}, \partial_s^i \vec{\kappa}\big\rangle \cdot \nabla_s^j \vec{\kappa}, \tag{3.56}$$

and, from (3.6),

$$\left|\partial_s^i \vec{\kappa}\right|^2 = \left|\nabla_s^i \vec{\kappa} + \sum_{\substack{[\![a,b]\!] \leq [\![i-2,3]\!] \\ c \leq i-2}} P_b^{a,c}(\vec{\kappa})\right|^2 + \left|\sum_{\substack{[\![a,b]\!] \leq [\![i-1,2]\!] \\ c \leq i-1}} P_b^{a,c}(\vec{\kappa})\right|^2.$$

Note that the term $\langle \vec{\lambda}, T\rangle \cdot \nabla_s^{m-1}\vec{\kappa}$, is the highest-order term on the R.H.S. of (3.56). By applying Gagliardo–Nirenberg inequalities and the uniform bound of $|\vec{\lambda}|$ proved in Lemma 11, we obtain

$$\left(\int_I \left|\nabla_s^{m-1}\big(\langle \vec{\lambda}, T\rangle \cdot \vec{\kappa}\big)\right|^2 ds\right)^{1/2} \leq C\big(\|\vec{\kappa}_0\|_{L^2}, L, |\Delta p|, m\big) \cdot \|\vec{\kappa}\|_{m+1,2}^{(m-1)/(m+1)}.$$

By adding $\frac{1}{2}\int_I |\nabla_s^m \vec{\kappa}|^2 ds$ on both sides of (3.55) and applying Gagliardo–Nirenberg inequalities, we have

$$\begin{aligned} &\frac{d}{dt}\int_I \left|\nabla_s^m \vec{\kappa}\right|^2 ds + \int_I \left|\nabla_s^m \vec{\kappa}\right|^2 ds + 2\int_I \left|\nabla_s^{m+1}\vec{\kappa}\right|^2 ds \\ &\quad \leq \epsilon \cdot \|\vec{\kappa}\|_{m+1,2}^2 + C\big(\epsilon, \|\vec{\kappa}_0\|_{L^2}, L, |\Delta p|, m\big). \end{aligned}$$

for any chosen $\epsilon > 0$. From the definition of the norm $\|\vec{\kappa}\|_{m+1,2}^2$, we may choose a sufficiently small $\epsilon > 0$, depending only on m and L, such that

$$\frac{d}{dt}\int_I \left|\nabla_s^m \vec{\kappa}\right|^2 ds + \int_I \left|\nabla_s^m \vec{\kappa}\right|^2 ds \leq C\big(\|\vec{\kappa}_0\|_{L^2}, L, |\Delta p|, m\big),$$

as desired in (3.53), which implies (3.54).

As we let $T = (\cos\varphi, \sin\varphi)$ represent the unit tangent vector of a *planar* curve, $\vec{\kappa} = \partial_s T$ is its curvature vector, and $k = \langle \vec{\kappa}, \vec{n} \rangle = \partial_s \varphi$ its signed curvature. Thus we may write that, for any $m \in \mathbb{N}$,

$$\nabla_s^m \vec{\kappa} = \partial_s^m k \cdot \vec{n}, \tag{3.57}$$

and

$$\partial_s^{m-1} k(s) - \partial_s^{m-1} k(\sigma) = \int_\sigma^s \partial_s^m k. \tag{3.58}$$

From (3.58) and applying Lemma 5, we have

$$\partial_s^{m-1} k(s) = \int_{\sigma^*}^s \partial_s^m k, \tag{3.59}$$

for some $\sigma^* \in \bar{I}$. To derive (3.59), we can simply choose $\sigma^* \in \partial \bar{I}$ as m is odd. As m is even, we can apply the mean value theorem and Lemma 5 to conclude that $\partial_s^{m-1} k(\sigma^*) = 0$ for some $\sigma^* \in \bar{I}$.

From (3.59), we obtain

$$\left\| \partial_s^{m-1} k \right\|_{L^\infty} \leq \left\| \partial_s^m k \right\|_{L^1}, \quad \forall m \in \mathbb{N}.$$

Thus, together with (3.54) and (3.57), we conclude that $\|\partial_s^m k\|_{L^\infty}$ is also uniformly bounded for all $m \in \mathbb{N}_0$. Now the proof of long-time existence is completed by a contradiction argument.

To analyze the asymptotic behavior of the flow, we choose a subsequence of $T(t, \cdot)$, which converges smoothly to $T_\infty(\cdot)$ as $t \to \infty$. In addition, we rewrite the energy identity in (2.12) as

$$u(t) := \int_I |\partial_t T|^2 ds = -\frac{d}{dt} \mathcal{F}_{L,\Delta p}[T], \tag{3.60}$$

which implies that $u \in L^1([0,\infty))$. On differentiating (3.60) and applying (1.1), (3.54), we obtain

$$\left| u'(t) \right| \leq C\big(\|\vec{\kappa}_0\|_{L^2}, L, |\Delta p| \big).$$

Therefore, $u(t) \to 0$ as $t \to \infty$. In other words, $T_\infty(\cdot)$ is independent of t and thus, by (1.1), is an equilibrium configuration. □

Acknowledgements During working on this project, C.-C. Lin was supported by the National Science Council of Taiwan NSC 101-2115-M-003-002 and the National Center for Theoretical Sciences in Taipei, Taiwan; and Y.-K. Lue was supported by MOST 103-2811-M-003-008. The authors also want to thank the referees for useful comments and corrections toward improving the presentation of this article.

Appendix

Lemma 13 *Let $\ell \in \mathbb{N}$, and*

$$\gamma_\ell(s,t) := \int_0^t \int_{-\infty}^{\infty} \partial_s^\ell K(s-\xi, t-\tau) \cdot (\hat{n} \circ \varphi)(\xi, \tau) d\xi d\tau.$$

Then, under the assumption in Lemma 10, $\gamma_\ell : D_0^b \to \mathbb{R}$ is well-defined for any $\ell \in \{1, 2, \ldots, m+1\}$. Moreover, for any $\delta \in (0, t)$, the following properties hold,

$$|\gamma_\ell(s,t)| \le C_\ell''(b,\delta), \quad \forall (s,t) \in D_\delta^b, \tag{4.1}$$

$$|\gamma_\ell(s_2,t) - \gamma_\ell(s_1,t)| \le C_\ell''(b,\delta) \cdot |s_2 - s_1|, \quad \forall (s_1,t),(s_2,t) \in D_\delta^b, \tag{4.2}$$

$$\gamma_\ell \in C^0\big(D_0^b\big), \tag{4.3}$$

where $C_\ell''(b,\delta)$ is a constant.

Proof Note that since $\varphi \in \mathfrak{B}_b^L(d_0, M_0)$, we may apply (3.34), (3.32), (3.33), and (3.20), and verify easily that, for any $\ell \in \{1, 2, \ldots, m\}$,

$$\partial_\xi^\ell (h \circ \varphi) \in C^0\big(D_0^b\big),$$

$$\big|\partial_\xi^\ell (h \circ \varphi)(\xi,\tau)\big| \le C(\ell, b, \delta), \tag{4.4}$$

$$\big|\partial_\xi^\ell (h \circ \varphi)(\xi_2,\tau) - \partial_\xi^\ell (h \circ \varphi)(\xi_1,\tau)\big| \le C(\ell, b, \delta) \cdot |\xi_2 - \xi_1|, \tag{4.5}$$

for all $(\xi,\tau) \in D_\delta^b$, $(\xi_i,\tau) \in D_\delta^b$, $i \in \{1,2\}$, and some positive constants $C(\ell,b,\delta)$. For convenience, we may assume without loss of generality that, for fixed ℓ and b, $C(\ell,b,\delta) \nearrow +\infty$ as $\delta \searrow +0$.

From the definition of fundamental solution K in (3.12), it is easy to verify that K is infinitely differentiable in $\mathbb{R} \times (0,\infty)$ and its derivatives of any order is uniformly bounded in $\mathbb{R} \times [\epsilon,\infty)$ for any fixed $\epsilon > 0$. In fact, for any $k \in \mathbb{N}$,

$$\partial_x^k K(x,t) = P_t^k(x) \cdot K(x,t), \tag{4.6}$$

where $P_t^k(x)$ is a polynomial of degree k whose coefficients depend continuously on t and could tend to infinity only if $t \to +0$. From using (4.6), it is easy to verify that, for any fixed $t > 0$, $s \in \mathbb{R}$, and $t_1, t_2 \ge 0$,

$$\int_{-\infty}^{\infty} \big|\partial_s^k K(s-\xi,t)\big| d\xi < C_K(k,t), \tag{4.7}$$

$$\int_{t_1}^{t_2} \int_{-\infty}^{\infty} \big|\partial_s K(s-\xi,\tau)\big| d\xi d\tau = \frac{2(\sqrt{t_2} - \sqrt{t_1})}{\sqrt{\pi}}, \tag{4.8}$$

where the constant $C_K(k,t) \to +\infty$ as $t \to +0$.

To show (4.1) when $\ell \in \{1, 2, \ldots, m+1\}$, observe that, with a change of variables and an integration by parts with respect to the space variables,

$$\begin{aligned} \gamma_\ell(s,t) = &\int_0^{\epsilon'} \int_{-\infty}^{\infty} \partial_s^\ell K(s-\xi, t-\tau) \cdot (h \circ \varphi)(\xi,\tau) d\xi d\tau \\ &+ \int_{\epsilon'}^{t} \int_{-\infty}^{\infty} \partial_s^{\ell-1} K(s-\xi, t-\tau) \cdot \frac{\partial (h \circ \varphi)}{\partial \xi}(\xi,\tau) d\xi d\tau \\ &- \int_{\epsilon'}^{t} \big[\partial_s^{\ell-1} K(s-\xi, t-\tau) \cdot (h \circ \varphi)(\xi,\tau)\big]_{-\infty}^{\infty} d\tau, \end{aligned} \tag{4.9}$$

for any fixed $\epsilon' \in (0,t)$. Note that, for any fixed $\tau \in (\epsilon', t)$, one can infer from (4.6) and (3.20) that the integrand in the last term is null. As we continue to integrate by parts in (4.9), using (4.4) and the property of exponentially vanishing rate of $[\partial_s^j K](x,t')$, we obtain that

$$\gamma_\ell(s,t) = \int_0^{\epsilon'} \int_{-\infty}^{\infty} \partial_s^\ell K(s-\xi, t-\tau) \cdot (h \circ \varphi)(\xi,\tau) d\xi d\tau \\ + \int_{\epsilon'}^{t} \int_{-\infty}^{\infty} \partial_s K(s-\xi, t-\tau) \cdot \left[\partial_\xi^{\ell-1}(h \circ \varphi)\right](\xi,\tau) d\xi d\tau \tag{4.10}$$

for any $\ell \in \{1, 2, \dots, m+1\}$ and any fixed $\epsilon' \in (0,t)$. Now, (4.1) follows on applying (3.20), (4.4), (4.7), and (4.8). Furthermore, since (4.1) holds for any $\delta \in (0,b)$, $\gamma_\ell : D_0^b \to \mathbb{R}$ is well-defined.

To show (4.2), we use the formula (4.10) and write

$$\begin{aligned} &\gamma_{m+1}(s_2,t) - \gamma_{m+1}(s_1,t) \\ &= \int_0^{\epsilon'} \int_{-\infty}^{\infty} \left(\partial_s^{m+1} K(s_2-\xi, t-\tau) - \partial_s^{m+1} K(s_1-\xi, t-\tau)\right) \\ &\quad \cdot (h \circ \varphi)(\xi,\tau) d\xi d\tau \\ &\quad + \int_{\epsilon'}^{t} \int_{-\infty}^{\infty} \left(\partial_s K(s_2-\xi, t-\tau) - \partial_s K(s_1-\xi, t-\tau)\right) \\ &\quad \cdot \partial_\xi^m (h \circ \varphi)(\xi,\tau) d\xi d\tau, \end{aligned} \tag{4.11}$$

for some fixed $\epsilon' \in (0,t)$. The first integral in (4.11) can be rewritten as

$$\int_0^{\epsilon'} \int_{-\infty}^{\infty} \int_{s_1}^{s_2} \partial_s^{m+2} K(s-\xi, t-\tau) \cdot (h \circ \varphi)(\xi,\tau) ds d\xi d\tau. \tag{4.12}$$

For fixed t, τ satisfying $t-\tau \geq t-\epsilon' > 0$, one can conclude from (4.6) and (3.20) that the function $K_{(t,\tau)}^{m+2}(\xi,s) := \partial_s^{m+2} K(s-\xi, t-\tau) \cdot (h \circ \varphi)(\xi,\tau)$ is uniformly bounded, integrable over the subset $\mathbb{R} \times [s_1, s_2] \subset \mathbb{R}^2$ and

$$\iint_{\mathbb{R}\times[s_1,s_2]} \left|K_{(t,\tau)}^{m+2}(\xi,s)\right| d(\xi,s) < +\infty. \tag{4.13}$$

By using Fubini's Theorem, the integral in (4.12) can be written as

$$\int_0^{\epsilon'} \int_{s_1}^{s_2} \int_{-\infty}^{\infty} \partial_s^{m+2} K(s-\xi, t-\tau) \cdot (h \circ \varphi)(\xi,\tau) d\xi ds d\tau. \tag{4.14}$$

The second integral in (4.11) can be written as

$$\int_{\epsilon'}^{t} \int_{-\infty}^{\infty} \partial_s K\left(s_1-\xi', t-\tau\right) \\ \cdot \left[\partial_{\xi'}^m (h \circ \varphi)\left(\xi'+\Delta s, \tau\right) - \partial_{\xi'}^m (h \circ \varphi)\left(\xi', \tau\right)\right] d\xi' d\tau, \tag{4.15}$$

by the change of variables, $\xi' = \xi - \Delta s$, where $\Delta s := s_2 - s_1 \geq 0$. Thus, by applying (3.20), (4.7) to the integral (4.14), and by applying (4.5), (4.8) to the integral (4.15), we obtain

$$\begin{aligned} &\left|\gamma_{m+1}(s_2,t) - \gamma_{m+1}(s_1,t)\right| \\ &\quad \leq \Delta s \cdot \left(C_K\left(m+2, t-\epsilon'\right) \cdot 2C_3 \cdot \epsilon' + C(m,t,\delta) \cdot \left(t-\epsilon'\right) \cdot \frac{2(\sqrt{t}-\sqrt{\epsilon'})}{\sqrt{\pi}} \right). \end{aligned} \tag{4.16}$$

Therefore, (4.16) implies the continuity of γ_{m+1} in the domain D_0^b along the s-direction, i.e., we have proved (4.2).

To show (4.3), we only need to prove the continuity of γ_{m+1} at any point in D_0^b along the t-direction, because of the result in (4.2). Now, for a fixed $t_1 \in (0,b)$, we assume that $\epsilon > 0$ satisfies

$$b > t_1 + \epsilon > t_1 - \epsilon > 0 \tag{4.17}$$

and choose $t_2 \neq t_1$ so that

$$|t_2 - t_1| < \epsilon/2. \tag{4.18}$$

From using (4.10), we write

$$\begin{aligned} &\gamma_{m+1}(s,t_2) - \gamma_{m+1}(s,t_1) \\ &\quad = \int_0^{t_1-\epsilon} \int_{-\infty}^{\infty} \left(\partial_s^{m+1} K(s-\xi, t_2-\tau) - \partial_s^{m+1} K(s-\xi, t_1-\tau)\right) \\ &\qquad \cdot (h \circ \varphi)(\xi,\tau) d\xi d\tau \\ &\qquad + \int_{t_1-\epsilon}^{t_2} \int_{-\infty}^{\infty} \partial_s K(s-\xi, t_2-\tau) \cdot \partial_\xi^m (h \circ \varphi)(\xi,\tau) d\xi d\tau \\ &\qquad - \int_{t_1-\epsilon}^{t_1} \int_{-\infty}^{\infty} \partial_s K(s-\xi, t_1-\tau) \cdot \partial_\xi^m (h \circ \varphi)(\xi,\tau) d\xi d\tau. \end{aligned} \tag{4.19}$$

By the property $\partial_t K(x,t) = \partial_x^2 K(x,t)$, the first integral in (4.19) can be written as

$$\int_0^{t_1-\epsilon} \int_{-\infty}^{\infty} \int_{t_1}^{t_2} \partial_s^{m+3} K(s-\xi, t-\tau) \cdot (h \circ \varphi)(\xi,\tau) dt d\xi d\tau. \tag{4.20}$$

For fixed $s \in \mathbb{R}$, $\tau \in [0, t_1 - \epsilon]$, one can conclude from (4.6) and (3.20) that the function $L^{m+3}_{(s,\tau)}(\xi,t) := \partial_s^{m+3} K(s-\xi, t-\tau) \cdot (h \circ \varphi)(\xi,\tau)$ is uniformly bounded, integrable over the subset $\mathbb{R} \times [t_1, t_2] \subset \mathbb{R}^2$ and

$$\iint_{\mathbb{R}\times[t_1,t_2]} \left|L^{m+3}_{(s,\tau)}(\xi,t)\right| d(\xi,t) < +\infty. \tag{4.21}$$

By using Fubini's Theorem, we may rewrite the integral in (4.20) as

$$\int_0^{t_1-\epsilon} \int_{t_1}^{t_2} \int_{-\infty}^{\infty} \partial_s^{m+3} K(s-\xi, t-\tau) \cdot (h \circ \varphi)(\xi,\tau) d\xi dt d\tau =: I_1. \tag{4.22}$$

By applying (4.7), (4.17), (4.18), we note that $t-\tau \geq \epsilon/2$ in the integrand of (4.22) and thus

$$|I_1| \leq |t_2-t_1| \cdot (t_1-\epsilon) \cdot 2C_3 \cdot \sup_{t\in[\epsilon/2,b]} \big\{C_K(m+3,t)\big\}. \tag{4.23}$$

Denote the second integral in (4.19) by I_2. Then, by applying (4.4), (4.8), (4.18), we obtain

$$|I_2| \leq \big(|t_2-t_1|+\epsilon\big) \cdot \frac{2}{\sqrt{\pi t_1}} \cdot C(m,b,t_1-\epsilon). \tag{4.24}$$

Similarly, we may denote the third integral in (4.19) by I_3 and obtain

$$|I_3| \leq \epsilon \cdot \frac{2}{\sqrt{\pi t_1}} \cdot C(m,b,t_1-\epsilon). \tag{4.25}$$

Notice that $\sup_{t\in[\epsilon/2,b]}\{C_K(m+3,t)\} \to +\infty$, as $\epsilon \to +0$, while $C(m,b,t_1-\epsilon)$ in (4.24) and (4.25) would remain bounded as $\epsilon \to +0$. Now, for any sufficiently small $\epsilon > 0$, we choose any $t_2 \in (t_1-\epsilon/2, t_1+\epsilon/2)$ such that

$$|t_2-t_1|^{1/2} \cdot \sup_{t\in(\epsilon/2,b)} \big\{C_K(m+3,t)\big\} \leq 1.$$

Then, from (4.23)–(4.25), we obtain

$$\begin{aligned} &\big|\gamma_{m+1}(s,t_2)-\gamma_{m+1}(s,t_1)\big| \\ &\quad < \left\{|\Delta t|^{\frac{1}{2}} \cdot t_1 \cdot 2C_3 + \big(|\Delta t|+2\epsilon\big) \cdot \frac{2}{\sqrt{\pi t_1}} \cdot C(m,b,t_1-\epsilon)\right\}, \end{aligned}$$

where $|\Delta t| := |t_2-t_1| < \epsilon/2$. Now, we have the continuity of γ_{m+1} along t-direction at any point in D_0^b. Thus, the proof of (4.3) is obtained. □

Proof of Lemma 10 We want to show that

$$\partial_s^\ell H_0^\varphi(\cdot,t) = \gamma_\ell(\cdot,t), \quad \forall t \in (0,b],\ \ell \in \{1,2,\ldots,m+1\}. \tag{4.26}$$

Let

$$J_\epsilon(s,t) := \int_0^{t-\epsilon}\int_{-\infty}^{\infty} K(s-\xi,t-\tau)\cdot(h\circ\varphi)(\xi,\tau)d\xi d\tau.$$

Note that, for fixed $t \in (0,b]$ and fixed $\epsilon \in (0,t)$, $J_\epsilon(\cdot,t)$ is smooth over $\mathbb{R}$, and we may exchange the operation of differentiation and integration below,

$$\partial_s^\ell J_\epsilon(s,t) = \int_0^{t-\epsilon}\int_{-\infty}^{\infty} \partial_s^\ell K(s-\xi,t-\tau)\cdot(h\circ\varphi)(\xi,\tau)d\xi d\tau.$$

Then, for any fixed $t \in (0,b]$ and $\ell \in \{1,2,\ldots,m+1\}$, we first show that, on any bounded and closed interval of $\mathbb{R}$,

$$\partial_s^\ell J_\epsilon(\cdot,t) \to \gamma_\ell(\cdot,t) \text{ uniformly as } \epsilon \to +0.$$

Observe that, from (4.10), choosing sufficiently small $\epsilon > 0$ so that $t - \epsilon > \delta > 0$, and applying (4.8), we have

$$\begin{aligned}&\left|\gamma_\ell(s,t) - \partial_s^\ell J_\epsilon(s,t)\right|\\&= \left|\int_{t-\epsilon}^{t}\int_{-\infty}^{\infty} \partial_s K(s-\xi, t-\tau)\cdot \partial_\xi^{\ell-1}(h\circ\varphi)(\xi,\tau)d\xi d\tau\right|\\&\le C(\ell-1,b,\delta)\cdot \frac{2(\sqrt{t}-\sqrt{t-\epsilon})}{\sqrt{\pi}}.\end{aligned}$$

Thus, over any compact subset $[s_0, s] \subset \mathbb{R}$ and for any fixed $t \in (0, b]$,

$$\partial_s^\ell J_\epsilon(\cdot,t) \to \gamma_\ell(\cdot,t) \text{ uniformly}, \quad \forall \ell \in \{1,2,\ldots,m+1\}, \tag{4.27}$$

as $\epsilon \to +0$. Moreover, $\gamma_\ell(\cdot,t)$ is uniformly continuous on $[s_0, s]$, for all $\ell \in \{1, 2, \ldots, m+1\}$.

Now, we prove (3.37) by induction argument. Namely, we assume that

$$\partial_s^\ell H_0^\varphi(\cdot,t) = \gamma_\ell(\cdot,t), \quad \forall \ell \in \{0,1,\ldots,m\},\ \forall t \in (0,b]. \tag{4.28}$$

Notice that, as $\ell = 0$, the equality holds by the definition of function γ_ℓ. Since

$$\partial_s^\ell J_\epsilon(s,t) = \int_{s_0}^{s} \partial_s^{\ell+1} J_\epsilon(x,t)dx + \partial_s^\ell J_\epsilon(s_0,t), \tag{4.29}$$

$\forall \ell \in \{1, \ldots, m\}$, fixed $t \in (0, b]$, and sufficiently small $\epsilon > 0$. By letting $\epsilon \to +0$ and applying (4.27) in (4.29), we derive

$$\gamma_\ell(s,t) = \int_{s_0}^{s} \gamma_{\ell+1}(x,t)dx + \gamma_\ell(s_0,t), \quad \forall \ell \in \{1,\ldots,m\}. \tag{4.30}$$

By applying (4.28) to the L.H.S. of (4.30), we obtain

$$\partial_s^\ell H_0^\varphi(s,t) = \int_{s_0}^{s} \gamma_{\ell+1}(x,t)dx + \gamma_\ell(s_0,t), \quad \forall \ell \in \{1,\ldots,m\}.$$

Then, by applying the fundamental theorem of calculus above, we have proved (4.26).

The proof of Lemma 10 is now completed by using an induction argument on m, and applying Lemma 13 and (4.26). Note that when $m = 1$, (3.34), (3.32), (3.33) hold from the fact that $\varphi \in \mathfrak{B}_{t_0}$, for some $t_0 \in (0, b_0]$, and applying Lemma 6. □

References

1. Angenent, S.: On the formation of singularities in the curve shortening flow. J. Differ. Geom. **33**(3), 601–633 (1991)
2. Antman, S.S.: Nonlinear Problems of Elasticity, 2nd edn. Applied Mathematical Sciences, vol. 107. Springer, New York (2005)
3. Bryant Robert, R., Griffiths, P.: Reduction for constrained variational problems and $\int \frac{1}{2}\vec{\kappa}^2 ds$. Am. J. Math. **108**(3), 525–570 (1986)
4. Brunnett, G., Wendt, J.: Elastic splines with tension control. In: Mathematical Methods for Curves and Surfaces, II, Lillehammer, 1997. Innov. Appl. Math., pp. 33–40. Vanderbilt Univ. Press, Nashville (1998)

5. Cannon, J.R.: The One-Dimensional Heat Equation. Encyclopedia of Mathematics and Its Applications, vol. 23. Addison-Wesley, Reading (1984). Advanced Book Program. With a foreword by Felix E. Browder
6. Dall'Acqua, A., Pozzi, P.: A Willmore–Helfrich L^2-flow of curves with natural boundary conditions. Commun. Anal. Geom. **22**(4), 617–669 (2014)
7. Dall'Acqua, A., Lin, C.-C., Pozzi, P.: Evolution of open elastic curves in $\mathbb{R}^n$ subject to fixed length and natural boundary conditions. Analysis (Berlin) **34**(2), 209–222 (2014)
8. Dall'Acqua, A., Lin, C.-C., Pozzi, P.: A gradient flow for open elastic curves with fixed length and clamped ends. Preprint (2014)
9. Dziuk, G., Kuwert, E., Schätzle, R.: Evolution of elastic curves in $\mathbb{R}^n$, existence and computation. SIAM J. Math. Anal. **33**(5), 1228–1245 (2002)
10. Gage, M., Hamilton, R.S.: The heat equation shrinking convex plane curves. J. Differ. Geom. **23**(1), 69–96 (1986)
11. Golomb, M., Jerome, J.: Equilibria of the curvature functional and manifolds of nonlinear interpolating spline curves. SIAM J. Math. Anal. **13**, 421–458 (1982)
12. Hearst, J.E., Shi, Y.: The elastic rod provides a model for DNA and its functions. In: Mathematical Approaches to Biomolecular Structure and Dynamics, Minneapolis, MN, 1994. IMA Vol. Math. Appl., vol. 82, pp. 59–70. Springer, New York (1996)
13. Koiso, N.: On the motion of a curve towards elastica. In: Actes de la Table Ronde de Géométrie Différentielle, Luminy, 1992. Sémin. Congr., vol. 1, pp. 403–436. Soc. Math. France, Paris (1996)
14. Langer, J., Singer, D.A.: Curve straightening and a minimax argument for closed elastic curves. Topology **24**(1), 75–88 (1985)
15. Langer, J., Singer, D.A.: Lagrangian aspects of the Kirchhoff elastic rod. SIAM Rev. **38**(4), 605–618 (1996)
16. Lin, C.-C.: L^2-flow of elastic curves with clamped boundary conditions. J. Differ. Equ. **252**(12), 6414–6428 (2012)
17. Linnér, A.: Some properties of the curve straightening flow in the plane. Trans. Am. Math. Soc. **314**(2), 605–618 (1989)
18. Mumford, D.: Elastica and Computer Vision, West Lafayette, IN, 1990. Algebraic Geometry and Its Applications, pp. 491–506. Springer, New York (1994)
19. Novaga, M., Okabe, S.: Curve shortening–straightening flow for non-closed planar curves with infinite length. J. Differ. Equ. **256**(3), 1093–1132 (2014)
20. Oelz, D., Schmeiser, C.: Derivation of a model for symmetric lamellipodia with instantaneous cross-link turnover. Arch. Ration. Mech. Anal. **198**(3), 963–980 (2010)
21. Oelz, D.: On the curve straightening flow of inextensible, open, planar curves. SeMA Journal **54**, 5–24 (2011)
22. Polden, A.: Curves and surfaces of least total curvature and fourth-order flows. Ph.D. dissertation, Universität Tübingen, Tübingen, Germany (1996)
23. Starostin, E.L., van der Heijden, G.H.M.: The shape of a Möbius strip. Nat. Mater. **6**, 563–567 (2007)
24. Wen, Y.: L^2 flow of curve straightening in the plan. Duke Math. J. **70**(3), 683–698 (1993)
25. Wen, Y.: Curve straightening flow deforms closed plane curves with nonzero rotation number to circles. J. Differ. Equ. **120**(1), 89–107 (1995)
26. Willmore, T.: Curves. In: Handbook of Differential Geometry, vol. I, pp. 997–1023. North-Holland, Amsterdam (2000) (English summary)

DOI 10.1007/978-94-017-7300-3_15
Reprinted from *Journal of Elasticity* Journal, DOI 10.1007/s10659-015-9520-y

Buckling of Naturally Curved Elastic Strips: The Ribbon Model Makes a Difference

Basile Audoly[1] · **Keith A. Seffen**[2]

Received: 28 October 2014 / Published online: 7 March 2015

Abstract We analyze the stability of naturally curved, inextensible elastic ribbons. In experiments, we first show that a loop formed using a metallic strip can become unstable if its radius is larger than its natural radius of curvature (undercurved case): the loop then folds onto itself into a smaller, multiply-covered loop. Conversely, a multi-covered, overcurved metallic strip can unfold dynamically into a circular configuration having a lower covering index. We analyze these instabilities using a one-dimensional mechanical model for an elastic ribbon introduced recently (Dias and Audoly in J. Elast., 2014), which extends Sadowsky's developable elastic ribbon model in the presence of natural curvature. Combining linear stability analyses and numerical computations of the post-buckled configurations, we classify the equilibria of the ribbon as a function of the ratio of its natural curvature to its actual curvature. Our ribbon model is formulated in close analogy with classical rod models; this allows us to adapt classical stability methods for rods to the case of a ribbon. The stability of a ribbon is found to differ significantly from that of an anisotropic rod: we attribute this difference to the fact that the tangent twisting modulus of a ribbon can be negative, in contrast to what is possible in the well-studied case of linearly elastic rods. The specific stability properties predicted by the curved ribbon model are confirmed by a finite element analysis of cylindrical shells having a small height-to-radius ratio.

Keywords Elastic plates · Elastic rods · Energy minimization

Mathematics Subject Classification 74K20 · 74K10 · 74G65

✉ B. Audoly
audoly@lmm.jussieu.fr

K.A. Seffen
kas14@cam.ac.uk

1 UMR 7190 Institut Jean Le Rond d'Alembert, Sorbonne Universités, UPMC Univ Paris 06, CNRS, 75005 Paris, France

2 Department of Engineering, University of Cambridge, Trumpington Street, Cambridge CB2 1PZ, UK

1 Introduction

As their cross-section is much smaller than their length, slender rods can undergo significant rotations while the material strains remain in the elastic range. The stability of thin elastic rods has long been recognized as an important issue, and has developed into a vast, and now classical field [1]. To give just two examples, excessive deformation of slender rods can describe the failure of engineering components such as drill strings, masts and pipelines [24] as well as the conformations of biological molecules, in particular, DNA [12]. Thin rods are usually considered capable of bending about two orthogonal axes and twisting along their centerline, where natural curvature and torsion can be specified initially to capture different starting geometries. Three imperatives underpin the analysis of deformation: equilibrium, compatibility and material constitutive behavior in the form of generalized Hooke's laws; and for finite rotations, these become coupled within a nonlinear framework, for example, the well-known Kirchhoff rod equations of force and moment equilibrium on a given cross-section.

The cross-sectional shape of a rod determines its bending and torsional stiffness about its principal axes. For linearly elastic rods, these laws are linear themselves. When the cross-section tends towards being flat, *i.e.*, when the rod is more like a strip or ribbon, a beam-like viewpoint merely adjusts the stiffnesses according to their geometrical definitions. For a very thin ribbon, the stiffness ratio may be very large, which affects the physical deformation in some dominant way and, more importantly, the integrity of the assumptions on which the beam theory is founded. It is therefore imperative to reformulate the assumptions of cross-sectional influence. Typically, researchers assume that the ribbon behaves like a narrow and thin elastic plate or shell [20–22]. Care needs to be exercised in the context of highly deformed strips, where the levels of local deformation are large enough to warrant a developable response in which the Gaussian curvature all but disappears; this imposes further constraint upon the kinematical relationships between curvature and torsion.

To date, the two approaches to thin elastic strips have been developed separately, each one having its own advantages and drawbacks. On one hand, modelling strips as anisotropic rods makes it possible to reuse the large body of work available on the analysis and computation of equilibria, stability and dynamics of thin rods; however, accuracy can be compromised because rod models ignore the special role played by developable configurations of the midsurface of a thin strip. Such configurations have a much lower elastic energy than non-developable ones when the thickness is much smaller than the width. On the other hand, ribbon models based on plate theory, which appeared in the seminal works of Sadowsky [9, 10, 19, 20] and Wunderlich [25, 26], are mechanically accurate but were developed as a largely independent branch of elasticity. This branch is much less advanced—for instance, the equations of equilibrium of a ribbon were not derived until relatively recently [22], and no stability analysis based on the one-dimensional ribbon model has been carried out to date, to the best of our knowledge.

In this paper, we try to bring together the best of both worlds, and show that the ribbon model, when appropriately formulated, can be analyzed in close analogy to a rod model, without losing physical accuracy. To this end, we use the ribbon model introduced by Dias and Audoly in another paper of this special issue [3], which extends both the energy proposed by Sadowsky [20] and the equations of equilibrium derived by Starostin and van der Heijden [22] for a flat ribbon, to a naturally curved ribbon. This curved ribbon model is formulated mathematically as a constrained elastic rod: like rods, but unlike other ribbon models, it makes use of a frame of directors; in this model, the inextensibility[1] of the mid-surface of the ribbon is taken into account by specifying kinematical constraints, which gives rise to an effectively nonlinear constitutive law [3]. Being expressed in the same language as an elastic rod model, the ribbon model of [3] enables us in the present paper (i) to set up a stability analysis for a one-dimensional ribbon model for the first time, by a straightforward adaptation of the classical stability methods used for rods, and (ii) to carry out a detailed comparison of the predictions of the rod and ribbon models.

To illustrate this idea, we consider a specific geometry, namely the buckling instabilities of simple or multi-covered loops made of a naturally elastic ribbon, when there is a mismatch between their natural radius of curvature and their actual radius of curvature. In previous work, this problem has been addressed using rod models including the effect of cross-section anisotropy and natural curvature [6, 7, 15]; as our paper shows, the ribbon model yields markedly different results. This choice of geometry is inspired by physical models with ribbon-like cross-sections encountered during our research; industrial bandsaw blades packaged as continuous loops of three turns and foldable sun-shades for protecting vehicle passengers from sideways sunlight, both of which unfold into single loop, planar structures. Both have initial, or "intrinsic" curvature, which is revealed when the loop is severed; the ribbon has a different, unstressed natural curvature. By reconnecting the original loop, the ribbon is loaded internally and acquires pre-stress, and the levels of pre-stress and, hence, intrinsic curvature dictate that each loop can be uniquely folded into three planar hoops. The stability of this configuration relative to the open-loop state is studied here for each of the different cross-sectional models, with the objective of better understanding the influences of each set of assumptions. We find that stable loops with more than three turns are also possible, and we build a physical demonstrator.

The layout of this paper is as follows. Section 2 details our construction and testing of a multistable hoop from a bandsaw blade. This blade originally has stable single and triple ring states, and we adjust the natural curvature plastically to explore the effect upon stability; we determine the values at which the stability is lost or gained, leading us to a novel configuration of five stable rings. These results are later compared to theoretical values in Sect. 6, and we also describe a simple procedure for reliably folding the hoop between states. Section 3 introduces our three modelling scenarios: the isotropic rod, the flat rod and the ribbon. All of them assume inextensibility along the tangent line normal to each cross-section, which we impose via a constrained Cosserat framework. The constitutive, or generalized Hooke's, laws relating the internal moment and curvature-twist components on a given cross-section to each other are *linear* for the rod models and *non-linear* for the ribbon model. In Sect. 4, all of these laws are imported into the equations of motion linearized about initially circular configurations, so that we may begin to understand later the differences in properties of the multistable hoop. A key difference between the ribbon and rod models is identified at this stage: the tangent twisting modulus of both rod models is constant and positive, although that of the ribbon model varies and may be negative. In Sect. 5, the consequence of the positiveness or non-positiveness of this modulus is explored in a simple scenario of the unbending of an initially curved rod: we find that the ribbon model becomes unstable close to the point where its tangent twisting modulus passes through zero, but the rod models remain stable until a considerably larger deformation is applied. We carry out a finite element analysis of a shell model to confirm the predictions of the ribbon model. In Sect. 6, we return to the original problem of the stability of the multiply-covered strips, which we approach analytically

[1] An inextensible surface is a surface that preserves its metric, hence the length of all curves drawn onto it, upon deformation. We do not consider the case of surfaces that are inextensible along a given set of material directions only, such as surfaces made up of woven networks of inextensible fibres.

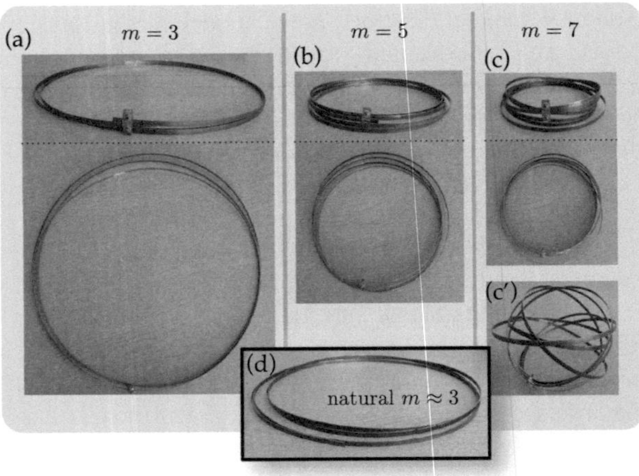

Fig. 1 Stable configurations of a bandsaw blade performing as a multistable hoop. The usual teeth have been removed from a commercially available product of initial length of around 3 m. The blade comes with two stable shapes; the open, single loop and the closed, folded triple loop. We adjusted the initial, or "natural" curvature by first cutting the loop, and then passing it through some mechanical rollers, which caused plastic curving about the cross-sectional width. The ends of cut loop are then re-connected by a simple clasp, visible here, before folding into higher-order configurations (see Fig. 2). Subfigures (**a**)–(**c**) show stable states with one, three and five loops (*top* and *bottom views*); in (**c**′) the five loop case has partially unraveled but is locked by frictional interplay between the loops. In (**d**), the naturally curved rod for each of these cases—the ends are free and it has just over three turns

and computationally using the ribbon model. We present a combination of dynamic stability analyses and numerical simulations of post-buckled equilibrium shapes. A global picture of the equilibria and their stability emerges, which is both consistent with our experiments and different from previous predictions based on rod models.

2 Experiments

Figure 1 shows a loop constructed from a 3 m length of bandsaw blade whose teeth have been ground down for ease of handling, resulting in a rectangular cross-section of width 5.65 mm and thickness 0.65 mm. The blade was passed in a controlled fashion through a set of rollers to impart uniform intrinsic curvature, κ_{n}, about the minor axis of bending of cross-section, and the final open section has around 3.33 turns as indicated, of radius $r_0 \approx 143$ mm. The ends of the blade are then connected together by a short bracket to form a continuous structure of three loops, which can be unfolded into a single loop or folded into five- and seven loops, as we soon describe. Apart from the single loop, each folded shape is almost but not quite planar owing to interference during the overlapping of turns; all have circular loops of the same radius and, crucially, the size of the initial curvature imparted by rolling allows each configuration to be locally stable, *i.e.*, to be relatively insensitive to sufficiently small out-of-plane perturbations.

Folding and obtaining each form is straightforward after some practice. Starting from a single loop, the ribbon may be twisted manually along any diameter in the out-of-plane direction, as displayed in Fig. 2, until it drives itself into three equal loops. To obtain five loops, the radius of one of the three loops is increased by feeding the arc-length of the two other loops into this larger loop by hand: when the larger loop is around 2–3 times as wide,

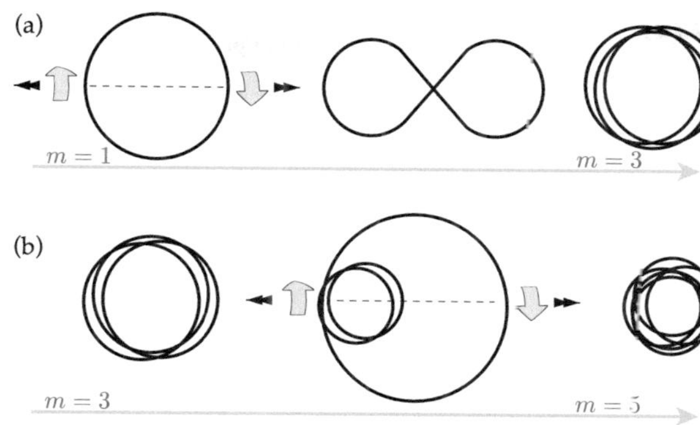

Fig. 2 Folding sequence for a loop with one, three and five stable loop configurations. (**a**) A single loop is twisted along a given diameter by hand to form a figure-of-eight loop in transition to the final three loop state. (**b**) Folding three loops into five is accomplished by, first, manually feeding arc-length into one of the three loops, at the expense of the other two, and then twisting this larger loop as per the single loop. The resulting five-looped state then adjusts itself automatically to the same radius everywhere even though we show them as being slightly different—purely to convey the number of loops

Table 1 Experimental regions of stability expressed in terms of $r_0\kappa_n$ (current radius × natural, open curvature) using a steel bandsaw blade, of rectangular cross-section 5.65 mm × 0.65 mm and $L \approx 3$ m

Covering index m	Stability bounds
1	$r_0\kappa_n < 3.50$
3	$0.28 < r_0\kappa_n <$ [beyond elastic limit]
5	$0.42 < r_0\kappa_n <$ [beyond elastic limit]

it can be twisted in a manner similar to the single loop, which forms three more loops in addition to the current pair of smaller loops. Once released, the final five loops adopt the same radii. Folding five- into seven loops is performed similarly provided plastic strains are not incurred. Each shape, or state, has different degrees of proximal stability: by far, the three loop configuration is most stable, and the other states move towards it in sequence if significantly perturbed or deliberately unfolded; from the single loop to three, or from seven- to five- to three loops. This happens to be the case for this physical model; if the intrinsic curvature is different, the stability character is altered.

A simple experiment is then performed to assess the limits at which stability is lost. For a given κ_n, the length of the blade is varied by feeding it through the bracket, thereby altering the closed length and r_0. The set-up is then laid on a horizontal table and gently released; stability is defined as being lost when the configuration either folds into more turns or opens up into fewer turns, and the critical data are reported in Table 1. In particular, it is found that a single loop is stable only if $r_0\kappa_n$ is less than 3.5. Above this value, a single loop cannot exist in an unloaded state but spontaneously collapses into three loops. The example in Fig. 1 has a ratio $r_0\kappa_n$ just below this value, and so is stable, but the degree of perturbation needed for destabilizing this shape is small. Three loops are just stable if $r_0\kappa_n$ is larger than 0.28, otherwise they open into a single loop for smaller values of this ratio. If the ratio is increased, then the system is increasingly resistant to perturbations. Unlike a single loop, three loops unfold, and the same is true for five loops, which open up into three loops if $r_0\kappa_n < 0.42$.

Other limits were considered, for example, for seven loops unfolding into five, but this was not recorded because there was too much contact between the loops, and friction played a significant role in preventing the transition. We also surmised that each of the three- and five loop scenarios were governed by an *upper bound* on stability in terms of $r_0\kappa_n$, where a higher ratio would cause each case to be unstable in the direction of more loops, *i.e.*, three loops would spontaneously collapse into five loops, *etc*. However, this was not observable without incurring plastic deformation because the original loop radius needs to be very small to unfold into a radius at least five-to-tenfold larger. In addition, higher loop configurations are reached only by folding in a very precise way, which is unlikely to happen spontaneously.

3 Models: Elastic Ribbon Versus Elastic Rods

In this section, we introduce three different mechanical models that can be used to represent a naturally curved elastic strip, such as the previously discussed bandsaw blade. The *ribbon* model is the only one that can accurately capture situations where the thickness h, the cross-sectional width w and the curvilinear length L are all very different, with $h \ll w \ll L$: this model is derived from a two-dimensional inextensible plate model by considering the limit $w \ll L$. The two other models are given merely for the sake of comparison; they are less accurate, but are somewhat simpler and have been extensively used in prior work. One model is the *isotropic rod* model: this is the standard model for a thin elastic rod having a circular cross-section—it simply ignores the aspect-ratio of the cross-section. The other one is the *flat rod* model, obtained by extrapolating the thin elastic rod model for a rectangular cross-section with dimensions $w \times h$ to the limit of a flat cross-section, *i.e.*, by letting $h/w \to 0$. This limit $h/w \to 0$, however, is incompatible with Kirchhoff kinematical hypothesis, which underpins the derivation of all rod models. The flat rod model is therefore mathematically defined, but ill-justified from a mechanical perspective.

3.1 Inextensible Cosserat Curve

We start with a kinematical description, common to all three models. Let S denote the arc-length in a reference configuration, and let L denote the curvilinear length of the rod. We use $S \in [0, L]$ as a Lagrangian coordinate to follow, or track, the location of material cross-sections. We will mostly consider static problems. However, to analyze dynamic stability, we will occasionally introduce a time variable, which we denote by τ. For any function $f(S,\tau)$, the derivative with respect to position in the reference configuration is denoted by $f'(S,\tau) = \partial f/\partial S$ and the temporal derivative by $\dot{f}(S,\tau) = \partial f/\partial \tau$.

Vectors are denoted by boldface characters. A configuration is specified by a Cosserat curve, *i.e.*, by the centerline $\mathbf{x}(S)$ and by three orthonormal directors $\mathbf{d}_i(S)$, with $i = 1, 2, 3$. We consider only *inextensible* and *unshearable* materials, for which the following two kinematical constraints are applicable. First, one of the directors, which we choose to be $\mathbf{d}_3$, coincides with the tangent to the curve:

$$\mathbf{x}'(S) = \mathbf{d}_3(S). \tag{1}$$

Second, the director's frame is orthonormal,

$$\mathbf{d}_i(S) \cdot \mathbf{d}_j(S) = \delta_{ij}, \tag{2}$$

which must hold for any $0 \le S \le L$ and for any combination of indices (i, j), where δ_{ij} denotes Kronecker's symbol. By convention, $\mathbf{d}_1$ is along the width of the strip and $\mathbf{d}_2$ is perpendicular to the plane tangent to strip, see Fig. 3. Note that inextensibility follows from projecting Eq. (1) along the tangent direction $\mathbf{d}_3$: this yields $|\mathbf{x}'| = 1$, *i.e.*, $|\mathrm{d}\mathbf{x}| = |\mathrm{d}S|$.

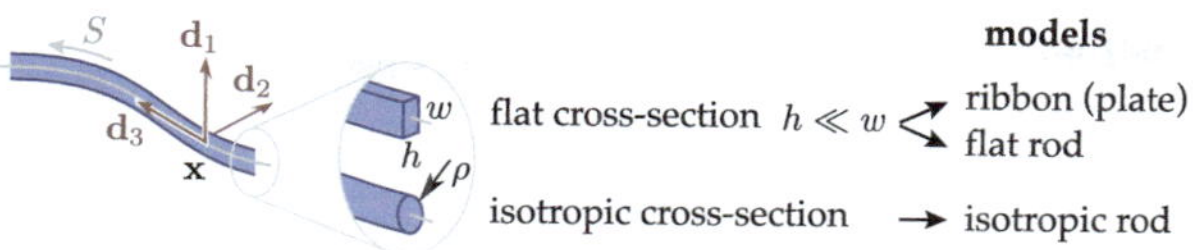

Fig. 3 An elastic ribbon is represented as a Cosserat curve $\mathbf{x}(S)$ with directors $\mathbf{d}_i(S)$. Throughout this paper, we compare the predictions of three mechanical models based on this kinematical description: the accurate ribbon model (which extends Sadowsky's narrow plate model to account for natural curvature), the classical rod model for a rectangular cross-section in the limit where the latter has a small aspect ratio, $h \ll w$ (flat rod model), and for a circular cross-section (isotropic rod model)

For closed ribbons, the centerline and directors are periodic, such that $\mathbf{x}(L) = \mathbf{x}(0)$ and $\mathbf{d}_i(L) = \mathbf{d}_i(0)$ (the equilibrium requires that the internal stress is periodic as well, and this will be imposed later).

The curvature strains (ω_1, ω_2) and the twisting strain ω_3 are defined as the components of the rotation gradient $\boldsymbol{\omega}$ relative to the basis formed by the directors:

$$\omega_i(S) = \frac{1}{2}\sum_{j=1}^{3}\sum_{k=1}^{3} \epsilon_{ijk}\, \mathbf{d}'_j(S) \cdot \mathbf{d}_k(S), \tag{3}$$

where ϵ_{ijk} is the alternating symbol. The rotation gradient, also known as the "Darboux vector", is represented by $\boldsymbol{\omega} = \sum_{i=1}^{3} \omega_i\, \mathbf{d}_i$ and obeys $\mathbf{d}'_i = \boldsymbol{\omega} \times \mathbf{d}_i$.

3.2 Kirchhoff's Equations of Motion

To state the equations of motion for ribbons and rods, we introduce the time variable τ. The internal stress is represented by the resultant $\mathbf{R}(S, \tau)$ and by the moment $\mathbf{M}(S, \tau)$. In the absence of external forces, the linear and angular momentum balances are expressed by Kirchhoff's equations of motion,

$$\mathbf{R}'(S, \tau) = \rho\, \ddot{\mathbf{x}}(S, \tau), \tag{4a}$$

$$\mathbf{M}'(S, \tau) + \mathbf{x}'(S, \tau) \times \mathbf{R}(S, \tau) = \mathbf{0}, \tag{4b}$$

where ρ is the mass per unit length. Since we have ignored the rotational inertia of the cross-section, the right-hand side of the equation for moment equilibrium is zero; as we explain later, this has implications for the growth rate of twisting-mode instabilities.

3.3 Constitutive Laws

The components $M_i = \mathbf{M} \cdot \mathbf{d}_i$ of the internal moment $\mathbf{M}$ relative to the basis of the directors are given by constitutive laws in terms of the strains ω_i. The three mechanical models (ribbon, isotropic rod, flat rod) arise from different constitutive laws, all of which are denoted generically by $\mathbf{f} = \mathbf{0}$,

$$\mathbf{f}(M_1, M_2, M_3, \omega_1, \omega_2, \omega_3) = \mathbf{0}. \tag{5}$$

Expressions for $\mathbf{f}$ corresponding to the different models are proposed below.

3.3.1 Constitutive Law for a Ribbon

The ribbon model was introduced in the seminal work of Sadowsky [20–22]. He derived the energy of an elastic ribbon by dimensional reduction by starting from the energy of a narrow, two-dimensional elastic plate, which he reduced to a one-dimensional model using the developability condition. His model correctly captures the inextensibility of the midsurface of the ribbon, which therefore remains a developable (hence ruled) surface after deformation.[2] Sadowsky explicitly parameterizes the ribbon's midsurface using the inverse cotangent η of the angle between the generatrices and the tangent $\mathbf{d}_3$ to the centerline.

Using a variational method, Starostin and van der Heijden [22] derived the equilibrium equations found by Sadowsky [9, 11, 18, 20], which they applied to geometries such as that in [23]. Even though their variational method departs from the standard treatment of thin elastic rods, the equilibrium equations they obtain are very similar to the classical Kirchhoff equations for thin rods, as they note.

In another paper in this special issue [3], Dias and Audoly extend Sadowsky's model to handle natural curvature. They also derive equilibrium equations for ribbons by adapting the classical approach, *i.e.*, by writing a principle of virtual work for elastic ribbons analogous to elastic rods. This viewpoint makes evident the strong similarities between the governing equations for rods and ribbons. It also has a practical advantage: it makes ribbons appear as a special case of rods. We adopt this view in the present paper, which enables us to analyze the stability of elastic ribbons by adapting of the classical stability analyses for elastic rods in a straightforward way.

We temporarily drop the time variable τ, as it does not enter into the constitutive laws. The direction of generatrices is parameterized by the quantity $\eta(S)$ in the Sadowsky model, and developability implies the constraint

$$\omega_3(S) = \eta(S)\,\omega_1(S). \tag{6a}$$

Following [3], we refrain from eliminating the parameter η, and view instead η and ω_3 as two variables linked by a kinematical constraint. The ribbons we consider here have zero geodesic curvature, *i.e.*, $\omega_2 = 0$ in the reference configuration, and their midsurface is inextensible, as previously noted. The geodesic curvature is preserved upon isometric deformations of the midsurface, hence the kinematical constraint

$$\omega_2 = 0 \tag{6b}$$

persists during deformation.

As written in Sect. 2, κ_n denotes the natural out-of-plane curvature. For simplicity, we assume κ_n is constant, and the reference configuration is taken to be a section of open cylinder with height w and natural radius, ${\kappa_n}^{-1}$. The strain energy stored in the deformed ribbon is derived in Eq. (34) of reference [3] from the energy of an inextensible cylindrical shell as

$$\mathcal{E}_{\text{rib}} = \frac{B_{\text{rib}}}{2}\int_0^L \left[{\omega_1}^2\left(1+\eta^2\right)^2 - 2\,\kappa_n\,\omega_1\left(1+\nu\,\eta^2\right)\right]\mathrm{d}S, \tag{7}$$

where $B_{\text{rib}} = D\,w$, is the product of the plate modulus $D = \frac{E\,h^3}{12\,(1-\nu^2)}$ and the width w of the ribbon; E is Young's modulus and ν is Poisson's ratio. The first term inside parenthesis in

[2] By contrast, Kirchhoff's kinematical hypothesis, which underpins rod models, valid when w and h are comparable, but it does not capture developable configurations well in the limit $h \ll w$.

the integrand is Sadowsky's energy, and the second term captures the effect of the natural curvature κ_n.

The first variation of the energy $\mathcal{E}_{\text{rib}}$ enables the following constitutive law to be found when the constraints in Eqs. (6a) and (6b) are applied:

$$M_1 = \frac{\partial \mathcal{E}_{\text{rib}}}{\partial \omega_1} - \eta\, M_3. \tag{8a}$$

Moreover, the condition of equilibrium for η writes:

$$\frac{\partial \mathcal{E}_{\text{rib}}}{\partial \eta} - \omega_1\, M_3 = 0. \tag{8b}$$

We refer the reader to [3] for details. Note that M_3 is the Lagrange multiplier associated with the constraint (6a): the principle of virtual work for ribbons is a constrained variational problem, and the terms $(-\eta\, M_3)$ and $(-\omega_1\, M_3)$ in Eqs. (8a) and (8b) are constraint terms, *i.e.*, are products of a Lagrange mutliplier by the gradient of a constraint.

Using Eq. (8b), we have $M_3 = 2\, B_{\text{rib}}\, (\omega_1\, (1+\eta^2)\, \eta - \kappa_n\, \nu\, \eta)$; Eq. (8a) then yields $M_1 = B_{\text{rib}}\, (\omega_1\, (1-\eta^4) - \kappa_n\, (1-\nu\, \eta^2))$. Eliminating η from Eq. (6a), the two constitutive equations and the constitutive constraint for a naturally curved ribbon are expressed altogether by $\mathbf{f}_{\text{rib}} = 0$, with

$$\mathbf{f}_{\text{rib}} = \begin{pmatrix} M_1 - B_{\text{rib}}\, [\omega_1 - \frac{{\omega_3}^4}{\omega_1^3} - \kappa_n (1 - \nu\, \frac{{\omega_3}^2}{\omega_1^2})] \\ \omega_2 \\ M_3 - 2\, B_{\text{rib}} [\omega_3 + \frac{{\omega_3}^3}{{\omega_1}^2} - \kappa_n\, \nu\, \frac{\omega_3}{\omega_1}] \end{pmatrix}. \tag{9}$$

Only the first and third components of $\mathbf{f}_{\text{rib}} = \mathbf{0}$ are genuine constitutive laws; the second component yields a constitutive constraint $\omega_2 = 0$, which is the geodesic constraint introduced earlier in Eq. (6b): in the presence of a kinematical constraint affecting a strain measure, $\omega_2 = 0$, this constraint indeed replaces the constitutive law for ω_2 and the bending moment M_2 is determined by solving the full equilibrium problem.

The ribbon model is fully captured by the expression of $\mathbf{f}_{\text{rib}}$ proposed in Eq. (9), so we do not need to use Eqs. (6a), (6b)–(8a), (8b) henceforth. It has been checked in reference [3] that the equations of equilibrium (4a)–(4b) and the constitutive law $\mathbf{f}_{\text{rib}} = \mathbf{0}$ are equivalent to the equilibrium equations derived by Starostin and van der Heijden [22] in the special case of a naturally straight and static ribbon ($\kappa_n = 0$ and $\ddot{\mathbf{x}} = \mathbf{0}$).

3.4 Constitutive Law for an Isotropic Rod

We define an 'isotropic' rod to be a linear elastic rod having equal principal bending moduli, as for a circular cross-section (note that isotropic rods may be naturally curved). The classical rod model representing an isotropic rod with natural curvature κ_n expresses $\mathbf{f}_{\text{iso}}$ as

$$\mathbf{f}_{\text{iso}} = \begin{pmatrix} M_1 - B_{\text{iso}}\, (\omega_1 - \kappa_n) \\ M_2 - B_{\text{iso}}\, \omega_2 \\ M_3 - C_{\text{iso}}\, \omega_3 \end{pmatrix}. \tag{10}$$

In its natural configuration, a section of this rod is a torus with cross-sectional radius ρ, enclosing a circular centerline with radius ${\kappa_n}^{-1}$. The bending and twisting moduli read $B_{\text{iso}} = E\, \frac{\pi\, \rho^4}{4}$, $C_{\text{iso}} = \frac{E}{2\,(1+\nu)}\, \frac{\pi\, \rho^4}{2}$, respectively, where $\mu = \frac{E}{2\,(1+\nu)}$ is the shear modulus of an isotropic Hookean material.

3.5 Constitutive Law for a Flat Rod

We call 'flat rod' the mathematical limit $w \gg h$ of the rod model with a rectangular cross-section, $w \times h$, and a natural curvature κ_n. When the aspect-ratio, h/w, of the cross-section tends to zero, the constitutive law is obtained in the limit:

$$\mathbf{f}_{\text{flat}} = \begin{pmatrix} M_1 - B_{\text{flat}}\,(\omega_1 - \kappa_n) \\ \omega_2 \\ M_3 - C_{\text{flat}}\,\omega_3 \end{pmatrix}, \tag{11}$$

where $B_{\text{flat}} = E\,\frac{w\,h^3}{12}$, $C_{\text{flat}} = \frac{E}{2\,(1+\nu)}\,\frac{w\,h^3}{3}$. Note that B_{flat} and C_{flat} are of the same order of magnitude, $\sim w\,h^3$. The other principal bending modulus is of order $\sim w^3\,h$, and is therefore much larger: this is why we replaced the corresponding constitutive law with the kinematical constraint $\omega_2 = 0$. For a rectangular rod, the torsional constant is obtained by solving a Laplace equation in a rectangular domain. The value of the torsional constant $\frac{w\,h^3}{3}$ used here corresponds to the solution of this Laplace problem for a thin strip of dimensions $w \times h$.

3.6 Not All Models Were Created Equal

As discussed at the beginning of Sect. 3, the ribbon model is the only accurate model if $h \ll w \ll L$. The two other models cannot be justified asymptotically in this limit and, worse, they are based on kinematical assumptions that cannot capture the inextensible modes of deformation of the center-surface (which are the lowest-energy modes when $h \ll w$). The rod models are given only for the sake of comparison.

4 Linearized Equations of Motion Near Circular Solutions

Without specifying boundary conditions, we linearize these equations near a circular configuration, as shown in Fig. 4. In so doing, we obtain the tangent bending and twisting moduli associated with the different constitutive laws, which play a key role in the stability analysis of the ribbon. The general linearized equations of motion derived in the present section are applied later to two specific geometries in Sect. 5 and Sect. 6.

4.1 Base State: Circular Solutions

We consider the twistless, circular state shown in Fig. 4, and analyze its stability for different boundary conditions. All quantities pertaining to the base circular state are marked with a subscript or superscript '0'. Let therefore ω_1^0 denote the curvature in the circular state, and $r_0 = 1/\omega_1^0$ the radius of curvature, as in Fig. 4. The centerline is parameterized by $\mathbf{x}_0(S) = r_0\,\mathbf{e}_r(\theta)$ where $\theta = \omega_1^0\,S$ is the polar angle. The directors read $\mathbf{d}_1^0(S) = \mathbf{e}_z$, $\mathbf{d}_2^0(S) = \mathbf{e}_r(\theta)$, $\mathbf{d}_3^0(S) = \mathbf{e}_\theta(\theta)$. The kinematical constraints in Eqs. (1–2) are satisfied, as can be checked. The rotation gradient $\boldsymbol{\omega}^0 = \omega_1^0\,\mathbf{e}_z$ has components $(\omega_1^0, \omega_2^0, \omega_3^0) = (\omega_1^0, 0, 0)$ in the directors basis $(\mathbf{d}_1^0, \mathbf{d}_2^0, \mathbf{d}_3^0)$.

In the present setting, the twisting strain is zero, $\omega_3 = 0$. Observe that the three constitutive laws all predict similar expressions for the bending and twisting moment, $M_1^0 = B\,(\omega_1^0 - \kappa_n)$ and $M_3^0 = 0$ when $\omega_3 = 0$. Here B denotes the bending modulus relevant to the model under consideration, $B = B_{\text{rib}}$, B_{iso} or B_{flat}. Now, let us calculate M_2^0. For the isotropic rod model, $M_2^0 = 0$ follows directly from the constitutive law. For the two other

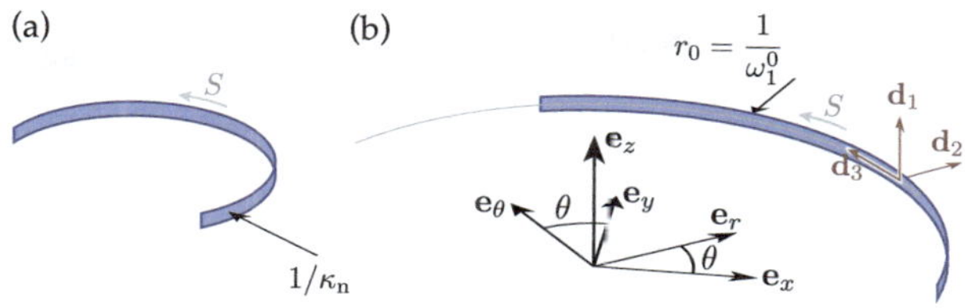

Fig. 4 (**a**) Reference configuration of a ribbon with natural curvature κ_n. (**b**) A circular actual configuration with curvature $\omega_1^0 = 1/r_0$

models, $M_2^0 = 0$ follows from equilibrium: the balance of moments in Eq. (4b) implies $0 = (\mathbf{M}_0' + \mathbf{d}_3^0 \times \mathbf{R}_0) \cdot \mathbf{d}_3^0 = \mathbf{M}_0' \cdot \mathbf{d}_3^0 = M_3^{0\prime} + \omega_1^0 M_2^0 = \omega_1^0 M_2^0$. Therefore, in all cases we have

$$\mathbf{M}_0(S) = M_1^0 \, \mathbf{d}_1^0, \quad \text{where } M_1^0 = B\left(\omega_1^0 - \kappa_n\right). \tag{12a}$$

The consequence of Eq. (12a) is that $\mathbf{M}_0' = \mathbf{0}$. Projecting now the balance of moments $\mathbf{M}_0' + \mathbf{d}_3^0 \times \mathbf{R}_0 = \mathbf{0}$ onto $\mathbf{d}_1^0$ and $\mathbf{d}_2^0$, we find that $R_1^0 = R_2^0 = 0$. The balance of forces is given by (4a) with $\ddot{\mathbf{x}} = \mathbf{0}$: this yields $(R_3^0 \, \mathbf{d}_3^0)' = \mathbf{0}$; hence $R_3^0 = 0$ and so

$$\mathbf{R}_0(S) = \mathbf{0}. \tag{12b}$$

By (12a)–(12b), the pre-stress in a circular configuration is a pure bending moment: $(\mathbf{R}_0(S), \mathbf{M}_0(S)) = (\mathbf{0}, M_1^0 \, \mathbf{d}_1^0)$. This bending moment drives the instabilities which we analyze later. The opposite case of instabilities driven by a twisting moment has been studied in the classical paper of Michell [16], see also [5].

4.2 Mode of Vibration Near a Circular Configuration

We consider a vibration mode near this circular configuration, *i.e.*, assume a small perturbation away from the circular state depending harmonically on the time variable τ. We use the notation of complex analysis and denote by $\hat{\mathbf{x}}$ the complex amplitude of displacement, by $\hat{\boldsymbol{\psi}}$ the complex infinitesimal rotation vector, by $\hat{\mathbf{R}}$ the complex amplitude of the internal force perturbation, and by $\hat{\mathbf{M}}$ the complex amplitude of the internal moment:

$$\mathbf{x}(S,t) = \mathbf{x}_0(S) + [r_0]\, \hat{\mathbf{x}}(\overline{S})\, e^{i\Omega\tau}, \tag{13a}$$

$$\mathbf{d}_i(S,t) = \mathbf{d}_i^0(S) + \left(\hat{\boldsymbol{\psi}}(\overline{S})\, e^{i\Omega\tau}\right) \times \mathbf{d}_i^0(S), \tag{13b}$$

$$\mathbf{R}(S,t) = \mathbf{0} + \left[B\left(\omega_1^0\right)^2\right] \hat{\mathbf{R}}(\overline{S})\, e^{i\Omega\tau}, \tag{13c}$$

$$\mathbf{M}(S,t) = \mathbf{M}_0 + \left[B\, \omega_1^0\right] \hat{\mathbf{M}}(\overline{S})\, e^{i\Omega\tau}. \tag{13d}$$

The perturbations $(\hat{\mathbf{x}}, \hat{\boldsymbol{\psi}}, \hat{\mathbf{R}}, \hat{\mathbf{M}})$ have been nondimensionalized by including the appropriate scaling factors in square brackets on the right-hand side. Our dimensionless units make use of the reference length $r_0 = 1/\omega_1^0$ (radius of the center-line the circular base state) and of the reference moment $B\,\omega_1^0$. All other reference quantities, such as forces and energies, are obtained by combining r_0 and $B\,\omega_1^0$. In Eqs. (13a)–(13d), the perturbations $(\hat{\mathbf{x}}, \hat{\boldsymbol{\psi}}, \hat{\mathbf{R}}, \hat{\mathbf{M}})$ are sought as a function of the rescaled arc-length defined by

$$\overline{S} = \omega_1^0\, S. \tag{14}$$

Note that the expression for the director $\mathbf{d}_i$ in Eq. (13b) in terms of the infinitesimal rotation $\hat{\boldsymbol{\psi}}$, automatically preserves the orthonormality between directors to a linear order in the perturbation.

Table 2 Values of the dimensionless tangent elastic moduli predicted by the different models near a circular configuration with radius r_0. These coefficients are obtained by linearizing Eqs. (9)–(11) and identifying the result with the form proposed in Eqs. (15a)–(15c)

	Contrast of bending moduli g	Dimensionless twisting modulus t
Ribbon	∞	$2(1-\nu\,\kappa_n\,r_0)$
Flat rod	∞	$\frac{2}{1+\nu}$
Isotropic rod	1	$\frac{1}{1+\nu}$

4.3 Linearized Constitutive Law

The constitutive law has been written in a generic form as $\mathbf{f}=0$ in Eq. (5), and special forms relevant to the different models have been obtained in Eqs. (9), (10) and (11). We now linearize these constitutive laws near the circular configuration

$$(M_1, M_2, M_3, \omega_1, \omega_2, \omega_3) = \left(M_1^0, 0, 0, \omega_1^0, 0, 0\right).$$

In terms of the increments of strain $\tilde{\omega}_i$ and increments of bending and twisting moments $\tilde{M}_i$, the linearized constitutive laws all take the form

$$\tilde{M}_1 = B\,\tilde{\omega}_1, \tag{15a}$$

$$\tilde{M}_2 = B\,g\,\tilde{\omega}_2, \tag{15b}$$

$$\tilde{M}_3 = B\,t\,\tilde{\omega}_3, \tag{15c}$$

where B is the model-dependent principal bending modulus introduced earlier, and (t, g) are the dimensionless second bending modulus and dimensionless twisting modulus, respectively. Both t and g are model-dependent as well: their expressions are listed in Table 2. The tangent moduli t and g have a strong influence on the stability of circular configurations: in particular, the tangent twisting moduli g predicted by the ribbon and flat rod models have different expression and this implies, as we shall confirm later, that these two models have different stability properties.

In Table 2, the statement of $g=\infty$ for the ribbon and flat rod models implies that Eq. (15b) should then be replaced by $\tilde{\omega}_2 = 0$ (which is indeed the incremental form of the kinematical constraint $\omega_2 = 0$).

Note that the increments $\tilde{M}_i$ of the bending and twisting moments $M_i = \mathbf{M}\cdot\mathbf{d}_i$ differ from the components $\hat{M}_i = \hat{\mathbf{M}}\cdot\mathbf{d}_i^0$ of the perturbed internal moment vector, $\hat{\mathbf{M}}$, onto the unperturbed directors basis: $\tilde{M}_i$ should not be confused with $\hat{M}_i$. For instance, for an infinitesimal rigid-body rotation, $\hat{\mathbf{M}}$ follows the rotation and its components $\hat{M}_i$ in the unperturbed directors basis are non-zero, although the quantities M_i are constant, so $\tilde{M}_i = 0$. Likewise, $\tilde{\omega}_i$ denote the increments of the strains ω_i and not the components of a vector $\hat{\boldsymbol{\omega}}$ (they are in fact the components of $\hat{\boldsymbol{\psi}}'$, as we note later).

4.4 Equations for the Modes

Linearizing the kinematical condition (1), the equations of motion (4a)–(4b), the definition of twist-curvature strains in Eq. (3) and using the linearized constitutive law in Eqs. (15a)–

(15c), we obtain

$$\hat{\mathbf{x}}'(\overline{S}) = \hat{\boldsymbol{\psi}} \times \mathbf{d}_3^0, \tag{16a}$$

$$\hat{\mathbf{R}}'(\overline{S}) = -\overline{\Omega}^2\, \hat{\mathbf{x}}, \tag{16b}$$

$$\hat{\mathbf{M}}'(\overline{S}) + \mathbf{d}_3^0 \times \hat{\mathbf{R}} = \mathbf{0}, \tag{16c}$$

$$\hat{\mathbf{M}} = \hat{\boldsymbol{\psi}} \times \overline{\mathbf{M}}_0 + \left(\mathbf{d}_1^0 \otimes \mathbf{d}_1^0 + g\, \mathbf{d}_2^0 \otimes \mathbf{d}_2^0 + t\, \mathbf{d}_3^0 \otimes \mathbf{d}_3^0\right) \cdot \hat{\boldsymbol{\psi}}'(\overline{S}), \tag{16d}$$

where we have defined the dimensionless pre-stress $\overline{\mathbf{M}}_0 = \overline{M}_1^0\, \mathbf{d}_1^0$ (which is identical for all models) and the dimensionless angular velocity $\overline{\Omega}$ by

$$\overline{M}_1^0 = \frac{M_1^0}{B\, \omega_1^0} = 1 - \kappa_{\mathrm{n}}\, r_0, \tag{17a}$$

$$\overline{\Omega} = \frac{\Omega}{\sqrt{\frac{B}{\rho}\, (\omega_1^0)^2}}. \tag{17b}$$

Equation (16d) can be interpreted by observing[3] that $\hat{\mathbf{M}} - \hat{\boldsymbol{\psi}} \times \overline{\mathbf{M}}_0 = \sum_i \tilde{M}_i\, \mathbf{d}_i^0$, and that $\hat{\boldsymbol{\psi}}' = \sum_i \tilde{\omega}_i\, \mathbf{d}_i^0$. When projected along any one of the undeformed directors $\mathbf{d}_i^0$, Eq. (16d) then appears to be equivalent to any one of the linearized constitutive laws in Eqs. (8a)–(8b), in nondimensional form, *i.e.*, with B set to 1.

To solve these equations, it will be convenient to represent the unknown amplitudes relative to the unperturbed director basis:

$$2\hat{\mathbf{x}}(\overline{S}) = \sum_{i=1}^{3} \hat{x}_i(\overline{S})\, \mathbf{d}_i^0(\overline{S}), \qquad \hat{\boldsymbol{\psi}}(\overline{S}) = \sum_{i=1}^{3} \hat{\psi}_i(\overline{S})\, \mathbf{d}_i^0(\overline{S}), \tag{18a}$$

$$\hat{\mathbf{R}}(\overline{S}) = \sum_{i=1}^{3} \hat{R}_i(\overline{S})\, \mathbf{d}_i^0(\overline{S}), \qquad \hat{M}(\overline{S}) = \sum_{i=1}^{3} \hat{M}_i(\overline{S})\, \mathbf{d}_i^0(\overline{S}). \tag{18b}$$

In dimensionless units, the unperturbed rotation gradient $\boldsymbol{\omega}_1^0 = \omega_1^0\, \mathbf{e}_z$ reads simply as $\overline{\boldsymbol{\omega}}_1^0 = \mathbf{e}_z$. Therefore, any one of the amplitudes $(\hat{\mathbf{x}}, \hat{\boldsymbol{\psi}}, \hat{\mathbf{R}}, \hat{\mathbf{M}})$, denoted generically by $\hat{\mathbf{v}}(\overline{S})$, is differentiated in the moving basis according to the formula

$$\hat{\mathbf{v}}'(\overline{S}) = \mathbf{e}_z \times \hat{\mathbf{v}} + \sum_{i=1}^{3} \hat{v}_i'\, \mathbf{d}_i^0 = \hat{v}_1'\, \mathbf{d}_1^0 + \left(\hat{v}_2' - \hat{v}_3\right) \mathbf{d}_2^0 + \left(\hat{v}_3' + \hat{v}_2\right) \mathbf{d}_3^0. \tag{19}$$

In the forthcoming sections, we analyze the stability of rods and ribbons in different circular geometries: we consider the unfolding of a semicircular ribbon in Sect. 5 and the stability of m-covered circles in Sect. 6. These stability analyses make use of the equations derived above: we insert the decompositions of the perturbations into components given in Eqs. (18a)–(18b) into the linearized equations of motion (16a)–(16d), making use of the differentiation rule in Eq. (19). In so doing, we obtain a set of twelve coupled linear ordinary differential equations—which we supplement by appropriate boundary conditions.

[3] To show these equalities, note that the left-hand side of the first equation is the derivative of $\hat{\mathbf{M}}$ in the moving basis following the virtual motion. The second equation is a classical formula for calculating the bending and twisting strain increments; see for instance [2].

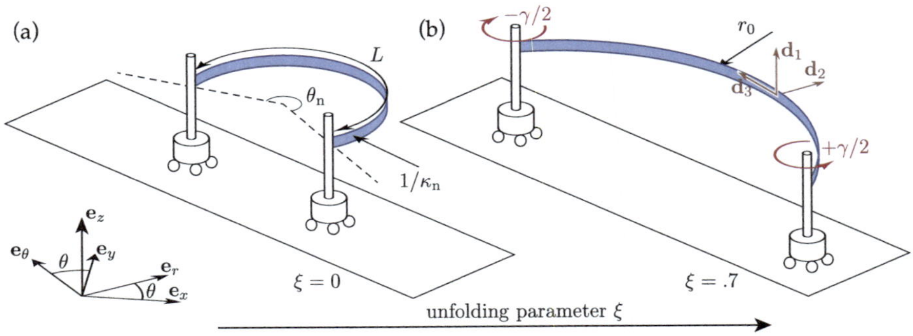

Fig. 5 Unfolding a semicircular ribbon. (**a**) Natural configuration: the case of a semicircle corresponds to $\theta_n = \pi$. (**b**) Undercurved, circular configuration with differential rotation of the ends $\gamma = \theta_n\,\xi$; its stability is analyzed in Sect. 5.3

5 Unfolding of a Semicircular Ribbon

Our analysis of the linearized constitutive laws has revealed a distinctive feature of the ribbon model: its tangent twisting modulus $t_{rib} = 2\,(1 - \nu\,\kappa_n\,r_0)$ in dimensionless form becomes negative when the curvature reaches the value $r_0^{-1} = \nu\,\kappa_n$. This suggests that a ribbon can become unstable when it is unbent, or *undercurved*, *i.e.*, when the curvature is decreased from its natural value κ_n. In this section, we consider the simplest geometry for which such an instability can appear: the unfolding of a semicircular strip. A linear stability analysis is first carried out, and predictions from the ribbon model are contrasted with those of the rod models, whose tangent twisting modulus is always positive, irrespective of the value of r_0^{-1}. A finite element analysis of a shell model is conducted, which confirms our findings.

5.1 Geometry

We consider a naturally curved ribbon of arclength L, subtending an initial angle $\theta_n = L\,\kappa_n$. It is progressively flattened by applying equal and opposite rotations to both ends normal to the (Oxy) plane; see Fig. 5. Out-of-plane displacements of the ends are completely restrained, but the ends are free to move without friction towards, or away from, each other in-plane. The end-wise rotations take place about the transverse direction $\mathbf{e}_z$ and the relative rotation, γ, of the two ends is expressed in terms of an unfolding parameter ξ as $\gamma = \theta_n\,\xi$; the initial, natural configuration corresponds to $\xi = 0$, and the flattened configuration corresponds to $\xi = 1$. We analyze the stability of the circular configuration during unfolding, *i.e.*, as $\xi \geq 0$ is increased.

5.2 Base Circular Solution

There exists a planar, circular solution to this problem, with curvature $\omega_1^0 = \kappa_n\,(1 - \xi)$, hence a radius of curvature $r_0 = (\omega_1^0)^{-1} = \frac{\kappa_n^{-1}}{(1-\xi)}$. By Eq. (17a), the bending prestress is identical for all three mechanical models, and is written as $\overline{M}_1^0 = 1 - \kappa_n\,r_0 = 1 - \frac{1}{1-\xi}$.

The values of the dimensionless moduli t and g are given in Table 2. In particular, for the ribbon, $t = t_{rib}$ is now a function of the unfolding parameter: $t_{rib} = 2\,(1 - \nu\,\kappa_n\,r_0) = 2\,(1 - \frac{\nu}{1-\xi})$. This modulus reaches zero when $\xi = 1 - \nu$, corresponding to a curvature $\omega_1^0 = \nu\,\kappa_n$. In view of this, we expect that the ribbon model predicts an instability before ξ reaches the value $(1-\nu)$. For materials having a positive Poisson's ratio ν where $1-\nu<1$, the instability is expected to occur well before the ribbon completely flattens.

5.3 Linear Stability Analysis of the Unfolding Semicircle

A linear stability analysis of the circular configuration is now carried out by specifying the linearized equations of motion derived in Sect. 4 for the particular geometry of Fig. 5 and for the static case, $\overline{\Omega}=0$. This yields the linearized equations of equilibrium in the interior, which are linear ordinary differential equations in the twelve amplitudes $(\hat{x}_i(\overline{S}), \hat{\psi}_i(\overline{S}), \hat{R}_i(\overline{S}), \hat{M}_i(\overline{S}))$ with $i=1,2,3$. These equations are supplemented by the relevant boundary conditions, namely $(\hat{x}_1, \hat{R}_2, \hat{R}_3, \hat{\psi}_1, \hat{\psi}_2, \hat{\psi}_3)_{S=0} = (\hat{x}_1, \hat{R}_2, \hat{R}_3, \hat{\psi}_1, \hat{\psi}_2, \hat{\psi}_3)_{\overline{S}=\overline{L}} = \mathbf{0}$. Here, the indices refer to the labelling of the material frame of the base circular solution, see Fig. 5: the indices 2 and 3 refer to the plane (Oxy), while the index 1 refers to the transverse direction (Oz).

The base solution is invariant by mirror symmetry with respect to the (Oxy) plane. As a result, the equations for the linear stability decouple into a set of equations for the quantities $(\hat{x}_2, \hat{x}_3, \hat{\psi}_1, \hat{R}_2, \hat{R}_3, \hat{M}_1)$ on one hand, and another set for the quantities $(\hat{x}_1, \hat{\psi}_2, \hat{\psi}_3, \hat{R}_1, \hat{M}_2, \hat{M}_3)$ on the other hand; see Sect. 5.8 of reference [2] for details. These two sets of equations characterize in-plane modes of instability involving bending, and out-of-plane modes involving twisting, respectively—the so-called 'ridge mode' and 'centerline mode' in reference [2], respectively. This decoupling is recovered in the analysis of m-covered circular solutions later on in Sect. 6.

A detailed analysis shows that the unfolding ribbon first becomes unstable by twisting out of plane. Therefore, to simplify the presentation, we assume that the quantities, $(\hat{x}_2, \hat{x}_3, \hat{\psi}_1, \hat{R}_2, \hat{R}_3, \hat{M}_1)$ are all identically zero. We cancel the corresponding terms in the decomposition given in Eqs. (18a)–(18b), and insert the result into the linearized equations of motion (16a)–(16d), making use of the differentiation rule in Eq. (19). Setting $\overline{\Omega}=0$, we find that

$$\hat{x}_1' - \hat{\psi}_2 = 0, \tag{20a}$$

$$g\,\hat{\psi}_2' + \left(\overline{M}_1^0 - g\right)\hat{\psi}_3 - \hat{M}_2 = 0, \tag{20b}$$

$$t\,\hat{\psi}_3' - \left(\overline{M}_1^0 - t\right)\hat{\psi}_2 - \hat{M}_3 = 0, \tag{20c}$$

$$\hat{M}_2' - \hat{M}_3 + \hat{R}_1 = 0, \tag{20d}$$

$$\hat{M}_3' + \hat{M}_2 = 0, \tag{20e}$$

$$\hat{R}_1' = 0, \tag{20f}$$

where $\hat{x}_1(\overline{S})$, $\hat{\psi}_2(\overline{S})$, $\hat{\psi}_3(\overline{S})$, $\hat{M}_2(\overline{S})$, $\hat{M}_3(\overline{S})$, $\hat{R}_1(\overline{S})$ are functions of the dimensionless arclength $\overline{S}=\frac{S}{L}$. The boundary conditions are

$$\left(\hat{x}_1(0), \hat{\psi}_2(0), \hat{\psi}_3(0), \hat{x}_1(\overline{L}), \hat{\psi}_2(\overline{L}), \hat{\psi}_3(\overline{L})\right) = \mathbf{0}, \tag{20g}$$

where the rescaled arclength is $\overline{L} = L\,\omega_1^0 = \theta_n\,(1-\xi)$.

When the expressions of $\overline{M}_1^0$, $\overline{L}$, g and t in terms of ξ and ν are inserted (here, we use the expressions of g and t relevant to any of the three mechanical models under consideration), Eqs. (20a)–(20g) become a linear eigenvalue problem for ξ depending on the parameters

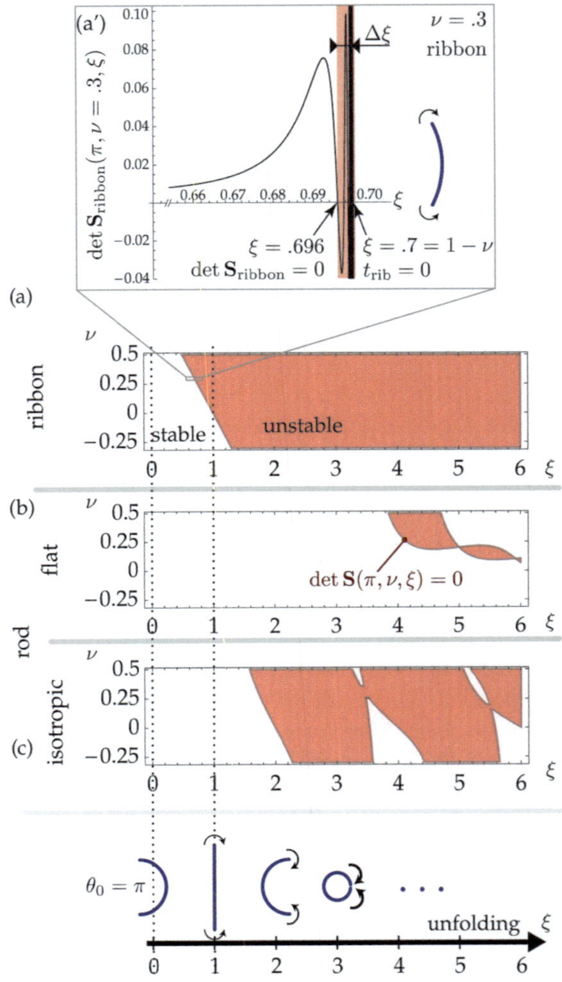

Fig. 6 Linear stability analysis of the unfolding semicircle ($\theta_n = \pi$), as predicted (**a**) by the ribbon model, (**b**) by the flat rod model, and (**c**) by the isotropic rod model. The ribbon model predicts a much lower instability threshold than the two other models. The *red domains* correspond to a change of sign of the determinant of the shooting matrix $\mathbf{S}(\theta_n, \nu, \xi)$: along its boundary, a marginally stable twisting mode exists. (**a′**) For the ribbon model, the first instability, corresponding to the first root of the determinant of S, occurs slightly below the value $\xi = 1 - \nu$ where the tangent twisting modulus cancels: the gap between the two is as small as $\Delta\xi = .004$ for $\nu = .3$

(θ_n, ν). For a non-trivial solution, these eigenvalues are marginally stable modes, and indicate a bifurcation from the unbuckled (planar, circular) solution branch. A closed-form solution is available because the coefficients of the equations are constant, but that solution turns out to be lengthy and tedious to present. Instead, we solve this problem numerically using a linear shooting method. We call $\mathbf{S}(\theta_n, \nu, \xi)$ the determinant of the shooting matrix; according to the shooting method [17], the entries in the shooting matrix are found by integrating an initial value problem for the differential system in Eqs. (20a)–(20g). The cancellation of the determinant $\det \mathbf{S}(\theta_n, \nu, \xi)$ indicates the existence of a marginally stable mode.

The numerical results for the linear stability of an unfolding semicircle are shown in Fig. 6. For the ribbon model, an instability occurs slightly before the tangent twisting modulus cancels, $t_{\text{rib}} = 0$, *i.e.*, slightly before ξ reaches the value $\xi = 1 - \nu$, as anticipated earlier. The critical value of ξ is numerically very close to $1 - \nu$: the difference is as small as $\Delta\xi = .004$ for $\nu = .3$, as shown in Fig. 6a′ (even though this difference is a function of Poisson's ratio and of the initial angular span, θ_n, it always stays of this order).

By contrast, the two rod models predict an instability at a much higher value of the unfolding parameter, as indicated in Figs. 6b, c. The instability typically does not take place

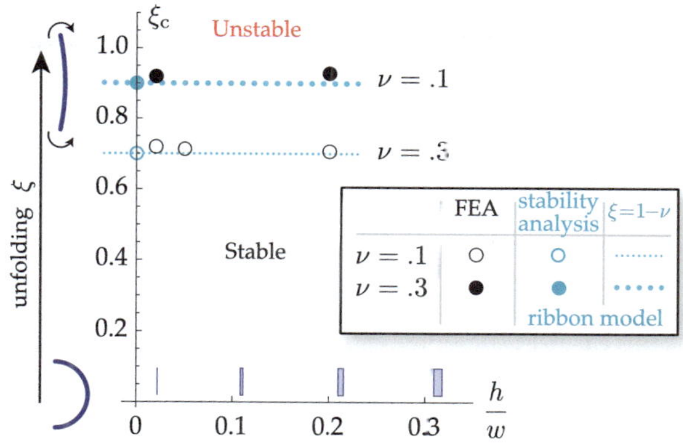

Fig. 7 Stability analysis of the unfolding semi-circular ribbon ($\theta_n = \pi$). Comparison of the finite element thin shell simulation (*black symbols*) to the ribbon model (*light blue*): full stability analysis (*symbols*), and approximate threshold $\xi = 1 - \nu$ where the twisting modulus cancels, $t_{rib} = 0$ (*dotted lines*)

before ξ reaches ≈ 2, corresponding to a circle that has been fully everted, *i.e.*, has been flattened out and then bent the other way around until it has become mirror-symmetric to the original one.

In summary, the out-of-plane instabilities are predicted at very different unfolding angles by the ribbon model on one hand, and by the rod models on the other hand. This is evident from the very different positions of the stability boundaries in the panel 'a' of Fig. 6, as compared to those in panels 'b' and 'c'. The early instability of the ribbon model is triggered by the change of sign of the tangent twisting modulus—no such change of sign takes place for the rod models.

5.4 Simulations by the Finite Element Method

In order to test our theoretical predictions, we performed computational simulations using a finite element analysis for thin elastic shells using the commercially available software package ABAQUS [8]. The semi-circle ($\theta_n = \pi$) is formed as a thin strip using four-noded S4R5 shell elements, with 20 elements across the width and 200 around its length. The previous boundary conditions are captured exactly: the ends are free to rotate about an axis parallel to the width direction, and no other rotations are permitted; one end is restrained against linear translation and the other end can only displace in a direction between the ends. The strip is loaded by imposing end-wise, in-plane rotations, and uniform flattening without end-wise reaction forces is accomplished by the ends freely moving relative to one another (they move apart for $\theta > \pi/2$, and then together when θ is larger). The shell elements can bend and stretch; we use a thickness parameter that is only a small fraction of the width, see Fig. 7, implying that the equilibrium configurations are close to inextensible. The high density of elements across the width of strip ensures that all stresses are accurately resolved.

The material is chosen to be linear elastic, with the Young's modulus of steel, 210 GPa, and three width-to-thickness ratios, 5, 20 and 50, are considered, along with two values, $\nu = .1$ and $\nu = .3$, of Poisson's ratio. The initial radius is 100 mm and the width is always $w = 5$ mm, giving a length-to-width ratio $L/w = 62.8$. The analysis follows two steps. First, the strip is unbent to a given end rotation, which is then held fixed with all other nodal displacements being free. Second, a linear perturbation analysis is performed where the value of the primary (lowest) eigenvalue is extracted. These steps are repeated for a range of end rotations where the aim is to capture the eigenvalue performance of the strip as a function of end rotation, width and Poisson's ratio: the value at which an eigenmode first becomes unstable is detected, and this critical value is plotted as a function of the aspect-ratio

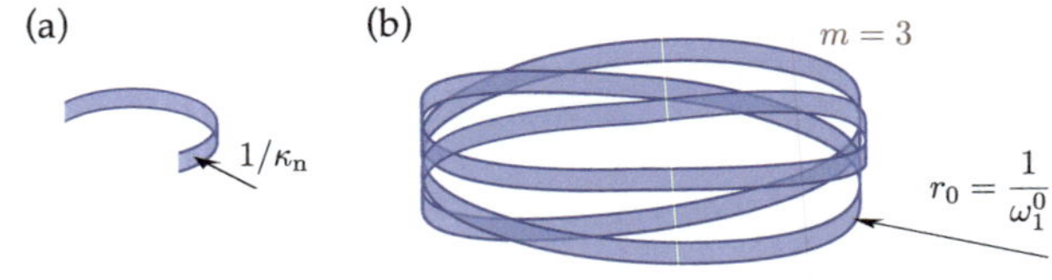

Fig. 8 A multiloop. (**a**) A small chunk of the loop shown in natural configuration, revealing the natural curvature κ_n. (**b**) An m-covered strip with covering index $m = 3$, exhibiting over-curvature ($\omega_1^0 > \kappa_n$) or under-curvature ($\omega_1^0 < \kappa_n$). The loop has been disentangled to help visualization but in our analysis self-penetration is allowed and a perfectly planar, self-intersecting configuration is considered

h/w and of Poisson's ratio ν. The results have been collated in Fig. 7. A good agreement is obtained with the linear stability analysis of the one-dimensional ribbon model. These finite element simulations confirm the existence of buckling instabilities of undercurved strips, triggered by the change of sign of their tangent twisting modulus.

6 Stability of Overcurved and Undercurved *m*-Covered Strips

6.1 Geometry

We return to the experiments of Sect. 2 concerning the stability of m-covered loops. The geometry is sketched in Fig. 8. The natural curvature is denoted by κ_n. The strip is closed and the single loop ($m = 1$) is used as the reference configuration. It is then bent into the m-covered strip sketched in Fig. 8b, using the folding sequence of Fig. 2. Topology shows that the single loop can be deformed by means of continuous deformations into an m-covered configuration *only if m is an odd integer*—see Figs. 1 and 2. We therefore restrict our attention to odd integers m; in addition, the m-covered configuration involves self-penetration if $m > 1$ but we ignore self-penetration entirely in our analysis.

We now analyze the stability of the m-covered strip, first by means of a linear stability analysis (Sect. 6.2), and then by calculating the post-bifurcation solutions numerically (Sect. 6.3). Let L be the length of the strip. The radius of the m-covered configuration is $r_0 = \frac{L}{2\pi m}$, which is said to be *undercurved* if $\omega_1^0 < \kappa_n$, *i.e.*, $m < \frac{\kappa_n L}{2\pi}$, and *overcurved* if $\omega_1^0 > \kappa_n$, *i.e.*, $m > \frac{\kappa_n L}{2\pi}$.

6.2 Linear Stability Analysis

The planar, m-covered configuration is a particular case of a circular solution: we use it as the base state of the general stability analysis carried out in Sect. 4.

Owing to the rotational invariance of the base state, we consider a vibration mode with dimensionless wavenumber $\overline{k}$, corresponding to a physical wavenumber $k = \omega_1^0 \overline{k}$. We denote by $\Re$ the real part of a complex number: the components of the perturbations in the directors basis are sought as

$$\begin{pmatrix} \hat{x}_j(\overline{S}) \\ \hat{\psi}_j(\overline{S}) \\ \hat{R}_j(\overline{S}) \\ \hat{M}_j(\overline{S}) \end{pmatrix} = \Re \left[e^{i\overline{k}\,\overline{S}} \begin{pmatrix} x_j^\dagger \\ \psi_j^\dagger \\ R_j^\dagger \\ M_j^\dagger \end{pmatrix} \right] \tag{21}$$

in terms of complex amplitudes

$$\hat{\mathbf{y}} = (\hat{x}_1, \hat{x}_2, \hat{x}_3, \hat{\psi}_1, \hat{\psi}_2, \hat{\psi}_3, \hat{R}_1, \hat{R}_2, \hat{R}_3, \hat{M}_1, \hat{M}_2, \hat{M}_3).$$

Since the ribbon is closed, the perturbations are periodic after m turns. This imposes a quantification condition for $\overline{k}$: there exists some integer n such that

$$\overline{k} = \frac{n}{m}. \tag{22}$$

The difference in phase after m turns, $\overline{k}\,\overline{L} = \frac{n}{m}\frac{L}{r_0} = 2\pi n$ is then a multiple of 2π.

We insert the harmonic form of the perturbations in Eq. (21) into Eqs. (18a)–(18b), and then into the linearized equations of motion (16a)–(16d), making use of the differentiation rule in Eq. (19). The resulting linearized equations of motion take the form

$$\mathbf{A}\left(\overline{M}_0^1, g, t; \overline{k}, \overline{\Omega}^2\right) \cdot \hat{\mathbf{y}} = \mathbf{0}, \tag{23}$$

where

$$\mathbf{A} = \left(\begin{array}{ccc|ccc|ccc|ccc}
i\overline{k} & 0 & 0 & 0 & -1 & 0 & & & & & & \\
0 & i\overline{k} & -1 & 1 & 0 & 0 & & \mathbf{0} & & & \mathbf{0} & \\
0 & 1 & i\overline{k} & 0 & 0 & 0 & & & & & & \\
\hline
\overline{\Omega}^2 & 0 & 0 & & & & i\overline{k} & 0 & 0 & & & \\
0 & \overline{\Omega}^2 & 0 & & \mathbf{0} & & 0 & i\overline{k} & -1 & & \mathbf{0} & \\
0 & 0 & \overline{\Omega}^2 & & & & 0 & 1 & i\overline{k} & & & \\
\hline
 & & & & & & 0 & -1 & 0 & i\overline{k} & 0 & 0 \\
 & \mathbf{0} & & & \mathbf{0} & & 1 & 0 & 0 & 0 & i\overline{k} & -1 \\
 & & & & & & 0 & 0 & 0 & 0 & 1 & i\overline{k} \\
\hline
 & & & -i\overline{k} & 0 & 0 & & & & & & \\
 & \mathbf{0} & & 0 & -ig\overline{k} & g - \overline{M}_1^0 & & \mathbf{0} & & & \mathbf{1} & \\
 & & & 0 & \overline{M}_1^0 - t & -i\overline{k}t & & & & & &
\end{array}\right). \tag{24}$$

Note that for $\overline{\Omega} = 0$ it is possible to write $\mathbf{A}$ as a diagonal block matrix by reordering the rows and columns—this is the decoupling discussed earlier in Sect. 5.3, arising from the symmetry of the base state.

The dispersion relation reads $\det \mathbf{A}(\overline{M}_0^1, g, t; \overline{k}, \overline{\Omega}^2) = 0$. Factoring the determinant, we obtain

$$\det \mathbf{A} = g\left(\left(1 + \overline{k}^2\right)\overline{\Omega}^2 - \overline{k}^2\left(1 - \overline{k}^2\right)^2\right)\left(c\,\overline{\Omega}^2 - b\,\overline{k}^2\left(1 - \overline{k}^2\right)\right), \tag{25a}$$

where

$$b = \frac{(\overline{M}_1^0)^2}{g} - \left(1 + \frac{t}{g}\right)\overline{M}_1^0 + t\left(1 - \overline{k}^2\right), \tag{25b}$$

$$c = 1 + \frac{\overline{k}^2 t - \overline{M}_1^0}{g}. \tag{25c}$$

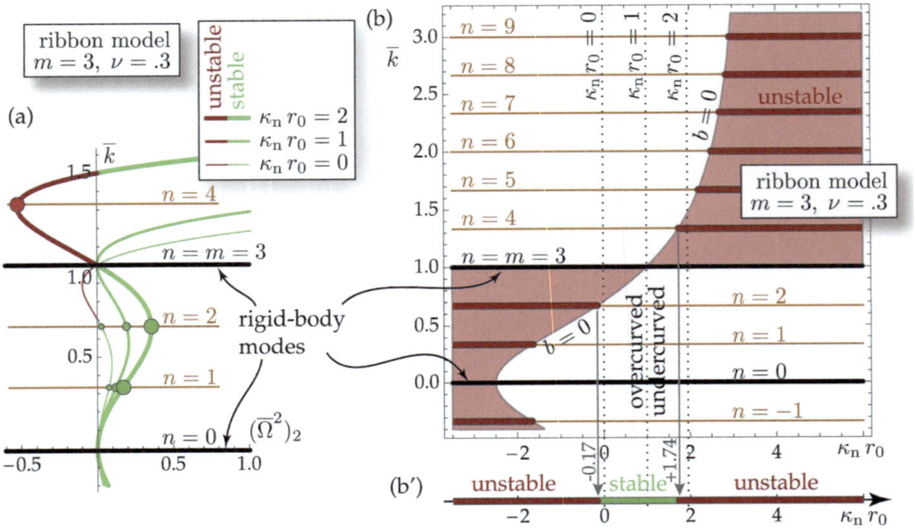

Fig. 9 Linear stability analysis of m-covered configurations with $m = 3$ and $\nu = .3$, based on the ribbon model. (**a**) Dependence of the most negative squared angular velocity $(\overline{\Omega}^2)_2$ on the wavenumber. (**b**) Graphical determination of the stability of m-covered ribbons for fixed $\nu = .3$ as a function of the parameter $\kappa_n r_0$ using the ribbon model. Unstable modes corresponds to the *thick, horizontal, dark red segments*. (**b′**) Configurations are stable when no unstable mode exists (*light green segment*), and unstable otherwise (*dark red segments*). See the text for details

The roots $\overline{\Omega}^2$ of the dispersion relation $\det \mathbf{A} = 0$ are the numbers $\overline{\Omega}^2 = (\overline{\Omega}^2)_1$ and $\overline{\Omega}^2 = (\overline{\Omega}^2)_2$ defined by

$$(\overline{\Omega}^2)_1 = \frac{\overline{k}^2 (1 - \overline{k}^2)^2}{1 + \overline{k}^2}, \tag{26a}$$

$$(\overline{\Omega}^2)_2 = \frac{b \overline{k}^2 (1 - \overline{k}^2)}{c}. \tag{26b}$$

The first root $(\overline{\Omega}^2)_1$ being always positive, the stability of the ribbon is governed by the sign of the second root. This sign changes when $\overline{k} = 0$, when $\overline{k} = \pm 1$, when $b = 0$, or when $c = 0$: these four equations yield the four possible types of boundaries surrounding the regions of stability in Figs. 9b and 10.

The quantity $(\overline{\Omega}^2)_2$ is a function of $\overline{k}$, b and c. On using Eqs. (25a)–(25c), b and c can be expressed in terms of $\overline{M}_1^0$, g and t. Using Table 2 and Eq. (17a), we can finally express $(\overline{\Omega}^2)_2$ as a function of $\overline{k}$, ν and $(\kappa_n r_0)$; there are three such expressions, one for each mechanical model.

In Fig. 9a, we plot $(\overline{\Omega}^2)_2$ predicted by the ribbon model for $\nu = .3$ and for $\kappa_n r_0 = 0$ (naturally straight ribbon, overcurved configuration, $\overline{M}_1^0 > 0$), $\kappa_n r_0 = 1$ (natural configuration, $\overline{M}_1^0 = 0$), and $\kappa_n r_0 = 2$ (undercurved configuration, $\overline{M}_1^0 < 0$). Stable modes corresponding to $(\overline{\Omega}^2)_2 > 0$ are shown in light green, and unstable modes corresponding $(\overline{\Omega}^2)_2 < 0$ are shown in dark red. For a given value of the covering index m (in this case, $m = 3$) the values of $\overline{k}$ satisfying the quantification condition (22) are shown as horizontal lines. The modes

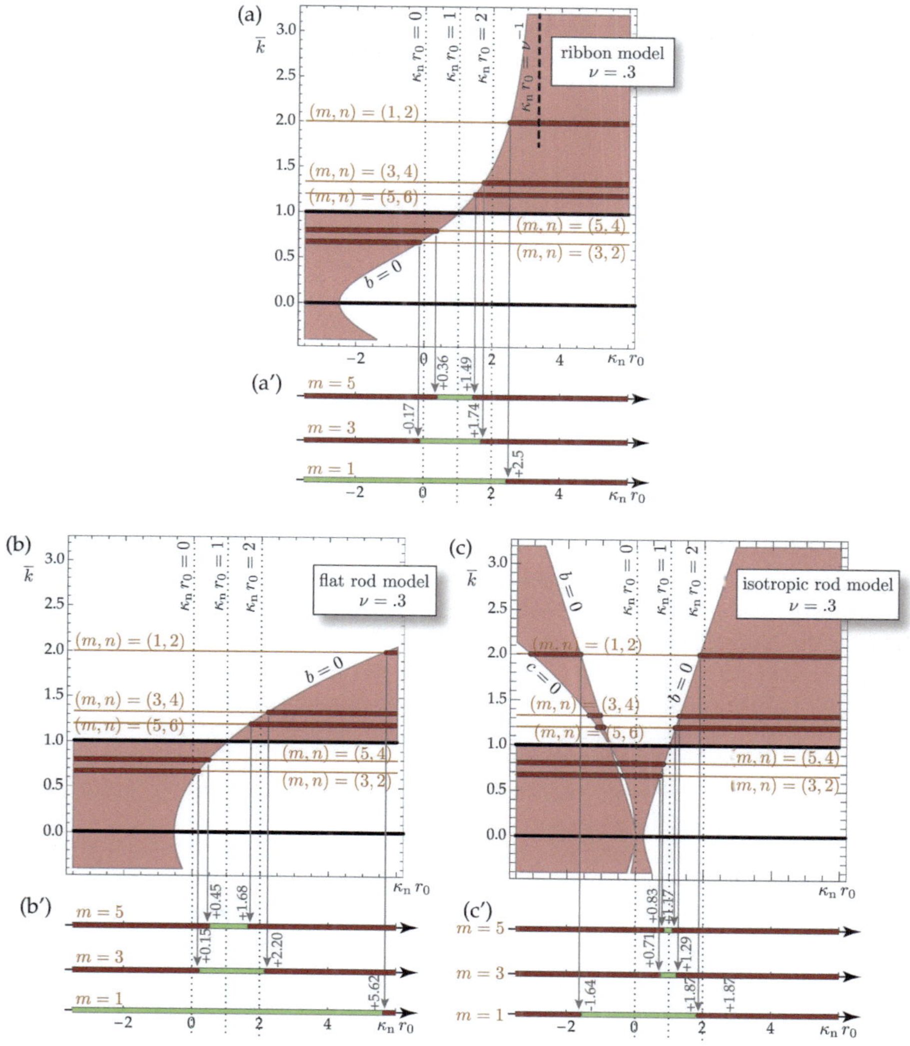

Fig. 10 Linear stability analysis of multiloop configurations, as predicted by the different mechanical models, for $\nu = .3$ and for various covering indices m. We recover the results of [15] for the rod models as particular cases

corresponding to $\overline{k} = 0$ and $\overline{k} = \pm 1$ (*i.e.*, $n = 0$ and $n = \pm m$) make both $(\overline{\Omega}^2)_1$ and $(\overline{\Omega}^2)_2$ vanish: these are rigid-body modes, which we represent by thick black lines. All other modes are non-rigid, and are shown using a thin brown line. Vibration modes are denoted by small disks: stable modes correspond to the intersection of a horizontal brown line with the green portion of a curve, and unstable vibration modes to an intersection with the red portion of a curve.

The influence of the parameter $\kappa_n r_0$ on the stability is grasped by the alternative presentation of the same data proposed in Fig. 9b. For a fixed value of ν and a specific mechanical model (in this case, $\nu = .3$ with the ribbon model), the region where $(\overline{\Omega}^2)_2 < 0$ is shaded in the plane $(\kappa_n r_0, \overline{k})$. The representative values $\kappa_n r_0 = 0, 1, 2$ used earlier now correspond to the vertical dotted lines. Unstable modes lie at the intersection of horizontal brown lines representing the quantization condition for $\overline{k}$ with the shaded regions where $(\overline{\Omega}^2)_2 < 0$. The m-covered ribbon is stable for a given value of $\kappa_{\rm n}\, r_0$ if, and only if, no unstable mode exists: this test is carried out graphically by projecting vertically the red segments onto the $(\kappa_{\rm n}\, r_0)$-axis in Fig. 9b′. For this particular set of parameters, the m-covered ribbon appears to be stable in a single interval, $-.17 < \kappa_{\rm n}\, r_0 < +1.74$ ($m = 3$, $\nu = .3$, ribbon model). This solves the linear stability problem of the m-covered ribbon, with $\nu = .3$ and $m = 3$. The same analysis is repeated in Fig. 10 for different values of m, and for the rod models as well. Note that we recover the results of [15] when we use the rod models, and the results of [6] in the particular case of a rod making a single loop, $m = 1$; the results concerning the stability of the ribbon model are novel.

We make a few comments on these results.

First, we check our analysis by confirming that the natural configuration is always stable: the stability regions in Figs. 10a′–c′ all contain the point $\kappa_{\rm n}\, r_0 = 1$, corresponding to a zero prestress $\overline{M}_1^0 = 0$.

Secondly, we observe that the ribbon model is, generally speaking, less stable than the flat rod model. This difference is particularly marked for the simple loop configuration ($m = 1$), which is stable for $\kappa_{\rm n}\, r_0 < 2.50$ with the ribbon model, and up to $\kappa_{\rm n}\, r_0 < 5.62$ with the flat rod model. This large difference in the upper bounds for $\kappa_{\rm n}\, r_0$ can be traced back to the presence of a vertical asymptote to the curve, $b = 0$, bounding the stability region for the ribbon model (shown as the vertical thick dashed line in Fig. 10a); by contrast, no such asymptote exists for the flat rod model and the upper part of the curve, $b = 0$, is much flatter in Fig. 10b than in Fig. 10a. Starting from Eq. (25b), it is easy to show that the vertical asymptote is where the tangent twisting modulus cancels, $t_{\rm rib} = 0$, and corresponds to $\kappa_{\rm n}\, r_0 = 1/\nu$. Therefore, the change of the sign of the tangent twisting modulus $t_{\rm rib}$ is responsible for the less stable behavior of the ribbon model as compared to the rod models—a similar conclusion as for the earlier unfolding semicircle in Sect. 5.1.

Thirdly, we observe that the isotropic rod model is less stable than the flat rod model. Recall that the difference between these models is that the flat rod is subjected to the kinematical constraint $\omega_2 = 0$, as captured by the statement $g_{\rm flat} = 0$. Unsurprisingly, the flat rod model has less flexibility to explore unstable modes, and is also more stable. As a particular case, we recover known results [15] concerning the stability of naturally straight isotropic rods, see the dotted line $\kappa_{\rm n}\, r_0 = 0$ in Figs. 10b–b′: for $\nu = .3$, m-covered configurations of a naturally straight isotropic rod are found to be unstable for all values of m, except for $m = 1$.

Finally, we found more boundaries to the stability domains in Fig. 10c for the isotropic rod model, which has a finite second bending modulus $g_{\rm iso} < \infty$, than in Figs. 10a–b for the flat rod and ribbon models for which $g_{\rm flat} = g_{\rm rib} = \infty$. This can be explained as follows. The equations $b = 0$ and $c = 0$ defining the boundaries are polynomial equations for $\kappa_{\rm n}\, r_0$ and $\overline{k}$ when the expressions of $\overline{M}_1^0 = 1 - \kappa_{\rm n}\, r_0$ and of t and g from Table 2 are inserted into Eqs. (25b)–(25c). Their order is decreased when g is infinite. The equation $b = 0$ is a quadratic equation for $\kappa_{\rm n}\, r_0$ when g is finite but a linear equation for $g = \infty$, and the equation $c = 0$ is a linear equation for $\kappa_{\rm n}\, r_0$ if g is finite and has no root if g is infinite (it does not depend on $\kappa_{\rm n}\, r_0$ then). Correspondingly, Figs. 10a–b (ribbon and flat rod models) have a single $b = 0$ branch and no $c = 0$ branch, and Fig. 10c has two $b = 0$ branches and another $c = 0$ branch (as in reference [15]).

Along the boundaries $b = 0$ and $c = 0$ marginally stable modes exist, which correspond to a loss of stability of the planar, m-covered configuration. There, the determinant $\det \mathbf{A}(\overline{M}_0^1, g, t; \overline{k}, \overline{\Omega}^2 = 0)$ nullifies and the marginally stable modes can be found by calculating the null vectors $\hat{\mathbf{y}}$ of $\mathbf{A}$. As noted earlier, these modes are rigid-body modes

Table 3 Upper and lower stability bounds for m-covered configurations based on the linear stability analysis of the ribbon model for $\nu = .3$, as predicted by Eq. (27). Values of $K_{\text{rib}}^{m,n} = \text{`}\mathit{1.}\text{'}$ in italics correspond to rigid-body modes ($n = m$) and should be ignored. Values in boldface ($n = m \pm 1$) correspond to critical values of $\kappa_{\text{n}}\, r_0$ where stability is lost. Other values correspond to higher-order modes appearing when the base circular solution is already unstable

$K_{\text{rib}}^{m,n}$	$n=1$	$n=2$	$n=3$	$n=4$	$n=5$	$n=6$	$n=7$
$m=1$	*1.*	**2.50**	2.93	3.10	3.18	3.23	3.26
$m=3$	−1.67	**−0.17**	*1.*	**1.74**	2.20	2.50	2.70
$m=5$	−2.17	−1.37	−0.45	**0.36**	*1.*	**1.49**	1.85
$m=7$	−2.33	−1.86	−1.24	−0.58	0.03	**0.56**	*1.*

for $\overline{k} = 0$ (translation along the axis $\mathbf{e}_z$, and rigid rotation about the axis $\mathbf{e}_z$) and $\overline{k} = \pm 1$ (translations along the directions $\mathbf{e}_x$ and $\mathbf{e}_y$, and rigid rotations about these directions); for $b = 0$, the marginally stable modes are out-of-plane modes,[4] similar to that studied earlier in Sect. 5.3; finally, the boundary $c = 0$, which is only encountered with the isotropic rod model ($g_{\text{iso}} < \infty$), corresponds to a marginal mode involving locally a pure-twist deformation about the tangent.[5]

Returning to the ribbon model, we note that an analytical expression of the stability boundaries can be obtained. Rewriting the equation $b = 0$ in terms of the parameter $\kappa_{\text{n}}\, r_0$ and using the quantification condition (22) for $\overline{k}$, we express the roots of $b = 0$ as $\kappa_{\text{n}}\, r_0 = K_{\text{rib}}^{m,n}$, where

$$K_{\text{rib}}^{m,n} = \frac{1}{\nu}\left(1 - \frac{1-\nu}{2\,\nu\,((\frac{n}{m})^2 - 1) + 1}\right) \quad \text{(ribbon model).} \tag{27}$$

All stability boundaries shown in Fig. 10a′ correspond to particular values of $K_{\text{rib}}^{m,n}$ listed Table 3; in fact, the values $K_{\text{rib}}^{m,m\pm 1}$ shown in boldface in the table are the critical values of the parameter $\kappa_{\text{n}}\, r_0$ found earlier in Fig. 10a′. We conclude that, according to the ribbon model, the m-covered ribbon is linearly stable for $\kappa_{\text{n}}\, r_0 < K_{\text{rib}}^{1,1} = 2.5$ for $m = 1$, and for $K_{\text{rib}}^{m,m-1} < \kappa_{\text{n}}\, r_0 < K_{\text{rib}}^{m,m+1}$ for $m \geq 3$. Similar expressions for the stability boundaries of the rod models have been obtained in reference [15].

In summary, we have analyzed the linear stability of m-covered configurations of an elastic strip, ignoring self-penetration. The ribbon model yields results that are qualitatively similar to those of the flat rod model. Quantitatively, however, the buckling threshold in the undercurved case (*i.e.*, for large positive values of $\kappa_{\text{n}}\, r_0$) are substantially lower with the ribbon model than with the flat rod model, especially for $m = 1$: this difference can be attributed mainly to the change of sign of the tangent twisting modulus, t_{rib}, at $\kappa_{\text{n}}\, r_0 = 1/\nu$ (thick dashed line in Fig. 10). All instabilities of the ribbon involve an out-of-plane mode.

[4] We just found that the stability boundaries for the ribbon model are either rigid-body modes, or out-of-plane modes ($b = 0$) but never in-plane modes ($c = 0$ has no root for $g = \infty$). This confirms the assumption made earlier in Sect. 5.3 that the stability of the ribbon model can be analyzed assuming that the perturbations associated with in-plane modes are zero.

[5] We have ignored the rotational inertia in Eq. (16c) for the balance of angular momentum and, as a result, pure twist modes are associated with zero inertia and an infinite frequency: this is why $c = 0$ makes $(\overline{\Omega}^2)_2$ infinite in Eq. (26a).

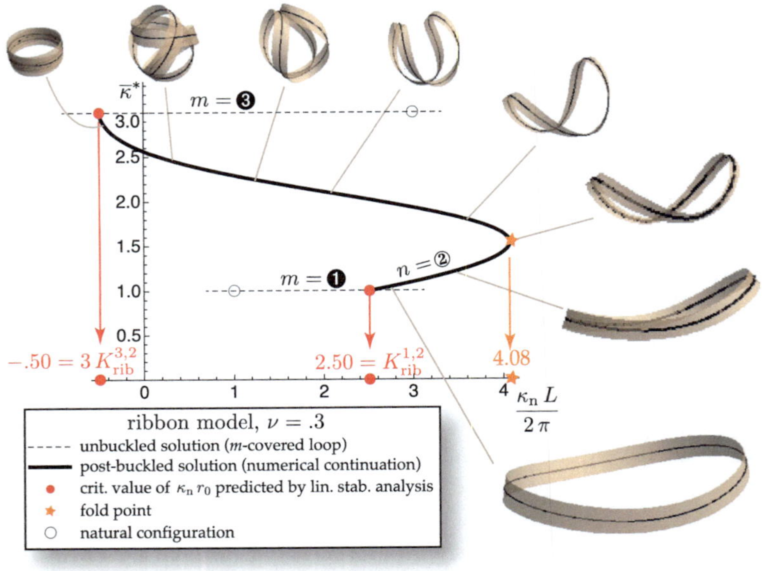

Fig. 11 A branch of post-bifurcated equilibrium solutions connecting the simple loop ($m = 1$) to the 3-covered loop ($m = 3$), as simulated by numerical continuation of the ribbon model with $\nu = .3$. The continuation parameter is the dimensionless natural curvature κ_n (*horizontal axis*); the quantity on the vertical axis is the conjugate quantity $\overline{\kappa}^*$

6.3 Numerical Post-bifurcated Solutions

To validate and complement this stability analysis, we solved numerically the post-bifurcation equilibrium solutions for a closed elastic ribbon. To do this, we solved Eqs. (1)–(4a), (4b) together with the constitutive law for an elastic ribbon in Eq. (9) and with periodic boundary conditions. We used AUTO-07p [4], a library which implements the arc-length continuation method [14]. We started from an m-covered planar configuration, detected bifurcations along this fundamental branch, and followed the various bifurcated solution branches emanating from it. The branches are plotted in a plane whose coordinates are the dimensionless natural curvature $\frac{\kappa_n L}{2\pi}$ and the conjugate quantity $\kappa^* = -\frac{\partial \mathcal{E}}{\partial \kappa_n}$: our diagram is distinguished in the sense of [13]. In dimensionless form, and after using the definition of the ribbon energy in Eq. (7), $\overline{\kappa}^*$ is given by

$$\overline{\kappa}^* = -\frac{1}{2\pi\, B}\frac{\partial \mathcal{E}}{\partial \kappa_n} = \frac{1}{2\pi}\int_0^L \omega_1 \left(1 + \nu \left(\frac{\omega_3}{\omega_1}\right)^2\right) \mathrm{d}S. \tag{28}$$

For planar configurations, note that $\omega_3 = 0$ and thus that $\overline{\kappa}^* = m$.

A typical branch is represented in Fig. 11; this is the branch starting off from the simple loop ($m = 1$), and extending the mode of instability $n = 2$ occurring for $\kappa_n r_0 = K_{\text{rib}}^{1,2}$, *i.e.*, $\kappa_n \frac{L}{2\pi} = K_{\text{rib}}^{1,2} = 2.5$. Above the buckling threshold, the dimensionless natural curvature $\frac{\kappa_n L}{2\pi}$ first increases, which is typical of a supercritical pitchfork bifurcation. It reaches a maximum value 4.08 at a fold point, and then decreases. Eventually, this branch connects to the triply-covered circular configuration ($m = 3$); the value of the control parameter where this connection occurs is $\frac{\kappa_n L}{2\pi} = \kappa_n\,(m\, r_0) = m\, K_{\text{rib}}^{m,m-1} = 3\, K_{\text{rib}}^{3,2} = -.5$. Therefore, the connections of the two endpoints of this branch with the planar, m-covered configurations are fully

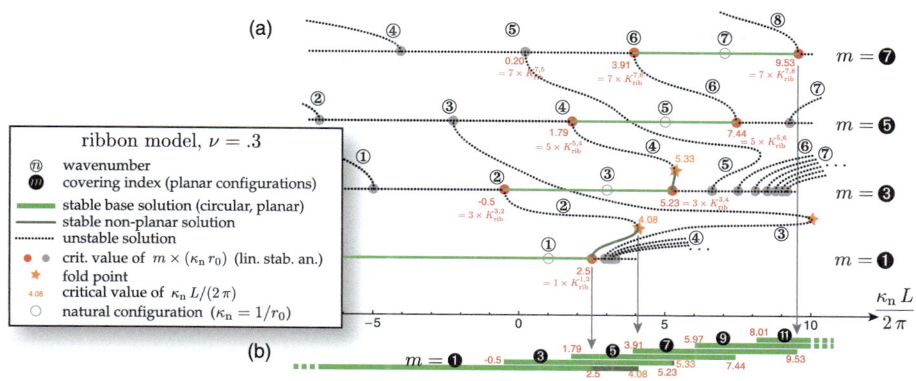

Fig. 12 (**a**) Sketch of the branches of non-linear equilibria of a closed elastic ribbon, for $\nu = .3$. The *horizontal axis* is the dimensionless natural curvature. The *stacked horizontal lines* represent the m-covered planar configurations, with m increasing from bottom to top. The *curves* represent the non-planar equilibrium branches connecting planar m-covered configurations. These curves were first plotted in the same set of conjugate coordinates as in Fig. 11, and they disentangled by hand to aid visualization, in such a way that their projection onto the horizontal axis is unaffected; as a result, the *vertical axis* loosely represents the parameter $\overline{\kappa}^*$. *Dots* denote connections of non-planar branches with planar (m-covered) configurations, as calculated by the linear stability analysis of Sect. 6.2. The stability of the planar solutions has been concluded based on the analysis of Sect. 6.2. The stability on non-planar solutions has been guessed based on the generic scenario of stability exchange at subcritical or subcritical pitchfork bifurcations and at fold points. (**b**) Domains of existence of stable planar (*green*) and stable non-planar (*dark green*) m-covered ribbons

consistent with the linear stability analysis of Sect. 6.2. Also, we note that the sequence of equilibria shown alongside the branch closely resemble the folding sequence shown in Fig. 2a.

For rods, there exists a similar branch of non-linear solutions $n = 2$ connecting the simple loop ($m = 1$) to the triply-covered loop ($m = 3$), but no fold point exists along this branch, see Fig. 2 in [15]: it is not surprising that the ribbon model, for which the constitutive laws are non-linear, gives rise to new bifurcations (here, a new fold point). As a result, the bifurcation $(m, n) = (1, 2)$ a supercritical (continuous) pitchfork bifurcation for rods, while it is a subcritical (discontinuous) pitchfork bifurcation for the ribbon model.

Repeating this continuation procedure for all the other branches emanating from the m-covered configurations, we constructed the diagram shown in Fig. 12. The branch studied in Fig. 11 appears there, along with many other branches. The stability of the non-planar branches has been established as follows: we first identified the bifurcations from the layout of the branches (such as a supercritical pitchfork bifurcation for $m = 1$ and $\frac{\kappa_n L}{2\pi} = 2.5$, a subcritical pitchfork bifurcation for $m = 3$ and $\frac{\kappa_n L}{2\pi} = -.5$, and fold point along the $n = 2$ branch at $\frac{\kappa_n L}{2\pi} = 4.08$), and assumed that the generic stability scenario associated with each of these bifurcations is applicable (*i.e.*, the bifurcated branches are stable past a supercritical pitchfork bifurcation, and unstable past a subcritical pitchfork bifurcation; an unstable mode appears across a fold point). The stable branches inferred in this way are summarized in part b of the figure. These stability results would need to be confirmed by a more thorough analysis, such as that done in [13] for the rod model.

A comparison with the bifurcation diagram for the anisotropic rod model, shown in Fig. 2 of [15], reveals some similarities concerning the general layout of the branches. There are two major differences, however. First, the evolution of the dimensionless curvature κ_n along the branches is different: on almost all branches the natural curvature varies monotonously for the anisotropic rod model [15], while along many branches of the ribbon model, the natural curvature first increases, passes through a maximum (fold points, orange stars in the diagram) and then decreases, see Fig. 12 above. Second, the bifurcations (and the equilibrium branches) for overcurved ribbons cluster near accumulation point, as happens for instance near $(\frac{\kappa_n L}{2\pi}, m) = (1, 3.33)$ and $(\frac{\kappa_n L}{2\pi}, m) = (3, 10.)$ in Fig. 12. This is consistent with the fact that the critical value $m\, K_{\text{rib}}^{m,n}$ of the quantity $\frac{\kappa_n L}{2\pi}$ converges to $\frac{m}{\nu}$ for $n \to \infty$ by Eq. (27). This value is precisely where the tangent twisting modulus cancels; when this happens, all wavenumbers become unstable, hence the accumulation of bifurcations. By contrast, the critical values and the branch of the rod models are evenly spaced [15].

To sum up, we have obtained a complete bifurcation diagram showing the non-linear equilibria of a closed ribbon. As for anisotropic rod models, branches corresponding to non-planar equilibria of a ribbon connect planar solutions of different covering indices m through bifurcations that are well predicted by the linear stability analysis of Sect. 6.2. However, both the critical values of the natural curvature for the onset of buckling from the planar configuration, and the aspect of the post-buckled branches differ significantly with the ribbon model, as compared to the rod model: the branches of the (more non-linear) ribbon model tend to cluster and they often include fold points where the natural curvature passes through a maximum.

6.4 Comparison to Experiments

In the experiments of Sect. 2, the single loop ($m = 1$) has been found to be stable for $\kappa_n r_0 < 3.5$; see Table 1. This value is above the threshold $K_{\text{ribb}}^{1,2} = 2.5$ predicted by the linear stability analysis of the ribbon model for $\nu = .3$, which is not surprising in view of the existence of a stable, non-planar branch past this bifurcation (dark green branch bifurcating from $(m, n) = (1, 2)$ in Fig. 12). The experimental threshold $\kappa_n r_0 = 3.5$ is reasonably close to the fold point terminating the stable part of the post-buckled branch at $\kappa_n r_0 = 4.08$. By contrast, the flat rod model predicts a much higher stability limit, namely $\kappa_n r_0 = 5.62$ (see Fig. 10b′).

Comparison of the experimental stability limits for $m \geq 3$ is more difficult because of the presence of self-contact and friction, both of which were ignored in the analysis. The experimental lower bounds are $\kappa_n r_0 = .28$ ($m = 3$) and $\kappa_n r_0 = .42$ ($m = 5$), see Table 1. This is consistent with the values $K_{\text{rib}}^{3,2} = -.17$ ($m = 3$) and $K_{\text{rib}}^{5,4} = .36$ ($m = 5$) found in the linear stability analysis for $\nu = .3$, see Table 3. Note that the bifurcation diagram predicts that a subcritical pitchfork bifurcation there, so these values are indeed the stability limits.

7 Conclusion

We presented experiments using metallic strips that had acquired residual curvature after being deformed plastically. These strips were formed into a loop shape, and then folded into multiply covered loops with a covering index equal to $m = 1$ (single loop), $m = 3$ and $m = 5$. Buckling instabilities were observed, both when the ratio of the natural curvature to the curvature of the loop is small (overcurved case) and when it is large (undercurved case).

Motivated by these experiments, we carried out stability analyses for elastic ribbons. We addressed the linear stability problem near m-covered circular configurations, and computed non-linear equilibrium branches numerically. We used the mathematical formulation of Dias and Audoly [3], who treat an elastic ribbon as a kinematically constrained rod governed by non-linear constitutive laws. This formulation allowed us to analyze the ribbon and rod models in a unified language, and to easily adapt stability analyses routinely performed for elastic rods to ribbons. By a detailed comparison of the models, we highlighted a number of differences in their stability properties.

Many of these differences were interpreted, based on a simple feature of naturally curved ribbons, which has not been reported so far: the tangent twisting modulus can become negative when a ribbon is undercurved. This change of sign induces buckling instabilities whose existence has been confirmed by finite element simulations of an elastic shell having a small height-to-radius ratio.

As a final remark, we would like to emphasize that these buckling instabilities are different from the classical lateral torsional buckling instabilities of anisotropic beams: the latter can be observed for linearly elastic rods, and they make use of twisting to transfer a bending moment initially applied in the stiffer direction (*i.e.*, involving the larger principal bending modulus) onto the weaker direction (*i.e.*, involving the smaller principal bending modulus); by contrast, the buckling instabilities for ribbons which we described in this paper require the twisting modulus to be non-constant, and have been demonstrated in geometries where the initial bending moment is along the weaker direction.

References

1. Antman, S.: Nonlinear Problems of Elasticity, 2nd edn. Springer, Berlin (2005)
2. Dias, M.A., Audoly, B.: A non-linear rod model for folded elastic strips. J. Mech. Phys. Solids **62**, 57–80 (2014)
3. Dias, M.A., Audoly, B.: "Wunderlich, meet Kirchhoff": a general and unified description of elastic ribbons and thin rods. J. Elast. (2014)
4. Doedel, E.J., Champneys, A.R., Fairgrieve, T.F., Kuznetsov, Y.A., Sandstede, B., Wang, X.J.: AUTO-07p: continuation and bifurcation software for ordinary differential equations (2007). http://indy.cs.concordia.ca/auto/
5. Goriely, A.: Twisted elastic rings and the rediscoveries of Michell's instability. J. Elast. **84**, 281–299 (2006)
6. Goriely, A., Shipman, P.: Dynamics of helical strips. Phys. Rev. E **61**(4), 4508–4517 (2000)
7. Haijun, Z., Zhong-can, O.Y.: Spontaneous curvature-induced dynamical instability of Kirchhoff filaments: application to DNA kink deformations. J. Chem. Phys. **110**, 1247 (1999)
8. Hibbitt, D., Karlsson, B., Sorensen, P.: Abaqus manual version 6.11. Tech. rep., Dassault Systèmes (2011)
9. Hinz, D.F., Fried, E.: Translation and interpretation of Michael Sadowsky's paper "Theory of elastically bendable inextensible bands with applications to the Möbius band". J. Elast., 1–11 (2014)
10. Hinz, D.F., Fried, E.: Translation of Michael Sadowsky's paper "An elementary proof for the existence of a developable Möbius band and the attribution of the geometric problem to a variational problem". J. Elast. (2014)
11. Hinz, D.F., Fried, E.: Translation of Michael Sadowsky's paper "The differential equations of the Möbius band". J. Elast., 1–4 (2014)
12. Hoffman, K.A.: Methods for determining stability in continuum elastic-rod models of DNA. Philos. Trans. R. Soc. Lond. A **362**, 1301–1315 (2004)
13. Hoffman, K.A., Manning, R.S., Maddocks, J.H.: Link, twist, energy and the stability of DNA minicircles. Biopolymers **70**(2), 145–157 (2003)
14. Keller, H.B.: Numerical solution of bifurcation and nonlinear eigenvalue problems. In: Rabinowitz, P.H. (ed.) Applications of Bifurcation Theory, pp. 359–384. Academic Press, San Diego (1977)
15. Manning, R.S., Hoffman, K.A.: Stability of n-covered circles for elastic rods with constant planar intrinsic curvature. J. Elast. **62**(1), 1–23 (2001)
16. Michell, J.H.: On the stability of a bent and twisted wire. Messenger Math. **11**, 181–184 (1889–1890)
17. Press, W.H., Teukolsky, S.A., Vetterling, W.T., Flannery, B.P.: Numerical Recipes, 3rd edn. Cambridge University Press, Cambridge (2007)
18. Sadowsky, M.: Die Differentialgleichungen des Möbiusschen Bandes. In: Jahresbericht der Deutschen Mathematiker-Vereinigung, vol. 39 (2. Abt. Heft 5/8), pp. 49–51 (1929)
19. Sadowsky, M.: Ein elementarer Beweis für die Existenz eines abwickelbaren Möbiusschen Bandes und die Zurückführung des geometrischen Problems auf ein Variationsproblem. In: Sitzungsberichte der Preussischen Akademie der Wissenschaften, physikalisch-mathematische Klasse, 17. Juli 1930, pp. 412–415 (1930). Mitteilung vom 26. Juni

20. Sadowsky, M.: Theorie der elastisch biegsamen undehnbaren Bänder mit Anwendungen auf das Möbiussche Band. In: Proceedings of the 3rd International Congress of Applied Mechanics, Stockholm, vol. 2, pp. 444–451 (1930)
21. Seffen, K.A., Pellegrino, S.: Deployment dynamics of tape springs. Proc. R. Soc. Lond., Ser. A, Math. Phys. Eng. Sci. **455**(1983), 1003–1048 (1999)
22. Starostin, E.L., van der Heijden, G.H.M.: The shape of a Möbius strip. Nat. Mater. **6**(8), 563–567 (2007)
23. Starostin, E.L., van der Heijden, G.H.M.: Tension-induced multistability in inextensible helical ribbons. Phys. Rev. Lett. **101**, 084301 (2008)
24. Thompson, J.M.T., Champneys, A.R.: From helix to localized writhing in the torsional post-buckling of elastic rods. Proc. R. Soc. A, Math. Phys. Eng. Sci. **452**(1944), 117–138 (1996)
25. Todres, R.E.: Translation of W. Wunderlich's "on a developable Möbius band". J. Elast., 1–12 (2014)
26. Wunderlich, W.: Über ein abwickelbares Möbiusband. Monatshefte Math. **66**(3), 276–289 (1962)

DOI 10.1007/978-94-017-7300-3_16
Reprinted from *Journal of Elasticity* Journal, DOI 10.1007/s10659-015-9525-6

Residual Stresses and Poisson's Effect Drive Shape Formation and Transition of Helical Structures

Zi Chen[1] · **Xiaomin Han**[1] · **Huang Zheng**[2]

Received: 4 February 2014 / Published online: 17 March 2015

Abstract Strained multilayer structures are extensively investigated because of their applications in microelectromechanical/nano-elecromechanical systems. Here we employ a finite element method (FEM) to study the bending and twisting of multilayer structures subjected to misfit strains or residual stresses. This method is first validated by comparing the simulation results with analytic predictions for the bending radius of a bilayer strip with given misfit strains. Then, the FEM simulations are used to study the deformation of a bilayer strip subjected to a certain residual stress to examine the influence of Poisson's effect. As predicted by elasticity theory, a nearly purely twisted ribbon results for a given mis-orientation angle, although the residual stress only has one non-zero principal component. Our results further show that for the same Poisson's ratio, a transition from a twisted ribbon to a nearly cylindrical helical shape can occur, either when the strip becomes wide and thin enough or when the driving force is large enough. The combined effects of the residual stress and the Poisson's ratio are also examined. Our work demonstrates the effective use of finite element simulations in controllable design of strained multilayer structures, which have broad potential applications in NEMS, sensors, drug delivery, morphing structures, active materials, optoelectronics, and bio-inspired robotics.

Keywords Misfit strain · Residual stress · Poisson's effect · Helices · Nanoribbon · Actuator

Mathematics Subject Classification (2010) 74B10 · 74G15 · 70C20

1 Introduction

Strained multilayer structures have received extensive attention because of their applications as sensors and actuators in microelectromechanical/nano-elecromechanical systems

✉ Z. Chen
Zi.Chen@Dartmouth.edu

1 Thayer School of Engineering, Dartmouth College, Hanover, NH 03755, USA

2 Fujian Radio and TV University, Fuzhou, China

(MEMS/NEMS) [1–14]. Multilayered thin films can also have tunable morphological and/or mechanical properties in response to the change of pH value or temperature [15, 16]. Stoney first analyzed the residual stress and deformation of a film-on-substrate bilayer structure [17]. His work applies to the scenario where the thickness of the film is relatively small compared to the substrate. Timoshenko investigated the bending of a bi-metal strip when heated uniformly and the related buckling problem [18]. Suo and co-workers further studied the deformation of film-on-substrate structures driven by differential thermal expansion [2]. Later, Tsui and Clyne developed an analytical model for calculating the residual stress in a bilayer strip due to misfit strain that can arise during deposition or due to mismatch in thermal contraction [19]. More recently, Huang and Zhang studied the relationship between residual stresses and the corresponding elastic deformation in bilayer cantilevers both analytically and numerically [20]. Nevertheless, the influence of Poisson's effect in these multilayer structures remain incompletely studied [2, 21, 22].

Helices are ubiquitous shapes in nature and engineering [23]. Often, helical shapes can arise because of misfit strains, residual stresses, surface stresses, or other sources of driving forces [13, 22, 24–26, 28–31]. The shape transitions among purely twisted ribbons (or helicoids), spiral ribbons, and tubular structures are found in a variety of physical systems, such as twist-nematic elastomers [30, 31], strained multilayer composites [22, 26–28, 33–35], peptides [34], germini surfactant [37], and nanoribbons[8]. Although there are some studies that involve the effects of Poisson's ratio on the spontaneous deformation of helical structures, few researchers have systematically investigated how Poisson's effect leads to the shaping of twisted ribbons.

In this work, we employed a finite element method (FEM) [33, 38] to model the bending and twisting of multilayer structures subjected to either misfit strain or residual stresses. In particular, our study shows that Poisson's effect can induce the formation of a purely twisted shape, even though the residual stress only has one non-zero principal component. Here we first show the validity of the finite element method employed in this work by comparing the simulation results with analytic predictions. Then, we use this FEM approach to study the influence of Poisson's effect on the spontaneous deformation of a multilayer strip subjected to a residual stress with only one non-zero principal component. As predicted, a nearly twisted ribbon forms when the mis-orientation angle satisfies a certain condition. We further demonstrated that for the same Poisson's ratio a transition from a twisted ribbon to a nearly cylindrical helical ribbon can occur when the strip becomes wide and thin enough or when the driving force is large enough. The combined effects of the residual stress and the Poisson's ratio have also been investigated. Our work complements the current studies on mechanical self-assembly of three-dimensional structures and can benefit the design and manufacturing of spontaneously bent and twisted structures for a number of engineering applications.

2 Self-scrolling of Strained Multi-layer Membranes

2.1 Theory

The theory for bending of a film-on-substrate is well established. Stoney first studied the residual stress and the bending radius of a strained bilayer structure [17]. Timoshenko further provided an improved solution for bilayer structures where the film thickness is comparable to the substrate. A recent treatment on subject, which we briefly recap here, was developed by Tsui and Clyne [19]. An undeformed bilayer strip is oriented relative to a right-handed

Fig. 1 Geometry of a bilayer strip (undeformed). The tetrahedral mesh is created in COMSOL Multiphysics 4.3a

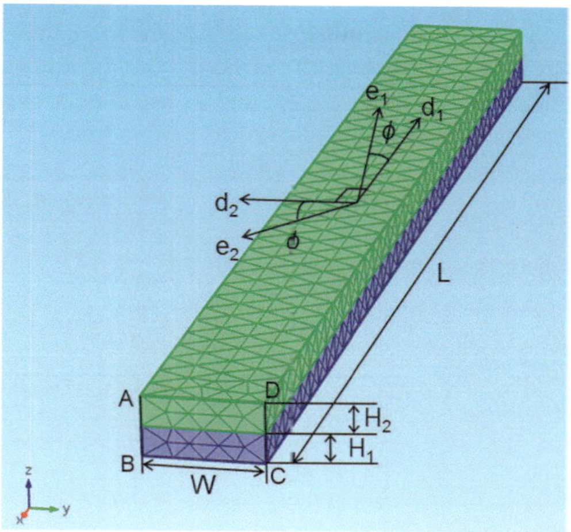

orthonormal basis $\boldsymbol{d}_i$, $i = 1, 2, 3$ as shown in Fig. 1, and the strip has a misfit strain $\Delta\varepsilon$ along the $\boldsymbol{d}_1$ direction and a zero mis-orientation angle is considered (i.e., $\phi = 0$ in Fig. 1). In [19], the radius of curvature is predicted as

$$R_t = \frac{1 + 4\alpha\beta + 6\alpha\beta^2 + 4\alpha\beta^3 + \alpha^2\beta^4}{6\alpha\beta(1+\beta)\Delta\varepsilon} H_1, \tag{1}$$

where $\alpha = E_2/E_1$ (E_1 and E_2 are Young's modulus of the bottom and top layer, respectively), $\beta = H_2/H_1$, and $\Delta\varepsilon$ is the misfit strain between the two layers.

2.2 Finite Element Simulations

We employ a model based on linear elasticity theory and stationarity principles [22, 28, 33], where both the bending and mid-plane stretching are taken into account, as well as geometric nonlinearity. Without making *a priori* assumptions about the deformed shape, the total potential energy density per unit area of the ribbon is $\prod = \int_{-E_1}^{H_1} (\frac{1}{2}\boldsymbol{\sigma} : \boldsymbol{\gamma}) dz$. The strain tensor ($\boldsymbol{\gamma}$) is obtained by superimposing elastic bending strains ($\boldsymbol{\gamma}_b$), an in-plane uniform "membrane" strain ($\boldsymbol{\gamma}_m$), internal residual strains ($\boldsymbol{\gamma}_0$), and strains along the direction perpendicular to the ribbon plane ($\boldsymbol{\gamma}_z$): $\boldsymbol{\gamma} = \boldsymbol{\gamma}_b + \boldsymbol{\gamma}_m + \boldsymbol{\gamma}_z + \boldsymbol{\gamma}_0$ [28].

The strain tensor can also be written as [32]

$$\boldsymbol{\gamma} = \gamma_{ij} \boldsymbol{d}_i \otimes \boldsymbol{d}_j, \tag{2}$$

where the components are

$$\begin{aligned}
\gamma_{xx} &= \varepsilon_{xx} + z(\kappa_1 \cos^2\phi + \kappa_2 \sin^2\phi) + \gamma_{xx}^0(z) \\
\gamma_{xy} &= \gamma_{yx} = \varepsilon_{xy} + z(\kappa_2 - \kappa_1)\sin\phi\cos\phi + \gamma_{xy}^0(z) \\
\gamma_{yy} &= \varepsilon_{yy} + z(\kappa_1 \sin^2\phi + \kappa_2 \cos^2\phi) + \gamma_{yy}^0(z) \\
\gamma_{zz} &= \varepsilon_{zz} + qz + \gamma_{zz}^0(z),
\end{aligned} \tag{3}$$

where γ_{ij} denotes the ij-component of the strain tensor [22, 28]. κ_1 and κ_2 denotes the principal curvatures along the principal bending axes, $\boldsymbol{e}_1$ and $\boldsymbol{e}_2$, respectively, and q is a constant required for compatibility.

If there is no residual stress (i.e., $\boldsymbol{\sigma}_0 = 0$), then $\boldsymbol{\sigma} = C : \boldsymbol{\gamma} + \boldsymbol{\sigma}_0 = C : \boldsymbol{\gamma}$. When the residual stress ($\boldsymbol{\sigma}_0$) is present, the principal components are denoted by σ_{0x}, σ_{0y}, and σ_{0z}, along the principal directions $\boldsymbol{e}_1$, $\boldsymbol{e}_2$, and $\boldsymbol{e}_3$ ($\boldsymbol{e}_3 \equiv \boldsymbol{e}_1 \times \boldsymbol{e}_2$), respectively. Thus, in terms of the global Cartesian coordinate system we may write

$$\boldsymbol{\sigma}_0 = \sigma_{0ij}\boldsymbol{d}_i \otimes \boldsymbol{d}_j, \tag{4}$$

where the non-zero components are

$$\begin{aligned}
\sigma_{0xx} &= \sigma_{0x}\cos^2\phi + \sigma_{0y}\sin^2\phi, \\
\sigma_{0xy} &= (\sigma_{0y} - \sigma_{0x})\sin\phi\cos\phi, \\
\sigma_{0yy} &= \sigma_{0x}\sin^2\phi + \sigma_{0y}\cos^2\phi, \\
\sigma_{0zz} &= \sigma_{0z}.
\end{aligned} \tag{5}$$

The finite element simulation used in this work is a full three-dimensional model using the structural mechanics module of COMSOL Multiphysics 4.3a [33, 38]. As shown in Fig. 1, the geometry is meshed using "Free Tetrahedral" elements in COMSOL. The boundary condition is set as follows. The point C is fixed in space, the point D has zero displacement in the y direction, while the points A and B have zero displacement both in the x and z directions. Other edges (or points) are free to move. In our model, we mainly consider two scenarios. In the first, there exists a uniform misfit strain between the two layers or the ribbon. In COMSOL Multiphysics, this is modeled by prescribing an initial strain that resides in one (or multiples) of the layers. In the other scenario, there exist residual stresses in one of the layers. In COMSOL, this is achieved by prescribing an initial stress tensor in one layer.

Verification of Finite Element Model Here we perform the following simulations to compare the results from this finite element model with the analytic solutions. From our finite element simulation, since the midplane should be strain-free, we can calculate the radius of curvature by considering the identity, $|OQ|^2 + |P'Q|^2 = |OP'|^2$, i.e., $(R - |w|)^2 + (L - |u|)^2 = R^2$ in Fig. 2(a). Therefore, the radius of curvature from the simulation is $R_s = [(L - u)^2 + w^2]/2|w|$. On the other hand, the analytical solution of the radius R in a strained bilayer structure is

$$R_t = \frac{4h}{3\Delta\varepsilon}, \tag{6}$$

when $\alpha = E_2/E_1 = 1$, $H_1 = H_2 = h$, and $\Delta\varepsilon$ is the misfit strain between the two layers.

Figure 2(b) shows very good agreement between the finite element simulations and theoretical predictions, as they almost completely overlap. This consistency validates the current finite element method in studying more complex, three-dimensional deformations driven by misfit strains or residual stresses (or a combination) in multilayered structures, for which analytical solutions may not be easy to obtain.

3 Formation of Twisted Ribbons due to Poisson's Effect

Helical shapes often arise because of mechanical anisotropy (the nature of the driving force being anisotropic) and geometric mis-orientation [22, 26, 28, 34, 38, 40], i.e., the principal

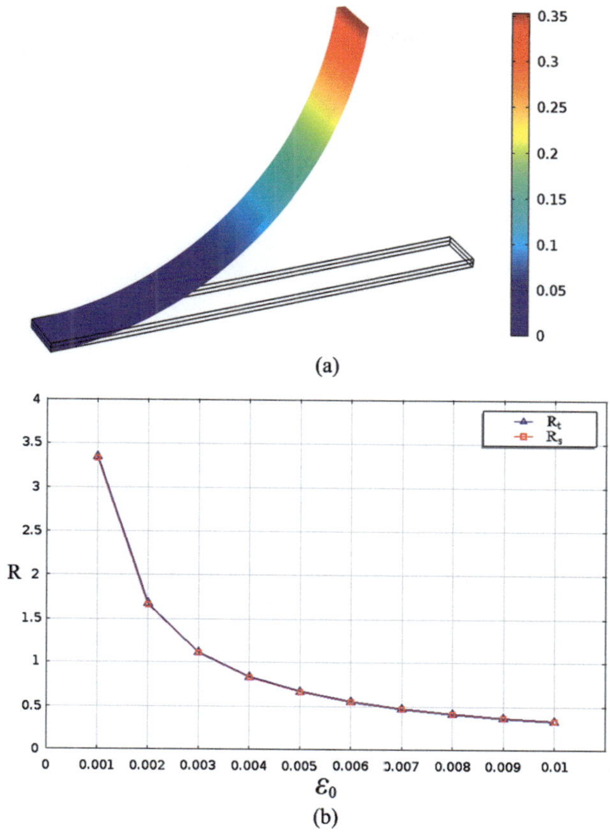

Fig. 2 (**a**) A stable circular shape driven by misfit strain in a bilayer ribbon. The initial strain in the top layer is $\varepsilon_t = -\varepsilon_0(e_1 \otimes e_1 + e_2 \otimes e_2)$. The misfit strain in the bottom layer is $\varepsilon_b = \varepsilon_0(e_1 \otimes e_1 + e_2 \otimes e_2)$, where $\varepsilon_0 = 0.01$. (**b**) The ratio between the radius of curvature from finite element simulation (R_s) and from theoretical calculation (R_t) for $\varepsilon_0 = 0.001n$ ($n = 1, 2, \ldots, 10$)

axes of the driving force form a non-trivial angle (neither 0 nor 90 degrees) with the geometric axes. As a result, the principal bending axes form a non-zero mis-orientation angle ϕ with the geometric axes. From a mathematical point of view [36], for example, a curvature tensor $\bar{\kappa}$ can be written as

$$\bar{\kappa} = \kappa_0(\boldsymbol{e}_1 \otimes \boldsymbol{e}_1 - \boldsymbol{e}_2 \otimes \boldsymbol{e}_2), \tag{7}$$

when the principal curvatures are κ_0 and $-\kappa_0$ in the basis $\{\boldsymbol{e}_1, \boldsymbol{e}_2\}$. From geometry, we have

$$\boldsymbol{e}_1 = \cos\phi \boldsymbol{d}_1 + \sin\phi \boldsymbol{d}_2, \tag{8}$$

and

$$\boldsymbol{e}_2 = -\sin\phi \boldsymbol{d}_1 + \cos\phi \boldsymbol{d}_2. \tag{9}$$

Then the curvature tensor in the new basis $\{\boldsymbol{d}_1, \boldsymbol{d}_2\}$ becomes

$$\begin{aligned} \bar{\kappa}' &= \kappa_0[(\cos^2\phi - \sin^2\phi)(\boldsymbol{d}_1 \otimes \boldsymbol{d}_1 - \boldsymbol{d}_2 \otimes \boldsymbol{d}_2) + 2\sin\phi\cos\phi(\boldsymbol{d}_1 \otimes \boldsymbol{d}_2 + \boldsymbol{d}_2 \otimes \boldsymbol{d}_1)] \\ &= \kappa_0(\boldsymbol{d}_1 \otimes \boldsymbol{d}_2 + \boldsymbol{d}_2 \otimes \boldsymbol{d}_1), \end{aligned} \tag{10}$$

when $\phi = \pi/4$ [29]. This result suggests that a narrow ribbon with principal curvatures κ_0 and $-\kappa_0$ in the basis $\{\boldsymbol{e}_1, \boldsymbol{e}_2\}$ indeed forms a purely twisted shape, because the curvature

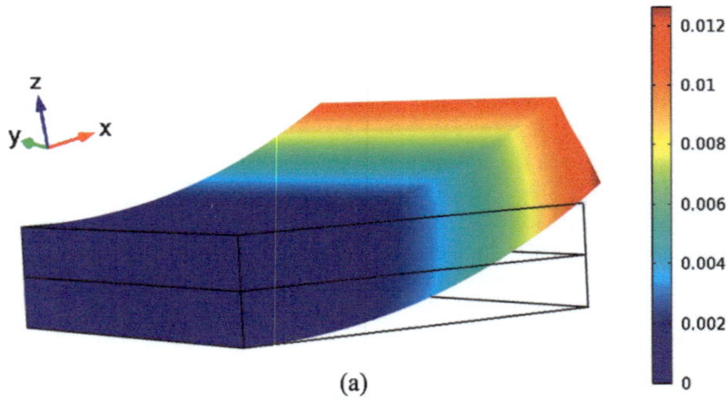

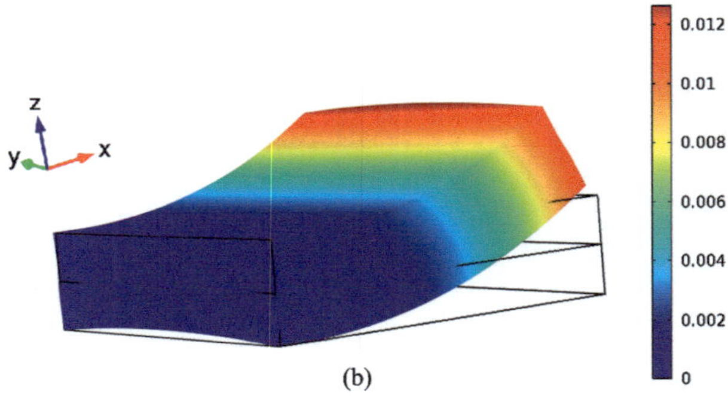

Fig. 3 Stable shapes driven by misfit strain in bilayer strips. (**a**) A cylindrical shape forms when the initial stress in the top layer is $\sigma_t = \sigma_0 e_1 \otimes e_1$, where $\sigma_0 = 10$ GPa. Here, both the width (W) and length (L) are 0.04 m. Young's modulus is $E = 100$ GPa and Poisson's ratio is $\nu = 0$. (**b**) A saddle shape forms for the same bilayer strip when the initial stress in the top layer is $\sigma_t = \sigma_0 e_1 \otimes e_1$. Young's modulus is $E = 100$ GPa and Poisson's ratio is $\nu = 0.3$. The mis-orientation angle is zero here ($\phi = 0$). The color indicates the total displacement

tensor only has shear (or torsional) components ($\boldsymbol{d}_1 \otimes \boldsymbol{d}_2$ and $\boldsymbol{d}_2 \otimes \boldsymbol{d}_1$) that are non-zero (when viewed from the basis $\{\boldsymbol{d}_1, \boldsymbol{d}_2\}$, the geometric axes of the ribbon). This example illustrates the fundamental reason of the formation of helical ribbons resulting from the presence of a geometric mis-orientation angle.

For the spontaneous bending of a multilayered structure driven by misfit strains, the previous studies suggest that Poisson's effect has little influence. For example, Tsui and Clyne showed that in a bilayer strip with given misfit strain, the radius of curvature does not depend on Poisson's ratios of the constituent layers [19]. However, some studies show the importance of Poisson's ratio [2, 21, 22, 28]. Here we investigated the possibility of a multilayer strip forming a purely twisted shape driven by residual stresses due to Poisson's effect.

First, we show that the resultant principal curvatures of a bilayer strip driven by a given residual stress become different when Poisson's effect is present. Here, the residual stress only has non-zero principal component along one axis. When Poisson's ratio is zero, the system deforms into a cylindrical shape with zero Gauss curvature ($\kappa_G = \kappa_1\kappa_2 = 0$, Fig. 3(a)). However, when Poisson's ratio is non-zero (e.g., $\nu = 0.3$), the Gauss curvature becomes negative (i.e., $\kappa_G = \kappa_1\kappa_2 < 0$, Fig. 3(b)). More specifically, when $\sigma_0 = 10$ GPa, the ratio between the principal curvature is $\kappa_2/\kappa_1 \approx -(1/14.9)/(1/4.2) \approx -0.28$. The absolute value of this ratio is very close to Poisson's ratio $\nu = 0.3$, supporting the hypothesis that Poisson's effect is the main cause of such deformation.

Our previous studies [22, 28, 34] have shown that a purely twisted ribbon forms when $\kappa_1 \cos^2\phi + \kappa_2 \sin^2\phi = 0$. So here we predict that a purely twisted shape occurs when $\cot^2\phi = -\kappa_2/\kappa_1 \approx \nu = 0.3$, i.e., $\phi \approx 61.3°$. Figure 4(a) shows the formation of a nearly twisted bilayer ribbon of width $W = 0.02$ m and thickness $H_1 = H_2 = H = 0.005$ m, when the mis-orientation angle is $\phi \approx 61.3°$ and the only non-zero principal component of the

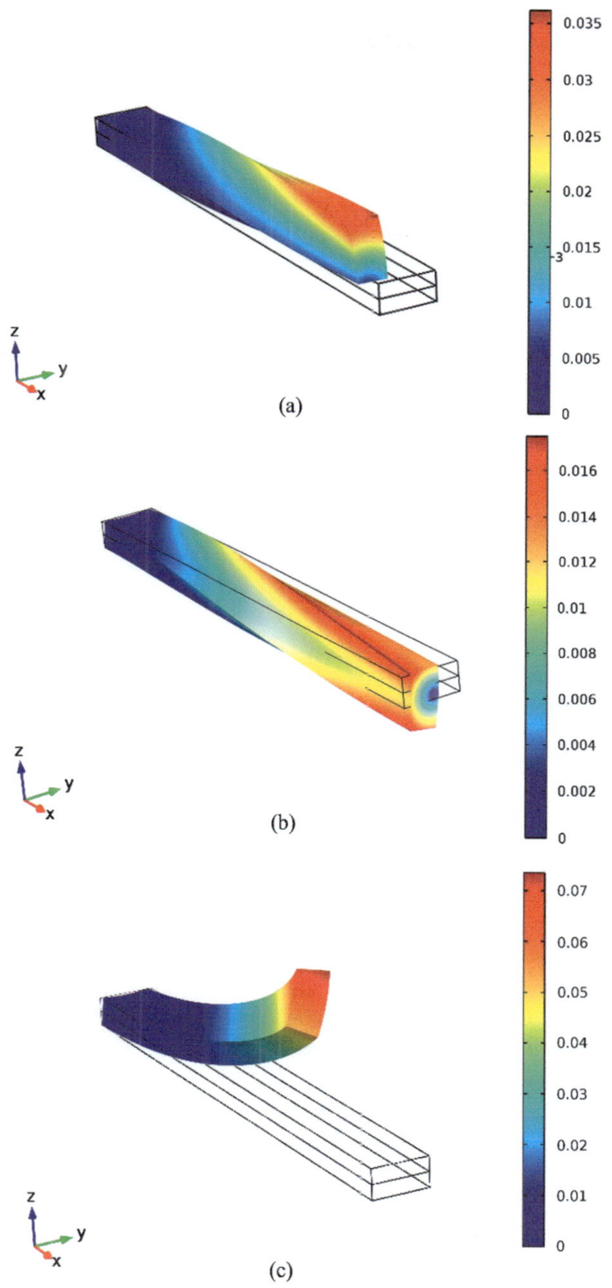

Fig. 4 Helical shapes driven by residual stress in one layer simulated by finite element method. The residual stress in the top layer is $\sigma_t = \sigma_0[e_1 \otimes e_1 \cos^2\phi + (e_1 \otimes e_2 + e_2 \otimes e_1)\cos\phi \sin\phi]$. Here, $\nu = 0.3$, $W = 0.02$ m, $H = 0.005$ m and $\sigma_0 = 10$ GPa. (**a**) A twisted shape forms when $\phi = 61.3°$, $\nu = 0.3$, $W = 0.02$ m, and $H = 0.005$ m. (**b**) A twisted shape forms when $\phi = 65.5°$, $\nu = 0.3$, $W = 0.02$ m, and $H = 0.005$ m. (**c**) A cylindrical helical shape forms when $\phi = 65.5°$, $\nu = 0$. Young's modulus is $E = 100$ GPa. The color indicates the total displacement

residual stress is $\sigma_0 = 10$ GPa along the e_1 direction. Through trial-and-error, we found that when $\phi \approx 65.5°$, the shape of the deformed bilayer ribbon is even closer to a purely twisted ribbon (the centerline stays straight, i.e., the helix radius is zero), as shown in Fig. 4(b). By contrast, when Poisson's effect is absent, the same driving force results in a cylindrical helical ribbon (Fig. 4(c)).

Fig. 5 $|\kappa_2/\kappa_1|$ as a function of ν for different width values

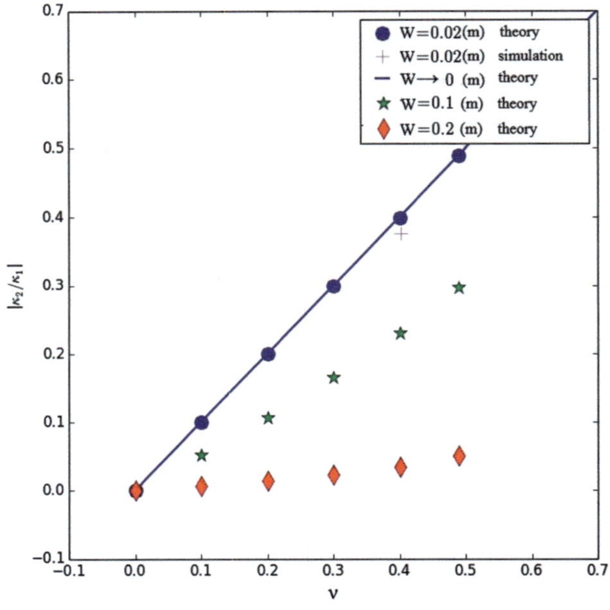

From a theoretical perspective, it should hold that $|\kappa_2/\kappa_1| = \nu$ [22] in the limit when the width vanishes (i.e., $W \to 0$), as shown by the blue line in Fig. 5. When the ribbon has a finite width, the theoretical values of the principal curvatures can also be numerically calculated [39]. Figure 5 shows the absolute value of the ratio between the principal curvatures ($|\kappa_2/\kappa_1|$) as a function of the Poisson's ratio for different width values. In the small width regime (e.g., $W = 0.02$ m), those values are in excellent agreement with the asymptotic values $|\kappa_2/\kappa_1| = \nu$, and the value from the previous finite element simulation also agree well with the corresponding theoretical value. It is worth noting that the ratio becomes very small when the width of the ribbon becomes large enough, demonstrating that a nearly cylindrical shape forms in that regime because of the competition between bending and stretching energy, consistent with the findings in previous studies [26, 39].

To see the effects of Poisson's ratio on the helical ribbon shape more intuitively, we plot the radius (theoretical value) of the helical ribbon as a function of ν for different width values in Fig. 6. In the small width regime (i.e., $W = 0.02$ m), the radius first decreases and then increases as the Poisson's ratio increases. When $\nu = 0.3$, the radius becomes zero, indicating a purely twisted ribbon forms. In the large width regime, by contrast, the radius of the helical ribbon does not change much as the Poisson's ratio increases, suggesting that the Poisson effect does not result in any dramatic shape changes when the width is large enough. In the intermediate regime when $W = 0.1$ m, the radius decreases monotonically with the increase of the Poisson's ratio, until a nearly helicoidal shape forms at a high Poisson's ratio ($\nu = 0.49$). We also plot the radius of the helical ribbon as a function of W for different Poisson's ratios in Fig. 7. As the width increases, the radius increases monotonically except for when $\nu = 0.49$, in which case the radius first decreases and then increases.

Moreover, we examine the effects of the residual stress on the stable geometric configurations. Figure 8(a) shows that the absolute value ($|\kappa_2/\kappa_1|$) of the ratio between the two principal curvatures is almost a constant when the width is relatively narrow ($W = 0.02$ m) or when $\nu = 0$ (in which case the ratio is zero because $\kappa_2 = 0$); however, when the width is relatively large (e.g., $W = 0.1$ m) the ratio decreases monotonically as the residual stress in-

Fig. 6 The radius of the helical ribbon as a function of ν for different width values. The mis-orientation angle is 61.3°

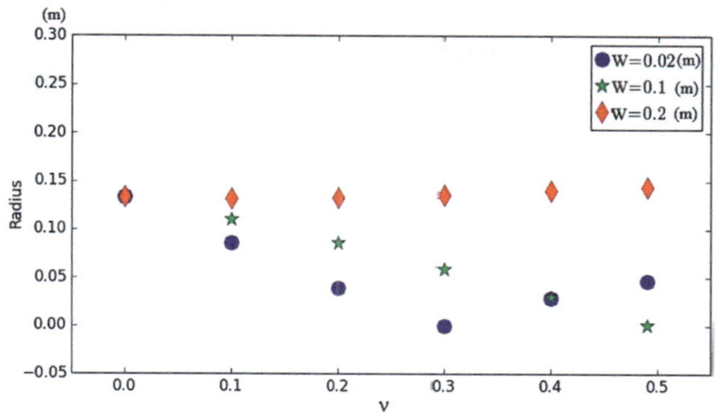

Fig. 7 The radius of the helical ribbon as a function of W for different Poisson's ratios

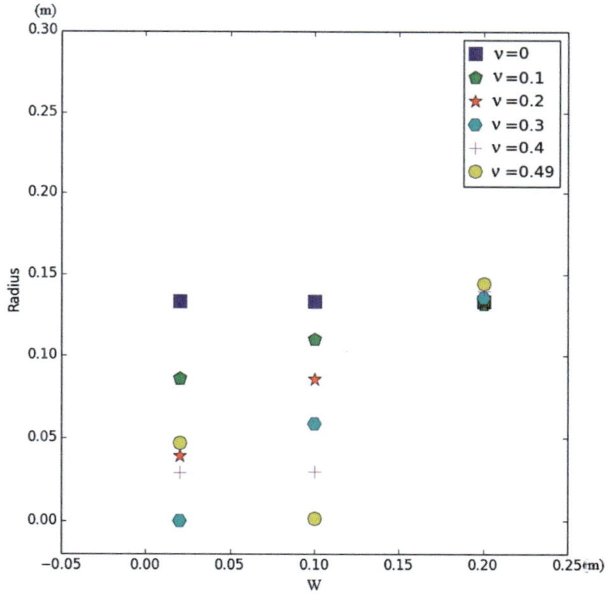

creases. The radius of the helical ribbon, on the other hand, always decreases monotonically as the residual stress increases as shown in Fig. 8(b). When the residual stress becomes sufficiently large (Fig. 9(b)), the system transitions from a twisted ribbon (Fig. 9(a)) to a nearly cylindrical helical shape. It is also worth mentioning that the finite element simulations provide convenient access to the strain components as well. In particular, the strain components along the three axes are found to be a linear function of the residual stress as expected. From the plane stress condition, it should hold that $\varepsilon_{33} = -\nu(\varepsilon_{11} + \varepsilon_{22})/(1 - \nu)$ [32]. The numerical simulation results indeed show good agreement with this prediction (Fig. 9(c)).

4 Shape Transition Between Twisted Ribbons and Cylindrical Helical Ribbons

It is interesting to note that a shape transition can occur from a purely twisted shape (Fig. 4(a), (b)) to a nearly cylindrical helical shape when Poisson's ratio changes (Fig. 4(c))

Fig. 8 Effects of residual stress. (**a**) The ratio between the two principal curvatures as a function of σ_0 for different combinations of Poisson's ratio and width. (**b**) The radius of the helical ribbon as a function of σ_0 for different combinations of Poisson's ratio and width

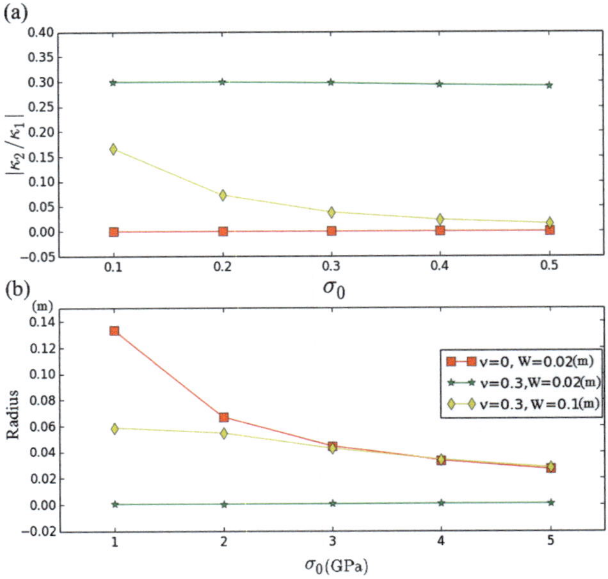

or when the strip becomes wide and thin enough (Fig. 5(a), (b)), even though all the other physical parameters stay the same. The main cause of the latter case is the competition between bending energy ($\sim \kappa^2 EWH^3$) and stretching energy ($\sim \kappa^4 EW^5 H$), as has been discussed in recent studies [26, 29, 34, 39]. On the other hand, when the driving force (i.e., residual stress) becomes strong enough, such transition can also occur, because the principal curvatures will become large enough. According to previous studies [26, 34], such shape transition occurs when $W\sqrt{\kappa/H}$ becomes large enough, which can be achieved by increasing W and κ, or decreasing H. It will be of further interest to investigate whether such system can also exhibit multi-stability, which has been studied in a number of related systems [26, 34, 41, 42, 44, 48]. This awaits future investigation, as the numerical convergent issue is currently preventing us from making a definite conclusion. There remain ample venues for future research on mechanical self-assembly and multi-stabilities of strained semiconductor nanostructures [8], liquid crystals [30, 31], peptides [34], and dielectric elastomer actuators [47].

5 Conclusion

In this work, we use a finite element method to model the large deformation (bending and twisting) of multilayer structures when subjected to misfit strains or residual stresses. We first compare the simulation results with analytic predictions for the radius of a bilayer strip subjected to given misfit strains. Afterwards, we use FEM simulations to study the spontaneous deformation of a multilayer strip subjected to a prescribed residual stress, in order to examine the influence of Poisson's effect. As predicted, a nearly twisted ribbon results for a calculated mis-orientation angle, even though the residual stress only has one non-zero principal component. Our results further show that for the same Poisson's ratio, a transition from a twisted ribbon to a nearly cylindrical helical shape can occur when the strip becomes wide and thin enough or when the driving force is large enough. Our work demonstrates the usefulness of finite element simulations in programmable design of strained multilayer

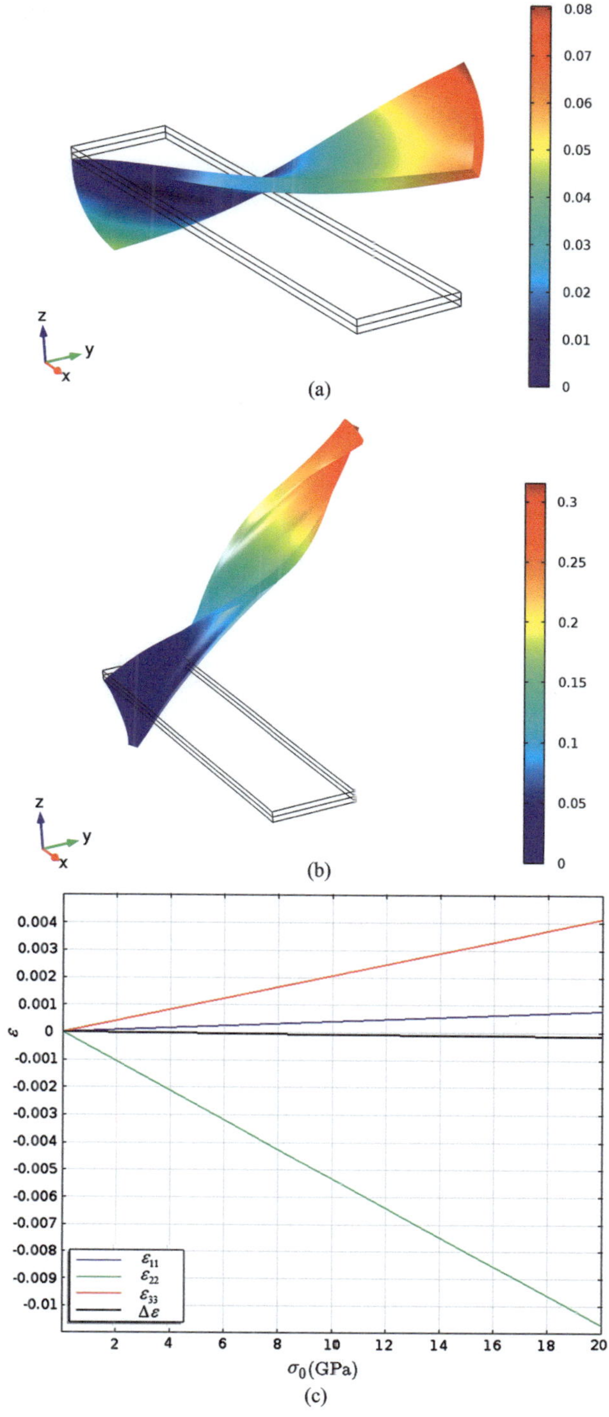

Fig. 9 Stable helical shapes driven by residual stress in the top layer ($\nu = 0.3 \neq 0$), $\sigma_t = \sigma_0[e_1 \otimes e_1 \cos^2\phi + e_2 \otimes e_2 \sin^2\phi + (e_1 \otimes e_2 + e_2 \otimes e_1) \times \cos\phi \sin\phi]$, where $\phi = 65.5°$. (**a**) A general helical shape forms when $W = 0.04$ m, $H = 0.0025$ m, and $\sigma_0 = 10$ GPa. (**b**) A nearly cylindrical helical shape forms when $W = 0.04$ m, $H = 0.0025$ m, and $\sigma_0 = 20$ GPa. Young's modulus is $E = 100$ GPa. The color indicates the total displacement. (**c**) Strain components as a function of residual stress: ε_{11}, ε_{22}, and ε_{33} are the strain components along x, y, z directions, respectively. $\Delta\varepsilon \equiv \varepsilon_{33} + \nu(\varepsilon_{11} + \varepsilon_{22})/(1 - \nu)$ is used to check whether the theoretical prediction $\varepsilon_{33} = -\nu(\varepsilon_{11} + \varepsilon_{22})/(1 - \nu)$ holds

structures with large elastic deformations. These structures have a range of potential applications in NEMS, sensors, drug delivery, morphing structures [43–46], active materials [49], optoelectronics [50], and bio-inspired robotics.

Acknowledgements Z.C. acknowledges the support from the Society in Science-Branco Weiss fellowship, administered by ETH Zürich. The authors thank the anonymous reviewers for helpful comments. The authors also thank Eric Dai and Jayne I. Hanlin for comments.

References

1. Salamon, N.J., Masters, C.B.: Bifurcation in isotropic thin film/substrate plates. Int. J. Solids Struct. **32**, 473–481 (1995)
2. Suo, Z., Ma, E.Y., Gleskova, H., Wagner, S.: Mechanics of rollable and foldable film-on-foil electronics. Appl. Phys. Lett. **74**, 1177–1179 (1999)
3. Freund, L.B.: Substrate curvature due to thin film mismatch strain in the nonlinear deformation range. J. Mech. Phys. Solids **48**, 1159–1174 (2000)
4. Hubbard, T., Wylde, J.: Residual strain and resultant deflection of surface micromachined structures. J. Vac. Sci. Technol. A **18**, 734–737 (2000)
5. Klein, C.A.: How accurate are Stoney's equation and recent modifications. J. Appl. Phys. **88**, 5487–5489 (2000)
6. Hsueh, C.: Modeling of elastic deformation of multilayers due to residual stresses and external bending. J. Appl. Phys. **91**, 9652–9656 (2002)
7. Feng, X., Huang, Y., Jiang, H., Ngo, D., Rosakis, A.J.: The effect of thin film/substrate radii on the Stoney formula for thin film/substrate subjected to nonuniform axisymmetric misfit strain and temperature. J. Mech. Mater. Struct. **1**, 1041–1053 (2006)
8. Zhang, L., Ruh, E., Grützmacher, D., Dong, L., Bell, D.J., Nelson, B.J., Schönenberger, C.: Anomalous coiling of SiGe/Si and SiGe/Si/Cr helical nanobelts. Nano Lett. **6**, 1311–1317 (2006)
9. Huang, S., Zhang, X.: Extension of the Stoney formula for film-substrate systems with gradient stress for MEMS applications. J. Micromech. Microeng. **16**, 382–389 (2006)
10. Kim, S., Boyd, J.G., Mani, S.: Mechanical behavior of mismatch strain-driven microcantilever. Microelectron. J. **38**, 371–380 (2007)
11. Li, X.: Strain induced semiconductor nanotubes: from formation process to device applications. J. Phys. D, Appl. Phys. **41**, 193001 (2008)
12. Singamaneni, S., LeMieux, M.C., Lang, H.P., Gerber, C., Lam, Y., Zauscher, S., Datskos, P.G., Lavrik, N.V., Jiang, H., Naik, R.R., Bunning, T.J., Tsukruk, V.V.: Bimaterial microcantilevers as a hybrid sensing platform. Adv. Mater. **20**, 653–680 (2008)
13. Li, W., Huang, G., Yan, H., Wang, J., Yu, Y., Hu, X., Wu, X., Mei, Y.: Fabrication and stimuli-responsive behavior of flexible micro-scrolls. Soft Matter **8**, 7103–7107 (2012)
14. Xu, D., Zhang, L., Dong, L., Nelson, B.J.: Nanorobotics for NEMS using helical nanostructures. In: Encyclopedia of Nanotechnology, pp. 1715–1721 (2012)
15. Childers, W.S., Anthony, N.R., Mehta, A.K., Berland, K.M., Lynn, D.G.: Phase networks of cross-β peptide assemblies. Langmuir **28**, 6386–6395 (2012)
16. Gong, X.: Controlling surface properties of polyelectrolyte multilayers by assembly pH. Phys. Chem. Chem. Phys. **15**, 10459–10465 (2013)
17. Stoney, G.G.: The tension of metallic films deposited by electrolysis. Proc. R. Soc. A **82**, 172–175 (1909)
18. Timoshenko, S.: Analysis of bi-metal thermostats. J. Opt. Soc. Am. **11**, 233–255 (1925)
19. Tsui, Y.C., Clyne, T.W.: An analytical model for predicting residual stresses in progressivelydeposited coatings part 1: planar geometry. Thin Solid Films **306**, 23–33 (1997)
20. Huang, S., Zhang, X.: Gradient residual stress induced elastic deformation of multilayer MEMS structures. Sens. Actuators A **134**, 177–185 (2006)
21. Pureza, J.M., Lacerda, M.M., De Oliveira, A.L., Fragalli, J.F., Zanon, R.A.S.: Enhancing accuracy to Stoney equation. Appl. Surf. Sci. **255**, 6426–6428 (2009)
22. Chen, Z., Majidi, C., Srolovitz, D.J., Haataja, M.: Tunable helical ribbons. Appl. Phys. Lett. **98**, 011906 (2011)
23. Chouaieb, N., Goriely, A., Maddocks, J.H.: Helices. Proc. Natl. Acad. Sci. USA **103**, 9398–9403 (2006)
24. Wang, J., Feng, X., Wang, G., Yu, S.: Twisting of nanowires induced by anisotropic surface stresses. Appl. Phys. Lett. **92**, 191901 (2008)

25. Majidi, C., Chen, Z., Srolovitz, D.J., Haataja, M.: Theory for the spontaneous bending of piezoelectric nanoribbons: mechanics, spontaneous polarization, and space charge coupling. J. Mech. Phys. Solids **58**, 73–85 (2010)
26. Armon, S., Efrati, E., Kupferman, R., Sharon, E.: Geometry and mechanics in the opening of chiral seed pods. Science **333**, 1726–1730 (2011)
27. Gerbode, S.J., Puzey, J.R., McCormick, A.G., Mahadevan, L.: How to cucumber tendril coils and overwinds. Science **337**, 1087–1091 (2012)
28. Chen, Z., Majidi, C., Srolovitz, D.J., Haataja, M.: Continuum elasticity theory approach for spontaneous bending and helicity of ribbons induced by mechanical anisotropy. arXiv:1209.3321
29. Armon, S., Aharoni, H., Moshe, M., Sharon, E.: Shape selection in chiral ribbons: from seed pods to supramolecular assemblies. Soft Matter (2014)
30. Sawa, Y., Ye, F., Urayama, K., Takigawa, T., Gimenez-Pinto, V., Selinger, R.L.B., Selinger, J.V.: Shape selection of twist-nematic-elastomer ribbons. Proc. Natl. Acad. Sci. USA **108**, 6364–6368 (2011)
31. Sawa, Y., Urayama, K., Takigawa, T., Gimenez-Pinto, V., Mbanga, B.L., Ye, F., Selinger, J.V., Selinger, R.L.B.: Shape and chirality transitions in off-axis twist nematic elastomer ribbons. Phys. Rev. E **88**, 022502 (2013)
32. Landau, L.D., Lifshitz, E.M.: Theory of Elasticity, 3rd edn. Pergamon, Elmsford (1986)
33. Guo, Q., Zheng, H., Chen, W., Chen, Z.: Finite element simulations on mechanical self-assembly of biomimetic helical structures. J. Mech. Med. Biol. **13**, 1340018 (2013)
34. Guo, Q., Mehta, A.K., Grover, M.A., Chen, W., Lynn, D.G., Chen, Z.: Shape selection and multi-stability in helical ribbons. Appl. Phys. Lett. **104**, 211901 (2014)
35. Guo, Q., Chen, Z., Li, W., Dai, P., Ren, K., Lin, J., Taber, L.A., Chen, W.: Mechanics of tunable helices and geometric frustration in biomimetic seashells. Europhys. Lett. **105**, 64005 (2014)
36. Knowles, J.K.: Linear Vector Spaces and Cartesian Tensors. Oxford University Press, New York (1998)
37. Oda, R., Huc, I., Schmutz, M., Candau, S.J., MacKintosh, F.C.: Tuning bilayer twist using chiral counterions. Nature **399**, 566–569 (1999)
38. Guo, Q., Zheng, H., Chen, W., Chen, Z.: Modeling bistable behaviors in morphing structures through finite element simulations. Bio-Med. Mater. Eng. **24**, 557–562 (2014)
39. Chen, Z., Guo, Q., Majidi, C., Chen, W., Srolovitz, D.J., Haataja, M.: Nonlinear geometric effects in bistable morphing structures. Phys. Rev. Lett. **109**, 114302 (2012)
40. Efrati, E., Irvine, W.T.M.: Orientation-dependent handedness and chiral design. Phys. Rev. X **4**, 011003 (2014)
41. Chen, Z.: Geometric nonlinearity and mechanical anisotropy in strained helical nanoribbons. Nanoscale **6**, 9443–9447 (2014)
42. Chen, Z.: Shape transition and multi-stability of helical ribbons: a finite element method study. Arch. Appl. Mech. **85**(3), 331–338 (2015)
43. Seffen, K.A., Guest, S.D.: Pre-stressed morphing bistable and neutrally stable shells. J. Appl. Mech. **78**, 011002 (2011)
44. Lachenal, X., Weaver, P.M., Daynes, S.: Multi-stable composite twisting structure for morphing applications. Proc. R. Soc. Lond. **468**, 1230–1251 (2012)
45. Pirrera, A., Lachenal, X., Daynes, S., Weaver, P.M., Chenchiah, I.V.: Multi-stable cylindrical lattices. J. Mech. Phys. Solids **61**, 2087–2107 (2013)
46. Lachenal, X., Weaver, P.M., Daynes, S.: Influence of transverse curvature on the stability of pre-stressed helical structures. Int. J. Solids Struct. **468**, 1230–1251 (2012)
47. Wissman, J., Finkenauer, L., Deseri, L., Majidi, C.: Saddle-like deformation in a dielectric elastomer actuator embedded with liquid-phase gallium-indium electrodes. J. Appl. Phys. **116**, 144905 (2014)
48. Giomi, L., Mahadevan, L.: Multi-stability of free spontaneously curved anisotropic strips. Proc. R. Soc. Lond. **468**, 511–530 (2012)
49. Ge, Q., Qi, H.J., Dunn, M.L.: Active materials by four-dimension printing. Appl. Phys. Lett. **103**, 131901 (2013)
50. Hwang, G., Dockendorf, C., Bell, D., Dong, L., Hashimoto, H., Poulikakos, D., Nelson, B.: 3-D InGaAs/GaAs helical nanobelts for optoelectronic devices. Int. J. Optomechatron. **2**, 88–103 (2008)

DOI 10.1007/978-94-017-7300-3_17
Reprinted from *Journal of Elasticity* Journal, DOI 10.1007/s10659-015-9521-x

Representation for a Smooth Isometric Mapping from a Connected Planar Domain to a Surface

Yi-Chao Chen[1] · Roger Fosdick[2] · Eliot Fried[3]

Received: 25 January 2015 / Published online: 7 March 2015

Abstract A representation theorem for a smooth isometric mapping of a flat, connected domain $\mathcal{D}$ in two-dimensional Euclidean point space $\mathcal{E}^2$ into a surface $\mathcal{S}$ in three-dimensional Euclidean point space $\mathcal{E}^3$ is presented. The form of the mapping is shown to be necessary and sufficient to describe any such smooth isometry. Importantly, this work is not based upon the hypothesis that the mapped surface is ruled. In general, a mapping from a flat planar domain into a ruled surface is far from being isometric, and the property of being ruled is a partial consequence of our representation theorem.

Keywords Isometric embedding · Rigidity · Representation theorem · 2-d to 3-d

Mathematics Subject Classification 53A05 · 74K15 · 74K35 · 57R40 · 53A45

1 Introduction

The representation for an isometric embedding of a three-dimensional rigid body into three-dimensional point space is of distinct simplicity and is well-known. However, nowhere in the existing literature have we been able to find an analogous representation for an isometric embedding of an arbitrary open, connected subset $\mathcal{D}$ of two-dimensional Euclidean point space $\mathcal{E}^2$ into three-dimensional point space $\mathcal{E}^3$. Here, we establish such a representation for the case of three-times continuously differentiable embeddings. This undertaking amounts

✉ E. Fried
eliot.fried@oist.jp

Y.-C. Chen
chen@uh.edu

R. Fosdick
fosdick@aem.umn.edu

[1] Department of Mechanical Engineering, University of Houston, Houston, TX 77204-4006, USA

[2] Department of Aerospace Engineering and Mechanics, University of Minnesota, Minneapolis, MN 55455-0153, USA

[3] Mathematical Soft Matter Unit, Okinawa Institute of Science and Technology Graduate University, Onna, Okinawa 904-0495, Japan

to finding all possible shapes of a material surface that is 'rigid' in its tangent plane and is directly relevant to any problem in which the shape of an unstretchable ribbon or band must be determined.

Prior to stating our main result, we mention some essential features of our notation. The vector spaces corresponding to $\mathcal{E}^2$ and $\mathcal{E}^3$ are respectively denoted by $\mathcal{V}^2$ and $\mathcal{V}^3$. Also, $\mathcal{U}^2$ and $\mathcal{U}^3$ denote the subsets consisting of all unit vectors belonging respectively to $\mathcal{V}^2$ and $\mathcal{V}^3$. The dyadic product $\boldsymbol{e} \otimes \boldsymbol{f}$ of vectors $\boldsymbol{e}$ and $\boldsymbol{f}$ in $\mathcal{V}^2$ is a linear transformation of $\mathcal{V}^2$ to itself. Analogously, the dyadic product $\boldsymbol{u} \otimes \boldsymbol{v}$ of a vector $\boldsymbol{u}$ in $\mathcal{V}^3$ and a vector $\boldsymbol{v}$ in $\mathcal{V}^2$ is a linear transformation of $\mathcal{V}^2$ to $\mathcal{V}^3$. Analogous interpretations go for dyadic products involving only elements of $\mathcal{V}^3$. Further, given another vector $\boldsymbol{w}$ in $\mathcal{V}^2$, $\boldsymbol{u} \otimes \boldsymbol{v} \otimes \boldsymbol{w}$ can be viewed as a linear transformation $\boldsymbol{u} \otimes (\boldsymbol{f} \otimes \boldsymbol{w})$ of linear transformations of $\mathcal{V}^2$ to itself to vectors in $\mathcal{V}^3$ or, alternatively, as a linear transformation $(\boldsymbol{u} \otimes \boldsymbol{v}) \otimes \boldsymbol{w}$ of $\mathcal{V}^2$ to linear transformations of $\mathcal{V}^2$ to $\mathcal{V}^3$.

Theorem *A three-times continuously differentiable mapping $\boldsymbol{r}$ from an open, connected subset $\mathcal{D}$ of two-dimensional Euclidean point space $\mathcal{E}^2$ into three-dimensional Euclidean point space $\mathcal{E}^3$ is isometric if and only if there exists a sufficiently smooth curve $\mathcal{C}$ in $\mathcal{E}^3$ with unit speed parametrization $\boldsymbol{f}$ such that, for each point $\boldsymbol{x}$ in $\mathcal{D}$,*

$$\boldsymbol{r}(\boldsymbol{x}) = \boldsymbol{f}(u(\boldsymbol{x})) + v(\boldsymbol{x})\big(\cos\phi(u(\boldsymbol{x}))\boldsymbol{p}(u(\boldsymbol{x})) + \sin\phi(u(\boldsymbol{x}))\boldsymbol{b}(u(\boldsymbol{x}))\big), \tag{1}$$

where

$$\boldsymbol{p} = \frac{\boldsymbol{t}'}{|\boldsymbol{t}'|} \quad \textit{and} \quad \boldsymbol{b} = \boldsymbol{t} \times \boldsymbol{p}, \quad \textit{with}\ \boldsymbol{t} = \boldsymbol{f}', \tag{2}$$

denote the unit normal and unit binormal elements of the Frenet frame $\{\boldsymbol{t}, \boldsymbol{p}, \boldsymbol{b}\}$ of $\mathcal{C}$, where ϕ is defined according to

$$\phi(u) = \phi_0 - \int_0^u \tau(s)\,\mathrm{d}s, \tag{3}$$

with

$$\tau = \boldsymbol{b} \cdot \boldsymbol{p}' \tag{4}$$

being the torsion of $\mathcal{C}$, and where u and v are mappings from $\mathcal{D}$ to $\mathbb{R}$ defined implicitly in terms of $\boldsymbol{x}$ by the equation

$$\boldsymbol{x} = \boldsymbol{x}_0 + \int_0^u \boldsymbol{e}(\theta(s))\,\mathrm{d}s + v\boldsymbol{e}_\perp(\theta(u)), \tag{5}$$

with $\boldsymbol{x}_0$ being a point in $\mathcal{D}$ and with $\boldsymbol{e}$ and $\boldsymbol{e}_\perp$ being mappings from $\mathcal{D}$ to $\mathcal{U}^2$ defined consistent with

$$\boldsymbol{e}(\theta) = \cos\theta\,\boldsymbol{\imath} + \sin\theta\,\boldsymbol{\jmath} \quad \textit{and} \quad \boldsymbol{e}_\perp(\theta) = -\sin\theta\,\boldsymbol{\imath} + \cos\theta\,\boldsymbol{\jmath} \tag{6}$$

for some fixed orthonormal basis $\{\boldsymbol{\imath}, \boldsymbol{\jmath}\}$ for $\mathcal{V}^2$ and θ defined according to

$$\theta(u) = \theta_0 + \int_0^u \kappa(s)\cos\phi(s)\,\mathrm{d}s, \tag{7}$$

with

$$\kappa = |\boldsymbol{t}'| \tag{8}$$

being the curvature of $\mathcal{C}$.

Due to the assumed smoothness of the mapping $\boldsymbol{r}$, its status as an isometry admits a characterization in terms of its gradient. Specifically, $\boldsymbol{r}$ is isometric if and only if its gradient

$$\boldsymbol{Q} = \nabla \boldsymbol{r}, \tag{9}$$

which is a linear transformation of $\mathcal{V}^2$ to $\mathcal{V}^3$, preserves lengths and angles in the sense that

$$|\boldsymbol{Q}\boldsymbol{a}| = |\boldsymbol{a}| \tag{10}$$

holds for every $\boldsymbol{a}$ in $\mathcal{V}^2$. With this, we have

$$\begin{aligned} \boldsymbol{Q}\boldsymbol{a} \cdot \boldsymbol{Q}\boldsymbol{b} &= \frac{1}{2}\big(|\boldsymbol{Q}(\boldsymbol{a}+\boldsymbol{b})|^2 - |\boldsymbol{Q}\boldsymbol{a}|^2 - |\boldsymbol{Q}\boldsymbol{b}|^2\big) \\ &= \frac{1}{2}\big(|\boldsymbol{a}+\boldsymbol{b}|^2 - |\boldsymbol{a}|^2 - |\boldsymbol{b}|^2\big) \\ &= \boldsymbol{a} \cdot \boldsymbol{b} \end{aligned} \tag{11}$$

for all $\boldsymbol{a}$ and $\boldsymbol{b}$ in $\mathcal{V}^2$. Equivalent to (10) we thus have

$$\boldsymbol{Q}^{\top}\boldsymbol{Q} = \boldsymbol{I}, \tag{12}$$

where $\boldsymbol{I}$ denotes the identity transformation on $\mathcal{V}^2$.

In view of the foregoing observations, the proof of the theorem is divided into two parts: Sect. 2 is devoted to showing that every continuously differentiable mapping $\boldsymbol{r}$ from a connected subset $\mathcal{D}$ of $\mathcal{E}^2$ to $\mathcal{E}^3$ that is defined in accord with (1)–(8) satisfies (10) and Sect. 3 is devoted to showing that every three-times continuously differentiable mapping $\boldsymbol{r}$ from $\mathcal{D} \subset \mathcal{E}^2$ to $\mathcal{E}^3$ that satisfies (10) must be of the form (1), where $\boldsymbol{p}$ and $\boldsymbol{b}$ obey (2), ϕ is given by (3)–(4), and u and v are defined implicitly by (5) with $\boldsymbol{e}$ and $\boldsymbol{e}_\perp$ given by (6) and θ given by (7)–(8). An example that illustrates the utility of the representation contained in the theorem is provided in Sect. 4.

2 Sufficiency of the Representation

Let $\boldsymbol{r}$ be a continuously differentiable mapping from a subset $\mathcal{D}$ of $\mathcal{E}^2$ into $\mathcal{E}^3$ defined in accord with (1)–(8). Then, computing the gradient of (1) and using $(2)_3$, the Frenet–Serret relations

$$\boldsymbol{t}' = \kappa \boldsymbol{p}, \qquad \boldsymbol{p}' = -\kappa \boldsymbol{t} + \tau \boldsymbol{b}, \qquad \boldsymbol{b}' = -\tau \boldsymbol{p}, \tag{13}$$

and the identities

$$\phi' = -\tau \quad \text{and} \quad \theta' = \kappa \cos\phi \tag{14}$$

arising from differentiating (3) and (7), respectively, yields

$$\begin{aligned} \boldsymbol{Q} = {} & \boldsymbol{f}' \otimes \nabla u + (\cos\phi\, \boldsymbol{p} + \sin\phi\, \boldsymbol{b}) \otimes \nabla v \\ & - v\big(\phi'(\sin\phi\, \boldsymbol{p} - \cos\phi\, \boldsymbol{b}) - \cos\phi\, \boldsymbol{p}' - \sin\phi\, \boldsymbol{b}'\big) \otimes \nabla u \end{aligned}$$

$$= \big((1 - v\kappa\cos\phi)\boldsymbol{t} - v(\phi' + \tau)(\sin\phi\boldsymbol{p} - \cos\phi\boldsymbol{b})\big) \otimes \nabla u$$

$$+ (\cos\phi\boldsymbol{p} + \sin\phi\boldsymbol{b}) \otimes \nabla v$$

$$= (1 - v\theta')\boldsymbol{t} \otimes \nabla u + (\cos\phi\boldsymbol{p} + \sin\phi\boldsymbol{b}) \otimes \nabla v. \tag{15}$$

Next, computing the gradient of (5) leads to a representation,

$$\boldsymbol{I} = \boldsymbol{e} \otimes \nabla u + \boldsymbol{e}_\perp \otimes \nabla v - v\theta'\boldsymbol{e} \otimes \nabla u$$

$$= (1 - v\theta')\boldsymbol{e} \otimes \nabla u + \boldsymbol{e}_\perp \otimes \nabla v, \tag{16}$$

for the identity transformation on $\mathcal{V}^2$. Since $\boldsymbol{I}$ must satisfy $\boldsymbol{I}^\top\boldsymbol{a} = \boldsymbol{I}\boldsymbol{a} = \boldsymbol{a}$ for every $\boldsymbol{a}$ in $\mathcal{V}^2$ and since (6) ensures that $\boldsymbol{e}$ and $\boldsymbol{e}_\perp$ are orthonormal, (16) gives

$$\boldsymbol{e} = (1 - v\theta')\nabla u \quad \text{and} \quad \boldsymbol{e}_\perp = \nabla v. \tag{17}$$

Inserting (17) into (15) yields

$$\boldsymbol{Q} = \boldsymbol{t} \otimes \boldsymbol{e} + (\cos\phi\boldsymbol{p} + \sin\phi\boldsymbol{b}) \otimes \boldsymbol{e}_\perp. \tag{18}$$

Thus,

$$\boldsymbol{Q}^\top\boldsymbol{Q} = \boldsymbol{e} \otimes \boldsymbol{e} + \boldsymbol{e}_\perp \otimes \boldsymbol{e}_\perp = \boldsymbol{I}, \tag{19}$$

since the properties of the Frenet frame $\{\boldsymbol{t}, \boldsymbol{p}, \boldsymbol{b}\}$ ensure that $|\boldsymbol{t}| = 1$ and that $\boldsymbol{t} \cdot (\cos\phi\boldsymbol{p} + \sin\phi\boldsymbol{b}) = 0$. Finally, we conclude that any continuously differentiable mapping $\boldsymbol{r}$ from a connected subset $\mathcal{D}$ of $\mathcal{E}^2$ into $\mathcal{E}^3$ that is defined in accord with (1)–(8) satisfies (12) and is, thus, isometric. ■

Remark 1 Each term in the expression (18) for $\boldsymbol{Q}$ derived from (1) involves a dyadic product of the form $\boldsymbol{u} \otimes \boldsymbol{v}$, with a vector $\boldsymbol{u}$ belonging to $\mathcal{U}^3$ and a vector $\boldsymbol{v}$ belonging to $\mathcal{U}^2$, and thus has a structure analogous to that of an orthogonal linear transformation of $\mathcal{V}^3$ to itself. Indeed, $\boldsymbol{Q}$ can be viewed as an orthogonal linear transformation of $\mathcal{V}^2$ to $\mathcal{V}^3$, and (19) reflects this interpretation.

Remark 2 Although establishing the sufficiency of the representation requires only that $\boldsymbol{r}$ be continuously differentiable, additional smoothness is required to establish the necessity of our representation theorem.

3 Necessity of the Representation

Let $\boldsymbol{r}$ be a three-times continuously differentiable and isometric mapping from $\mathcal{D} \subset \mathcal{E}^2$ to $\mathcal{E}^3$ and define $\boldsymbol{Q}$ and $\boldsymbol{n}$ by

$$\boldsymbol{Q} := \nabla\boldsymbol{r} \quad \text{and} \quad \boldsymbol{n} := \boldsymbol{Q}\boldsymbol{\imath} \times \boldsymbol{Q}\boldsymbol{\jmath}, \tag{20}$$

with $\{\boldsymbol{\imath}, \boldsymbol{\jmath}\}$ being an arbitrary fixed orthonormal basis for $\mathcal{V}^2$. Then, by Lemmas 1 and 3 of the Appendix, there exist mappings k and $\boldsymbol{e}$ from $\mathcal{D}$ to $\mathbb{R}$ and $\mathcal{U}^2$, respectively, such that

$$\nabla\boldsymbol{Q} = k\boldsymbol{n} \otimes \boldsymbol{e} \otimes \boldsymbol{e}. \tag{21}$$

Given such a mapping $\boldsymbol{e}$, it is convenient in the following to use the alternative orthonormal basis $\{\boldsymbol{e}, \boldsymbol{e}_\perp\}$ for $\mathcal{V}^2$ introduced in Lemma 3 of the Appendix, where $\boldsymbol{e}_\perp$ is defined by

$$\boldsymbol{e}_\perp := \boldsymbol{W}\boldsymbol{e}, \tag{22}$$

with

$$\boldsymbol{W} := \boldsymbol{\jmath} \otimes \boldsymbol{\imath} - \boldsymbol{\imath} \otimes \boldsymbol{\jmath}. \tag{23}$$

Next, we consider the mapping $\bar{\boldsymbol{x}}$ from $\mathbb{R}^2$ to $\mathcal{E}^2$ defined as

$$\bar{\boldsymbol{x}}(u, v) := \boldsymbol{g}(u) + v\boldsymbol{e}_\perp(\boldsymbol{g}(u)), \tag{24}$$

where $\boldsymbol{g}$ is a mapping from $\mathbb{R}$ to $\mathcal{D}$ determined by solving the initial-value problem

$$\boldsymbol{g}' = \boldsymbol{e}(\boldsymbol{g}), \qquad \boldsymbol{g}(0) = \boldsymbol{x}_0, \tag{25}$$

with $\boldsymbol{x}_0$ being an arbitrary point in $\mathcal{D}$. Notice that, by $(25)_1$ and the chain rule,

$$\boldsymbol{g}'' = (\nabla \boldsymbol{e})\boldsymbol{g}' = (\nabla \boldsymbol{e})\boldsymbol{e}. \tag{26}$$

Further, by (22), $\nabla \boldsymbol{e}_\perp = \boldsymbol{W}\nabla \boldsymbol{e}$ and (26) gives

$$(\nabla \boldsymbol{e}_\perp)\boldsymbol{e} = \boldsymbol{W}(\nabla \boldsymbol{e})\boldsymbol{e} = \boldsymbol{W}\boldsymbol{g}''. \tag{27}$$

Thus, it is straightforward to show that the Jacobian determinant of the mapping $\bar{\boldsymbol{x}}$ has the form $\bar{\boldsymbol{x}}_v \cdot \boldsymbol{W}\bar{\boldsymbol{x}}_u = 1 - v\boldsymbol{g}'' \cdot \boldsymbol{e}_\perp$.

Granted that $\bar{\boldsymbol{x}}$ can be locally inverted, namely $1 - v\boldsymbol{g}'' \cdot \boldsymbol{e}_\perp \neq 0$, we may obtain u and v as mappings from $\mathcal{D}$ to $\mathbb{R}$ and it follows that

$$\boldsymbol{x} = \bar{\boldsymbol{x}}(u(\boldsymbol{x}), v(\boldsymbol{x})) = \boldsymbol{g}(u(\boldsymbol{x})) + v(\boldsymbol{x})\boldsymbol{e}_\perp(\boldsymbol{g}(u(\boldsymbol{x}))). \tag{28}$$

In view of $(25)_1$ and (27), computing the gradient of (28) yields an expression

$$\begin{aligned} \boldsymbol{I} &= (\boldsymbol{g}' + v(\nabla \boldsymbol{e}_\perp)\boldsymbol{g}') \otimes \nabla u + \boldsymbol{e}_\perp \otimes \nabla v \\ &= (\boldsymbol{e} + v\boldsymbol{W}\boldsymbol{g}'') \otimes \nabla u + \boldsymbol{e}_\perp \otimes \nabla v \\ &= \nabla u \otimes (\boldsymbol{e} + v\boldsymbol{W}\boldsymbol{g}'') + \nabla v \otimes \boldsymbol{e}_\perp, \end{aligned} \tag{29}$$

for the identity transformation $\boldsymbol{I}$ of $\mathcal{V}^2$ to itself, the latter step being a consequence of the symmetry of $\boldsymbol{I}$. Since

$$\boldsymbol{a} = \boldsymbol{I}\boldsymbol{a} = (\boldsymbol{e} \cdot \boldsymbol{a} - v\boldsymbol{g}'' \cdot \boldsymbol{W}\boldsymbol{a})\nabla u + (\boldsymbol{e}_\perp \cdot \boldsymbol{a})\nabla v \tag{30}$$

and since $\{\boldsymbol{e}, \boldsymbol{e}_\perp\}$ constitutes an orthonormal basis for $\mathcal{V}^2$, applying $\boldsymbol{I}$ to $\boldsymbol{e}$ and $\boldsymbol{e}_\perp$ while bearing in mind (22), (27), and the identity $(\nabla \boldsymbol{e}_\perp)^\top \boldsymbol{e}_\perp = \nabla(\frac{1}{2}|\boldsymbol{e}_\perp|^2) = \boldsymbol{0}$, gives

$$\boldsymbol{e} = (1 - v\boldsymbol{g}'' \cdot \boldsymbol{e}_\perp)\nabla u \quad \text{and} \quad \boldsymbol{e}_\perp = \nabla v. \tag{31}$$

Next, let $\bar{\boldsymbol{r}}$ be a mapping from $\mathbb{R}^2$ to $\mathcal{V}^3$ such that

$$\bar{\boldsymbol{r}}(u(\boldsymbol{x}), v(\boldsymbol{x})) = \boldsymbol{r}(\boldsymbol{x}) \tag{32}$$

for each $\boldsymbol{x}$ in $\mathcal{D}$. Then, by $(20)_1$ and (31), we obtain

$$\begin{aligned}\boldsymbol{Q} &= \nabla\boldsymbol{r} \\ &= \bar{\boldsymbol{r}}_u \otimes \nabla u + \bar{\boldsymbol{r}}_v \otimes \nabla v \\ &= \frac{\bar{\boldsymbol{r}}_u \otimes \boldsymbol{e}}{1 - v\boldsymbol{g}'' \cdot \boldsymbol{e}_\perp} + \bar{\boldsymbol{r}}_v \otimes \boldsymbol{e}_\perp.\end{aligned} \tag{33}$$

Now, knowing that $\boldsymbol{r}$ is an isometric mapping and using the orthogonality of $\boldsymbol{e}$ and $\boldsymbol{e}_\perp$, we note that $\boldsymbol{Q}\boldsymbol{e}$ and $\boldsymbol{Q}\boldsymbol{e}_\perp$ must satisfy the necessary conditions

$$|\boldsymbol{Q}\boldsymbol{e}| = 1, \qquad |\boldsymbol{Q}\boldsymbol{e}_\perp| = 1, \quad \text{and} \quad \boldsymbol{Q}\boldsymbol{e} \cdot \boldsymbol{Q}\boldsymbol{e}_\perp = 0, \tag{34}$$

whereby (33) yields

$$|\bar{\boldsymbol{r}}_u| = |1 - v\boldsymbol{g}'' \cdot \boldsymbol{e}_\perp|, \qquad |\bar{\boldsymbol{r}}_v| = 1, \quad \text{and} \quad \bar{\boldsymbol{r}}_u \cdot \bar{\boldsymbol{r}}_v = 0. \tag{35}$$

Additionally, because (22) and (23) yield the following equivalent representation for $\boldsymbol{n}$ defined in $(20)_2$

$$\boldsymbol{n} = \boldsymbol{Q}\boldsymbol{e} \times \boldsymbol{Q}\boldsymbol{e}_\perp, \tag{36}$$

we see from (33) that

$$\boldsymbol{n} = \frac{1}{1 - v\boldsymbol{g}'' \cdot \boldsymbol{e}_\perp}\bar{\boldsymbol{r}}_u \times \bar{\boldsymbol{r}}_v. \tag{37}$$

Next, let $\bar{\boldsymbol{n}}$ be a mapping from $\mathbb{R}^2$ to $\mathcal{U}^3$ such that

$$\bar{\boldsymbol{n}}(u(\boldsymbol{x}), v(\boldsymbol{x})) = \boldsymbol{n}(\boldsymbol{x}). \tag{38}$$

Since, by Lemmas 2 and 3 of the Appendix, we know that

$$\nabla\boldsymbol{n} = -k\,\boldsymbol{Q}\boldsymbol{e} \otimes \boldsymbol{e}, \tag{39}$$

then (31), (33), and (38) yield

$$\begin{aligned}\nabla\boldsymbol{n} &= \bar{\boldsymbol{n}}_u \otimes \nabla u + \bar{\boldsymbol{n}}_v \otimes \nabla v \\ &= \frac{1}{1 - v\boldsymbol{g}'' \cdot \boldsymbol{e}_\perp}\bar{\boldsymbol{n}}_u \otimes \boldsymbol{e} + \bar{\boldsymbol{n}}_v \otimes \boldsymbol{e}_\perp \\ &= -\frac{k}{1 - v\boldsymbol{g}'' \cdot \boldsymbol{e}_\perp}\bar{\boldsymbol{r}}_u \otimes \boldsymbol{e},\end{aligned} \tag{40}$$

from which it follows that

$$\bar{\boldsymbol{n}}_u = -k\bar{\boldsymbol{r}}_u \quad \text{and} \quad \bar{\boldsymbol{n}}_v = \boldsymbol{0}. \tag{41}$$

Next, by $(35)_{1,3}$, (37), and $(41)_1$, it can be seen that

$$\bar{\boldsymbol{n}}_u \times \boldsymbol{n} = -\frac{k}{1 - v\boldsymbol{g}'' \cdot \boldsymbol{e}_\perp}\bar{\boldsymbol{r}}_u \times (\bar{\boldsymbol{r}}_u \times \bar{\boldsymbol{r}}_v) = k(1 - v\boldsymbol{g}'' \cdot \boldsymbol{e}_\perp)\bar{\boldsymbol{r}}_v \tag{42}$$

or, equivalently,

$$\bar{\boldsymbol{r}}_v = \frac{1}{k(1 - v\boldsymbol{g}'' \cdot \boldsymbol{e}_\perp)} \bar{\boldsymbol{n}}_u \times \boldsymbol{n}. \tag{43}$$

However, from $(35)_2$ we have $|\bar{\boldsymbol{r}}_v| = 1$ so that

$$\bar{\boldsymbol{r}}_v = \pm \frac{\bar{\boldsymbol{n}}_u \times \boldsymbol{n}}{|\bar{\boldsymbol{n}}_u \times \boldsymbol{n}|} \quad \text{and} \quad \frac{|\bar{\boldsymbol{n}}_u \times \boldsymbol{n}|}{|k(1 - v\boldsymbol{g}'' \cdot \boldsymbol{e}_\perp)|} = 1. \tag{44}$$

Thus, since, by $(41)_2$, $\bar{\boldsymbol{n}}$, and therefore $\bar{\boldsymbol{n}}_u$, must be independent of v, it follows from $(44)_1$ that

$$\bar{\boldsymbol{r}}_v(u, v) = \boldsymbol{m}(u), \tag{45}$$

where $\boldsymbol{m}$ obeys

$$|\boldsymbol{m}| = 1. \tag{46}$$

As a consequence of (45), there exists a mapping $\boldsymbol{f}$ from $\mathbb{R}$ to $\mathcal{E}^3$ such that

$$\bar{\boldsymbol{r}}(u, v) = \boldsymbol{f}(u) + v\boldsymbol{m}(u). \tag{47}$$

Next, using (47) in the representation (33) we find that

$$\boldsymbol{Q} = \frac{(\boldsymbol{f}' + v\boldsymbol{m}') \otimes \boldsymbol{e}}{1 - v\boldsymbol{g}'' \cdot \boldsymbol{e}_\perp} + \boldsymbol{m} \otimes \boldsymbol{e}_\perp, \tag{48}$$

which, clearly, yields

$$\boldsymbol{Q}\boldsymbol{e} = \frac{\boldsymbol{f}' + v\boldsymbol{m}'}{1 - v\boldsymbol{g}'' \cdot \boldsymbol{e}_\perp} \quad \text{and} \quad \boldsymbol{Q}\boldsymbol{e}_\perp = \boldsymbol{m}. \tag{49}$$

Of course, since $\boldsymbol{r}$ is an isometric mapping and $(34)_1$ holds, then (49) gives

$$|\boldsymbol{f}' + v\boldsymbol{m}'| = |1 - v\boldsymbol{g}'' \cdot \boldsymbol{e}_\perp|, \tag{50}$$

consequences of which are

$$|\boldsymbol{f}'| = 1, \qquad \boldsymbol{f}' \cdot \boldsymbol{m}' = -\boldsymbol{g}'' \cdot \boldsymbol{e}_\perp, \quad \text{and} \quad |\boldsymbol{m}'| = |\boldsymbol{g}'' \cdot \boldsymbol{e}_\perp| = |\boldsymbol{f}' \cdot \boldsymbol{m}'|. \tag{51}$$

Notice also, by $(34)_3$ and (49), that $(\boldsymbol{f}' + v\boldsymbol{m}') \cdot \boldsymbol{m} = 0$. Thus, since $\boldsymbol{m}' \cdot \boldsymbol{m} = 0$ by (46), it follows that

$$\boldsymbol{f}' \cdot \boldsymbol{m} = 0. \tag{52}$$

From $(51)_1$, $\boldsymbol{f}$ defines a unit speed parametrization for a curve, say $\mathcal{C}$, in $\mathcal{E}^3$. Let

$$\boldsymbol{t} = \boldsymbol{f}', \qquad \boldsymbol{p} = \frac{\boldsymbol{f}''}{|\boldsymbol{f}''|} \quad \text{and} \quad \boldsymbol{b} = \boldsymbol{t} \times \boldsymbol{p} \tag{53}$$

denote the elements of the Frenet frame of $\mathcal{C}$. By (52) and $(53)_1$, the unit vector $\boldsymbol{m}$ may be written as

$$\boldsymbol{m} = (\boldsymbol{m} \cdot \boldsymbol{p})\boldsymbol{p} + (\boldsymbol{m} \cdot \boldsymbol{b})\boldsymbol{b}, \tag{54}$$

from which it follows that

$$\boldsymbol{m}\cdot\boldsymbol{p}=\cos\phi \quad \text{and} \quad \boldsymbol{m}\cdot\boldsymbol{b}=\sin\phi. \tag{55}$$

Moreover, differentiating (54) and using the Frenet–Serret relations (13), where $\kappa=|\boldsymbol{t}'|$ and $\tau=\boldsymbol{b}\cdot\boldsymbol{p}'$ denote the curvature and torsion of $\mathcal{C}$, we readily find that

$$\boldsymbol{m}'=-\kappa\cos\phi\boldsymbol{t}-(\tau+\phi')\sin\phi\boldsymbol{p}+(\tau+\phi')\cos\phi\boldsymbol{b}, \tag{56}$$

with

$$|\boldsymbol{m}'|^2=\kappa^2\cos^2\phi+(\tau+\phi')^2. \tag{57}$$

But, in accord with $(51)_3$, $(53)_1$, and (56), we find that

$$|\boldsymbol{m}'|^2=|\boldsymbol{t}\cdot\boldsymbol{m}'|^2=\kappa^2\cos^2\phi, \tag{58}$$

so that, by (57),

$$\phi'=-\tau, \tag{59}$$

and we see that

$$\boldsymbol{m}'=-\kappa\cos\phi\boldsymbol{t}. \tag{60}$$

Thus, (47) leads to the representation

$$\boldsymbol{r}(\boldsymbol{x})=\boldsymbol{f}(u(\boldsymbol{x}))+v(\boldsymbol{x})\big(\cos\phi(u(\boldsymbol{x}))\boldsymbol{p}(u(\boldsymbol{x}))+\sin\phi(u(\boldsymbol{x}))\boldsymbol{b}(u(\boldsymbol{x}))\big), \tag{61}$$

namely (1) in the statement of the theorem. Notice also that the relation (3) for ϕ of the theorem follows from (59).

Next, observe that since $\{\boldsymbol{e},\boldsymbol{e}_\perp\}$ is an orthonormal basis for $\mathcal{V}^2$, with $\boldsymbol{e}_\perp$ given by (22) and (23), we have

$$\boldsymbol{e}=\cos\theta\boldsymbol{\imath}+\sin\theta\boldsymbol{\jmath} \quad \text{and} \quad \boldsymbol{e}_\perp=-\sin\theta\boldsymbol{\imath}+\cos\theta\boldsymbol{\jmath}, \tag{62}$$

where θ is defined by

$$\theta:=\arccos(\boldsymbol{e}(\boldsymbol{g})\cdot\boldsymbol{\imath})=\arcsin(\boldsymbol{e}(\boldsymbol{g})\cdot\boldsymbol{\jmath}). \tag{63}$$

Evidently, (6) of the theorem holds. Thus, noting (25), we see that

$$\boldsymbol{g}''=(\boldsymbol{e}(\boldsymbol{g}))'=\theta'\boldsymbol{e}_\perp \tag{64}$$

and it follows from $(51)_2$, $(53)_1$, and (60) that

$$\theta'=\kappa\cos\phi, \tag{65}$$

which yields the expression (7) for θ in the theorem. Finally, the implicit relation (5) for u and v in terms of $\boldsymbol{x}$ is a consequence of (25) and (28). We therefore conclude that a three-times continuously differentiable isometric mapping $\boldsymbol{r}$ from a connected subset $\mathcal{D}$ of $\mathcal{E}^2$ into $\mathcal{E}^3$ must be of the form determined by (1)–(8). ■

Remark 3 By (34) and (36), $\{\boldsymbol{Qe}, \boldsymbol{Qe}_\perp\}$ provides at each point of $\mathcal{S}$ an orthonormal basis for the tangent space of $\mathcal{S}$ at that point. Further, by (49), (60), (64), and (65),

$$\boldsymbol{Qe} = \boldsymbol{t}, \tag{66}$$

and, thus, the representation (48) for $\boldsymbol{Q}$ takes the elementary form

$$\boldsymbol{Q} = \boldsymbol{t} \otimes \boldsymbol{e} + \boldsymbol{m} \otimes \boldsymbol{e}_\perp. \tag{67}$$

As a consequence of (67),

$$\begin{aligned}\boldsymbol{QQ}^\top &= \boldsymbol{t} \otimes \boldsymbol{t} + \boldsymbol{m} \otimes \boldsymbol{m} \\ &=: \boldsymbol{I}_{\mathcal{S}},\end{aligned} \tag{68}$$

where $\boldsymbol{I}_{\mathcal{S}}$ is the identity transformation of the tangent space of $\mathcal{S}$ to itself.

Now, let $\tilde{\boldsymbol{n}}$ be a mapping from $\mathcal{S}$ to $\mathcal{U}^3$ defined such that

$$\tilde{\boldsymbol{n}}(\boldsymbol{r}(\boldsymbol{x})) = \boldsymbol{n}(\boldsymbol{x}) \tag{69}$$

for each $\boldsymbol{x}$ in $\mathcal{D}$. Then, by the chain rule and $(20)_1$,

$$\begin{aligned}\nabla \boldsymbol{n} &= (\nabla_{\mathcal{S}} \tilde{\boldsymbol{n}}) \nabla \boldsymbol{r} \\ &= -\boldsymbol{L} \boldsymbol{Q},\end{aligned} \tag{70}$$

where $\nabla_{\mathcal{S}}$ denotes the gradient operator on $\mathcal{S}$ and the Weingarten map

$$\boldsymbol{L} := -\nabla_{\mathcal{S}} \tilde{\boldsymbol{n}} \tag{71}$$

of $\mathcal{S}$ is at each point of $\mathcal{S}$ a symmetric linear transformation of the tangent space of $\mathcal{S}$ at that point into itself. In view of (39), (66), (68), and (70),

$$\begin{aligned}\boldsymbol{LQQ}^\top &= -(\nabla \boldsymbol{n}) \boldsymbol{Q}^\top \\ &= (k \boldsymbol{Qe} \otimes \boldsymbol{e}) \boldsymbol{Q}^\top \\ &= k \boldsymbol{Qe} \otimes \boldsymbol{Qe} \\ &= k \boldsymbol{t} \otimes \boldsymbol{t}.\end{aligned} \tag{72}$$

With reference to (68) and (72), we conclude that $\boldsymbol{L}$ has the particular form

$$\boldsymbol{L} = k \boldsymbol{t} \otimes \boldsymbol{t} \tag{73}$$

from which it follows that k is the principal curvature of $\mathcal{S}$ in the direction of $\boldsymbol{Qe}$. Immediate but important consequences of this are that the mean curvature $H := \frac{1}{2} \mathrm{tr} \boldsymbol{L}$ of $\mathcal{S}$ is simply

$$H = \frac{1}{2} k \tag{74}$$

and that the Gaussian curvature $K := \det \boldsymbol{L}$ of $\mathcal{S}$ vanishes, viz.

$$K = 0. \tag{75}$$

The result (75) is consistent with the observation that a surface determined by an isometric mapping of a connected subset of $\mathcal{E}^2$ is also developable. However, nonisometric mappings of planar domains into developable surfaces certainly exist.

Remark 4 To determine k, the nontrivial principal curvature of $\mathcal{S}$, we first observe that, as a consequence of (18),

$$\boldsymbol{Q}\boldsymbol{e}_\perp = \cos\phi\,\boldsymbol{p} + \sin\phi\,\boldsymbol{b}, \tag{76}$$

which, with (36), (66), and the elementary properties of the Frenet frame, gives

$$\begin{aligned}\boldsymbol{n} &= \boldsymbol{t} \times (\cos\phi\,\boldsymbol{p} + \sin\phi\,\boldsymbol{b})\\ &= -\sin\phi\,\boldsymbol{p} + \cos\phi\,\boldsymbol{b}.\end{aligned} \tag{77}$$

Further, by (77), (17), (59), and (65), we find that

$$\begin{aligned}\nabla\boldsymbol{n} &= \kappa\sin\phi\,\boldsymbol{t}\otimes\nabla u\\ &= \frac{\kappa\sin\phi}{1 - v\kappa\cos\phi}\boldsymbol{t}\otimes\boldsymbol{e}.\end{aligned} \tag{78}$$

However, by (66), (70), and (73),

$$\begin{aligned}\nabla\boldsymbol{n} &= -\boldsymbol{L}\boldsymbol{Q}\\ &= -k(\boldsymbol{t}\otimes\boldsymbol{t})\boldsymbol{Q}\\ &= -k\boldsymbol{t}\otimes\boldsymbol{Q}^\top\boldsymbol{t}\\ &= -k\boldsymbol{t}\otimes\boldsymbol{e},\end{aligned} \tag{79}$$

which, when compared with (78), yields

$$k = -\frac{\kappa\sin\phi}{1 - v\kappa\cos\phi}. \tag{80}$$

From (80) it is evident that $k \to \infty$ as $v\kappa\cos\phi \to 1$, in precise correspondence with the vanishing of the Jacobian of the transformation (24). In the next section, we present an example for which this occurs.

Remark 5 Consistent with (73) and (75), the principal curvature in the direction $\boldsymbol{Q}\boldsymbol{e}_\perp = \boldsymbol{m}$ on $\mathcal{S}$ orthogonal to $\boldsymbol{Q}\boldsymbol{e} = \boldsymbol{t}$ vanishes. Thus, $\boldsymbol{Q}\boldsymbol{e}_\perp$ describes the orientation of straight lines on $\mathcal{S}$. With this in mind, (32), (47), and $(49)_2$ yield

$$\bar{\boldsymbol{r}}(u, v) = \boldsymbol{f}(u) + v\boldsymbol{Q}(u)\boldsymbol{e}_\perp(u), \tag{81}$$

from which it follows that $\mathcal{S}$ can be viewed as a ruled surface obtained by rotating a line of possibly variable length and orientation $\boldsymbol{e}_\perp$ along the curve $\mathcal{C}$ parametrized by $\boldsymbol{f}$. This property of being ruled is a particular necessary consequence of hypothesizing that $\boldsymbol{r}$ be isometric and three-times continuously differentiable. Nowhere in the proof of our theorem do we assume that $\mathcal{S}$ is ruled.

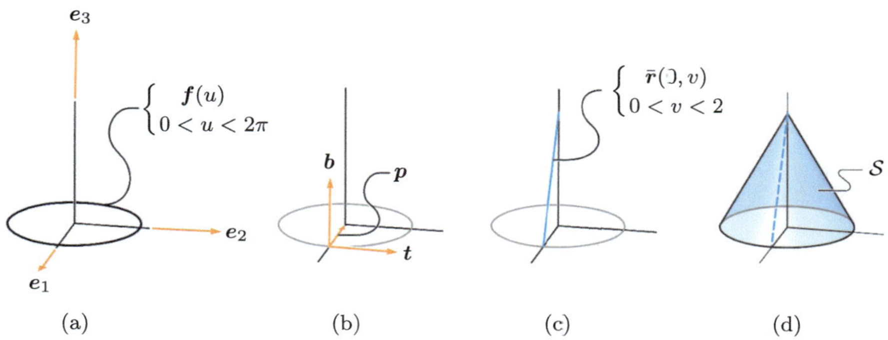

Fig. 1 Construction of a conical surface $\mathcal{S}$ according to (1)–(8): (**a**) The circle of unit radius parametrized by the mapping $\boldsymbol{f}$. (**b**) The Frenet frame corresponding to $\boldsymbol{f}$. (**c**) A generator of the cone. (**d**) The conical surface $\mathcal{S}$

4 Illustrative Example

The theorem provides a sufficient and necessary condition for a mapping $\boldsymbol{r}$ from a connected subset $\mathcal{D}$ of $\mathcal{E}^3$ to be isometric. If such a mapping is isometric, it must be representable in the form of (1). On the other hand, for a choice of a mapping $\boldsymbol{f}$ consistent with conditions (2)–(8), the representation (1) provides an isometric mapping.

Consider a circle Γ of unit radius, as depicted in Fig. 1(a) and define a mapping $\boldsymbol{f}$ from $[0, 2\pi]$ to $\mathcal{E}^3$ such that

$$\boldsymbol{f}(u) = \cos u\boldsymbol{e}_1 + \sin u\boldsymbol{e}_2, \tag{82}$$

where $\{\boldsymbol{e}_1, \boldsymbol{e}_2, \boldsymbol{e}_3\}$, with $\boldsymbol{e}_3 = \boldsymbol{e}_1 \times \boldsymbol{e}_2$, provides an orthonormal basis for $\mathcal{V}^3$. For the choice (82), (2) yields explicit expressions,

$$\left.\begin{aligned} \boldsymbol{t}(u) &= -\sin u\boldsymbol{e}_1 + \cos u\boldsymbol{e}_2, \\ \boldsymbol{p}(u) &= -\cos u\boldsymbol{e}_1 - \sin u\boldsymbol{e}_2, \\ \boldsymbol{b}(u) &= \boldsymbol{t}(u) \times \boldsymbol{b}(u) = \boldsymbol{e}_3, \end{aligned}\right\} \tag{83}$$

for the tangential, normal, and binormal elements of the Frenet frame of Γ, as depicted in Fig. 1(b) for $u = 0$. Further, using (83) in (4) and (8) shows that, consistent with intuitive expectations, the torsion and curvature of Γ are

$$\tau = 0 \quad \text{and} \quad \kappa = 1. \tag{84}$$

Thus, (3) and (7) of the theorem yield

$$\phi(u) = \phi_0 \quad \text{and} \quad \theta(u) = \theta_0 + u\cos\phi_0. \tag{85}$$

Specializing the representation (1) of the theorem in accord with (82), $(83)_{2,3}$, and $(85)_1$, while suppressing dependence of u and v on $\boldsymbol{x}$, yields

$$\begin{aligned} \bar{\boldsymbol{r}}(u, v) &= \boldsymbol{f}(u) - v\cos\phi_0(\cos u\boldsymbol{e}_1 + \sin u\boldsymbol{e}_2) + v\sin\phi_0\boldsymbol{e}_3 \\ &= (1 - v\cos\phi_0)(\cos u\boldsymbol{e}_1 + \sin u\boldsymbol{e}_2) + v\sin\phi_0\boldsymbol{e}_3. \end{aligned} \tag{86}$$

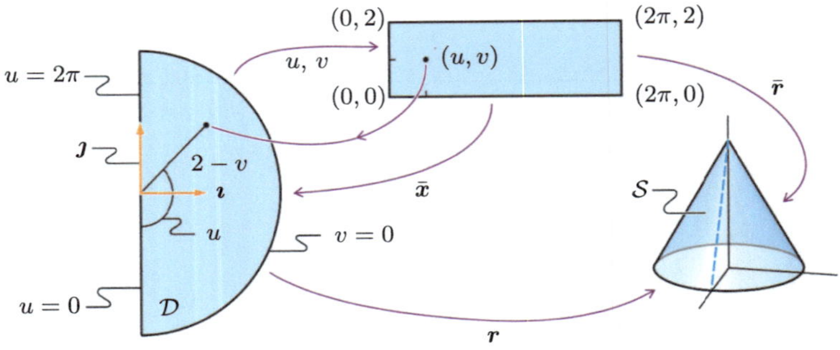

Fig. 2 Isometric mapping of a semicircle $\mathcal{D}$ into a conical surface $\mathcal{S}$ according to (1)–(8): The reference flat semicircular domain $\mathcal{D}$. The rectangular domain, with coordinates u and v, that $\bar{\boldsymbol{x}}$ maps into $\mathcal{D}$ is also shown

The line $\boldsymbol{r} = \bar{\boldsymbol{r}}(0, v)$ is depicted in Fig. 1(c) for $\phi_0 = \pi/3$ and $0 \le v \le 2$ and the surface $\boldsymbol{r} = \bar{\boldsymbol{r}}(u, v)$ is depicted in Fig. 1(d) for $\phi_0 = \pi/3$ and $0 \le u \le 2\pi$ and $0 \le v \le 2$.

Notice that, by (5) and (6) of the theorem and $(85)_2$,

$$\begin{aligned}\boldsymbol{x} = \boldsymbol{x}_0 &+ \left(\frac{1}{\cos\phi_0} - v\right)\big(\sin(\theta_0 + u\cos\phi_0)\boldsymbol{\imath} - \cos(\theta_0 + u\cos\phi_0)\boldsymbol{\jmath}\big)\\ &- \frac{1}{\cos\phi_0}(\sin\theta_0\boldsymbol{\imath} - \cos\theta_0\boldsymbol{\jmath}).\end{aligned} \tag{87}$$

For the particular choices $\boldsymbol{x}_0 = \boldsymbol{o} - 2\boldsymbol{\jmath}$, where $\boldsymbol{o}$ denotes a conveniently chosen origin, $\phi_0 = \pi/3$, and $\theta_0 = 0$ and for $0 \le u \le 2\pi$ and $0 \le v \le 2$, (87) describes a half disk as depicted in Fig. 2. This is the particular domain $\mathcal{D}$ which is mapped isometrically to the conical surface in Fig. 2.

The relation (87) determining $\boldsymbol{x}$ in terms of u and v is not globally invertible on the entire plane. To each $\boldsymbol{x}$, there exist infinitely many u and v. Given an arbitrary integer n, these can be represented by

$$\left.\begin{aligned}u(\boldsymbol{x}) &= \frac{1}{\cos\phi_0}\left(n\pi - \theta_0 - \arctan\frac{\cos\phi_0(\boldsymbol{x}-\boldsymbol{x}_0)\cdot\boldsymbol{\imath} + \sin\theta_0}{\cos\phi_0(\boldsymbol{x}-\boldsymbol{x}_0)\cdot\boldsymbol{\jmath} - \cos\theta_0}\right),\\ v(\boldsymbol{x}) &= \frac{1}{\cos\phi_0}\big(1 + (-1)^n|\cos\phi_0(\boldsymbol{x}-\boldsymbol{x}_0) + \sin\theta_0\boldsymbol{\imath} - \cos\theta_0\boldsymbol{\jmath}|\big).\end{aligned}\right\} \tag{88}$$

For u and v properly restricted, (87) can be solved to determine them in terms of $\boldsymbol{x}$. Suppose that, as before, $\boldsymbol{x}_0 = \boldsymbol{o} - 2\boldsymbol{\jmath}$, $\phi_0 = \pi/3$, and $\theta_0 = 0$. Then, if $0 \le u \le 2\pi$ and $0 \le v \le 2$, we must take $n = 1$ in (88), yielding

$$u(\boldsymbol{x}) = 2\left(\pi - \arctan\frac{(\boldsymbol{x}-\boldsymbol{o})\cdot\boldsymbol{\imath}}{(\boldsymbol{x}-\boldsymbol{o})\cdot\boldsymbol{\jmath}}\right), \qquad v(\boldsymbol{x}) = 2 - |\boldsymbol{x}-\boldsymbol{o}|. \tag{89}$$

On substituting (89) into (82) and $(83)_{2,3}$, we arrive at the relevant special version,

$$\begin{aligned}\boldsymbol{r}(\boldsymbol{x}) = &\left(\frac{|\boldsymbol{x}-\boldsymbol{o}|}{2} - \frac{|(\boldsymbol{x}-\boldsymbol{o})\cdot\boldsymbol{\imath}|^2}{|\boldsymbol{x}-\boldsymbol{o}|}\right)\boldsymbol{e}_1\\ &- \frac{(\boldsymbol{x}-\boldsymbol{o})\cdot\boldsymbol{\imath}\,(\boldsymbol{x}-\boldsymbol{o})\cdot\boldsymbol{\jmath}}{|\boldsymbol{x}-\boldsymbol{o}|}\boldsymbol{e}_2 + \sqrt{3}\left(1 - \frac{|\boldsymbol{x}-\boldsymbol{o}|}{2}\right)\boldsymbol{e}_3,\end{aligned} \tag{90}$$

of the representation (1) for an isometric mapping.

Finally, notice in accord with (80) that the nonvanishing principal curvature of $\mathcal{S}$ is

$$k = -\frac{\sin\phi_0}{1 - v\cos\phi_0} = -\frac{\sqrt{3}}{2 - v}, \tag{91}$$

whereby $k \to -\infty$ as $v \to 2$, which corresponds to the tip of the conical surface $\mathcal{S}$.

Appendix: Lemmas Invoked in Establishing the Necessity of the Representation

Consider a smooth mapping $\boldsymbol{r}$ from $\mathcal{D}$ to $\mathcal{E}^3$, the gradient $\boldsymbol{Q}$ of which obeys (10) for all $\boldsymbol{a}$ in $\mathcal{V}^2$. Let $\{\boldsymbol{\imath}, \boldsymbol{\jmath}\}$ be a fixed orthonormal basis for $\mathcal{V}^2$. Then, by (10),

$$|\boldsymbol{Q}\boldsymbol{\imath}| = 1, \qquad |\boldsymbol{Q}\boldsymbol{\jmath}| = 1, \tag{A.1}$$

and

$$2\boldsymbol{Q}\boldsymbol{\imath} \cdot \boldsymbol{Q}\boldsymbol{\jmath} = |\boldsymbol{Q}(\boldsymbol{\imath} + \boldsymbol{\jmath})|^2 - |\boldsymbol{Q}\boldsymbol{\imath}|^2 - |\boldsymbol{Q}\boldsymbol{\jmath}|^2 = |\boldsymbol{\imath} + \boldsymbol{\jmath}|^2 - 2 = 0. \tag{A.2}$$

Let $\boldsymbol{n}$ be a mapping from $\mathcal{D}$ to $\mathcal{U}^3$ defined in accord with $(20)_2$, so that

$$\boldsymbol{n} = \boldsymbol{Q}\boldsymbol{\imath} \times \boldsymbol{Q}\boldsymbol{\jmath}. \tag{A.3}$$

Importantly, (A.3) is but one of many equivalent ways by which $\boldsymbol{n}$ may be defined. For example, suppose that $\boldsymbol{b}$ and $\boldsymbol{b}_\perp$ are elements of $\mathcal{U}^2$ defined such that

$$\boldsymbol{b} := \cos\alpha\boldsymbol{\imath} + \sin\alpha\boldsymbol{\jmath} \quad \text{and} \quad \boldsymbol{b}_\perp := -\sin\alpha\boldsymbol{\imath} + \cos\alpha\boldsymbol{\jmath} \tag{A.4}$$

for some α in $[0, 2\pi]$. Then, $\boldsymbol{b} \cdot \boldsymbol{b}_\perp = 0$ and it is easy to see that

$$\boldsymbol{Q}\boldsymbol{b} \times \boldsymbol{Q}\boldsymbol{b}_\perp = \boldsymbol{n}. \tag{A.5}$$

In particular, choosing $\boldsymbol{b} = \boldsymbol{e}$ and $\boldsymbol{b}_\perp = \boldsymbol{e}_\perp$ in (A.5) yields (36).

We next establish some lemmas useful for the proof of the necessity of our representation contained in Sect. 3.

Lemma 1 *At each point in $\mathcal{D}$, there exists a symmetric linear transformation of $\mathcal{V}^2$ to itself, namely a member of* Sym_2, *such that*

$$\nabla\boldsymbol{Q} = \boldsymbol{n} \otimes \boldsymbol{S}. \tag{A.6}$$

Proof In view of (A.1)–(A.3), define mappings $\boldsymbol{m}_i$, $i = 1, 2, 3$, from $\mathcal{D}$ to $\mathcal{U}^3$ by

$$\boldsymbol{m}_1 = \boldsymbol{Q}\boldsymbol{\imath}, \qquad \boldsymbol{m}_2 = \boldsymbol{Q}\boldsymbol{\jmath}, \quad \text{and} \quad \boldsymbol{m}_3 = \boldsymbol{m}_1 \times \boldsymbol{m}_2 = \boldsymbol{n}. \tag{A.7}$$

Since the value of $\nabla\boldsymbol{Q} = \nabla\nabla\boldsymbol{r}$ at any point in $\mathcal{D}$ can be viewed as a linear transformation from Sym_2 to $\mathcal{V}^3$, it admits a representation of the form

$$\nabla\boldsymbol{Q} = \sum_{i=1}^{3} \boldsymbol{m}_i \otimes \boldsymbol{S}_i, \tag{A.8}$$

where $\boldsymbol{S}_i$ is a mapping of $\mathcal{D}$ to Sym_2 for each $i = 1, 2, 3$. Choose and fix arbitrary elements $\boldsymbol{a}$ and $\boldsymbol{b}$ of $\mathcal{V}^2$. Then, since

$$\begin{aligned}\nabla(\boldsymbol{Qa}\cdot\boldsymbol{Qb}) &= \big(\nabla(\boldsymbol{Qa})\big)^{\top}\boldsymbol{Qb} + \big(\nabla(\boldsymbol{Qb})\big)^{\top}\boldsymbol{Qa}\\ &= \big((\nabla\boldsymbol{Q})\boldsymbol{a}\big)^{\top}\boldsymbol{Qb} + \big((\nabla\boldsymbol{Q})\boldsymbol{b}\big)^{\top}\boldsymbol{Qa},\end{aligned} \tag{A.9}$$

(A.8) and the recognition that $\boldsymbol{Qa}$ and $\boldsymbol{Qb}$ are orthogonal to $\boldsymbol{m}_3$ for all $\boldsymbol{a}$ and $\boldsymbol{b}$ in $\mathcal{V}^2$ yield

$$\begin{aligned}\nabla(\boldsymbol{Qa}\cdot\boldsymbol{Qb}) &= \sum_{i=1}^{3}\big((\boldsymbol{m}_i\otimes\boldsymbol{S}_i\boldsymbol{a})^{\top}\boldsymbol{Qb} + (\boldsymbol{m}_i\otimes\boldsymbol{S}_i\boldsymbol{b})^{\top}\boldsymbol{Qa}\big)\\ &= \sum_{i=1}^{3}\big((\boldsymbol{S}_i\boldsymbol{a}\otimes\boldsymbol{m}_i)\boldsymbol{Qb} + (\boldsymbol{S}_i\boldsymbol{b}\otimes\boldsymbol{m}_i)\boldsymbol{Qa}\big)\\ &= \sum_{i=1}^{2}\big((\boldsymbol{Qb}\cdot\boldsymbol{m}_i)\boldsymbol{S}_i\boldsymbol{a} + (\boldsymbol{Qa}\cdot\boldsymbol{m}_i)\boldsymbol{S}_i\boldsymbol{b}\big)\\ &= \sum_{i=1}^{2}\boldsymbol{S}_i(\boldsymbol{a}\otimes\boldsymbol{b}+\boldsymbol{b}\otimes\boldsymbol{a})\boldsymbol{Q}^{\top}\boldsymbol{m}_i,\end{aligned} \tag{A.10}$$

which, with (11) and $(\mathrm{A}.7)_{1,2}$, implies that $\boldsymbol{S}_1$ and $\boldsymbol{S}_2$ must satisfy

$$\boldsymbol{S}_1(\boldsymbol{a}\otimes\boldsymbol{b}+\boldsymbol{b}\otimes\boldsymbol{a})\boldsymbol{\imath} + \boldsymbol{S}_2(\boldsymbol{a}\otimes\boldsymbol{b}+\boldsymbol{b}\otimes\boldsymbol{a})\boldsymbol{\jmath} = \boldsymbol{0} \tag{A.11}$$

for all $\boldsymbol{a}$ and $\boldsymbol{b}$ in $\mathcal{V}^2$. With the particular choices $\boldsymbol{a}=\boldsymbol{b}=\boldsymbol{\imath}$, $\boldsymbol{a}=\boldsymbol{b}=\boldsymbol{\jmath}$, and $\boldsymbol{a}=\boldsymbol{\imath}$ and $\boldsymbol{b}=\boldsymbol{\jmath}$ in sequence, (A.11) results in requirements,

$$\boldsymbol{S}_1\boldsymbol{\imath}=\boldsymbol{0}, \qquad \boldsymbol{S}_2\boldsymbol{\jmath}=\boldsymbol{0}, \quad \text{and} \quad \boldsymbol{S}_1\boldsymbol{\jmath}+\boldsymbol{S}_2\boldsymbol{\imath}=\boldsymbol{0}, \tag{A.12}$$

which, since $\boldsymbol{S}_1$ and $\boldsymbol{S}_2$ are mappings of $\mathcal{D}$ to Sym_2, can be met if and only if

$$\boldsymbol{S}_1=\boldsymbol{0} \quad \text{and} \quad \boldsymbol{S}_2=\boldsymbol{0}. \tag{A.13}$$

Finally, using $(\mathrm{A}.7)_3$ and (A.13) in (A.8) and setting $\boldsymbol{S}_3=\boldsymbol{S}$, we obtain

$$\nabla\boldsymbol{Q}=\boldsymbol{n}\otimes\boldsymbol{S}, \tag{A.14}$$

which completes the proof. ■

Lemma 2 *The gradient* $\nabla\boldsymbol{n}$ *of* $\boldsymbol{n}$ *defined by* (A.3) *admits a representation of the form*

$$\nabla\boldsymbol{n}=-\boldsymbol{QS}, \tag{A.15}$$

where $\boldsymbol{Q}$ *denotes the gradient of* $\boldsymbol{r}$ *and* $\boldsymbol{S}$ *is the mapping of* $\mathcal{D}$ *to* Sym_2 *encountered in Lemma* 1.

Proof Differentiating (A.3) and using (A.6), we find that

$$\begin{aligned}\nabla\boldsymbol{n} &= \big((\boldsymbol{n}\times\boldsymbol{Q\jmath})\otimes\boldsymbol{\imath} - (\boldsymbol{n}\times\boldsymbol{Q\imath})\otimes\boldsymbol{\jmath}\big)\boldsymbol{S}\\ &= -(\boldsymbol{Q\imath}\otimes\boldsymbol{\imath}+\boldsymbol{Q\jmath}\otimes\boldsymbol{\jmath})\boldsymbol{S}\end{aligned}$$

$$\begin{aligned} &= -\boldsymbol{Q}(\boldsymbol{\imath} \otimes \boldsymbol{\imath} + \boldsymbol{\jmath} \otimes \boldsymbol{\jmath})\boldsymbol{S} \\ &= -\boldsymbol{Q}\boldsymbol{S}, \end{aligned} \tag{A.16}$$

which completes the proof. ■

Lemma 3 *Let $\boldsymbol{Q}$ and $\boldsymbol{n}$ be given by $\boldsymbol{Q} = \nabla \boldsymbol{r}$ and* (A.3), *with $\boldsymbol{r}$ being three-times continuously differentiable. Then, there exists mappings k from $\mathcal{D}$ to $\mathbb{R}$ and $\boldsymbol{e}$ from $\mathcal{D}$ to $\mathcal{U}^2$ such that the mapping $\boldsymbol{S}$ of $\mathcal{D}$ to* Sym_2 *encountered in Lemma* 1 *can be expressed as*

$$\boldsymbol{S} = k\boldsymbol{e} \otimes \boldsymbol{e}. \tag{A.17}$$

Proof Let $\boldsymbol{W} = \boldsymbol{\jmath} \otimes \boldsymbol{\imath} - \boldsymbol{\imath} \otimes \boldsymbol{\jmath}$. Granted that $\boldsymbol{r}$ is three-times continuously differentiable, we differentiate (A.6) and use (A.15) to find that

$$\begin{aligned} \nabla\nabla\boldsymbol{Q}[\boldsymbol{W}] &= -(\nabla\boldsymbol{n})\boldsymbol{W}\boldsymbol{S} + \boldsymbol{n} \otimes \nabla\boldsymbol{S}[\boldsymbol{W}] \\ &= \boldsymbol{Q}\boldsymbol{S}\boldsymbol{W}\boldsymbol{S} + \boldsymbol{n} \otimes \nabla\boldsymbol{S}[\boldsymbol{W}]. \end{aligned} \tag{A.18}$$

Since $\boldsymbol{S}$ is in Sym_2 and $\boldsymbol{W}$ is skew, $(\boldsymbol{S}\boldsymbol{W}\boldsymbol{S})^{\top} = -\boldsymbol{S}\boldsymbol{W}\boldsymbol{S}$ and since $\{\boldsymbol{W}\}$ provides a basis for the collection of skew linear transformations of $\mathcal{V}^2$ to $\mathcal{V}^2$, there exists a scalar λ such that

$$\boldsymbol{S}\boldsymbol{W}\boldsymbol{S} = \lambda\boldsymbol{W}. \tag{A.19}$$

Applying the Cayley–Hamilton theorem to $\boldsymbol{S}\boldsymbol{W}$ while bearing in mind that $\mathrm{tr}(\boldsymbol{S}\boldsymbol{W}) = 0$ and that $\det \boldsymbol{W} = 1$, gives

$$\boldsymbol{S}\boldsymbol{W}\boldsymbol{S}\boldsymbol{W} + (\det \boldsymbol{S})\boldsymbol{I} = \boldsymbol{0}. \tag{A.20}$$

Next, taking the trace of (A.20) while invoking (A.19) yields

$$\begin{aligned} \det \boldsymbol{S} &= -\frac{1}{2}\mathrm{tr}(\boldsymbol{S}\boldsymbol{W}\boldsymbol{S}\boldsymbol{W}) \\ \det \boldsymbol{S} &= \frac{1}{2}\mathrm{tr}(\boldsymbol{S}\boldsymbol{W}\boldsymbol{S}\boldsymbol{W}^{\top}) \\ &= \frac{1}{2}\boldsymbol{S}\boldsymbol{W}\boldsymbol{S} \cdot \boldsymbol{W} \\ &= \frac{1}{2}\lambda\boldsymbol{W} \cdot \boldsymbol{W} \\ &= \lambda, \end{aligned} \tag{A.21}$$

with which (A.19) becomes

$$\boldsymbol{S}\boldsymbol{W}\boldsymbol{S} = (\det \boldsymbol{S})\boldsymbol{W}. \tag{A.22}$$

Thus,

$$\nabla\nabla\boldsymbol{Q}[\boldsymbol{W}] = (\det \boldsymbol{S})\boldsymbol{Q}\boldsymbol{W} + \boldsymbol{n} \otimes \nabla\boldsymbol{S}[\boldsymbol{W}] \tag{A.23}$$

and, because $\nabla\nabla\boldsymbol{Q}[\boldsymbol{W}] = \boldsymbol{0}$, we see that

$$\det \boldsymbol{S} = 0 \tag{A.24}$$

and that $\nabla \boldsymbol{S}[\boldsymbol{W}] = \boldsymbol{0}$. Finally, we recognize that (A.24) implies that there exist mappings k from $\mathcal{D}$ to $\mathbb{R}$ and $\boldsymbol{e}$ from $\mathcal{D}$ to $\mathcal{U}^2$ such that the spectral representation of $\boldsymbol{S}$ has the form

$$\boldsymbol{S} = k\boldsymbol{e} \otimes \boldsymbol{e}, \tag{A.25}$$

which completes the proof. ■

Remark 6 In the foregoing proof, $\nabla\nabla\boldsymbol{Q}$ is a linear transformation which can be viewed as a linear transformation of linear transformations of $\mathcal{V}^2$ to itself to linear transformations of $\mathcal{V}^2$ to $\mathcal{V}^3$.

DOI 10.1007/978-94-017-7300-3_18
Reprinted from *Journal of Elasticity* Journal, DOI 10.1007/s10659-015-9533-6

ERRATUM

Erratum to: Four Papers Published in the Journal of Elasticity (2015) 119(1–2)

Roger Fosdick[1] **· Eliot Fried**[2]

Published online: 21 May 2015

Erratum to: J Elast (2015) 119:23–34
DOI 10.1007/s10659-014-9489-y

The translation of Wunderlich's article "Über ein abwickelbaresMöbiusband", which appeared in the Monatshefte für Mathematik 66 (1962), 276–289, was undertaken with the kind permission of Springer, its publisher.

Erratum to: J Elast (2015) 119:19–22
DOI 10.1007/s10659-014-9491-4

This article was originally published in German in Jahresbericht der Deutschen Mathematiker-Vereinigung 39 (2. Abt. Heft 5/8, Jahresversammlung vom 16. bis 23. September), 49–51 (1929) and has been published in English with kind permission of the current editor in chief of this journal and the Deutschen Mathematiker-Vereinigung.

Erratum to: J Elast (2015) 119:3–6
DOI 10.1007/s10659-014-9490-5

There are instances where we have been unable to trace or contact the copyright holder of the translated articles. If notified the publisher will be pleased to rectify any errors or omissions at the earliest opportunity.

The online version of the original articles can be found under doi:10.1007/s10659-014-9489-y, doi:10.1007/s10659-014-9491-4, doi:10.1007/s10659-014-9490-5, doi:10.1007/s10659-014-9492-3.

✉ R. Fosdick
fosdick@aem.umn.edu

E. Fried
eliot.fried@oist.jp

1 Department of Aerospace Engineering and Mechanics, University of Minnesota, Minneapolis, MN 55455-0153, USA

2 Mathematical Soft Matter Unit, Okinawa Institute of Science and Technology, Onna, Okinawa 904-0495, Japan

Erratum to: J Elast (2015) 119:7–17
DOI 10.1007/s10659-014-9492-3

There are instances where we have been unable to trace or contact the copyright holder of the translated articles. If notified the publisher will be pleased to rectify any errors or omissions at the earliest opportunity.

MIX
Papier aus verantwortungsvollen Quellen
Paper from responsible sources
FSC® C105338

If you have any concerns about our products,
you can contact us on
ProductSafety@springernature.com

In case Publisher is established outside the EU,
the EU authorized representative is:
Springer Nature Customer Service Center GmbH
Europaplatz 3, 69115 Heidelberg, Germany

Printed by Libri Plureos GmbH
in Hamburg, Germany